Hi Charlotte

It was

doing ... book. !!

Ian

A Guide to
Synchrotron Radiation Science

A Guide to
Synchrotron Radiation Science

Editors
Makoto Watanabe
Shigeru Sato
Ian Munro
G. S. Lodha

Narosa Publishing House
New Delhi Chennai Mumbai Kolkata

A Guide to Synchrotron Radiation Science

394 pgs. | 211 figs. | 03 tbls.

Editors
M. Watanabe
S. Sato
Tohoku University
Sendai, Japan

I. Munro
University of Manchester
Manchester Science and Technology Facilities Council
London, U.K.

G. S. Lodha (*Retd.*)
Raja Ramanna Centre for Advanced Technology (RRCAT)
Indore, India

NAROSA PUBLISHING HOUSE PVT. LTD.

22 Delhi Medical Association Road, Daryaganj, New Delhi 110 002
35-36 Greams Road, Thousand Lights, Chennai 600 006
306 Shiv Centre, Sector 17, Vashi, Navi Mumbai 400 703
2F-2G Shivam Chambers, 53 Syed Amir Ali Avenue, Kolkata 700 019

www.narosa.com

ISBN 978-81-8487-373-3

Published by N.K. Mehra for Narosa Publishing House Pvt. Ltd.,
22 Delhi Medical Association Road, Daryaganj, New Delhi 110 002

Printed in India

Preface

Synchrotron radiation is the name given to the brilliant electromagnetic radiation emitted from electrons moving along circular or undulating orbits and traveling with a velocity almost equal to that of light. The radiation is emitted tangentially to the direction of motion and synchrotron radiation can be either in continuum or in quasi-monochromatic spectral form. It is used widely around the world as one of the most brilliant sources of radiation from the infrared to the soft and hard X-ray regions. Synchrotron radiation offers exceptional brightness while still covering such a very wide photon energy and wavelength region. The benefits are particularly striking in the soft and hard X-ray regions where there are few alternative bright continuum light sources. Synchrotron radiation has been used extensively to determine the physico-chemical characteristics of many materials from an atomic and molecular view point and it can be readily applied to topics ranging from electronic structure analyses and simple crystal structure studies to protein crystallography, trace element mapping, high resolution microscopy and many others. Synchrotron radiation is also ideal for the investigation of the microscopic characteristics of materials which have been newly synthesized or extracted. Therefore it has now become an indispensable tool for structural studies from the materials sciences to the life sciences. This situation can be fully appreciated knowing that there are now more than 60 synchrotron radiation sources are working or under construction around the world.

This book is a new Guide presenting a totally revised and wholly rewritten version of a Japanese book produced by Tohoku University Press in 2004. The book is intended to be a guide for young scientists about to undertake experiments using synchrotron radiation or who are considering entering the field of accelerator design or of optical engineering. This guide offers an introduction to these expanding areas of science starting from their basic concepts. It presents in an easily assimilable form, the fundamentals of synchrotron radiation from its generation principles, through source design and operation to the principles of instruments used for synchrotron radiation exploitation. This is followed by the basic theory of the interaction between light and matter. Also included is a survey of important applications in some selected research fields, including absorption spectroscopy, photoelectron spectroscopy and emission spectroscopy in the VUV and soft X-ray regions, microfabrication, X-ray diffractometry, EXAFS, X-ray fluorescence analysis, inelastic X-ray scattering, infrared spectroscopy and imaging. All the subject areas described are well documented and the reader can easily access the latest information through the concise descriptions together with the references provided. This book is intended to be at post graduate level for students in physics, chemistry, biology and engineering. Readers will require some previous knowledge of electromagnetism, relativity theory, optics and quantum mechanics. Most synchrotron light source facilities are freely open to all potential users and we hope that many readers will feel encouraged to join this important and exciting field of synchrotron radiation science now expanding around the world.

Acknowledgement

We wish to express our grateful thanks to many people for their help in preparing the earlier and the current versions of this book. They include H. Akazawa, S. Aoki, D. Attwood, G. L. Carr, C. T. Chen, M. Fujita, K. Fukui, Y. Hatano, K. Hayashi, S. Hayakawa, U. Hergenhahn, A. Iida, G. Isoyama, K. Ishii, T. Ishii, A. Kakizaki, H. Kihara, H. Kimura, S. Kimura, T. Kinoshita, S. Koda, K. Kubota Y. Kuroiwa, A. R. Lang, G. Margaritondo, L. M. Miller, T. Murata, H. Nakagawa, I. Nakai, T. Nanba, H. Ogawa, H. Ohkuma, V. Saile, K. Sakurai, Y. Sakurai, H. Sato, V. Schmidt, K. Seki, R. A. Senin, Y. Shibata, Y. Shimizu, K. Shinoda, S. Suga, M. Suzuki, Takashi Takahashi, Toshiharu Takahashi, K. Takemoto, A. Tanaka, M. Taniguchi, N. Ueno, R. I. G. Uhrberg, T. Urisu, Y. Uspenski, N. Watanabe, G. Williams, M. Yamamoto and S. Yamamoto.

For generous permission to use various Figures, we are indebted to the following societies, publishers and organizations, who reserve copyright for the Figures. The American Physical Society, American Institute of Physics, American Chemical Society, The American Society for Biochemistry and Molecular Biology, International Union of Crystallography, The Korean Physical Society, The Physical Society of Japan, The Japan Society of Applied Physics, The Japan Institute of Metals and Materials, The Japanese Society of Synchrotron Radiation Research, The Crystallographic Society of Japan, IOP (Institute of Physics) Publising, Elsevier, John Wiley and Sons, AGNE Gijutsu Center, SPring-8 (Japan Synchrotron Radiation Research Institute), UVSOR (Institute for Molecular Science) and INSION.

We also thank Tohoku University Press for permission to produce this English version and CROSS (Comprehensive Research Organization for Science and Society) for their support.

Editors
M. Watanabe
S. Sato
I. Munro
G.S. Lodha

Contents

Contributors

EJIMA, Takeo 7.5.3, 14.2.1
ESASHI, Masaki 8.2
FUKUDA, Hiroshi 14.2.4
HAMA, Hiroyuki 3.5.2, App. E
HATANO, Tadashi 5.2.2, 5.5
HAYASHI, Hisashi 12
ISHII, Hisao 7.7
ITO, Tasuku 10.3.3
KAJITANI, Tsuyoshi 9.1, 9.2.1, 9.2.2
KAMISHIMA, Osamu 10.3.4
KONDO, Yasuhiro 3.5.1, 5.7, 13.1, 13.2, 13.3.1.c, 14.2.3, App. A
KONO, Shozo 7.9
KURIHARA, Kazue 9.7
MATSUBARA, Eiichiro 9.8, 10.4, 11.2
MURAKAMI, Youichi 9.4
NIWANO, Michio 13.3.3
NODA, Yukio 9.2.3, 9.3, 9.5.3
OHTANI, Eiji 9.5.1, 9.5.2
SAITO, Masao 9.6, 10.5
SATO, Shigeru 2, 7.1, 7.2, 7.5, 7.8
SUZUKI, Shoji 5.1, 5.2.1, 5.2.3, 5.2.4, 5.3.1, 5.4, 5.8, 7.3.2, 14.2.1
TANAKA, Akinori 7.8
TAKAKUWA, Yuji 8.1
TOHJI, Kazuyuki 10.3.1
TSUJI, Kouichi 11.1, 14.2.2
UEDA, Kiyoshi 6, Apps. G, H
UNNO, Masaki 9.6
WASEDA, Yoshio 9.8
WATANABE, Makoto 1, 3, 5.3.2, 5.6, 7.1, 7.2, 7.3.1, 7.4, 7.5, 7.6, 9.1, 9.2.1, 9.2.2, 10.1, 10.2, 13.3.1, 13.3.2, 14.1, 14.2.1, 14.2.2, Apps. A, B, C, D, F, I, J, K, L
YANAGIHARA, Mihiro 4, 7.4, 14.2.2
YONENAGA, Ichiro 14.3

(All contributors were staff of Tohoku University, Sendai, Japan in 2004.)

CHAPTER 1

Introduction
The Characteristics of Synchrotron Radiation and Its Importance

There is visible light (wavelengths from 750 to 400 nm; photon energies from 2 to 3 eV) with which we are very familiar. There also exists the invisible, but nevertheless exceedingly important light across the whole range of the electromagnetic spectrum with wavelengths both longer and shorter than that of visible light, in the regions of the infrared and also the ultraviolet (from about 400 to 200 nm; energies from 3 to 6 eV), the vacuum ultraviolet (VUV) and soft X-ray regions (from 200 to 0.4 nm, with photon energies from about 6 eV to 3 keV) and finally the hard X-ray region which extends from around 0.4 to 0.01 nm; from about 3 keV up to 100 keV. Synchrotron radiation (often abbreviated to SR) manifests itself as a light source of considerable brilliance, or brightness across this enormous wavelength and energy range. Synchrotron radiation is generated by high energy electrons moving with almost exactly the velocity of light along curved (circular or undulating) orbits in a direction tangential to their curved orbit. From the 1970s it has become an essential tool used to establish the physico-chemical characteristics of materials from an atomic view point from the basic physical sciences to the life sciences.[1–9] This book introduces the reader to the essentials of synchrotron radiation science commencing with its generation principles, source design and operation and the instrumentation required for its exploitation. Finally, the book leads to a survey of its many applications in selected research fields including spectroscopy, diffractometry, microanalysis, imaging and microfabrication. The basic theories needed for a good understanding the principles of each application are provided. In this preliminary chapter, the history of this field describing the advances in light sources and measurements is briefly reviewed and followed by explanations of the characteristics of synchrotron radiation. Overviews of its many applications are given for readers to appreciate fully the importance of synchrotron radiation. A comparison of synchrotron radiation with laboratory sources is provided to enable potential users to recognize the benefits and merits in the use of synchrotron radiation.

1.1 BRIEF HISTORY

Synchrotron radiation was first observed in 1947 from an electron synchrotron of the General Electric Company in USA.[10] Subsequently, its characteristics were investigated at other synchrotron facilities around the world (for early history, see Ref. 1). The generation mechanism for synchrotron radiation is the same as that for all electromagnetic radiation generated by electrical oscillations (the alternating movements of charges) and emitted from an antenna. There is one major difference between them however, in that the movement of charge is relativistic in the former, while it is non-relativistic in the latter (see Chap. **2**). Synchrotron radiation was originally considered to be a serious nuisance and a major source of energy loss from high energy electron synchrotrons therefore reducing their efficiency because of the extra energy required to accelerate the electrons. However, synchrotron radiation has many unique and important characteristics, which include a strong continuum spectrum of the highest brilliance across virtually the entire electromagnetic spectrum and with polarization characteristics which can be selected to be from linear to circular at any wavelength. Furthermore, the peak intensity and maximum brightness lie in the X-ray region where alternative sources are difficult to construct and to operate. Its unique characteristics were first exploited in the early half of the 1960s by researchers who were making spectroscopic studies of atoms, molecules and solids in the VUV and soft X-ray regions where there were no effective light sources with a good continuum emission spectrum. The first crucial experiment to demonstrate the potential of synchrotron radiation was through the observation of two-electron excitation absorption in rare gases. This work was undertaken at NIST in Gaithersburg, USA using the radiation from a 180 MeV electron synchrotron (SURF I). This study was rapidly followed by work at LNF (Frascati, Italy), INS-SOR (Tokyo, Japan), DESY (Hamburg, Germany), NINA (Daresbury, UK) and others. Synchrotron radiation is essential for any study of the electronic structure of materials, because the response to VUV and soft X-ray radiation is very large. Hard X-rays have a high transmission through matter and can easily be exploited to provide X-ray, or Röntgen, photographs. Since X-ray wavelengths are of the same order as the interatomic distances in crystals, their atomic structures can be analyzed. Hard X-rays can excite core (inner shell) electrons, whose binding energies are element-specific as is the case also in the VUV and soft X-ray regions. These wavelengths are very much shorter than those of visible light, so that any diffuseness at edges of microscopic images, caused by diffraction of light is much reduced. The usefulness of synchrotron radiation was quickly recognized for hard X-ray research and intensively exploited from the latter half of the 1960s.

Synchrotron accelerators are used to accelerate electrons up to high energy to enable them to collide with targets for high energy physics studies. The acceleration frequency used in synchrotrons is from a few to several 10s of Hz and the intensities of the electron beam fluctuate from acceleration to acceleration. This means that the intensities of the emitted synchrotron radiation are not stable causing an unsatisfactory situation for any precise experiments. Therefore users of synchrotron radiation in the 1970s began to use storage rings as shown in Fig. 1.1, in which electrons, or positrons, once injected and accumulated can be contained for very long times. Details are given in Chap. **3**. This type of accelerator was originally used in experiments to create and observe high energy collisions between charged particles, *e.g.* in the LHC at Geneva in Switzerland. Their working principles are similar to those of a synchrotron. There are storage rings with electron energies ranging from a few to several 100s of MeV. Examples include Tantalus (USA), ACO (France), SOR-RING (Japan) and those with energies above 1 GeV include SPEAR (USA), DORIS (Germany), VEPP-3 and VEPP-4 (Russia), DCI (France) and CESR (USA). The SOR-RING in Tokyo, Japan, was the first one in the world to be designed specifically for use as a spectroscopic synchrotron radiation light source.[11)] Throughout this period, photoelectron spectroscopy, photoionization spectroscopy, X-ray diffraction (XRD) and EXAFS (extended X-ray absorption fine structure) continued to be developed using stable synchrotron radiation from storage rings. The electron beam does not actually circulate as a uniform beam, but rather as a bunched beam. By exploiting this fact, synchrotron radiation has also been used as a pulsed light source for the measurement of nanosecond decay times of fluorescence and many other kinds of time-resolved

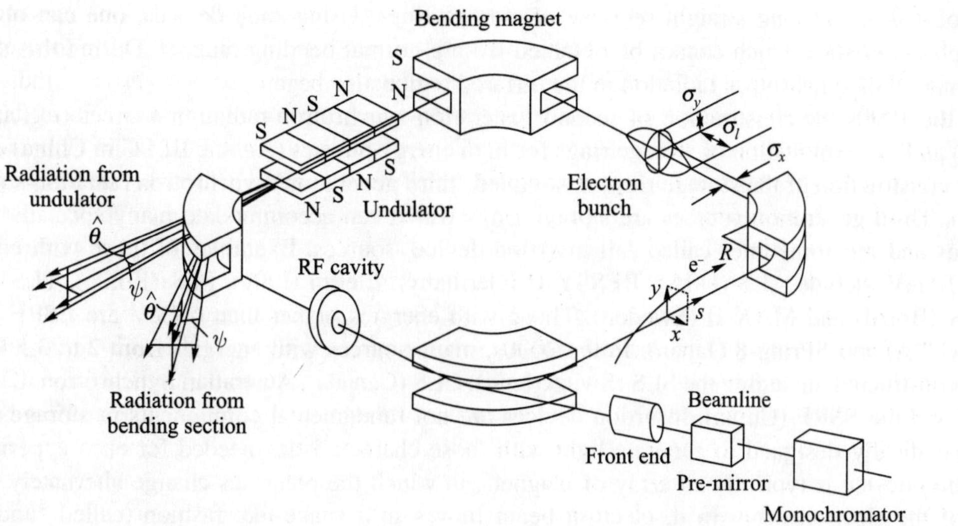

Figure 1.1 Schematic of an electron storage ring used as a synchrotron radiation light source equipped with beamlines. The motion of the electrons is illustrated with position coordinates and radiation from bending sections and undulators, with angular coordinates.

spectroscopy. In the 1970s, synchrotrons and small storage rings were exploited as rings for synchrotron radiation use, while very large storage rings were still used primarily for high energy physics with any synchrotron radiation users carrying out their experiments parasitically. This early period when synchrotron radiation research was undertaken using synchrotrons and small storage rings exclusively, but large storage rings parasitically, has been designated the 'first generation'.

During the 1980s, synchrotron radiation users recognized that it was difficult to carry out experiments satisfactorily under the operating conditions and schedules needed for high energy physics, so that many projects were soon established to construct the new storage rings designed exclusively for synchrotron radiation research. These storage rings were the SRS (UK), NSLS, Aladdin (USA), Photon Factory: PF, UVSOR (Japan), BESSY (Germany), Super ACO (France), Indus-1 (India), NSRL (China) and others. They are now called the 'second generation' synchrotron radiation sources. In second generation sources the principal source parameters and machine operating conditions are optimized specifically for synchrotron radiation experiments and users. Methods to achieve small beam cross sections and small angular spread of electron beams had already been developed by this time and using these new synchrotron radiation sources, much more challenging experiments such as angle-resolved photoelectron spectroscopy for the determination of band dispersion, spin-resolved photoelectron spectroscopy, photoelectron spectroscopy on clean and adsorbed surfaces, X-ray structure analysis for complicated materials such as on proteins and X-ray fluorescence analysis and so on, became readily feasible. The opportunities for research were subsequently extended from basic and industrial materials to biological materials. New optical elements and systems, and detectors specifically designed for synchrotron radiation use, also became available. Circular dichroism experiments had commenced using circularly-polarized synchrotron radiation emitted from above and below the median plane of the storage ring. At a later stage, novel insertion devices such as undulators and wigglers were developed and located in the long straight sections of storage rings. Using such devices, one can obtain light with characteristics which cannot be obtained from a normal bending magnet. During this period the exploitation of synchrotron radiation in the infrared region also began.

In the 1990s the construction of second generation synchrotron radiation sources (*e.g.* Indus-2 in India) and the exploitation of storage rings for high energy physics (*e.g.* the BEPC in China) continued while construction of the storage rings now called 'third generation' synchrotron radiation sources had begun. Third generation sources are storage rings which can accommodate many specialist insertion devices and are sometimes called 'all insertion device' sources. Examples of those with energy less than 2 GeV include ALS (USA), BESSY II (Germany), Elettra (Italy), PLS (Korea), TLS (Taiwan), LNLS (Brazil) and MAX II (Sweden). Those with energies higher than 6 GeV are ESRF (Europe), APS (USA) and SPring-8 (Japan). In the 2000s, many sources with energies from 2 to 3.5 GeV have been constructed, including the SLS (Switzerland), CLS (Canada), Australian Synchrotron, DIAMOND (UK) and the SSRF (China). Insertion devices are not fundamental components in storage rings, but are specifically designed to generate light with those characteristics needed for each experiment. An insertion device is typically an array of magnets, in which the polarities change alternately along the line of the electron orbit. In it, electron beam moves in a snake-like fashion (called 'undulation'). If the magnets are superconducting, for example in a superconducting wiggler, where the magnetic field is much stronger than that in a bending magnet, one can obtain light with wavelengths much shorter than those from a bending section. This device is called a 'wavelength shifter'. On the other hand, if the angles of undulation are small, one can obtain brilliant quasi-monochromatic light arising from an interference effect between the light from each individual undulation, or bend (see Sect. **2.4**). Third generation synchrotron radiation sources are those storage rings where the emittance, defined to be approximately equal to the product of beam size multiplied by angular spread of the electron beam, is made very small in order to generate highly brilliant quasi-monochromatic light. In general the horizontal emittance defines the overall performance of any storage ring and is much larger than vertical emittance. Therefore in this book the term emittance always refers to the horizontal emittance. The emittance of most third generation synchrotron radiation sources is from a few nm · rad to a few 10s nm · rad. The availability of such sources has made feasible some very advanced experiments such as high resolution spectroscopy, microscopic measurements on small samples or on very small volumes in samples, measurements on very dilute samples, multi-coincidence measurements of electron and ions in atoms and molecules, measurements under externally applied fields such as magnetic fields, high pressures and many others. Different types of two-photon experiments using lasers combined with synchrotron radiation have also become important. The third generation sources have led to the development of many new X-ray optical elements such as superpolished mirrors, multilayers bearing high heat loads and effective cooling methods for optical element together with high resolution grating monochromators and also high resolution crystal monochromators.

From the early 1960s to the early 2000s, the levels of technology associated with synchrotron radiation sources[12,13] and their applications advanced very rapidly.[7,12] Synchrotron radiation is now essential and important not only in the basic sciences, but particularly in the materials and life sciences. It is noteworthy that SORTEC (which had operated in Tsukuba in Japan) was redesigned after closure and has been revived as a synchrotron radiation source with much improved characteristics and is now called the Siam Photon Source and has been relocated to Nahkon Ratchasima in Thailand. In a similar manner the BESSY storage ring originally from Berlin in Germany has been used so as to create the SESAME source at Allaan in Jordan. Table 1.1 lists those synchrotron radiation facilities in the world, which are open to users or under construction in 2014. Much more information is available on the relevant source home pages.[14)] The storage ring with the smallest energy is that of NIST (200 MeV), while the largest one is SPring-8 (8 GeV). In terms of circumference, the smallest one is a few m, while the longest is 2.3 km. The normal stored current levels are from 100 to 500 mA. All these synchrotron radiation sources are surrounded by radiation shields, so that beamlines for most experiments are situated outside the shields. The longest beamline is currently about 1 km long. About 55 synchrotron radiation sources are now in operation and about 25 sources are of the third generation.

Table 1.1 List of synchrotron radiation facilities in the world†

Country	Location	Name	Electron energy (GeV)	Critical photon energy (keV)	Horizontal emittance (nm · rad)
Spain	Cerdanyola del Valles	ALBA[3)]	3		4.3
UK	Didcot	DIAMOND[3)]	3	8.39	2.74
France	Gif sur Yvette	SOLEIL[3)]	2.75	8.6	3.9
	Grenoble	ESRF[3)]	6.03	9.6, 20.6	4
Switzerland	Villigen	SLS[3)]	2.4	5.4	4.8
Italy	Frascati	DAΦNE[1)]	0.51	0.2	1000
	Trieste	Elettra[3)]	(2) 2.4	(3.2) 5.5	(7.0) 9.7
Germany	Karlsruhe	ANKA[2)]	2.5	6.23	50
	Bonn	ELSA[1)]	1.6 ~ 2.7		
	Dortmund	DELTA[3)]	1.5		11
	Hamburg	PETRA III[3)]	6		1
	Berlin	BESSY II[3)]	1.7 (1.9)	2.6	6
		MLS	0.105 ~ 0.65	0.017 ~ 0.364	
Denmark	Aarhus	ASTRID[1)]	0.58	0.358	140
		ASTRID 2[3)]	0.58	0.238	12
Sweden	Lund	MAX I[2)]	0.55	0.31	
		MAX II[3)]	1.5		8.8
		MAX III[3)]	0.7		13.3
		+MAX IV[3a)]	3		0.2 ~ 0.3
Porland	Krakow	+Solaris[3)]	1.5		6
Ukraine	Kharkov	NESTOR[2)]	0.04 ~ 0.225		
Armenia	Yerevan	+CANDLE[3)]	3		8.4

Country	Location	Name	Electron energy (GeV)	Critical photon energy (keV)	Horizontal emittance (nm · rad)
Russia	Moscow	Siberia-1[2)]	0.45	0.2	880
		Siberia-2[2)]	2.5	1.775, 7.1	76
	Dubna	+DELSY[3)]	1.2	1.16	11.4
	Novosibirsk	VEPP-3[1)]	2.2	2.95	
		VEPP-4[1)]	6		
Jordan	Allaan	+SESAME[3)]	2.5	6.06	24.9
India	Indore	Indus-1[2)]	0.45	0.2	73
		Indus-2[2)]	2.5	6.3	58.1
Thailand	Nakhon Ratchasima	Siam Photon Source[2)]	1.2		41
Singapore	Singapore	Helios 2[2)]	0.7	#1.47	1370
China	Beijing	BEPC II[1)]	2.5		
	Hefei	HLS-II[2)]	0.8	0.52	40
	Shanghai	SSRF[3)]	3.5		3.9
Taiwan	Hsinchu	TLS[3)]	1.51		25
		+TPS[3)]	3		1.6
Korea	Pohang	PLS-II[3)]	3		5.9
Japan	Tsukuba	PF[2)]	2.5	4.0	34.6
		AR[2)]	6.5	26.4	293
	Okazaki	UVSOR-III[3)]	0.75	0.43	15
	Nagoya	AichiSR[2)]	1.2	1.34, #4.8	53
	Kusatsu	AURORA[2)]	0.575	#0.83	
	Nishiharima	SPring-8[3)]	8	28.9	2.8
		NewSUBARU[2)]	1.0 (1.5)	0.7	37
	Higashi-Hiroshima	HiSOR[2)]	0.7	0.87	400
	Saga	SAGA LS[3)]	1.4	1.9	25
Australia	Melbourne	Australian Synchrotron[3)]	3	1.9	12
Canada	Saskatoon	CLS[3)]	2.9	7.57	18.3
USA	Stanford	SPEAR 3[3)]	3	7.6	10
	Berkeley	ALS[3)]	1.9		2
	Stoughton	Aladdin[2)]	(0.8) 1.0	(0.55) 1.1	(120) 187

Country	Location	Name	Electron energy (GeV)	Critical photon energy (keV)	Horizontal emittance (nm · rad)
	Ithaca	CESR[1)]	5.5	11.0	
	Argonne	APS[3)]	7	19.5	2.5
	Gaithersburg	SURF III[2)]	0.38	0.146	
	Upton	NSLS/VUV[2)]	0.808	0.612	160
		NSLS/X-Ray[2)]	2.8	7.1	66
		+NSLS-II[3a)]	3		0.55
	Baton Rouge	CAMD[2)]	1.5	2.56	200
Brazil	Campinas	LNLS[2)]	1.37	2	100
		+Sirius[3a)]	3		0.28

†: Open or to be open to common users. + : under construction, # : superconducting bending magnet, 1): first generation, 2): second generation, 3): third generation, 3a): third generation (emittance<1 nm · rad).

Their emittances are very small. Some of them have been upgraded (see numbers attached behind the names of rings in Table 1.1). Some storage rings with extremely small emittances less than 1 nm · rad (see Sect. **3.2.5.e**) are under construction.[6,12,13)]

Since early 2000s, even 'fourth generation' synchrotron radiation sources have been seriously considered. An example of such a source would be a 'free electron laser' (FEL), which can generate coherent light using an undulator (see Sect. **3.5.2**). In an FEL, electron beams interact with the light in an undulator and emit exceptionally brilliant light. In 1977 in the USA, the first successful FEL experiment operated in the infrared region using electron beams from a linear accelerator (LINAC) and by making the undulator light traverse many times within an optical cavity consisting of two mirrors, thus causing it to interact effectively with the electron beam itself (a 'cavity-FEL'). Subsequently, FEL experiments have succeeded in the visible and ultraviolet regions using electron beams in storage rings. However in the wavelength regions shorter than ultraviolet, normal incidence mirror reflectivity decreases, so that it is very difficult to create an efficient optical cavity. Therefore, a new technique called SASE (self-amplified spontaneous emission) has been developed, in which the electron beam from the LINAC passes through the undulator only once and yet is still able to amplify the light.[15)] This type of FEL ('SASE-FEL') was first demonstrated in the VUV region at DESY in 2000. Later, a SASE-FEL in the VUV region was successfully operated at SPring-8 and recently ones in the X-ray region have operated at SLAC (USA) and SPring-8.[12,13)] Currently a major FEL facility, called the European XFEL is under construction in Germany and another type of FEL, a 'laser-seeded FEL' is operational at Elettra and at NSLS.[14)]

A different type of fourth generation source is the 'energy recovery LINAC' (ERL) type. Other advances have yielded electron beams with small emittance (<1 nm · rad) and with short bunches (<100 fs) from LINACs. These are properties which cannot be obtained from usual third generation storage rings. The use of such electron beams and undulators may give even more brilliant and shorter pulses of light in the future. However if the acceleration frequency of the LINAC is increased to provide large currents, similar to those in a storage ring, then the consumed power becomes very large. In order to solve this problem the ERL has been proposed. In an ERL, the electron beam is accelerated to its final energy, then passes through bending magnets situated in a ring and is returned back to the entrance of the accelerator structure of the LINAC and passed yet again through the acceleration structures. This time however the recycled electron beam generates a high frequency field in the accelerator structure, thus losing kinetic energy, which results in beam deceleration and it is finally discarded into a beam dump. The high frequency field so generated is used to accelerate the succeeding electron beam. In this manner, the energy of the electron beam can be substantially preserved and the energy losses of the system effectively minimized. Users can then exploit brilliant light obtained from any undulators situated in the straight sections between the bending magnets which are used to make the electron beam return to the entrance of accelerator structure. Of course the synchrotron radiation from the bending magnets can be used also. ERLs of a few 10s MeV have been successfully operated so far.[16,17] The most recent such facility is operational at Daresbury Lab. (UK).[18] They are used to generate coherent synchrotron radiation (CSR), infrared FEL and high energy light using inverse Compton scattering employing an optical laser. However it has not proved possible yet to exploit the technique in the GeV region. At present, ERLs in the GeV region have been proposed for Nobosibirsk (Russia),[19] BNL,[20] Cornel Univ.,[21] SLAC (USA) (Ref. 6 in Chap. **3**) and PF (Japan).

1.2 THE CHARACTERISTICS OF SYNCHROTRON RADIATION

1.2.1 The shape of electron beam and intensity units of synchrotron radiation

Figure 1.1 shows a schematic drawing of a storage ring. It illustrates the motions of electrons with their position coordinates and that of radiation from bending and undulator magnets with their angular coordinates. Electrons with nominal energy E circulate, oscillating with small amplitudes around an equilibrium orbit which coincides with a central orbit. These are called betatron oscillations. Electrons of energy $E + \Delta E$ circulate, oscillating similarly around another equilibrium orbit, a little apart from the central orbit. The ratio of energy spread and nominal energy σ_E/E is $5 \times 10^{-4} \sim 1 \times 10^{-3}$, where σ_E represents the standard deviation in a Gaussian distribution. These betatron oscillations and the associated energy spread result in a finite cross section and angular spread for the electron beam. They are expressed using σ_x and σ_x' in the horizontal direction, and σ_y and σ_y', in the vertical direction. If the contribution of energy spread in these values is small, the emittance in each direction is approximately equal to $\sigma_x\sigma_x'$ and $\sigma_y\sigma_y'$, respectively. The shape of the cross section is nearly elliptical. In typical third generation synchrotron radiation sources, $\sigma_x = 0.05 \sim 0.5$ mm, $\sigma_y = 0.01 \sim 0.05$ mm, $\sigma_x' = 0.01 \sim 0.05$ mrad and $\sigma_y' = 0.005 \sim 0.01$ mrad. In second generation sources, these values are larger by one or two orders of magnitude. The energy lost through the emission of synchrotron radiation is replenished using high frequency (RF) cavities which operate typically at frequencies from about 100 to 500 MHz. As a result the electron beams circulate in bunches created by this accelerating RF field. The bunch length is $\sigma_l = 1 \sim 30$ cm, which corresponds to a bunch width of from 30 ps to 1 ns. The inter-bunch interval is from 2 to 10 ns. When only one bunch ('single bunch operation') or a few bunches separated by equidistant intervals circulate in a storage ring, one can obtain short pulses of light separated by intervals from 0.1 to 1.0 μs.

The intensity of synchrotron radiation is expressed as the number of photons emitted in 1 s from the stored current in the storage ring. In Chap. **2**, the expression for the intensity is given per unit of stored current I (mA). Let us define the number of emitted photons in 1 s to be the photon flux. This definition is the standard one except that it is now applied to a specific, 0.1% relative band width (bw), or $\Delta\lambda/\lambda = \Delta\omega/\omega = 10^{-3}$. Let us now define the number of emitted photons from a bending section in 1 s in a horizontal angle of $\Delta\psi = 1$ mrad to be the angular flux. This is the result of the integration of intensity with respect to the vertical angle $\hat{\theta}$. This is sometimes called the flux or flux density. This definition is useful only for light from a dipole bending section. Finally, let us define the number of photons emitted in 1 s in $\Delta\psi = \Delta\hat{\theta} = 1$ mrad divided by the cross section of electron beam to be the brilliance (which is sometimes called the brightness). Of course, synchrotron radiation is pulsed light and so the peak brilliance has to be calculated by taking into account the peak intensity. The unit of brilliance is useful for the characterization of radiation from bending magnets and from undulators. The units of synchrotron radiation intensity are summarized as follow.

$$\begin{aligned} \text{flux} &= \text{photons/s/0.1\% bw} \\ \text{angular flux} &= \text{photons/s/mrad/0.1\% bw} \\ \text{brilliance} &= \text{photons/s/mrad}^2/\text{mm}^2/0.1\%\text{ bw} \end{aligned} \tag{1.1}$$

There are several software programs for the calculation of synchrotron radiation intensities, which are freely available.[22)]

1.2.2 Light from bending sections

In Fig. 1.2, the angular fluxes of synchrotron radiation from various sources are plotted against photon energy on a log-log scale. The light from each bending section has a smooth continuum spectrum, characterized by a critical energy u_c(keV) = 2.22E^3 (GeV)/R(m) (see Sect. **2.3.3**), where R is the radius of the circular electron orbit in the bending magnet. The angular flux shows a maximum at around 0.3 u_c. As can be seen in this figure, the intensity decreases gradually on the low energy side and rapidly on the high energy side in this plot. The photon flux can be obtained from product of angular flux and acceptance angle $\Delta\psi$. The magnetic field of the bending magnet is given by B(T) = 3.34E(GeV)/R(m) (see Sect. **2.1.2**). Therefore, the critical energy is proportional to the magnetic field, for fixed electron energy. The bending magnets of BESSY II, PF and SPring-8 in Fig. 1.2 from which synchrotron radiation is emitted are of about 1 T. On the other hand, superconducting magnets can generate magnetic fields of more than 4 to 10 T, so that by using them one can obtain photons with energy several times higher than those from bending magnets. Some superconducting wigglers are used specifically for this purpose for example in BESSY II SW. They are the wavelength shifters mentioned above. There are two storage rings, in which the bending magnets themselves are superconducting. They are AURORA and Helios 2. In the light source at Nagoya, Japan, both normal and superconducting magnets are used.

The angular spread of light from a bending section in the vertical direction is small around the critical energy u_c which is about $1/\gamma$ (see Sect. **2.3.1**). Here $\gamma = E/mc^2$ which is called Lorentz factor and mc^2 = 0.511 MeV which is the rest energy of the electron. Actually E is the total energy, or the sum of the acceleration energy and the rest energy, but in the case of such high energy electrons, the total energy is almost equal to the acceleration energy. The angular spread in the vertical direction depends on the photon energy and becomes greater with decreasing photon energy. The light from bending sections emitted in the median plane is totally linearly-polarized and that emitted above or below the median plane is, in general, elliptically-polarized. When the vertical observation angle is large, the light is essentially circularly-polarized.

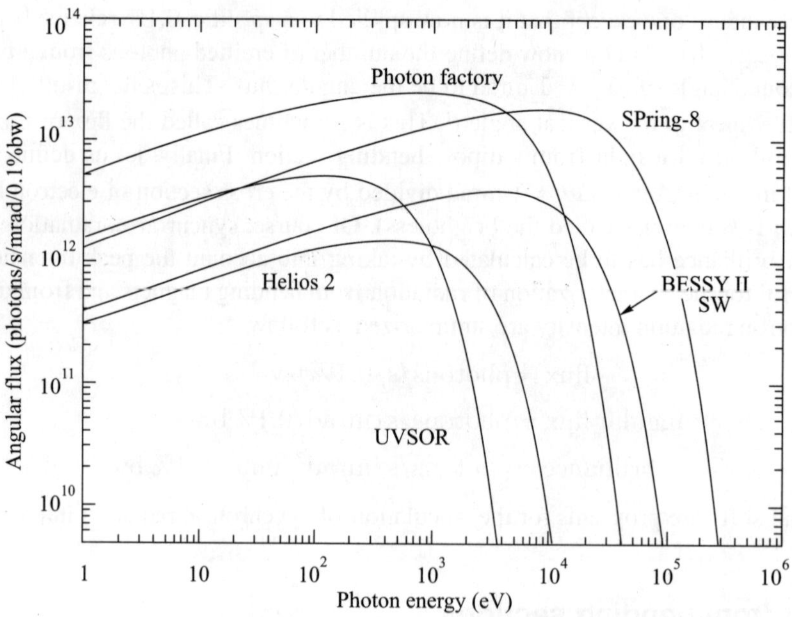

Figure 1.2 Angular flux spectra of light from bending sections. SW: superconducting wiggler.

1.2.3 Light from undulators

In undulators, the magnetic field changes sinusoidally along the orbit. The period λ_u is usually longer than 1 cm. The number of periods can be chosen to be from several to more than 100. Figure 1.1 shows the schematic of a planar undulator, in which magnetic fields are directed vertically and the undulation of electron beam is in the horizontal plane and the angle coordinates of the radiation are shown. The radiation from a planar undulator is a spread of quasi-monochromatic light beams consisting of the fundamental and its higher harmonics. Their wavelengths are given by

$$\lambda = \frac{\lambda_u}{2n\gamma^2}\left(1 + \frac{K^2}{2} + \gamma^2\theta^2\right) \tag{1.2}$$

(see Sect. **2.4.1**), where n is integer expressing the order of the harmonics, $\theta^2 = \hat{\theta}^2 + \psi^2$ and $K = 93.4\lambda_u(\mathrm{m})B_0(\mathrm{T})$ is called deflection parameter. The maximum of the magnetic field is B_0. When the magnetic fields are not too large and the deflection angle is small and comparable to $1/\gamma$, quasi-monochromatic light is obtained. In this case, $K \approx 1$. The relative band width of the quasi-monochromatic light is given by $\Delta\lambda/\lambda \approx 1/nN_u$, where N_u is number of periods. By the use of planar undulators, one can obtain totally linearly-polarized light. By the use of helical undulators in which the electrons move on helical orbits, one can generate circularly- or elliptically-polarized light. In the case of circularly-polarized light, the second term $K^2/2$ in parenthesis in Eq. (1.2) has to be replaced by K^2 and n takes only 1 (see Sect. **2.4.2.b**). The wavelength of the quasi-monochromatic light can be altered by changing K, that is, $B_0(\mathrm{T})$, without changing the electron energy. In many cases, the undulators are composed of permanent magnets, so that $B_0(\mathrm{T})$ can be changed mechanically by altering the magnet gaps.

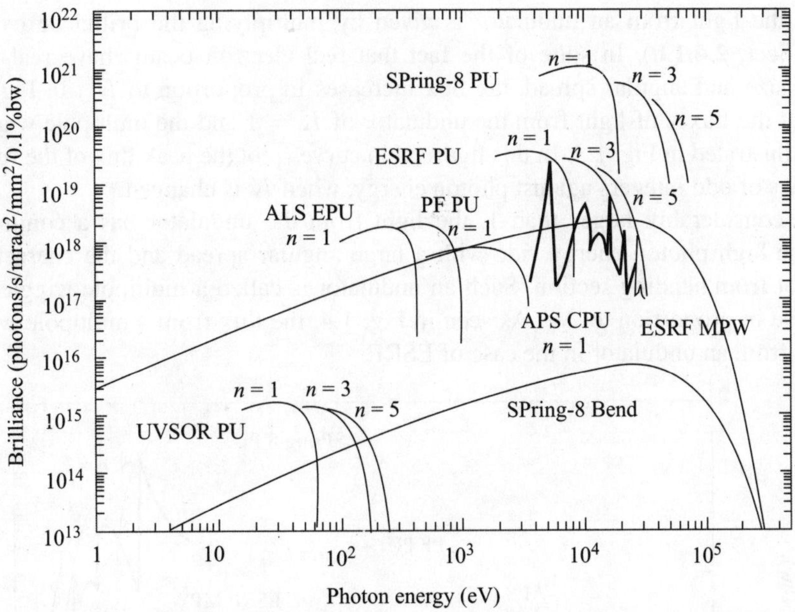

Figure 1.3 Brilliance spectra of light from an undulator (PU: planar, EPU: elliptical-linear polarization changeable, CPU: circular-linear polarization changeable), bending section and multipole wiggler (MPW).

The radiation emitted by an electron has an intensity distribution with close to axial symmetry, even when it passes through a planar undulator. Its angular spread $\sigma'_{\rm r}$ is smaller than the angular spread of the light from bending section and when $K \approx 1$, $\sigma'_{\rm r} \approx 1/\gamma\sqrt{nN_{\rm u}} \approx \sqrt{\lambda/L_{\rm u}}$, where $L_{\rm u}$ is the total length of the undulator (see Sect. **2.4.1.b**). When the light has the minimum product of cross section and angular spread (Gaussian beam)[23], $\sigma_{\rm r}\sigma'_{\rm r} \approx \lambda/4\pi$, where $\sigma_{\rm r}$ is the radius of the light beam.[24] Then the light from undulator has the finite size of $\sigma_{\rm r} \approx \sqrt{\lambda L_{\rm u}}/4\pi$. The total beam size and angular spread of the light in which both the contribution of electron beam and photon beam are taken into account, are defined to be Σ_x, Σ'_x (in the horizontal direction) and Σ_y, Σ'_y (in the vertical direction). These quantities are given by $\Sigma_x = \sqrt{\sigma_{\rm r}^2 + \sigma_x^2}$, $\Sigma_y = \sqrt{\sigma_{\rm r}^2 + \sigma_y^2}$, $\Sigma'_x = \sqrt{\sigma_{\rm r}'^2 + \sigma_x'^2}$ and $\Sigma'_y = \sqrt{\sigma_{\rm r}'^2 + \sigma_y'^2}$.

The light output from undulators is best defined by their brilliance. Their brilliance is proportional to $N_{\rm u}^2$ in ideal case. In Fig. 1.3, thick curves represent the brilliance of light from a planar undulator of $K = 1$ at $\theta = 0$, from a bending section and from a multipole wiggler mentioned later. As can be seen in this figure, the higher harmonics of odd integers are stronger than the adjacent higher harmonics of even integers in the planar undulator. Thin curves plot the peak brilliances of the fundamental and higher harmonics of odd integers against photon energy, when K is changed ($0 \sim 2.5$). For circularly-polarized light from a helical undulator, only the fundamental is strong. The brilliances of the fundamentals can be more than 10^3 times larger than those of the light from a bending section. The brilliance increases with increase in $N_{\rm u}$, but it is not proportional to $N_{\rm u}^2$ when $N_{\rm u}$ becomes very large. This is due to the energy spread of the electron beam, which prevents the relative band width from becoming small. The longest undulator inserted in any storage ring is one of 25 m at SPring-8. The brilliance of the fundamental from this undulator exceeds 10^{20} photons/s/mrad^2/mm^2/0.1% band width in the 8 to 18 keV region.

The flux of the light from an undulator is given by multiplying the brilliance by $2\pi\Sigma_x\Sigma_y$ and $2\pi\Sigma'_x\Sigma'_y$ (see Sect. **2.4.1.b**). In spite of the fact that real electron beams have real energy spread, cross-sectional size and angular spread, the flux increases in proportion to N_u. In Fig. 1.4, the thick curves represent the fluxes of light from the undulator of $K = 1$ and the multipole wiggler which are the same ones illustrated in Fig. 1.3. In this figure, thin curves plot the peak flux of the fundamental and higher harmonics of odd integers against photon energy, when K is changed.

When K is considerably larger than 1, the light from the undulator has a continuum spectrum extending to the high photon energy side with a large angular spread and the character of the light is similar to that from bending section. Such an undulator is called a multipole wiggler. Its brilliance and flux increase in proportion to N_u. As seen in Fig. 1.4, the flux from a multipole wiggler is much larger than that from an undulator in the case of ESRF.

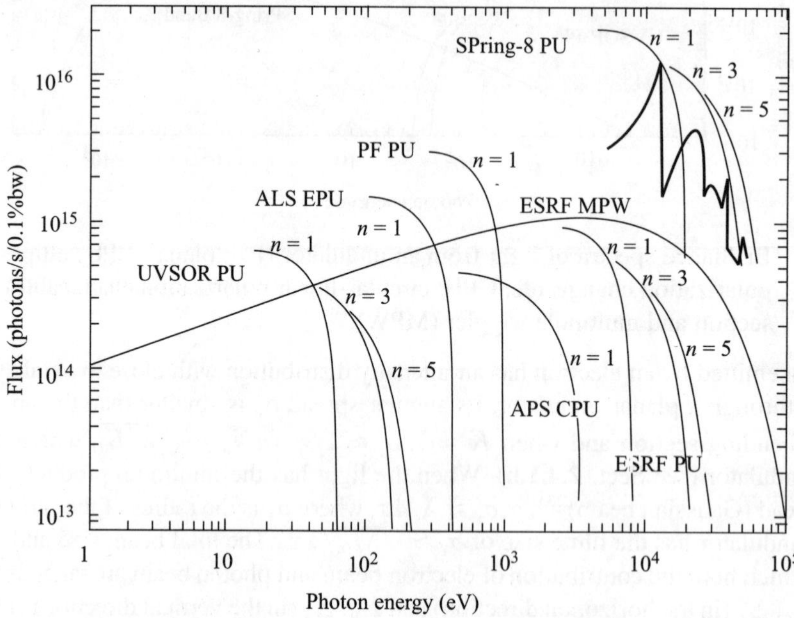

Figure 1.4 Flux spectra of light from an undulator (PU: planar, EPU: elliptical-linear polarization changeable, CPU: circular-linear polarization changeable) and multipole wiggler (MPW).

1.2.4 Coherence

There are two types of coherence of light, both transverse and longitudinal. Transverse coherence expresses the probability of interference of light at two different points in a plane perpendicular to the direction of propagation. Generally speaking, xx' which is the product of the size of object (x) and divergence angle (x') is conserved at the focal point in imaging optics. The invariant is called the Lagrange invariant (see App. **A**). The invariant cannot be smaller than $\lambda/4\pi$ mentioned above. This value is called the diffraction limit and the light having this value interferes completely. When the emittance of the electron beam is smaller than the diffraction limit, the size and angular spread are those defined by the source and light at the diffraction limit can be obtained. In third generation synchrotron radiation sources, light at the diffraction limit can be obtained in the VUV range, but not in the soft and hard X-ray regions. Longitudinal coherence expresses the possibility of interference of light at two different points along the line of propagation. The longitudinal coherence length is given by $l_c = \lambda^2/\Delta\lambda$, where $\Delta\lambda$ is band width. In the case of the light from an undulator, $l_c = nN_u\lambda$, because $\Delta\lambda/\lambda \approx 1/nN_u$. The light from objects in this length l_c can interfere. Coherence can be enhanced by the use of optical elements, but the intensity is then decreased. The light from any undulator is partially coherent and its coherence is higher than that of laboratory sources, so that it always has some advantages. The degree of coherence, however is much lower than that of a laser. Photon degeneracy is one measure of coherence, which is given by $\delta = B_p\lambda^3/4c$, where B_p is peak brilliance given in units using m, rad and 100% band width, and c is the velocity of light.[25] This value is less than 1 for light from an undulator, while it is much larger than 1 for a laser. The degeneracy δ expresses the density of light which can interfere. For interference experiments, the larger δ becomes, the better.

1.2.5 The light available at beamlines

Beamlines are used to transfer synchrotron radiation to the experiments. A standard beamline is composed of a front end, pre-mirrors, a monochromator, refocusing mirrors (post-mirror) and instruments for measurements. A schematic layout is illustrated in Fig. 1.1. In VUV and soft X-ray beamlines, the vacuum chambers are connected directly to the vacuum chamber of the storage rings without windows, while for hard X-ray beamlines, the experimental vacuum chambers are separated from the storage ring vacuum by Be windows. Pre-mirrors are often used to collect a wide angular range of synchrotron radiation for beamlines used below about 30 keV. In some cases, multilayer coated mirrors are used as the pre-mirror. For monochromatization of synchrotron radiation, gratings are used in the region below 1.5 keV and monochromatizing crystals, above 1 keV. Figure 1.5 shows schematics of grating and crystal monochromators. For advanced experiments, high resolution monochromators equipped with concave gratings of large radii or plane gratings with varied groove space are used. Even four-crystal monochromators utilizing back scattering with asymmetric reflection have been developed.

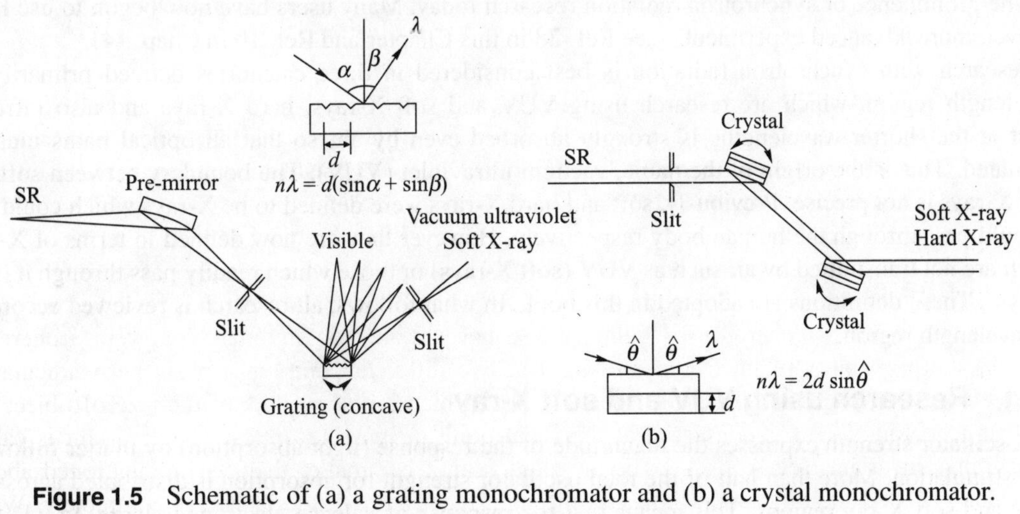

Figure 1.5 Schematic of (a) a grating monochromator and (b) a crystal monochromator.

When synchrotron radiation reaches the sample after passing through the collecting or mochromatizing optical elements, its intensity will be decreased because of their limited efficiency. Intensity is also diminished by any misalignment of optical elements. Therefore high precision mirrors, efficient monochromatizing elements and precise alignment of all optical elements are essential in the assembly of beamlines to obtain maximum intensity.

Examples of the intensity obtained at sample positions are as follow. At 0.4 keV, 5×10^8/s photons are obtained with a band width of 40 meV, at 10 keV, 4×10^{12}/s with a relative band width of 0.02% and at 14 keV for a special use, 1.6×10^9/s with a band width of 2.5 meV. Photon beams with cross sections from 1 to a few mm^2 can be obtained without the use of post-mirrors behind monochromators. With post-mirrors, even smaller cross sections are obtained. By using microscopic mirrors or a Fresnel zone plate, microbeams of diameter from a few 10s to a few 100s of nm can be obtained (see Sect. **14.1**). When the wavelength of the fundamental or of higher harmonics of the light from an undulator is tuned to the wavelength chosen by the monochromator (by adjusting the magnet gap), one can always obtain very intense monochromatic light. Note that the polarization characteristics of synchrotron radiation change as it passes along the beamline and will not necessarily be maintained at the sample position. For details of optical techniques, see Chap. **5**.

1.3 RESEARCH USING SYNCHROTRON RADIATION

In general, when light interacts with materials the phenomena of refraction, absorption and scattering are the primary events which take place. Such phenomena are well understood and described macroscopically using optical constants of materials, which are specific to each material (see Sects. **4.1** and **4.2**). Additional processes such as photoelectron emission and fluorescence emission may also be stimulated. By studying these phenomena, one can discover information about the electronic and atomic structure of any material. Synchrotron radiation provides such exceptional brilliance across such an extensive wavelength range that it has been used in all the basic sciences, including materials sciences,[26)] life sciences[27)] and environmental sciences over many decades, resulting in the significance and the prominence of synchrotron radiation research today. Many users have now begun to use FELs for even more advanced experiments (see Ref. 28 in this Chapter and Ref. 10 in Chap. **14**).

Research with synchrotron radiation is best considered in three categories defined primarily by wavelength region, which are research using VUV and soft X-rays, hard X-rays and also infrared. Light at the shorter wavelengths is strongly absorbed even by air, so that all optical paths must be evacuated. This is the origin of the name, vacuum ultraviolet (VUV). The boundary between soft and hard X-rays is not precise. Previously, soft and hard X-rays were defined to be X-rays which could not, or could pass through the human body respectively. However they are now defined in terms of X-rays which are not transmitted by air such as VUV (soft X-rays) or those which readily pass through it (hard X-rays.) These definitions are adopted in this book. In what follows, all research is reviewed according to wavelength region.

1.3.1 Research using VUV and soft X-ray

The oscillator strength expresses the magnitude of the response (light absorption) by matter following light stimulation. More than half of the total oscillator strength for absorption is distributed across the VUV and soft X-ray regions. This means that the response of valence electrons is large. In addition, the response of electrons in shallow cores is also important. In the hard X-ray region, the response still exists, but absorption of light becomes weaker. In the visible region, there are many opaque materials such as metals. This does not mean all materials have a large absorption only in the VUV and soft X-ray regions, but means that many materials have large absorption at lower energies. Given that absorption is large in the VUV and soft X-ray regions, absorption spectroscopy using VUV and soft X-ray radiation enables us to determine energy levels in many materials, so that it can provide information on their electronic structures.[29] Such light can excite electrons above the vacuum level and expel them from the material as photoelectrons. By measuring the energy distributions of such photoelectrons, the electronic structure of the ground state can be understood. Such studies of electronic structure are indispensable not only in basic research, but also for the characterization of new materials and an understanding of their functions.

a. Atoms and molecules: The first important experiment using synchrotron radiation was the measurement of absorption spectra originating from a two-electron excitation in rare gases. This state is an autoionization state mixed with a one-electron ionization state, which had previously been investigated only by inelastic electron scattering experiments. From absorption measurements using VUV and soft X-rays, excited states of valence electrons in atoms and molecules have now been observed, including highly excited states of valence electrons, two-electron excitation states, superexcited states and others, and excited states of core electrons as well. The core electrons are located inside the valence electrons and they do not contribute directly to bonding in molecules. However the excitation of core electrons can give considerable information about excited states. Using photoelectron spectroscopy, ion yield spectroscopy and coincidence measurements of excited electrons and ions, the dynamical processes in photodissociation have been studied. Recently, high resolution monochromators have become available, so that the Auger and photodissociation processes in molecules consisting of light elements have been investigated by identifying even the vibrational levels (see Chap. **6**). Further experiments have been made on highly excited states in molecules aligned by lasers with probing by synchrotron radiation. Basic studies have been made on various photochemical processes associated with the upper atmosphere.

b. Solids and surfaces: Spectroscopic studies on solids started with simple absorption measurements. From absorption spectra due to excitations of valence and core electrons, the justification for band calculations was made for metals, semiconductors and ionic crystals. Core exciton absorption has also been studied and subsequently photoelectron spectroscopy has flourished. By angle-integrated photoelectron spectroscopy, the density of states of valence bands can be studied. From angle-resolved experiments, the determination of band dispersion in the valence band has become possible for semiconductors and many other materials. Photoelectron diffraction experiments are now feasible. By the use of both these techniques, the structures of clean and adsorbed surfaces have been clarified. In addition, spin-resolved photoelectron spectroscopy has been developed and the density of states of majority and minority spins can now be obtained separately for magnetic materials. Recently, high resolution photoelectron spectroscopy has become available with band widths of a few meV. Using excitation light of from about 500 eV to a few keV, information concerning the bulk, rather than the surface properties can be obtained. Materials of choice for research have now been extended from simple compounds to complex novel materials, such as high T_c superconductors, heavy electron systems, nano structure compounds and so on. They include not only inorganic materials, but also organic materials whose applications have been hugely extended. Using circularly-polarized light, magnetic circular dichroism (MCD) experiments have become straightforward. Even the MCD measurement of core absorption and the determination of the orbital and spin magnetic momenta have become possible. This exploits the fact that the symmetry of the core level with respect to angular momentum is well defined. The method is also popular in the hard X-ray region. Since such intense light is available at all sample positions, even core emission spectroscopy has become possible where light (fluorescence) is emitted when a core hole is buried with an electron at higher levels. It is possible to observe normal fluorescence whose spectral shapes are independent of excitation energy and also anomalous emission whose spectral shape changes and emission bands shift with scanning excitation energy around absorption edges. They are spectator emission and inelastically scattered (Raman) light. (Inelastic scattering in the hard X-ray region is considered in Chap. **12**). More details are given in Chap. **7**.

c. Photochemical processes: Because the absorption coefficients of materials in the VUV and soft X-ray regions are so large, excitation in this region offers the advantage to obtain chemically active specimens *via* the ionization or by the dissociation of isolated molecules. These are molecular ions at highly excited states and radicals. When allowed to react with other molecules, one can obtain materials which cannot be synthesized by normal chemical processes. This kind of reaction is a called a 'photochemical process' (see Sect. **8.1**). Experiments which comprise the making of thin films on solid surfaces with simultaneous irradiation of exposing gases and the solid surfaces can be carried out. Conversely, etching of solids using photo-activated ions is possible. There exist other irradiation experiments using synchrotron radiation to make thin metal films by dissociating organometallic molecules deposited on solids and by making hydrogen-passivated Si surfaces clean by desorbing hydrogen.

d. Lithography: When light shines a small object, the light is diffracted and the edge of its shadow becomes diffuse. The amount of the diffuseness is proportional to wavelength of the light. The wavelength is smaller in VUV and soft X-ray ranges than in the visible. Therefore one can transcribe finer patterns using VUV and soft X-rays than is feasible in the visible. This is called X-ray lithography. Mask patterns with line width of a few 10s of nm are made initially by electron beam lithography and then transcribed with high fidelity using irradiation with synchrotron radiation. By superimposing many such fine patterns on the same substrates, one can fabricate ultra-large-scale integrated circuits (ULSI) or even micromachines such as micro gears, actuators and so on. Especially for the latter, the advantage of soft and hard X-rays is that these rays penetrate deeply into the photoresist. At present, ULSIs are made using ultraviolet lasers. Lithography using light other than synchrotron radiation called extreme ultraviolet (EUV) lithography has been developed, but 'micromachining' to create miniature devices for practical use are now made using synchrotron radiation (see Sect. **8.2**).

e. Microscopy: Microscopes with improved spatial resolution can readily be made (< 50 nm) using VUV and soft X-ray rather than using visible, for the same reasons mentioned above. The light can excite core electrons of any materials, so that element-selected excitation is possible. Therefore one can obtain images which distinguish between different elements by choosing the appropriate photon energy (called 'element mapping' or 'chemical mapping'). The binding energies of core levels differ from compound to compound (by their chemical shifts), so that one can obtain microscopic images to identify any chosen element or compound as described in Chap. **14**. This is of course valid also in microscopes using hard X-rays which have been intensively developed by the use of mirrors at grazing incidence or Fresnel zone plates (often abbreviated as zone plate). Using such microscopes, structures of materials composed of small grains or biological materials have been observed. For the study of biological materials consisting of light elements C, N, O and H with good contrast in water, observations in the 'water window' region is most appropriate, whose energy region is higher than the C K absorption edge and lower than the O K absorption edge.

1.3.2 Research using hard X-rays

The most prominent characteristic of X-rays is their large transmissibility through matter. However, their transmittance differs from element to element, which offers contrast information, so that the contrasting transmission images can easily be obtained. This is the principle of Röntgen photographs. The core absorption spectra in the hard X-ray region can also provide information concerning the electronic structures of materials as in the VUV and soft X-ray regions and information to clarify local structures around selected atoms. Following photoabsorption, materials revert to their ground states, emitting characteristic (fluorescent) X-rays or Auger electrons, which have element-specific energies and are used for element analysis. Those X-rays not absorbed by matter may still be scattered by electrons. There are two main scattering processes. The first one is elastic scattering without energy loss and the other inelastic scattering in which a part of energy of X-ray is lost (see Sects. **4.2.3** and **4.3.3**). The former is called 'Thomson scattering', which is used for crystal structure analyses or crystallography.[30)] The latter includes 'Compton scattering' and 'X-ray Raman scattering', which are used for studies of electronic structures.

a. Structure analysis: X-ray structure analysis is a powerful tool to determine the lattice type and lattice constants in periodic structures such as crystals. This method utilizes elastic X-ray scattering from periodic lattice planes, in which X-rays of certain wavelengths are reflected at specific angles given by the Bragg diffraction condition. Instruments for structure analysis using characteristic X-rays were normally used in many laboratories. However the merits of synchrotron radiation have brought great advances to structure analysis research (see Chap. **9**). The foremost advantage is that of tuneability, giving the ability to select any photon energy. Using characteristic X-rays, the photon energy is of course fixed. If one measures diffraction patterns by choosing a specific X-ray core absorption edge of an element in any compound, one can obtain structural information about the compound and readily discover new features such as superlattices *etc*. A second advantage is simply that of high brilliance. With such benefits, diffractometry becomes possible on very small crystals and permits the use of small diamond anvil cells for high pressure instruments and so on. Structure analyses have been made on micro samples such as powders freshly synthesized, dusts from the cosmos and so on. X-ray diffractometry gives not only structure parameters such as lattice distances, but also the electronic distributions. Utilizing the latter, one can measure the positions of atoms even in extremely large unit cells. With this information, the structures of many proteins have been visualized to identify, for example, the positions of functional units in proteins where specific biological functions are exhibited and to observe the changes of their conformation. By the use of the high parallelism of synchrotron radiation, the structures of muscles, collagens and other fibrous materials which have large lattice constants have been investigated by especially small angle diffraction experiments. Elastic X-ray scattering provides the structural information of not only periodic systems, but also non-periodic systems. Clarified are for instance, the shapes of small objects such as macromolecules of biological materials by small angle scattering, the surface layer structures by grazing incidence scattering, the distribution of atoms in amorphous solids and liquids, and so on.

b. EXAFS: In core absorption spectra in the hard X-ray region, strong and complicated features are found around the core absorption edges, as is also seen in VUV and soft X-ray regions. These X-ray structures are called XANES (X-ray absorption near edge structure) with which one can identify unknown materials such as biological materials, materials in pharmaceutical processes, waste remnants in incinerators and so on. When the photon energy becomes large and remote from the absorption edge, the intensity of the core absorption decreases with a weakly oscillating structure. These spectral features are called EXAFS. Both XANES and EXAFS are called XAFS (X-ray absorption fine structure). The EXAFS structure results from the interference between a wave of an electron emitted from an excited atom and that scattered by the surrounding atoms. Therefore the analysis of the EXAFS structure can yield detailed information about the local structure around the excited atom. This method can also be applied to samples which do not have any periodicity (see Chap. **10**). Therefore it has now become widely used in the analysis of local structures in complex molecules, amorphous solids and liquids.

c. Fluorescence: 'X-ray fluorescence analysis' is a method to measure the energies of fluorescent X-rays emitted from any material which then allows element identification *via* their characteristic X-ray spectra. It is a powerful tool to identify trace elements contained in materials. In commercial instruments, X-ray tubes are usually used. The advantage in using synchrotron radiation is due to very high sensitivity (see Sect. **11.1**). Microscopic measurements can be made using X-ray microbeams as the excitation light (see Sect. **14.2.2**). Currently, many researchers are competing in the detection of elements at exceedingly small concentrations. By the use of synchrotron radiation sources of large energy, one can use high energy X-rays. In this case, the K shell electrons of heavy elements can be excited easily. It has the advantage that one can measure fluorescent X-rays of specific elements with minimal overlap from signals from fluorescent X-rays of other elements. This cannot be achieved by L shell excitation. The method is used in many fields including the identification of materials in archaeological artifacts, heritage science experiments and also in criminal forensic investigation *etc.* Recently, 'X-ray fluorescence holography' in which the angle dependence of the intensity of fluorescent X-rays has been measured, has been developed to determine the local structure around the atom emitting the fluorescence (see Sect. **11.2**).

d. Inelastic scattering: In inelastic scattering the incident X-ray photon is scattered by an electron and gives energy to it. In this case the momentum of the scattered X-ray differs from that of the incident X-ray. The spectra of the scattered X-rays are obtained to measure the given energy (energy loss) with scanning the scattering angle or the excitation energy to change the transfer momentum (see Chap. **12**). This experimental method is similar to that of fluorescent X-ray spectroscopy. One type of inelastic scattering is that of 'Compton scattering'. Using this principle, one can study electronic structure through the measurement, the momentum distribution of electrons in metals, for instance, which have a free-electron like behavior. Another example is that of 'X-ray Raman scattering', in which the X-rays are scattered not by such free electrons, but by tightly bound electrons. These experiments give results which reflect the absorption spectra. In this case, the measurement is of scattered intensity as a function of the energy lost from incident light. Using this technique, soft X-ray spectra can also be obtained using hard X-rays.

e. Imaging: Since the date of their discovery the most universally important application of X-rays has been that of imaging, as can be seen from the many health checks taking, for example, X-ray transmission photographs of breasts. Transmission imaging is used not only for diagnosis in medicine and dentistry, but also for inspection of defects in metals and many other fabricated devices in industry. The advantages of synchrotron radiation are its ability to see very small objects with good contrast and with good spatial resolution. Using a well collimated X-ray beam from a brilliant synchrotron radiation source, the transmission images can be obtained with spatial resolution of about 1 μm. Furthermore, the microscopic technologies developed in the soft X-ray region mentioned above have been extended to the hard X-ray region and magnified images of small objects can now be obtained with a resolution of much less than 1 μm. Using the coherence of synchrotron radiation, holography and study of speckles are also in progress (see Sect. **14.1**). The very high brilliance has enabled X-ray topography to be carried out, in which diffraction patterns of Bragg or Laue types are of crystals taken using a collimated X-ray beam to observe images originating from defects in crystals. It is a powerful tool to investigate the structural integrity of crystals (see Sect. **14.2**).

1.3.3 Research using the Infrared

Synchrotron radiation from bending sections is important not only in the short wavelength regions, but also in the very long wavelength region. The advantage is that in the infrared and far infrared regions, synchrotron radiation is much more brilliant than normal blackbody radiation. This is due to the fact that synchrotron radiation is emitted from such a small source size and with a small angle of vertical divergence, so that its brilliance is considerably higher than blackbody radiation which is emitted isotropically. Even though the angular spread of synchrotron radiation is larger in infrared than in VUV, soft X-ray and hard X-ray regions, the directionality is better defined than in the usual infrared light sources. In the infrared region, absorptions due to molecular vibrations and lattice vibrations, and free electrons and electrons at shallow levels can be studied. Magnetic circular dichroism experiments have been made using synchrotron radiation on magnetic metals and also the analyses of structures of adsorbed molecules on surfaces have been studied (see Chap. **13**). Nowadays, many beamlines for infrared microscopy have been constructed and even microscopic images of biological materials have been taken in order to map the distribution of special molecules in the sample (for the microscope, see Sect. **14.2.3**). The resolution is usually larger than μm in the infrared range.

1.4 COMPARISON WITH OTHER LIGHT SOURCES

In order to compare synchrotron radiation with alternative light sources, one needs to use common units of intensity. In the following, the average brilliance in time is used. Figure 1.6 shows brilliances of some synchrotron radiation sources, SASE-FELs,[31] X-ray tubes,[32] and cavity-FEL and other various light sources.[33]

The light sources in the VUV and soft X-ray ranges in the laboratory fall into two groups, first, line sources (not shown in the figure) and second, continuum sources. Line sources are, for example, hydrogen and rare gas discharge lamps. These sources are not useful in comparison with synchrotron radiation for absorption spectroscopy which needs wavelength scanning, but the rare gas discharge lamps will continue to be used as excitation sources for photoelectron spectroscopy. Many types of VUV and soft X-ray lasers have been developed utilizing higher harmonic generation from visible and ultraviolet lasers and excitation by the use of excimer and other lasers.[24] These lasers emit intense light with a line spectrum with high coherence and have been used to take speckle patterns by shining them on solid surfaces and more recently to undertake high resolution photoelectron spectroscopy. The intensities of some soft X-ray lasers per pulse are very large although the repetition periods may be

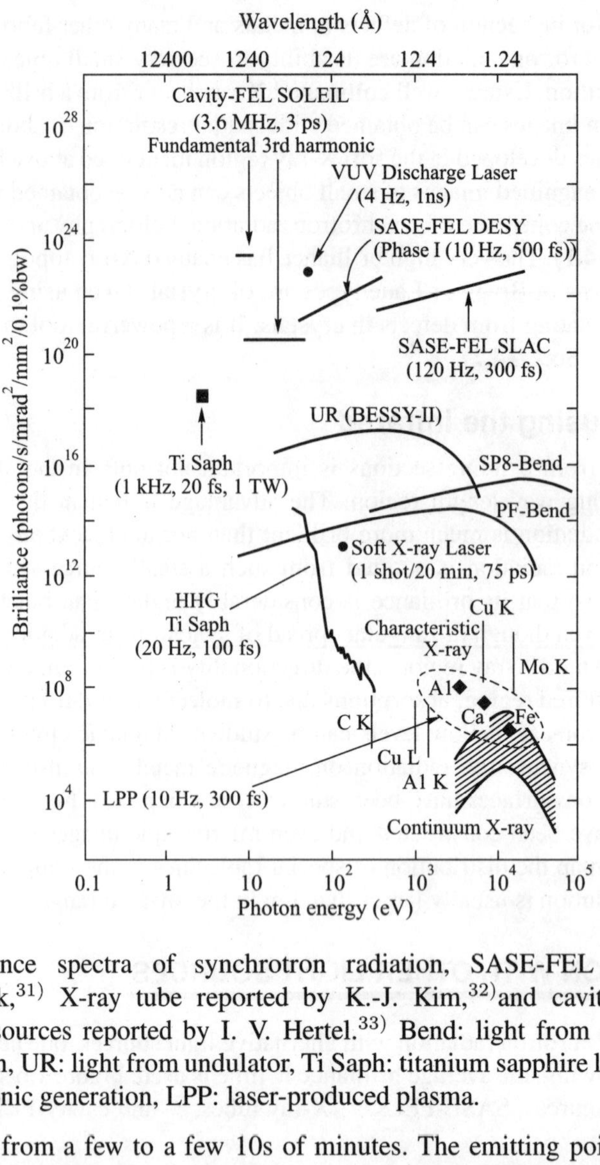

Figure 1.6 Brilliance spectra of synchrotron radiation, SASE-FEL reported by H. Winick,[31)] X-ray tube reported by K.-J. Kim,[32)] and cavity-FEL and other light sources reported by I. V. Hertel.[33)] Bend: light from a normal bending section, UR: light from undulator, Ti Saph: titanium sapphire laser, HHG: higher harmonic generation, LPP: laser-produced plasma.

very long, for instance from a few to a few 10s of minutes. The emitting point and angle may both change slightly from shot to shot. The continuum sources are laser-produced plasma light sources.[24)] In them, infrared radiation from a Nd:YAG laser is used, for instance to irradiate a metal to produce plasma on its surface which emits VUV and soft X-rays. When a light metal element is irradiated, a line spectrum is available, while a heavy element yields a continuum spectrum mixed with line spectrum. The intensity from the Nd:YAG laser fluctuates pulse by pulse, so that monitors are essential for intensity normalization. This source can be used in the laboratory, though the brilliance and the stability are much worse than those of synchrotron radiation.

Sources in hard X-ray in laboratories are X-ray tubes, which emit characteristic X-rays with weak continuum X-rays. Using characteristic X-rays, diffractometry can be made and using continuum

X-rays, even some EXAFS measurements are possible. However, synchrotron radiation is always required for diffractometry on small samples or with changing photon energy, and for all precise EXAFS measurements, especially for short timescales. Recently, however, time-resolved structure analyses using pulsed X-rays emitted from targets irradiated by femtosecond lasers (laser-produced plasma) have been achieved.[34)] There are alternative kinds of pulsed X-rays. The shortest pulse width of synchrotron radiation (*i.e.* the electron bunch width), it is about 30 ps, so that femtosecond experiments are not feasible with synchrotron radiation, but a novel technique has been developed to obtain femtosecond pulsed synchrotron radiation (100 fs) as mentioned in Sect. **3.3.3**.

REFERENCES

1. C. Kunz ed.: "Synchrotron Radiation. Techniques and Applications" (Springer-Verlag, 1979).

2. H. Winick and S. Doniach eds.: "Synchrotron Radiation Research" (Plenum Press, 1980).

3. E. E. Koch ed.: Series of "Handbook on Synchrotron Radiation" (North-Holland, 1983∼1991).

4. "Oxford Series on Synchrotron Radiation" (Oxford University Press, 1996∼).

5. D. H. Bilderback, K. O. Hodgson, M. P. Kiskinova and R. Rosei eds.: Series on "Synchrotron Radiation Techniques and Applications" (World Scientific, 1998∼).

6. "Synchrotron Radiation News" (Taylor & Francis Inc., 1988∼).

7. "Journal of Synchrotron Radiation" (International Union of Crystallography, 1994∼).

8. G. Margaritondo: "Elements of Synchrotron Light: For Biology, Chemistry and Medical Research" (Oxford University Press, 2002).

9. P. Willmott: "An Introduction to Synchrotron Radiation –Techniques and Applications–" (John Wiley & Sons, 2011).

10. J. P. Blewett: J. Synch. Rad. **5** (1998) 135.

11. T. Miyahara, H. Kitamura, S. Sato, M. Watanabe, S. Mitani, E. Ishiguro, T. Fukushima, T. Ishii, Shigeo Yamaguchi, M. Endo, Y. Iguchi, H. Tsujikawa, T. Sugiura, T. Katayama, Y. Yamakawa, Seitaro Yamaguchi and T. Sasaki: Particle Accelerators **7** (1976) 163.

12. Int. Conf. on Synchrotron Radiation Instrumentation has been held every 3 years. The proceedings are published in Nucl. Instr. Meth. A **467-468** (2001), AIP Conf. Proc. **705** (2004), AIP Conf. Proc. **879** (2007), AIP Conf. Proc. **1234** (2010), J. Phys.: Conf. Ser. **425** (2013) *et al.*

13. Int. Particle Accelerator Conf. has been held every year. The proceedings are published from each organizing institution.

14. Home pages of synchrotron facilities in the world can be accessed through http://www.lightsources.org.

15. H.-D. Nuhn and J. Rossbach: Synch. Rad. News **13** (2000) No.1, 18.

16. G. R. Neil, C. L. Bohn, S. V. Benson, G. Biallas, D. Douglas, H. F. Dylla, R. Evans, J. Fugitt, A. Grippo, J. Gubeli, R. Hill, K. Jordan, R. Li, L. Merminga, P. Piot, J. Preble, M. Shinn, T. Siggins, R. Walker and B. Yunn: Phys. Rev. Lett. **84** (2000) 662.

17. R. Hajima, T. Shizuma, M. Sawamura, R. Nagai, N. Nishimori, N. Kikuzawa and E. J. Minehara: Nucl. Instr. Meth. A **507** (2003) 115.
18. http://www.stfc.ac.uk/1320.aspx.
19. F. N. Kulipanov, A. N. Skrinsky and N. A. Vinokurov: J. Synch. Rad. **5** (1998) 176.
20. I. Ben-Zvi and S. Krinsky: Synch. Rad. News **14** (2001) No.2, 20.
21. D. Bilderback, I. Bazarov, K. Finkelstein, S. Gruner, G. Krafft, L. Merminge, H. Padamsee, Q. Shen, C. Sinclair, R. Talman and M. Tigner: Synch. Rad. News. **14** (2001) No.3, 12.
22. For example, T. Tanaka and H. Kitamura: Synchrotron Radiation Calculation Software SPECTRA, http://radiant.harima.riken.go.jp/spectra/index.html.
23. R. Guenther: "Modern Optics" (John Wiley and Sons, 1990).
24. D. T. Attwood: "Soft X-Ray and Extreme Ultraviolet Radiation. Principles and Applications" (Cambridge University Press, 1999).
25. E. Gluskin, I. McNulty. P. J. Viccaro and M. R. Howells: Nucl. Instr. Meth. A **319** (1992) 213.
26. Int. Conf. on Synchrotron Radiation in Materials Science has been held every 2 years. The proceedings are published in Nucl. Instr. Meth. B **199** (2003) 1, Nucl. Instr. Meth. B **238** (2005) 1, Proc. SRMS-5 Conf. (Chicago, 2006), AIP Conf. Proc. **1092** (2009) *et al.*
27. Int. Confs. on Biology and Synchrotron Radiation (BSR) and on Medical Applications of Synchrotron Radiation (MASR) have been held every 2 or 3 years. The proceedings of 6th MASR is published in AIP Conf. Proc. **1266** (2010).
28. P. Bucksbaum, T. Moeller and K. Ueda eds.: J. Phys. B: At. Mol. Opt. Phys. **46** (2013) 160201.
29. Int. Conf. on Vacuum Ultraviolet Radiation Physics (Vacuum Ultraviolet and X-Ray Physics from 2010) has been held every 3 years. The proceedings are published in Surf. Rev. Lett. **9** (2002) 1, J. Electr. Spectr. Rel. Phenom. **144-147** (2005) 1, Appl. Phys. A **92** (2008) 421 (invited papers only), J. Electr. Spectr. Rel. Phenom. **184** (2011) 63 *et al.*
30. International Union of Crystallography (IUCr) has published the journal of Acta Crystallographica since 1948, which covers whole field of crystallography.
31. H. Winick: J. Synch. Rad. **5** (1998) 168.
32. K.-J. Kim: in "X-Ray Data Booklet" A. C. Thompson and D. Vaughan eds., (LBNL/PUB-490 Rev.2, Center for X-Ray Optics and Advanced Light Source, 2001), p.2-15.
33. I. V. Hertel: in Abstract of Satellite Workshop on "Lasers and Short Wavelength Applications" to the 7th Int. Conf. on Synchrotron Radiation Instrumentation (Berlin, 2000).
34. K. Kinoshita, H. Harano, K. Yoshii, T. Ohkubo, A. Fukasawa, K. Nakamura and M. Uesaka: Laser and Particle Beams **19** (2001) 125.

CHAPTER 2

Synchrotron Radiation is Emitted by Moving Electrons

Synchrotron radiation is emitted by moving electrons. Their energy is extremely high and therefore their velocity is almost exactly equal to the velocity of light. The consequences of the theory of relativity are considered first and later the general theory for the generation of electromagnetic waves emitted by moving charged particles is explained using Maxwell's equations. Both theories are applied to the case of the relativistic electrons traversing both circular and undulating orbits. The spectral shape and the polarization characteristics of synchrotron radiation are described in detail. The derivations of all these formulae are given in Apps. **B**, **C** and **D**.

2.1 THE RELATIVISTIC DYNAMICS OF MOVING ELECTRONS

2.1.1 The special theory of relativity

Einstein proposed a novel relation between time and space called the 'special theory of relativity', in 1905. His theory was based on the following two postulates.

(1) The laws of nature in a given inertial system of reference are independent of the translational motion of the system.

(2) The velocity of light is a constant in all frames. The velocity of light is independent of the motion of the source and is given the constant value of c.

We will now review the Lorentz transformation between space and time coordinates in different inertial frames on the basis of the special theory of relativity.

Consider two inertial reference frames K and K′ in vacuum. The space and time coordinates of a point are $(\boldsymbol{x}, t)$ in K and $(\boldsymbol{x}', t')$ in K′ where $\boldsymbol{x}$ and $\boldsymbol{x}'$ are position vectors and give three-dimensional coordinates (x, y, z) and (x', y', z') respectively. The coordinate axes in the two frames are oriented parallel so that the frame K′ is moving in the positive x direction with velocity $\boldsymbol{v}$, as seen from K. In this case, t is independent of t' according to postulate (1). For simplicity, let the origin of coordinates in K and K′ be coincident at $t = t' = 0$. If a light source at rest at the origin in K and K′ is illuminated at $t = t' = 0$, the postulate (2) predicts that observers in K and K′ will see a spherical shell of radiation expanding outward from the respective origins with velocity c. The wave front reaches a point $\boldsymbol{x}$ (x, y, z) in the frame K at a time t given by the equation,

$$\boldsymbol{x}^2 - c^2t^2 = 0. \tag{2.1}$$

Similarly, in the frame K′, the wave front reaching a point $\boldsymbol{x}'$ (x', y', z') is specified by

$$\boldsymbol{x}'^2 - c^2t'^2 = 0. \tag{2.2}$$

The relation between Eqs. (2.1) and (2.2) is given by

$$\boldsymbol{x}^2 - c^2t^2 = \boldsymbol{x}'^2 - c^2t'^2. \tag{2.3}$$

After solving Eq. (2.3), the Lorentz transformation is found to be

$$x' = \frac{x - vt}{\sqrt{1-\beta^2}}, \quad y' = y, \quad z' = z, \quad t' = \frac{t - vx/c^2}{\sqrt{1-\beta^2}}, \tag{2.4}$$

where $\beta = v/c$. According to Eq. (2.4) the coordinates perpendicular to the direction of relative motion are unchanged while parallel coordinates and the time are transformed. This consequence is to be contrasted with the Galilean transformation. From this first part in Eq. (2.4), the Lorentz contraction is derived. When we see a bar with length l in K standing on K′, it appears to be contracted to a length $l\sqrt{1-\beta^2} = l/\gamma$, where $\gamma = 1\Big/\sqrt{1-\beta^2}$.

Suppose that there is a moving point whose velocity vector $\boldsymbol{u}$ takes the values (u_x, u_y, u_z) in the inertial frame K, while the frame K′ is moving with velocity $\boldsymbol{v} = c\boldsymbol{\beta}$ in the positive x direction with respect to the inertial frame K. The components of the velocity $\boldsymbol{u}'(u'_x, u'_y, u'_z)$ of the point seen from K′ are given by

$$u'_x = \frac{u_x - v}{1 - vu_x/c^2}, \quad u'_y = \frac{u_y\sqrt{1-\beta^2}}{1 - vu_x/c^2}, \quad u'_z = \frac{u_z\sqrt{1-\beta^2}}{1 - vu_x/c^2}. \tag{2.5}$$

2.1.2 The equation of motion for an electron in an electromagnetic field

In the theory of special relativity, the momentum $\boldsymbol{p}$ and the energy E of a particle with rest mass m and velocity $\boldsymbol{v}$ are related by

$$\boldsymbol{p} = m\boldsymbol{v}\Big/\sqrt{1-\beta^2} = \gamma m\boldsymbol{v}, \tag{2.6}$$

$$E = mc^2\Big/\sqrt{1-\beta^2} = \gamma mc^2. \tag{2.7}$$

Equation (2.7) gives the total energy including the static energy of the particle. In the presence of an electromagnetic field, the derivatives of the momentum $\boldsymbol{p}$ and the energy E with respect to t are given by

$$\frac{d\boldsymbol{p}}{dt} = \boldsymbol{F}, \tag{2.8}$$

$$\frac{dE}{dt} = \boldsymbol{v} \cdot \frac{d\boldsymbol{p}}{dt} = \boldsymbol{v} \cdot \boldsymbol{F}. \tag{2.9}$$

If we consider the case of a single electron under the same conditions as above, the Lorentz force, $\boldsymbol{F}$, applied to the electron is given by

$$\boldsymbol{F} = -e(\boldsymbol{E} + \boldsymbol{v} \times \boldsymbol{B}), \tag{2.10}$$

where e, $\boldsymbol{E}$, $\boldsymbol{v}$ and $\boldsymbol{B}$ are the electric charge ($e > 0$), electric field, velocity of the electron and the magnetic field, respectively. For simplicity consider the magnetic field $\boldsymbol{B}$ only. The equation of motion for an electron is then given by

$$\frac{d\boldsymbol{p}}{dt} = -e(\boldsymbol{v} \times \boldsymbol{B}). \tag{2.11}$$

When $\boldsymbol{B}$ is uniform and perpendicular to the orbit, Eq. (2.11) shows that the electron must rotate in a circle of radius $R = vE/ec^2B$ as well as with an angular frequency $\omega = ec^2B/E$. If v is close to c, the radius R of the circular orbit is approximately equal to the constant value, $R = E/ecB$. In addition, we apply an electric field $\boldsymbol{E}$ to the rotating electron. Then the derivative of the total energy E with respect to time t is given by

$$\frac{dE}{dt} = -e\boldsymbol{v} \cdot \boldsymbol{E} \tag{2.12}$$

2.2 THE ELECTROMAGNETIC WAVE EMITTED FROM MOVING ELECTRONS

2.2.1 Lienard-Wiechert potentials

When electric charges move, magnetic fields are induced giving rise to electric fields, and *vice versa*. These time-varying fields spread out through space as electromagnetic waves. We can treat electricity and magnetism independently in the case of steady-state problems. However we must also consider the relationship between electric and magnetic phenomena in circumstances when they are treated time-dependent. For these cases, we commence with Maxwell's equations compatible with the relativity, which provide a basis for an understanding of all classical electromagnetic phenomena.

Consider the position and time dependent electric and magnetic vectors of $\boldsymbol{E}(\boldsymbol{r}, t)$ and $\boldsymbol{H}(\boldsymbol{r}, t)$ together with related electric flux density $\boldsymbol{D}(\boldsymbol{r}, t)$ and magnetic flux density $\boldsymbol{B}(\boldsymbol{r}, t)$. The Maxwell's equations are then given by

$$\text{rot}\, \boldsymbol{E} = -\frac{\partial \boldsymbol{B}}{\partial t}, \tag{2.13}$$

$$\text{rot}\, \boldsymbol{H} = \frac{\partial \boldsymbol{D}}{\partial t} + \boldsymbol{j}, \tag{2.14}$$

$$\text{div}\, \boldsymbol{D} = \rho, \tag{2.15}$$

$$\text{div}\, \boldsymbol{B} = 0, \tag{2.16}$$

where $\boldsymbol{j}$ and ρ are current density and true charge density, respectively. Next, by using the scalar potential $\Phi(\boldsymbol{r}, t)$ and vector potential $\boldsymbol{A}(\boldsymbol{r}, t)$, Maxwell's equations can be rewritten as follow,

$$\boldsymbol{E} = -\frac{\partial \boldsymbol{A}}{\partial t} - \text{grad}\, \Phi, \tag{2.17}$$

$$\boldsymbol{B} = \text{rot}\, \boldsymbol{A}, \tag{2.18}$$

$$\left(\nabla^2 - \frac{1}{c^2}\frac{\partial^2}{\partial t^2}\right)\Phi = -\frac{1}{\varepsilon_0}\rho, \tag{2.19}$$

$$\left(\nabla^2 - \frac{1}{c^2}\frac{\partial^2}{\partial t^2}\right)\boldsymbol{A} = -\mu_0 \boldsymbol{j}. \tag{2.20}$$

Here ε_0 and μ_0 represent the dielectric constant and permeability in vacuum. In addition, the Lorentz gauge

$$\text{div}\,\boldsymbol{A} + \frac{1}{c^2}\frac{\partial \Phi}{\partial t} = 0 \tag{2.21}$$

is chosen. By using the Lorentz gauge, the wave equations Eqs. (2.19) and (2.20), in which Φ and $\boldsymbol{A}$ are on an equivalent footing, are obtained. Moreover, Eq. (2.21) gives the concept of the independence of the arbitrarily chosen coordinate system and fits in the postulate (1) of special relativity.

At first the scalar potential is derived from the wave equation Eq. (2.19). Since it is difficult to solve wave equations directly, the electric charge density $\rho(\boldsymbol{r}, t)$ and the scalar potential $\Phi(\boldsymbol{r}, t)$ are transformed to Fourier integrals as follow,

$$\rho(\boldsymbol{r}, \omega) = \frac{1}{2\pi}\int_{-\infty}^{\infty} \rho(\boldsymbol{r},\, t)\, e^{i\omega t}\, dt, \tag{2.22}$$

$$\Phi(\boldsymbol{r}, \omega) = \frac{1}{2\pi}\int_{-\infty}^{\infty} \Phi(\boldsymbol{r},\, t)\, e^{i\omega t}\, dt. \tag{2.23}$$

By substituting Eqs. (2.22) and (2.23) for Eq. (2.19), the inhomogeneous Helmholtz equation is obtained as

$$\left(\nabla^2 + \frac{\omega^2}{c^2}\right)\Phi(\boldsymbol{r},\, \omega) = -\frac{1}{\varepsilon_0}\,\rho(\boldsymbol{r},\, \omega). \tag{2.24}$$

Next, we consider the case where an electric point charge is located at the origin. This situation resembles the Poisson equation in electrostatics. In this case, Eq. (2.24) is simplified to become

$$\left(\nabla^2 + \frac{\omega^2}{c^2}\right) G(\boldsymbol{r}) = -\delta(\boldsymbol{r}) \tag{2.25}$$

and its solution is obtained as

$$G(\boldsymbol{r}) = \frac{1}{4\pi}\,\frac{e^{\pm i(\omega/c) r}}{r}, \tag{2.26}$$

where $|\boldsymbol{r}| = r$. When $\omega = 0$, $G(\boldsymbol{r})$ gives the Coulomb potential for a single point charge. Once the solution for a single point charge is obtained, we can extend it to the case of a continuous charge density $\rho(\boldsymbol{r}, t)$. In Fig. 2.1, P and P′ show the observation and the charge points. The corresponding position vectors to P and P′ measured from the origin O are $\boldsymbol{r}_{\mathrm{P}}$ and $\boldsymbol{r}_{\mathrm{P}'}$, respectively. $\Phi(\boldsymbol{r}_{\mathrm{P}}, \omega)$ is obtained by superposition of particular solutions as

$$\Phi(\boldsymbol{r}_{\mathrm{P}}, \omega) = \frac{1}{\varepsilon_0}\int G(\boldsymbol{r}_{\mathrm{P}} - \boldsymbol{r}_{\mathrm{P}'})\,\rho\,(\boldsymbol{r}_{\mathrm{P}'},\, \omega)\, d\boldsymbol{r}_{\mathrm{P}'}. \tag{2.27}$$

Moreover we may rewrite Eq. (2.27) by changing the variable ω to t,

$$\begin{aligned}\Phi(\boldsymbol{r}_{\mathrm{P}}, t) &= \frac{1}{\varepsilon_0}\int_{-\infty}^{\infty} e^{-i\omega t} d\omega \int G(\boldsymbol{r}_{\mathrm{P}} - \boldsymbol{r}_{\mathrm{P}'}) d\boldsymbol{r}_{\mathrm{P}'}\, \frac{1}{2\pi}\int_{-\infty}^{\infty} \rho(\boldsymbol{r}_{\mathrm{P}'},\, t')\, e^{i\omega t'} dt' \\ &= \frac{1}{4\pi\varepsilon_0}\int \frac{1}{|\boldsymbol{r}_{\mathrm{P}} - \boldsymbol{r}_{\mathrm{P}'}|}\, d\boldsymbol{r}_{\mathrm{P}'} \int_{-\infty}^{\infty} \rho(\boldsymbol{r}_{\mathrm{P}'},\, t')\, dt'\, \frac{1}{2\pi}\int_{-\infty}^{\infty} e^{-i\omega\left(t - t' \mp \frac{1}{c}|\boldsymbol{r}_{\mathrm{P}} - \boldsymbol{r}_{\mathrm{P}'}|\right)} d\omega. \end{aligned} \tag{2.28}$$

Since the last integral is equal to the delta function, the scalar potential Eq. (2.28) becomes

$$\Phi(\boldsymbol{r}_\mathrm{P}, t) = \frac{1}{4\pi\varepsilon_0}\int \frac{\rho(\boldsymbol{r}_{\mathrm{P}'}, t')}{|\boldsymbol{r}_\mathrm{P} - \boldsymbol{r}_{\mathrm{P}'}|}\,\delta\left(t' - t + \frac{|\boldsymbol{r}_\mathrm{P} - \boldsymbol{r}_{\mathrm{P}'}|}{c}\right) d\boldsymbol{r}_{\mathrm{P}'}dt'. \tag{2.29}$$

By repeating the procedure, the vector potential $A(\boldsymbol{r}_\mathrm{P}, t)$ is written as

$$\boldsymbol{A}(\boldsymbol{r}_\mathrm{P}, t) = \frac{\mu_0}{4\pi}\int \frac{\boldsymbol{j}(\boldsymbol{r}_{\mathrm{P}'}, t')}{|\boldsymbol{r}_\mathrm{P} - \boldsymbol{r}_{\mathrm{P}'}|}\,\delta\left(t' - t + \frac{|\boldsymbol{r}_\mathrm{P} - \boldsymbol{r}_{\mathrm{P}'}|}{c}\right) d\boldsymbol{r}_{\mathrm{P}'}dt'. \tag{2.30}$$

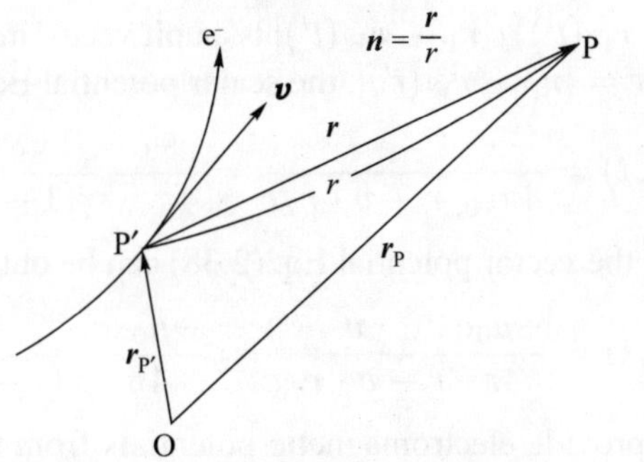

Figure 2.1 Relation between the position of an electron P′ and the observation point P. The electromagnetic wave is observed at point P.

Generally $t' = t \pm |\boldsymbol{r}_\mathrm{P} - \boldsymbol{r}_{\mathrm{P}'}|/c$, however a negative sign due to postulate (2) of the special theory of relativity is chosen, which means there is no phenomenon faster than the velocity of light in vacuum. Electromagnetic potentials given in Eqs. (2.29) and (2.30) are called 'retarded potentials'.

Consider the case where an electron is moving along an orbit $\boldsymbol{r}_{\mathrm{P}'} = \boldsymbol{r}_{\mathrm{P}'}(t')$. Also, by using velocity $d\boldsymbol{r}_{\mathrm{P}'}/dt' = \boldsymbol{v}(t')$, the charge density and the current density are given by

$$\rho(\boldsymbol{r}_{\mathrm{P}'}, t') = -e\delta\left(\boldsymbol{r}_{\mathrm{P}'}(t') - \boldsymbol{r}_{\mathrm{P}'}\right), \tag{2.31}$$

$$\boldsymbol{j}(\boldsymbol{r}_{\mathrm{P}'}, t') = -e\boldsymbol{v}(t')\delta\left(\boldsymbol{r}_{\mathrm{P}'}(t') - \boldsymbol{r}_{\mathrm{P}'}\right). \tag{2.32}$$

After substituting Eq. (2.31) for Eq. (2.29), and Eq. (2.32) for Eq. (2.30), we integrate them with $\boldsymbol{r}_{\mathrm{P}'}$ and obtain the simplified forms below,

$$\Phi(\boldsymbol{r}_\mathrm{P}, t) = \frac{-e}{4\pi\varepsilon_0}\int \frac{\delta\left(t - t' - \dfrac{|\boldsymbol{r}_\mathrm{P} - \boldsymbol{r}_{\mathrm{P}'}(t')|}{c}\right)}{|\boldsymbol{r}_\mathrm{P} - \boldsymbol{r}_{\mathrm{P}'}(t')|}\,dt', \tag{2.33}$$

$$\boldsymbol{A}(\boldsymbol{r}_\mathrm{P}, t) = \frac{-\mu_0 e}{4\pi}\int \frac{\boldsymbol{v}(t')\,\delta\left(t - t' - \dfrac{|\boldsymbol{r}_\mathrm{P} - \boldsymbol{r}_{\mathrm{P}'}(t')|}{c}\right)}{|\boldsymbol{r}_\mathrm{P} - \boldsymbol{r}_{\mathrm{P}'}(t')|}\,dt'. \tag{2.34}$$

In order to carry out further integration in Eqs. (2.33) and (2.34), a new expression $f(t') = t' + |\boldsymbol{r}_\mathrm{P} - \boldsymbol{r}_{\mathrm{P}'}(t')|/c$ is introduced, and integral calculus is extended in t'. As a result, we can consider the electromagnetic waves on a scalar potential with the simplified form,

$$\Phi(\boldsymbol{r}_{\mathrm{P}},t)=\frac{-e}{4\pi\varepsilon_0}\frac{1}{|\boldsymbol{r}_{\mathrm{P}}-\boldsymbol{r}_{\mathrm{P'}}(t')|}\left(\frac{df(t')}{dt'}\right)^{-1}. \tag{2.35}$$

Here, $t'=t-|\,\boldsymbol{r}_{\mathrm{P}}-\boldsymbol{r}_{\mathrm{P'}}(t')\,|/c$ is called the 'emission time' or the 'retarded time'. Furthermore, by using $\boldsymbol{\beta}=\boldsymbol{v}/c$, $df(t')/dt'$ can be rewritten as follows.

$$\frac{df(t')}{dt'}=1-\boldsymbol{n}(t')\cdot\boldsymbol{\beta}(t')=\alpha(t'). \tag{2.36}$$

In Eq. (2.36), $\boldsymbol{n}(t')=(\boldsymbol{r}_{\mathrm{P}}-\boldsymbol{r}_{\mathrm{P'}}(t'))/|\boldsymbol{r}_{\mathrm{P}}-\boldsymbol{r}_{\mathrm{P'}}(t')|$ is a unit vector in the direction from $\boldsymbol{r}_{\mathrm{P'}}(t')$ to $\boldsymbol{r}_{\mathrm{P}}$. Since $\boldsymbol{r}=\boldsymbol{r}_{\mathrm{P}}-\boldsymbol{r}_{\mathrm{P'}}(t')$ and $r=|\boldsymbol{r}_{\mathrm{P}}-\boldsymbol{r}_{\mathrm{P'}}(t')|$, the scalar potential Eq. (2.35) can be written as

$$\Phi(\boldsymbol{r}_{\mathrm{P}},t)=\frac{-e}{4\pi\varepsilon_0}\frac{1}{r-\boldsymbol{v}\cdot\boldsymbol{r}/c}=\frac{-e}{4\pi\varepsilon_0}\frac{1}{r(1-\boldsymbol{n}\cdot\boldsymbol{\beta})}. \tag{2.37}$$

Following similar procedure, the vector potential Eq. (2.38) can be obtained as

$$\boldsymbol{A}(\boldsymbol{r}_{\mathrm{P}},t)=\frac{-\mu_0 e}{4\pi}\frac{\boldsymbol{v}}{r-\boldsymbol{v}\cdot\boldsymbol{r}/c}=\frac{-\mu_0 ec}{4\pi}\frac{\boldsymbol{\beta}}{r(1-\boldsymbol{n}\cdot\boldsymbol{\beta})}. \tag{2.38}$$

Equations (2.37) and (2.38) provide electromagnetic potentials from moving electrons and are called 'Lienard-Wiechert potentials'.

2.2.2 Electromagnetic waves from moving electrons

In order to characterize the electromagnetic waves emitted from a moving electron, we calculate the electric field $E(\boldsymbol{r}_{\mathrm{P}},t)$ and magnetic flux density $\boldsymbol{B}(\boldsymbol{r}_{\mathrm{P}},t)$ by partially differentiating the Lienard-Wiechert potentials with respect to $\boldsymbol{r}_{\mathrm{P}}$ and t. However, because $\boldsymbol{r}_{\mathrm{P}}$ is not a variable independent of t as well as of t', and it is difficult to differentiate the potentials directly. Since the integrand in Eq. (2.33) (and also in Eq. (2.34)) has a delta function, t' must be equal to $t'=t-r/c$. By using t', we are able to choose $\boldsymbol{r}=\boldsymbol{r}(t')$, $r=r(t')$, $\boldsymbol{v}=\boldsymbol{v}(t')=\partial\boldsymbol{r}_{\mathrm{P'}}(t')/\partial t'=-\partial\boldsymbol{r}(t')/\partial t'$ and $\boldsymbol{\beta}=\boldsymbol{\beta}(t')=\boldsymbol{v}(t')/c$. The distance r between the position of an electron P′ and the observation point P at the emission time t' is then defined to be

$$r(t')=c\,(t-t'). \tag{2.39}$$

Partially differentiating Eq. (2.39) with respect to the observation time t', the following relation is obtained.

$$\frac{\partial r(t')}{\partial t}=\frac{\partial r(t')}{\partial t'}\frac{\partial t'}{\partial t}=-\frac{\boldsymbol{r}(t')\cdot\boldsymbol{v}(t')}{r(t')}\frac{\partial t'}{\partial t}=c\left(1-\frac{\partial t'}{\partial t}\right). \tag{2.40}$$

Repeating the procedure, using the identity $r^2(t')=\boldsymbol{r}^2(t')$ with respect to the emission time t', $2r(t')\partial r(t')/\partial t'=2\boldsymbol{r}(t')\cdot\partial\boldsymbol{r}(t')/\partial t'$ is known and the relation is given below,

$$\frac{\partial r(t')}{\partial t'}=\frac{\boldsymbol{r}(t')}{r(t')}\cdot\frac{\partial\boldsymbol{r}(t')}{\partial t'}=-\frac{\boldsymbol{r}(t')}{r(t')}\cdot\boldsymbol{v}(t'). \tag{2.41}$$

Since Eq. (2.41) was inserted into Eq. (2.40), the partial derivative of t' with respect to t is given below,

$$\frac{\partial t'}{\partial t} = \frac{1}{1 - \boldsymbol{v}(t') \cdot \boldsymbol{r}(t')/r(t')c} = \frac{r(t')}{s(t')} = \frac{1}{1 - \boldsymbol{n} \cdot \boldsymbol{\beta}(t')}, \tag{2.42}$$

where

$$s(t') = r(t') - \boldsymbol{v}(t') \cdot \boldsymbol{r}(t')/c = r(t')(1 - \boldsymbol{n} \cdot \boldsymbol{\beta}(t')). \tag{2.43}$$

Likewise from Eq. (2.39),

$$\text{grad } t' = -\frac{1}{c} \text{ grad } r(t') = -\frac{1}{c} \left(\frac{\partial r(t')}{\partial t'} \text{grad } t' + \boldsymbol{n} \right).$$

Hereafter and for simplicity, t' in r, $\boldsymbol{r}$, $\boldsymbol{v}$ and $\boldsymbol{\beta}$ is abbreviated, thus

$$\text{grad } t' \left(1 + \frac{1}{c} \frac{\partial r}{\partial t'} \right) = -\frac{1}{c} \boldsymbol{n}. \tag{2.44}$$

By substituting Eq. (2.41) for Eq. (2.44),

$$\text{grad } t' = \frac{-\boldsymbol{r}}{c(r - \boldsymbol{r} \cdot \boldsymbol{v}/c)} = \frac{\boldsymbol{r}}{cs}. \tag{2.45}$$

Next, we will calculate the changing ratio of the source distance s with respect to the observation time t as

$$\frac{\partial s}{\partial t} = \frac{\partial s}{\partial t'} \frac{\partial t'}{\partial t} = \frac{r}{s} \left(\frac{\partial r}{\partial t'} - \frac{1}{c} \frac{\partial \boldsymbol{v}}{\partial t'} \cdot \boldsymbol{r} - \frac{1}{c} \boldsymbol{v} \cdot \frac{\partial \boldsymbol{r}}{\partial t'} \right) = \frac{r}{s} \left(\frac{-\boldsymbol{r} \cdot \boldsymbol{v}}{r} - \dot{\boldsymbol{\beta}} \cdot \boldsymbol{r} + c\beta^2 \right). \tag{2.46}$$

Here $\dot{\boldsymbol{\beta}}$ is equal to $\partial \boldsymbol{\beta}/\partial t'$, in addition, the changing ratio of s with respect to the space is given by

$$\begin{aligned} \text{grad } s &= \text{grad}(r - \boldsymbol{r} \cdot \boldsymbol{\beta}) = \text{grad } r - \text{grad } (\boldsymbol{r} \cdot \boldsymbol{\beta}) \\ &= \boldsymbol{n} - \boldsymbol{\beta} + \frac{\boldsymbol{r}}{cs} \left(\frac{\boldsymbol{r} \cdot \boldsymbol{v}}{r} + \boldsymbol{r} \cdot \dot{\boldsymbol{\beta}} - c\beta^2 \right). \end{aligned} \tag{2.47}$$

The following equations which are related to Eq. (2.17) are calculated as

$$\frac{\partial \boldsymbol{v}}{\partial t} = \dot{\boldsymbol{v}} \frac{r}{s}, \quad \text{rot } \boldsymbol{v} = -\dot{\boldsymbol{v}} \times \text{grad } t' = -\boldsymbol{r} \times \boldsymbol{\beta}/s. \tag{2.48}$$

Using these expressions as well as Eqs. (2.17) and (2.18), the electric field $\boldsymbol{E}$ can be calculated to be

$$\boldsymbol{E} = \frac{-e}{4\pi\varepsilon_0} \left\{ -\text{grad} \left(\frac{1}{s} \right) - \frac{1}{c^2} \frac{\partial}{\partial t} \frac{\boldsymbol{v}}{s} \right\} = \frac{-e}{4\pi\varepsilon_0} \left\{ \left(\frac{1}{s^2} \right) \text{grad } s - \frac{1}{cs} \dot{\boldsymbol{\beta}} + \frac{\boldsymbol{\beta}}{cs^2} \frac{\partial s}{\partial t} \right\}. \tag{2.49}$$

After inserting Eqs. (2.43), (2.46), (2.47) and (2.48) into Eq. (2.49), the expression below is finally obtained.

$$\begin{aligned} \boldsymbol{E}(\boldsymbol{r}_{\mathrm{P}}, t) &= \frac{-e}{4\pi\varepsilon_0} \left[\frac{1 - \boldsymbol{\beta}^2}{s^3} (\boldsymbol{r} - r\boldsymbol{\beta}) + \frac{1}{s^3 c^2} \boldsymbol{r} \times \{ (\boldsymbol{r} - r\boldsymbol{\beta}) \times \dot{\boldsymbol{v}} \} \right] \\ &= \frac{-e}{4\pi\varepsilon_0} \left[\frac{(\boldsymbol{n} - \boldsymbol{\beta})(1 - \beta^2)}{r^2 (1 - \boldsymbol{n} \cdot \boldsymbol{\beta})^3} + \frac{\boldsymbol{n} \times \{ (\boldsymbol{n} - \boldsymbol{\beta}) \times \dot{\boldsymbol{\beta}} \}}{cr (1 - \boldsymbol{n} \cdot \boldsymbol{\beta})^3} \right], \end{aligned} \tag{2.50}$$

$$B(r_P, t) = \frac{1}{c} n \times E(r_P, t). \tag{2.51}$$

In Eq. (2.50), the first term falls off inversely as the square of the distance r. This term originates from the velocity of the electron, v, and gives the Coulomb force when $v = 0$. In contrast, the second term arises from the accelerating motion of electrons and slowly falls off as $1/r$. Since the emitted electromagnetic waves are usually observed at long distances, it is clear that the first term in Eq. (2.50) shows a rapid decrease for short distances, however the effect of the second term extends to large distances. Generally, electromagnetic waves from accelerating charged particles are called 'bremsstrahlung'. We can see several examples in electron and atomic scattering. In the present case, therefore, only the characteristics of the second term in Eq. (2.51) are considered. Then, electromagnetic fields produced by bremsstrahlung are given by

$$E(r_P, t) = \frac{-e}{4\pi\varepsilon_0} \frac{n \times \{(n - \beta) \times \dot{\beta}\}}{cr(1 - n \cdot \beta)^3}, \tag{2.52a}$$

$$B(r_P, t) = \frac{1}{c} n \times E(r_P, t). \tag{2.52b}$$

Next we wish to derive the energy flow carried by bremsstrahlung per unit time and per unit area at the observing position r_P and at time t. In this case, bremsstrahlung was emitted at the emission time t', where the Poynting vector S is the amount of energy crossing unit area perpendicular to the vector, per unit time, and is defined to be $S = E \times B$. Then we have

$$\begin{aligned} S(r_P, t) &= \frac{1}{\mu_0 c} \{E(r_P, t)\}^2 n \\ &= \frac{1}{\mu_0 c} \left(\frac{e}{4\pi\varepsilon_0}\right)^2 \frac{n}{c^2 r^2} \frac{[n \times \{(n - \beta) \times \dot{\beta}\}]^2}{(1 - n \cdot \beta)^6}. \end{aligned} \tag{2.53}$$

The energy emitted at time t by an electron per unit time and per unit solid angle toward the surface of the sphere with the radius of r is given by

$$\frac{dP}{d\Omega} = \frac{d^2 W}{dt d\Omega} = r^2 S(r_P, t) \cdot n, \tag{2.54}$$

where the solid angle $d\Omega = \sin\theta d\theta d\phi$. From Eq. (2.42), the power emitted at t' per unit solid angle is

$$\frac{dP}{d\Omega} = \frac{d^2 W}{dt' d\Omega} = S(r_P, t) \cdot n\,(1 - n \cdot \beta)\, r^2. \tag{2.55}$$

In Eq. (2.55), P is equal to dW/dt' . After substitution of Eq. (2.53) for Eq. (2.55) and integration of Eq. (2.55) with solid angle over the whole area, the power radiated as bremsstrahlung can be written as

$$\frac{dW}{dt'} = \frac{1}{\mu_0 c^3} \left(\frac{e}{4\pi\varepsilon_0}\right)^2 \int \frac{[n \times \{(n - \beta) \times \dot{\beta}\}]^2}{(1 - n \cdot \beta)^5} d\Omega. \tag{2.56}$$

When $\beta \ll 1$, that is, the velocity of the electron is very small in comparison with c, the total instantaneous power emitted appears as the familiar 'Larmor radiation' where

$$\frac{dW}{dt'} = \frac{e^2}{16\pi^2\varepsilon_0 c^3}\dot{\boldsymbol{v}}^2 \int \sin^2\theta d\Omega = \frac{e^2}{6\pi\varepsilon_0 c^3}\dot{\boldsymbol{v}}^2. \tag{2.57}$$

Here θ is the angle between $\boldsymbol{n}$ and $\dot{\boldsymbol{\beta}}$. In addition, $\varepsilon_0\mu_0 = c^{-2}$ is used. It is interesting to compare the change in the angular distribution pattern for the radiation between slow and fast electrons. Knowing the radiation pattern from slow electrons in Eq. (2.57), we can proceed to the case of fast electrons accelerated in a linear accelerator. In this case, as $\boldsymbol{\beta}$ is parallel to $\dot{\boldsymbol{\beta}}, \boldsymbol{n} \times \{(\boldsymbol{n} - \boldsymbol{\beta}) \times \dot{\boldsymbol{\beta}}\} = \boldsymbol{n} \times (\boldsymbol{n} \times \dot{\boldsymbol{\beta}})$. This relation is inserted into Eq. (2.56) and we obtain

$$\frac{dW}{dt'} = \frac{e^2\dot{\boldsymbol{v}}^2}{16\pi^2\varepsilon_0 c^3} \int \frac{\sin^2\theta}{(1 - \beta\cos\theta)^5} d\Omega. \tag{2.58}$$

From Eq. (2.58), we can see the denominator in the integrand reduces as the velocity of the electron $\boldsymbol{v}$ increases, resulting in an increase in the angle integration. This means the angular distribution of radiation from the accelerated electron in a linear accelerator is confined to a very narrow cone in the direction of motion. Figure 2.2 shows the radiation pattern for an accelerated electron in its direction of motion both for the non-relativistic ($\beta \approx 0$) and the relativistic ($\beta \approx 1$) cases. This behavior is also predicted from the Lorentz transformation Eq. (2.5). It is pointed out that the radiated power from an electron in a linear accelerator is much smaller than that from one in a circular accelerator.

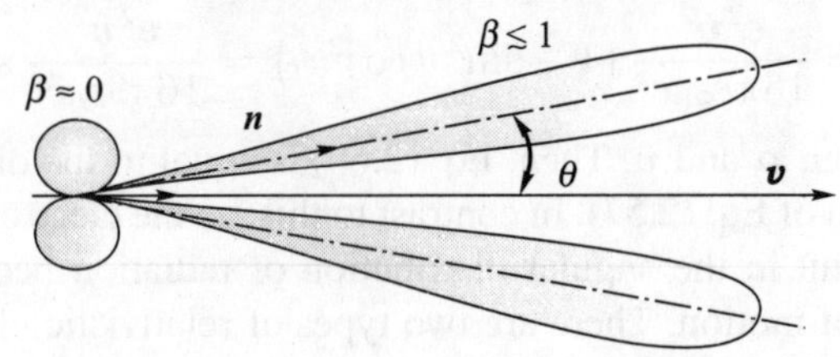

Figure 2.2 The angular distribution pattern of electromagnetic radiation from linearly accelerating electrons, for the non-relativistic ($\beta \approx 0$) and the relativistic ($\beta \lesssim 1$) cases.

2.3 LIGHT PRODUCED BY AN ELECTRON IN CIRCULAR MOTION

2.3.1 The directionality of radiation from an electron in circular motion

In the bending magnets of the storage ring, electrons move on a part of a circular orbit. Suppose an electron is moving in a circle with radius R. A tangential line which touches the circle at source point O is selected as the z-axis, and $\boldsymbol{n}$ is the unit vector in the direction from O to the observing position. The electron is moving circularly with its acceleration $\dot{\boldsymbol{v}}$ perpendicular to its velocity $\boldsymbol{v}$. We choose a coordinate system such that instantaneously $\boldsymbol{v}$ is in the z-axis and $\dot{\boldsymbol{v}}$ is in the x-axis. In addition, we define that polar angle of $\boldsymbol{n}$ from the direction of the velocity vector $\boldsymbol{v}$ is θ and azimuth angle of $\boldsymbol{n}$ in the xy-plane, ϕ as shown in Fig. 2.3. Accordingly, $\boldsymbol{n} \cdot \boldsymbol{\beta} = \beta\cos\theta, \boldsymbol{n} \cdot \dot{\boldsymbol{\beta}} = \dot{\beta}\sin\theta\cos\phi, \boldsymbol{\beta} \cdot \dot{\boldsymbol{\beta}} = 0$.

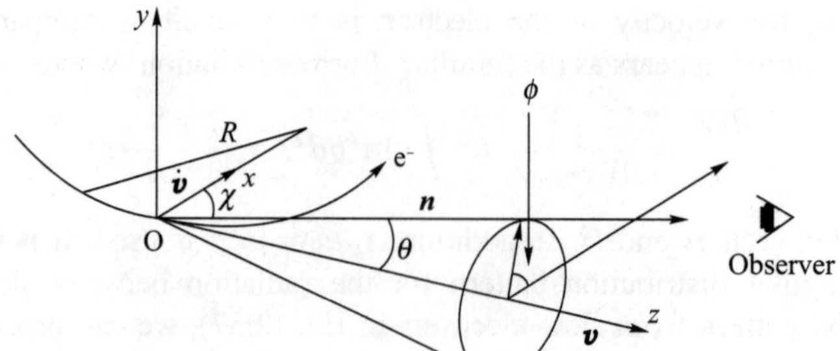

Figure 2.3 The relationship between the coordinates of the electron orbit and the observation direction.

By using the formula of the outer product, $\boldsymbol{A} \times (\boldsymbol{B} \times \boldsymbol{C}) = \boldsymbol{B}(\boldsymbol{A} \cdot \boldsymbol{C}) - \boldsymbol{C}(\boldsymbol{A} \cdot \boldsymbol{B})$, we rewrite Eq. (2.55) and derive the power, $P = P(t') = dW/dt'$, radiated per solid angle $d\Omega$ as follows.

$$\frac{dP}{d\Omega} = \frac{e^2\dot{v}^2}{16\pi^2\varepsilon_0 c^3}\frac{1}{(1-\beta\cos\theta)^3}\left\{1 - \frac{\sin^2\theta\cos^2\phi}{\gamma^2(1-\beta\cos\theta)^2}\right\}. \tag{2.59}$$

In attempting to understand Eq. (2.59) qualitatively, note that when an electron is moving slowly, as $\beta \approx 0$ and $\gamma \approx 1$, Eq. (2.59) is given by

$$\frac{dP}{d\Omega} = \frac{e^2\dot{v}^2}{16\pi^2\varepsilon_0 c^3}(1 - \sin^2\theta\cos^2\phi) = \frac{e^2\dot{v}^2}{16\pi^2\varepsilon_0 c^3}\sin^2\chi, \tag{2.60}$$

where χ is the angle between $\boldsymbol{n}$ and $\dot{\boldsymbol{v}}$. Then, Eq. (2.60) is equal to the dipole radiation as well as the Larmor radiation expression of Eq. (2.57). In contrast to this, as the electron moves with velocity closer to c, relativistic effects result in the angular distribution of radiation becoming increasingly focused into the forward direction of motion. There are two types of relativistic effect. One is the effect of the specific spatial relationship between $\boldsymbol{\beta}$ and $\dot{\boldsymbol{\beta}}$. The other is a general, relativistic effect arising from the transformation from the rest frame of the electron to the observer's frame and then manifesting itself by the presence of $(1 - \boldsymbol{n}\cdot\boldsymbol{\beta})$ in the denominator of Eq. (2.53). In relativistic limits, $\beta \approx 1$, $dP/d\Omega$ is tilted forward very steeply, showing its maximum intensity around $\theta = 0$. After expanding Eq. (2.59) in θ around $\theta = 0$, we substitute $\beta \approx 1$ and $1 - \beta \approx 1/2\gamma^2$ for Eq. (2.59), then

$$\frac{dP}{d\Omega} = \frac{e^2\dot{v}^2\gamma^6}{2\pi^2\varepsilon_0 c^3(1+\gamma^2\theta^2)^3}\left(1 - \frac{4\gamma^2\theta^2\cos^2\phi}{(1+\gamma^2\theta^2)^2}\right). \tag{2.61}$$

Equation (2.61) shows that the radiated power is concentrated within a very narrow cone with a small polar angle θ. This situation is shown in Fig. 2.4(a). Furthermore, mean value of θ^2 is given by

$$\langle\theta^2\rangle = \int \theta^2\frac{dP}{d\Omega}d\Omega \bigg/ \int \frac{dP}{d\Omega}d\Omega \approx \frac{1}{\gamma^2}, \tag{2.62}$$

where θ is not the angle of elevation from z-axis but is the polar angle around the z-axis. In fact, the curving electron emits an extremely focused beam in tangential directions in the horizontal plane all around the circular orbit. For this reason, we are able to regard the root mean square of Eq. (2.62) as giving the divergence angle of $\hat{\theta}_r$ from the horizontal plane. Finally, the angular divergence of the radiation is given by

$$\langle\theta^2\rangle^{1/2} \approx \hat{\theta}_r = 1/\gamma \approx mc^2/E \approx 0.511/E(\text{GeV}) \text{ (mrad)}. \tag{2.63}$$

2.3.2 Radiated power

The total power radiated from an accelerating relativistic electron will now be calculated. Eq. (2.59) must be integrated with respect to solid angle, Ω. At first, integration with respect to ϕ is carried out, and $P = P(\theta)$ is obtained as

$$\begin{aligned} P &= \int \frac{dP}{d\Omega} d\Omega = \frac{e^2\dot{v}^2}{16\pi^2\varepsilon_0 c^3}\int_0^{\pi/2}\frac{\sin\theta d\theta}{(1-\beta\cos\theta)^3}\int_0^{2\pi}\left\{1-\frac{\sin^2\theta\cos^2\phi}{\gamma^2(1-\beta\cos\theta)^2}\right\}d\phi \\ &= \frac{e^2\dot{v}^2}{16\pi\varepsilon_0 c^3}\int_0^{\pi/2}\frac{\sin\theta d\theta}{(1-\beta\cos\theta)^3}\left\{2-\frac{\sin^2\theta}{\gamma^2(1-\beta\cos\theta)^2}\right\}. \end{aligned} \tag{2.64}$$

Replacing $1-\beta\cos\theta$ in Eq. (2.64) by x, integration with respect θ is made. The total power emitted from the accelerating relativistic electron per turn is given by

$$P = \frac{e^2\dot{v}^2}{16\pi\varepsilon_0 c^3}\int_{1-\beta}^{1}\frac{dx}{\beta x^3}\left(2-\frac{\beta^2-1+2x-x^2}{\gamma^2\beta^2x^2}\right) \approx \frac{e^2\dot{v}^2\gamma^4}{6\pi\varepsilon_0 c^3}. \tag{2.65}$$

In the final term in Eq. (2.65), we replace $\dot{v}^2 \approx (c^2/R)^2$ and rewrite the total power per turn as

$$U_0 = \frac{e^2c\gamma^4}{6\pi\varepsilon_0 R^2}\frac{2\pi R}{c} = \frac{e^2\gamma^4}{3R\varepsilon_0}. \tag{2.66}$$

In practical units,

$$U_0(\text{keV}) = 88.5E^4(\text{GeV})/R(\text{m}), \tag{2.67}$$

where the radiation power U_0 is expressed in keV, the acceleration energy E in GeV and bending radius R in m.

2.3.3 The emitted frequency spectrum

As shown in Fig. 2.4(a), an accelerating electron emits synchrotron radiation in a direction tangential to its curved orbit as a narrow light cone with the full solid angle of $2\gamma^{-1}$. The observer at a point P on the tangential line can see brilliant synchrotron radiation for the period during which the electron moves round on the circumference corresponding to the solid angle $2\gamma^{-1}$.

We suppose that the observer could observe one edge of the cone at time $t_1 = t_1' + r(t_1')/c$ and the other side of the cone at $t_2 = t_2' + r(t_2')/c$, respectively. During the time interval between t_1' and t_2', the electron moves a distance $2R/\gamma$ along the circular orbit. As the velocity of the electron is $c\beta$, the migration distance is $c\beta(t_2' - t_1') = 2R/\gamma$. From this relation, the time interval is obtained as

$$t_2' - t_1' = \frac{2R}{\gamma c\beta}. \tag{2.68}$$

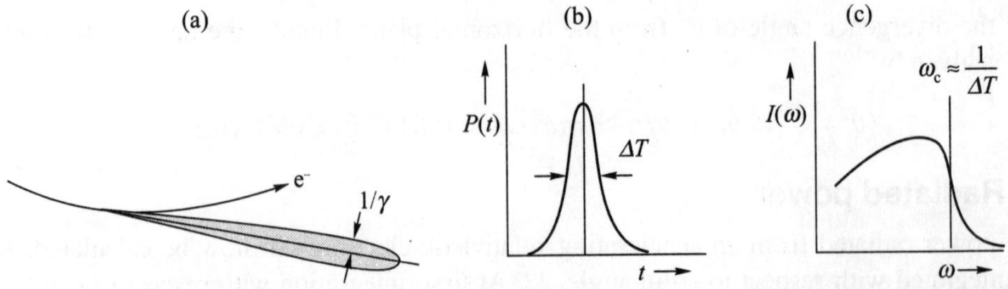

Figure 2.4 The emission of light from an electron in a bending section. (a) Angular spread of synchrotron radiation, (b) observation as an electromagnetic pulse, and (c) qualitative spectral distribution.

Moreover $r(t_2') = r(t_1' + t_2' - t_1') \approx r(t_1') - c\boldsymbol{n} \cdot \boldsymbol{\beta}\,(2R/\gamma c\beta)$ is attainable. Since the full solid angle of the light cone is $2/\gamma$, the time interval of observation is

$$\Delta T = t_2 - t_1 = (1 - \boldsymbol{n} \cdot \boldsymbol{\beta})\frac{2R}{\gamma c\beta}. \tag{2.69}$$

Because we observe synchrotron radiation on the tangential direction, $\boldsymbol{n}||\boldsymbol{\beta}$ and $\boldsymbol{n} \cdot \boldsymbol{\beta} = \beta$. Therefore, Eq. (2.69) is rewritten as

$$\Delta T = \frac{2R}{c\beta}\frac{1}{\gamma}(1 - \beta) \approx \frac{R}{c}\frac{1}{\gamma^3}. \tag{2.70}$$

Here, the denominator of $c\gamma^3$ is huge quantity, as a result, ΔT is extremely short.

Let us qualitatively explain the reason why such a short electromagnetic pulse of synchrotron radiation provides a continuous frequency spectrum. A conceptual explanation for the generation of a continuous spectrum is given in Figs. 2.4(b) and 2.4(c). The synchrotron radiation pulse is approximated to be a rectangular-shape with time width of ΔT and unit intensity. It is well known that any short time pulse is equivalent to the superposition of many waves with differing angular frequencies. Therefore, the Fourier transform of this pulse will provide a continuous frequency spectrum of synchrotron radiation emitted from the bending section. If we square the absolute value of the Fourier integral, the intensity distribution is qualitatively given as the continuous frequency spectrum

$$I(\omega) \propto \frac{\sin^2(\Delta T\omega/2)}{(\omega/2)^2}. \tag{2.71}$$

In this case, the critical angular frequency $\omega_{\rm m}$ is given by

$$\omega_{\rm m} = 2\pi/\Delta T = 2\pi c\gamma^3/R. \tag{2.72}$$

In fact, the synchrotron radiation pulse measured at the observing point is not a simple rectangular wave. In order to get the correct spectral function, it is necessary to make a frequency-transformation of the electromagnetic fields of synchrotron radiation arriving at the observation point. The Fourier transform for the total power of synchrotron radiation is performed by using the scalar potential Eq. (2.29), vector potential Eq. (2.30), charge density Eq. (2.31) and current density Eq. (2.32) (see App. **B**).

The total power radiated by an electron moving with velocity of light, in a circular orbit per turn and per $\Delta\omega$ can be calculated to be

$$P(\omega)\Delta\omega = \left\{\frac{\sqrt{3}e^2\gamma}{8\pi^2\varepsilon_0 R}\frac{\omega}{\omega_c}\int_{\omega/\omega_c}^{\infty} K_{5/3}(\eta)d\eta\right\}\Delta\omega,\ \ \omega_c = 3\gamma^3 c/2R. \tag{2.73}$$

Here, $K_{5/3}$ is the modified Bessel function of the second kind of order 5/3, and ω_c is called characteristic or critical angular frequency. This gives the spectrum of the light from bending sections. If we compare $1/\Delta T$ in Eq. (2.70), ω_m in Eq. (2.72) and ω_c in Eq. (2.73), they are all proportional to $c\gamma^3/R$. The asymptotic form of ranging from smaller or larger frequencies than ω_c is given as follows,

$$P(\omega) \approx \frac{3^{1/6}\Gamma(2/3)}{4\pi\varepsilon_0}\frac{e^2}{R^{2/3}}\frac{\omega^{1/3}}{c^{1/3}} \quad (\omega \ll \omega_c),$$

$$P(\omega) \approx \frac{1}{8\pi\varepsilon_0}\left(\frac{3}{2\pi}\right)^{1/2}\frac{e^2}{R}\gamma\left(\frac{\omega}{\omega_c}\right)^{1/2} e^{-\omega/\omega_c} \quad (\omega \gg \omega_c). \tag{2.74}$$

For the case where the electron energy is expressed as E (GeV) and the bending radius as R(m), we can derive a critical wavelength $\lambda_c(= 2\pi c/\omega_c)$ and also a corresponding photon energy in keV, respectively to be

$$\lambda_c = 4\pi R/3\gamma^3 = 5.59R(\text{m})/E^3(\text{GeV}) \ \ (\text{Å}),$$

$$u_c = 12.4/\lambda_c = 2.22E^3(\text{GeV})/R(\text{m}) \ \ (\text{keV}). \tag{2.75}$$

Equation (2.73) defines the total power radiated by an electron moving around the circular orbit. It is convenient to express Eq. (2.73), in practical units of photons/s/mrad/0.1% band width. In order to express the total power in the units, first we must divide it by $2\pi \times 10^3$ (mrad). Next, it is divided by $\hbar\omega$ to find the photon flux of synchrotron radiation per turn. In this case, the specific frequency resolution *i.e.*, the relative band width $\Delta\omega/\omega$, is chosen to be 0.1%, then $\Delta\omega$ is equal to $1 \times 10^{-3}\omega$. Using the substitution described above, we obtain the angular flux emitted by I(mA) of circulating electrons with an accelerating energy of E (GeV) per second, per 0.1% band width in practical units to be

$$\frac{d^3N_p}{dtd\psi(\Delta\omega/\omega)} = 2.457 \times 10^{10}E(\text{GeV})I(\text{mA})G_1(\omega/\omega_c)$$

$$(\text{photons/s/mrad/0.1\% band width}),$$

$$G_1(\omega/\omega_c) = \frac{\omega}{\omega_c}\int_{\omega/\omega_c}^{\infty} K_{5/3}(\eta)d\eta. \tag{2.76}$$

Figure 2.5 exhibits the $G_1(\omega/\omega_c)$ function which gives the spectral distribution of synchrotron radiation emitted from a bending section. This is presented in a log-log scale plot, (a), and in a linear scale plot, (b). Notice that Fig. 2.5 is obtained by calculation from Eq. (2.76).

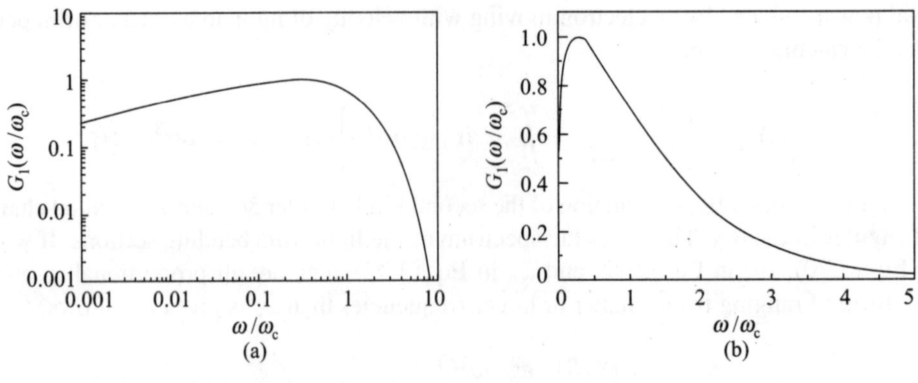

Figure 2.5 The spectral distribution function $G_1(\omega/\omega_c)$ for synchrotron radiation from bending section, (a) the distribution as a function of ω/ω_c in log-log scale and (b) in linear scale.

2.3.4 The angular distribution

Figure 2.6 shows the intensity angular distributions for synchrotron radiation where the electric vector is parallel to the orbital plane, $I_{\parallel}$ (parallel (σ) component, solid lines) and also for synchrotron radiation where the electric vector is perpendicular to the orbital plane, $I_{\perp}$ (perpendicular (π) component, dashed lines). Here $\hat{\theta}$ is an observation angle relative to the orbital plane. It is convenient to normalize $\hat{\theta}$ by $1/\gamma$. The abscissa of the figure is the relative observation angle $\gamma\hat{\theta}$, and the ordinate is the relative intensity of synchrotron radiation normalized by the peak height in the orbital plane. In the figure, the normalized photon energy ω/ω_c is chosen as a parameter. From Fig. 2.6, it is remarkable that synchrotron radiation lying within the orbital plane ($\gamma\hat{\theta} = 0$) consists solely of parallel components and is totally linearly-polarized. In contrast to this, perpendicular components appear in any off-plane direction ($\gamma\hat{\theta} \neq 0$). The general expression for the electric field for any observation angle $\hat{\theta}$ includes phase relation of $\pi/2$

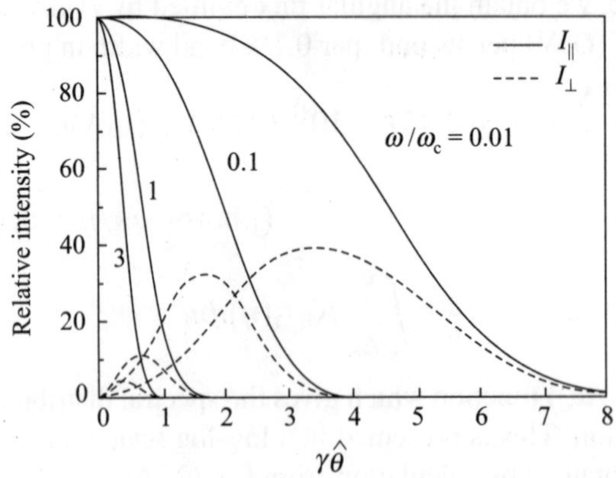

Figure 2.6 Angular distribution of spectral intensity emitted from a bending section. $\hat{\theta}$: observation angle relative to the orbital plane, ω: angular frequency of the light and ω_c: the critical angular frequency.

between parallel and perpendicular components. For small $\hat{\theta}$, the ratio of $I_{\perp}/I_{||}$ is considerably less than unity which means that elliptically-polarized light is obtained in general. However, when we observe synchrotron radiation at large angles of $\hat{\theta}$, approximately circularly-polarized light will be observed. Detailed calculations of the angular distribution are presented in App. **C**. The final result is as follows.

$$\frac{d^2P(\omega)}{d\omega d\Omega} = \frac{e^2}{12\pi^3\varepsilon_0 c}\left(\frac{\omega R}{c}\right)^2\left(\frac{1}{\gamma^2}+\hat{\theta}^2\right)\left\{K_{2/3}^2(\xi)+\frac{\gamma^2\hat{\theta}^2}{1+\gamma^2\hat{\theta}^2}K_{1/3}^2(\xi)\right\}. \tag{2.77}$$

In Eq. (2.77), $K_{1/3}(\xi)$ and $K_{2/3}(\xi)$ are modified Bessel functions of the second kind, where $\xi = (\omega R/3c)(1/\gamma^2+\hat{\theta}^2)^{3/2}$. The two terms in Eq. (2.77) represent the contributions from the two directions of polarization. The first one describes the contribution of the parallel component and the second one, the perpendicular component. The brilliance is given by $\dfrac{d^4N_{\mathrm{p}}}{dtd\Omega(\Delta\omega/\omega)dS}$, whose practical units are photons/s/mrad²/mm²/0.1% band width. Here dS is the cross section of the electron beam. If we substitute 0 for $\hat{\theta}$ in Eq. (2.77), the brilliance on the orbital plane can be calculated.

2.4 LIGHT FROM ELECTRONS MOVING IN AN UNDULATOR

2.4.1 Light from a planar undulator

a. The movement of the electron: An undulator is a linear array of magnets with alternating polarities. It can be installed in straight sections of a storage ring. The pole faces of a planar undulator are parallel to the horizontal plane and the vertical strength of magnetic field along the forward direction periodically changes sinusoidally. When high velocity electrons traverse the undulator, they 'wiggle' their way through the magnetic fields. We can describe the motion of electrons in the planar undulator (see Fig. 1.1) in the following way. Figure 2.7 schematically shows the direction of magnetic field, electron orbit, the maximum deflection angle of electrons ψ_0 and the radiation cone angle $1/\gamma$. In the figure, the magnetic field is parallel to the y-axis, and the direction of electron motion is chosen as the z-axis. The vertical magnetic field B_y along the z-axis is considered to be a sine wave with periodic length λ_{u} as

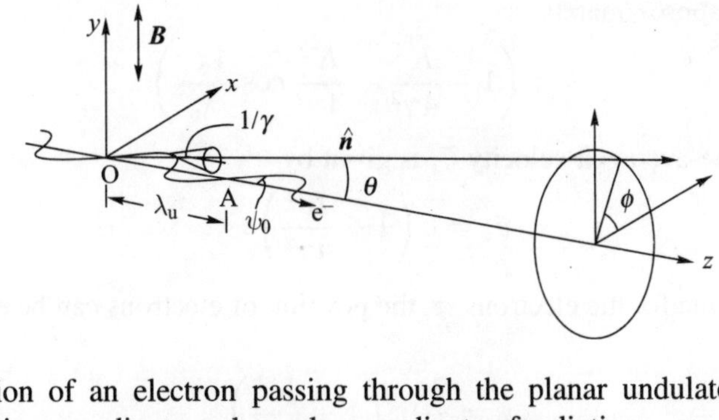

Figure 2.7 Motion of an electron passing through the planar undulator illustrated with position coordinate and angular coordinate of radiation.

$$B_y = -B_0 \sin \frac{2\pi z}{\lambda_u}. \tag{2.78}$$

Here, B_0 is the maximum value of the sinusoidal magnetic field. In the laboratory frame, where the light from undulator, undulator radiation is observed, the velocity and acceleration of electrons in the xz-plane are expressed by $v_x = dx/dt'$, $\dot{v}_x = d^2x/dt'^2$, $v_z = dz/dt'$, $\dot{v}_z = d^2z/dt'^2$. By using these relations and Eq. (2.11), equations of motion for an electron are written as

$$\gamma m \dot{v}_x = -ev_z B_0 \sin \frac{2\pi z}{\lambda_u}, \tag{2.79}$$

$$\gamma m \dot{v}_z = ev_x B_0 \sin \frac{2\pi z}{\lambda_u}. \tag{2.80}$$

Since we can rewrite differentiation by t' to that by z in Eq. (2.79),

$$\dot{v}_x = \frac{dv_x}{dz}\frac{dz}{dt'} = v_z \frac{dv_x}{dz}. \tag{2.81}$$

Insertion of Eq. (2.81) to Eq. (2.79) yields the equation,

$$\gamma m v_z \frac{dv_x}{dz} = -ev_z B_0 \sin \frac{2\pi z}{\lambda_u}.$$

After integration with respect to z, the important result,

$$v_x = \frac{e\lambda_u B_0}{2\pi m\gamma} \cos \frac{2\pi z}{\lambda_u} = \frac{Kv}{\gamma} \cos \frac{2\pi z}{\lambda_u} \tag{2.82}$$

is obtained. $K = e\lambda_u B_0/2\pi mv \approx e\lambda_u B_0/2\pi mc$ and is called the deflection parameter of the undulator. In practical units, K is simply given by

$$K = 93.4\lambda_u(\mathrm{m})B_0(\mathrm{T}). \tag{2.83}$$

As the maximum value of v_x is Kv/γ, the corresponding maximum deflection angle ψ_0 is K/γ. In other words, $K = \gamma\psi_0$. Subsequently v_z can be calculated from the following equation

$$v^2 = v_x^2 + v_z^2. \tag{2.84}$$

When $K/\gamma \ll 1$, v_z is approximately

$$v_z = v\left(1 - \frac{K^2}{4\gamma^2} - \frac{K^2}{4\gamma^2}\cos\frac{4\pi z}{\lambda_u}\right). \tag{2.85}$$

From Eq. (2.85), the time-averaged velocity $\overline{v}_z$ is given by

$$\overline{v}_z = v\left(1 - \frac{K^2}{4\gamma^2}\right). \tag{2.86}$$

If t' is the observation time for the electrons, z, the position of electrons can be expressed by $z = \overline{v}_z t'$. A new parameter

$$\omega_0 = \frac{2\pi}{\lambda_u} v \left(1 - \frac{K^2}{4\gamma^2}\right) \tag{2.87}$$

is introduced, where $2\pi z/\lambda_u = \omega_0 t'$. Using these relations, Eqs. (2.82) and (2.85) are integrated, and the following approximate solutions,

$$x = (\lambda_u K/2\pi\gamma)\sin\omega_0 t',$$
$$y = 0,$$
$$z = c\bar{\beta}_z t' - (\lambda_u K^2/16\pi\gamma^2)\sin 2\omega_0 t' \tag{2.88}$$

are obtained, where $\overline{\beta}_z = \overline{v}_z/c = \beta(1 - K^2/4\gamma^2)$.

b. The spectrum of undulator radiation: What is the nature of the light emitted by a high-velocity electron moving through such an undulator with small oscillation amplitude ($\psi_0 \approx 1/\gamma$)? In the case for light from a bending section, only one light pulse per turn is observed at the tangential direction from each bending section. In contrast, the time dependence of electromagnetic fields of undulator radiation is not of single pulse nature but is of almost sinusoidal high frequency periods. Therefore, it is clear that the frequency spectrum from the Fourier transform of time-dependent undulator radiation will produce quasi-monochromatic light and the fundamental wave therefore exhibits radiation of greatly enhanced brilliance.

The remarkable enhancement of brilliance is caused by constructive interference between the radiation emitted from every undulation (bend) in the orbit. The electron moves in z direction with the average velocity $\overline{v}_z$ sinusoidally oscillating in horizontal plane with angular frequency of ω_0. One period of oscillation T' or 'passing time' for the electron to move unit length of undulator is given by

$$T' = 2\pi/\omega_0. \tag{2.89}$$

We provide that the electron emits the light at O point and emits the light at A point after moving by one period length of the undulator λ_u (see Fig. 2.7), and we observe the light at an angle of observation θ measured form z direction. The condition of interference is that the path difference between the light emitted at O and that at A is the multiple of wavelength of undulator radiation, which is given by

$$n\lambda_n = \frac{2\pi c}{\omega_0} - \lambda_u \cos\theta = \lambda_u\left(\frac{1}{\overline{\beta}_z} - \cos\theta\right), \tag{2.90}$$

where λ_u stands for the wavelength of n-th order higher harmonic undulator radiation. For the case of a high-velocity electron, v is so large and the angle of observation θ naturally becomes small. Therfore, $\cos\theta \approx 1 - \theta^2/2$. Furthermore, in Eq. (2.90) one can approximate $\overline{\beta}_z \approx \beta(1 - K^2/4\gamma^2)$ and $1 - \beta \approx 1/2\gamma^2$ because $1/\gamma = \sqrt{1-\beta^2}$ is small. As a result, Eq. (2.90) is modified to the familiar form,

$$\lambda_n \approx \frac{\lambda_u}{n}\left\{\frac{1}{(1-1/2\gamma^2)(1-K^2/4\gamma^2)} - \left(1 - \frac{\theta^2}{2}\right)\right\} \approx \frac{\lambda_u}{2n\gamma^2}(1 + \frac{K^2}{2} + \gamma^2\theta^2) \tag{2.91}$$

The angular frequency is given by

$$\omega_n = 2\pi\nu_n = 2\pi c/\lambda_n = \frac{4\pi c n\gamma^2}{\lambda_u(1 + K^2/2 + \gamma^2\theta^2)}. \tag{2.92}$$

This result is generally derived using the relation between retarded time and observing time given by Eq. (2.42). The relation between passing time in observing time T and that in retarded time T' is given by

$$\frac{T}{T'} = 1 - \boldsymbol{n}(T') \cdot \boldsymbol{\beta}(T') \approx 1 - \overline{\beta}_z \cos\theta. \tag{2.93}$$

Using the same approximation used to obtain Eq. (2.91), T is calculated as

$$T \approx \frac{\lambda_u}{\beta c}\left\{1 - \beta\left(1 - \frac{K^2}{4\gamma^2}\right)\left(1 - \frac{\theta^2}{2}\right)\right\} \approx \frac{\lambda_u}{2\gamma^2 c}\left(1 + \frac{1}{2}K^2 + \gamma^2\theta^2\right). \tag{2.94}$$

Calculating $\lambda_n = (c/n)T$, one can also obtain Eq. (2.91).

Now undulator radiation is overviewed using relativity theory. In the moving frame with the average velocity of the high-velocity electron, the movement of the electron is seen as a harmonic oscillation with constant angular frequency. In other words, the succession of light sources at each small bend in orbit is effectively concentrated at one place, which is the origin of the enhancement of high brilliance. The light emitted in this frame is equivalent to the Larmor radiation of Eq. (2.57) with an angular dependence on $\sin^2\theta$. When we observe the periodic length of the undulator, it becomes a shortened relativistic length due to Lorentz contraction. Taking into account this and the fact that the magnetic field of the undulator moves at the average velocity of the electron, one can obtain the angular frequency of the light originating from the electron on this frame. Next, we transfer to the observer's frame. In this frame, the angular distribution of the light is confined to a very narrow cone in the direction of motion by relativistic effect and the angular frequency of the light is Doppler-shifted from that in the moving frame, which is expressed by the same equation as Eq. (2.92).

Equation (2.91) illustrates two important facts. First, that undulator generates longer wavelength light for larger K values with stronger magnetic field. Second, that the shortest wavelength is observed at the angle of $\theta = 0$, along the z-axis. As the angle of observation θ increases, longer wavelength light is observed.

We now calculate the electric field for the case of $\theta = 0$ which corresponds to the z-axis, where the electric field consists of parallel components only, shown by

$$|\hat{E}_{nx}(0)| = K|J_{(n+1)/2}(Y_0) - J_{(n-1)/2}(Y_0)|,$$
$$Y_0 = nK^2/(4 + 2K^2), \tag{2.95}$$

where J is Bessel function (see App. **D**). Furthermore the radiated power by an electron, per unit angular frequency, and per unit solid angle on the z-axis is given by

$$\left.\frac{d^2P(\omega)}{d\omega d\Omega}\right|_{\theta=0} = \frac{e^2\gamma^2}{4\pi\varepsilon_0 c}\sum_n F_n^{(p)}(K)H_n(\omega/\omega_1). \tag{2.96}$$

In Eq. (2.96), $H_n(\omega/\omega_1)$ exhibits the spectral shape function, and is expressed in the explicit form,

$$H_n(\omega/\omega_1) = \frac{\sin^2[N_u\pi\{(\omega/\omega_1) - n\}]}{[\pi\{(\omega/\omega_1) - n\}]^2}. \tag{2.97}$$

When $\omega/\omega_1 = n$, $H_n(\omega/\omega_1)$ is equal to N_u^2. Full width at half maximum (FWHM) of the spectrum is $\Delta\omega/\omega \approx 1/nN_u$. Figure 2.8 shows the dependence of the spectral shape function $H_n(\omega/\omega_1)$ on N_u, where N_u is the number of undulator periods. As N_u increases, the spectral width becomes narrower and closer to a delta function. If the angular distribution of emitted undulator radiation on the z-axis is approximately Gaussian with FWHM of $1/nN_u$, then the angular spread on the z-axis, σ'_r, is estimated to be $\sigma'_r \approx \sqrt{1 + K^2/2}/\gamma\sqrt{2nN_u} \approx 1/\gamma\sqrt{nN_u}$. From this expression, it is clear that

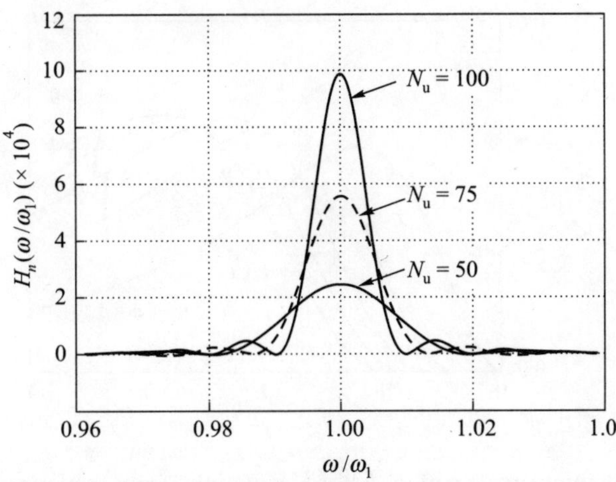

Figure 2.8 The spectral shape function $H_n(\omega/\omega_1)$ of light from a planar undulator for the case of $n = 1$ (fundamental wave). The periodic number N_u is chosen as a parameter.

undulator radiation emitted along the z-axis is highly concentrated in the forward direction. $F_n^{(p)}(K)$ gives the spectral intensity. On the z-axis ($\theta = 0$) for the case of a planar undulator,

$$F_n^{(p)}(K) = \frac{n^2}{(1+K^2/2)^2}|\hat{E}_{nx}(0)|^2 \quad (n : \text{odd number}), \tag{2.98a}$$

$$F_n^{(p)}(K) = 0 \quad (n : \text{even number}). \tag{2.98b}$$

In Fig. 2.9 the intensity variation of $F_n^{(p)}(K)$ as a function of K is shown for n = 1, 3, 5, 7 and 9. As shown in the figure, the radiation power of the fundamental wave (n = 1) is strong in the region near to $K = 1$. According to increase in the value of K, the maximum intensity of $F_n^{(p)}(K)$ shifts successively from lower to higher order waves. $F_n^{(p)}(K)$ for even orders is zero for the case of $\theta = 0$. However electrons effectively move with small oscillations $2\omega_0$ along the direction of motion through the undulator as seen in Eq. (2.88) with some angular spread around the z-axis. Thus, higher harmonic waves of even order may appear in real insertion devices (see Fig. 2.2). In the region with larger K values, Eq. (2.96) is qualitatively closer to the expression for the continuous spectrum. The spectral shape also resembles that of radiation from a bending magnet, but the flux is N_u times larger than from a single bending magnet. This is the obvious advantage of a multipole wiggler. In the planar undulator case, the light in the median plane will be linearly-polarized.

In the case when the stored current is I (mA) and all the electrons are moving parallel to the orbital plane, the emitted photon numbers per second, per mrad for both horizontal and vertical directions, and per 0.1% band width is given by

$$\left.\frac{d^4N_p}{dtd\psi d\hat{\theta}(\Delta\omega/\omega)}\right|_{\theta=0} = 1.744\times10^{11}N_u^2E^2(\text{GeV})I(\text{mA})F_n^{(p)}(K)$$

(photons/s/mrad^2/0.1% band width). (2.99)

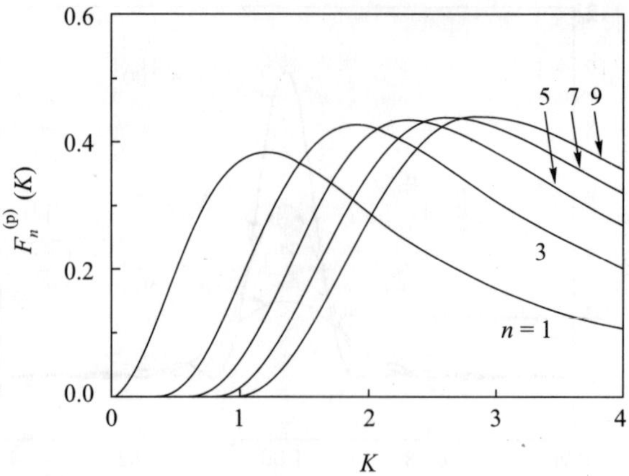

Figure 2.9 Spectral intensity function $F_n^{(\mathrm{p})}(K)$ of light from a planar undulator. Higher harmonic number n is chosen as a parameter.

Here angles $\hat{\theta}$ and ψ are defined as those in Fig. 1.1 and the relation between them is $\theta^2 = \hat{\theta}^2 + \psi^2$. If we assume the angular spread of undulator radiation is to be a Gaussian distribution with standard deviation of σ'_{r}, the photon flux within the radiation cone is approximately calculated to be Eq. (2.99) times $2\pi\sigma_{\mathrm{r}}'^2$. Finally the brilliance of undulator radiation is approximately given by the photon flux divided by $(2\pi)^2\, \Sigma_x \Sigma'_x \Sigma_y\, \Sigma'_y$ (see Sect. **1.2.3**). In the case of Fig. 1.3, the angular divergence of electron beam is introduced exactly for the calculation of spectral intensity.

2.4.2 Light from a helical undulator

a. The movement of an electron: By passing an electron through a helical magnetic field, produced by an alternating change in the polarity of magnetic fields between the vertical and horizontal directions, an electron is induced to execute a spiral motion around the central axis. In the case when magnetic field rotates clockwise along the direction of electron motion, the helical magnetic field $\boldsymbol{B}$ consists of the combination of successive changes from the horizontal field B_x to the vertical field B_y shifted by $\pi/2$ phase as given by

$$\boldsymbol{B} = -\boldsymbol{i} B_{x0} \sin\left(\frac{2\pi z}{\lambda_{\mathrm{u}}} + \frac{\pi}{2}\right) - \boldsymbol{j} B_{y0} \sin\left(\frac{2\pi z}{\lambda_{\mathrm{u}}}\right). \tag{2.100}$$

The equation of motion $d\boldsymbol{p}/dt = -e\boldsymbol{v} \times \boldsymbol{B}$ ($\boldsymbol{p} = \gamma m \boldsymbol{v}$) for an electron in this field is solved. The deflection parameters K_x and K_y are defined to be same as that in Eq. (2.83) for each direction. The derived solutions are as follow.

$$\boldsymbol{v} = \boldsymbol{i}\frac{K_y v}{\gamma}\cos\omega_0 t' + \boldsymbol{j}\frac{K_x v}{\gamma}\sin\omega_0 t' + \boldsymbol{k} v\left\{1 - \frac{K_x^2 + K_y^2}{4\gamma^2} + \frac{K_x^2 - K_y^2}{4\gamma^2}\cos 2\omega_0 t'\right\}, \tag{2.101}$$

$$\boldsymbol{r} = \boldsymbol{i}\frac{K_y \lambda_{\mathrm{u}}}{2\pi\gamma}\sin\omega_0 t' - \boldsymbol{j}\frac{K_x \lambda_{\mathrm{u}}}{2\pi\gamma}\cos\omega_0 t'$$

$$+ \boldsymbol{k}\left\{\left(1 - \frac{K_x^2 + K_y^2}{4\gamma^2}\right)\beta c t' + \frac{\left(K_x^2 - K_y^2\right)\lambda_{\mathrm{u}}}{16\pi\gamma^2}\sin 2\omega_0 t'\right\}. \tag{2.102}$$

Under the helical magnetic field of Eq. (2.100), an electron executes helical motion as a result of the superposition of the sinusoidal oscillation in xz-plane and that in yz-plane shifted by one fourth period. In this case, as we can see the rotational motion of an electron from the observation point, undulator radiation generated by the electron is right-handed elliptically-polarized light (for definition, see App. **J**).

b. The spectrum: The absolute value of the electric field of undulator radiation on the z-axis ($\theta = 0$) is given for the case of $K_y \geq K_x$ as

$$|\hat{E}_{nx}(0)| = K_y |J_{(n+1)/2}(Y_0) - J_{(n-1)/2}(Y_0)|, \tag{2.103}$$

$$|\hat{E}_{ny}(0)| = K_x |J_{(n+1)/2}(Y_0) + J_{(n-1)/2}(Y_0)|, \tag{2.104}$$

$$Y_0 = \frac{n}{1 + (K_x^2 + K_y^2)/2} \frac{K_y^2 - K_x^2}{4}. \tag{2.105}$$

Then the degree of circular polarization is given by

$$P_\text{c} = \frac{P_\text{circle}}{P_\text{total}} = 2\frac{|E_{nx}|\,|E_{ny}|}{|E_{nx}|^2 + |E_{ny}|^2}. \tag{2.106}$$

When an electron passes through the helical magnetic field with the period of N_u, the emitted power of undulator radiation per unit angular frequency and per unit solid angle in the z direction is

$$\left.\frac{d^2 P(\omega)}{d\omega d\Omega}\right|_{\theta=0} = \frac{e^2\gamma^2}{4\pi\varepsilon_0 c} \sum_n F_n^{(\text{h})}(K_x, K_y)\, H_n(\omega/\omega_1). \tag{2.107}$$

In Eq. (2.107), ω_1 exhibits the fundamental angular frequency on the z-axis as

$$\omega_1 = \frac{4\pi c\gamma^2}{\lambda_\text{u}\{1 + (K_x^2 + K_y^2)/2\}}. \tag{2.108}$$

Similarly, $F_n^{(\text{h})}(K_x, K_y)$ gives the spectral intensity on z-axis ($\theta = 0$) for the case of helical undulator as

$$F_n^{(\text{h})}(K_x, K_y) = \left\{\frac{n}{1 + (K_x^2 + K_y^2)/2}\right\}^2 \{|\hat{E}_{nx}(0)|^2 + |\hat{E}_{ny}(0)|^2\}. \tag{2.109}$$

When $K_x = K_y$, $Y_0 = 0$ is known from Eq. (2.105). In this case the Bessel function has a finite value only for $n = 1$ (the fundamental wave), and undulator radiation is circularly-polarized light.

CHAPTER 3

The Light Source

The normal type of source dedicated to synchrotron radiation research is the storage ring, which is one particular type of electron circular accelerator. First we outline the general features of a storage ring light source, then consider its magnetic structure and the theory of electron orbits, followed by a description of the theory of high (radio) frequency (RF) acceleration and RF cavities, how the beam lifetime is governed by vacuum pressure and intra-beam interaction, and beam instability. These are the fundamental parameters of every light source. In addition, coherent synchrotron radiation (CSR) generation and free electron lasers (FEL) are described. All recent developments of light sources including ERL are included in Refs. 10 and 11 in Chap. **1**.

3.1 OVERVIEW

All synchrotron radiation sources contain an injector and a storage ring which is schematically drawn in Fig. 1.1. A storage ring accumulates electrons and therefore needs an electron injector. The injector may be either a linear accelerator (LINAC), a microtron or a synchrotron which itself uses a LINAC or microtron as a pre-injector. The electrons accelerated in injectors are usually thermally derived electrons emitted either from metal filaments or oxides. Recently, electrons generated by laser irradiation have been used. The acceleration tube of a LINAC is composed of an array of high frequency (RF) cavities resonating at 2856 MHz for instance, in which electrons are accelerated by the traveling wave field. A synchrotron is an approximately circular accelerator enclosing an electron orbit composed of arcs and straight sections. Within the orbit one or more RF cavities are incorporated to supply energy to the electrons. When the magnetic field increases, the electron's energy is increased synchronously by feeding energy from the RF cavity as they pass through it. In this process the orbit remains fixed. The microtron is also a circular accelerator and in this case electrons also pass through an RF cavity but the orbit increases with increase in energy because the magnetic field is kept constant. These accelerators will not be considered further, but the storage ring as a light source will be discussed in detail.[1–3]

The operating principle of a storage ring is similar to that of a synchrotron. However, the synchrotron is a machine which accelerates low energy electrons up to high energy quickly, while a storage ring can be used to store electrons at a fixed energy for a long time. When the energy of the injected electrons is less than that of the operating energy, the electrons are accumulated at the injection energy and are subsequently accelerated to the operational energy and stored. When the injection energy is the same as the operational energy, any subsequent acceleration is unnecessary. In many cases, the stored current is from about 100 mA to 500 mA.

Among dedicated light sources, the lowest electron energy is 200 MeV, while the highest is 8 GeV. The principal components of any storage ring are the electromagnets which define the electron orbit, the RF cavities which supply the power and the vacuum orbit chambers, or the evacuated beam ducts in which the electrons circulate. In addition, there will be an electron injection system and a control system with monitors. Computers are now used routinely to control the entire light source and operate all the functions mentioned above.

Now, we consider the orbit of the electron beam, RF acceleration, the beam lifetime and instability, and generation of coherent radiation.

3.2 THE ORBIT OF THE ELECTRON BEAM

An array of magnets is called a lattice. An example of a plan view of the electron storage ring dedicated to synchrotron radiation is given in Fig. 3.1. The lattice is designed so that bending and quadrupole magnets are aligned to enable electrons to circulate stably and to minimize the emittance of the lattice. In general, storage rings are assembled from magnetic units, called cells which repeat periodically around the ring lattice. In what follows the equations of motion for electrons in a magnetic field and the methods to solve them are explained together with descriptions of emittance, beam cross section, magnet design and others.

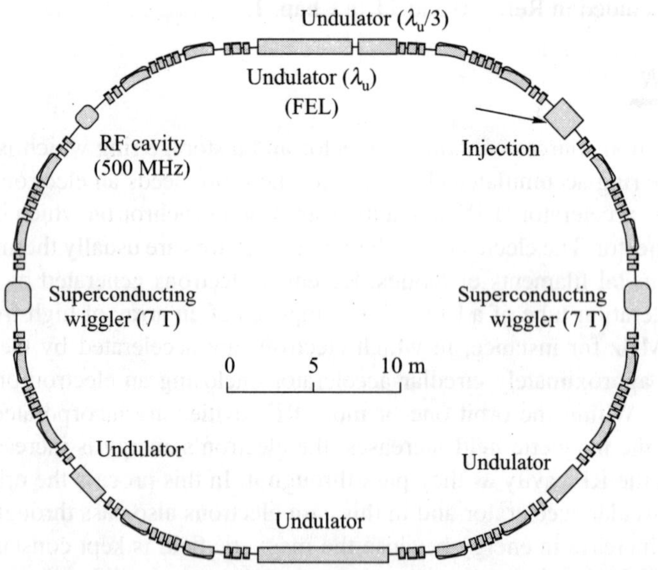

Figure 3.1 The plan view of 1.5 GeV electron storage ring proposed by Tohoku University in 2003 as synchrotron radiation source. Its injector is 1.2 GeV synchrotron with pre-injector of 300 MeV LINAC. Circumference: 99.5 m, emittance; 26.9 nm • rad.

3.2.1 The role of magnets in defining electron motion

When an electron moves in magnetic field it experiences the Lorentz force. Figures 3.2(a) and 3.2(b) show sections across bending and quadrupole magnets, respectively. In the bending (dipole) magnets the electron undergoes circular motion. However in quadrupole magnets the four poles are located 90° apart with the polarities of adjacent poles of opposite sign. Its magnetic field is 0 at the central orbit and becomes stronger with increase in distance from the central orbit to the outside. Close to the center of the ring, it increases but its sign is opposite to that outside. As shown in Fig. 3.2(b), the further the electron moves outward from the central orbit, the stronger the force directing it inward becomes, therefore any electron moving away from the central orbit always receives a force directing it back. The electron does not experience any force on the central orbit itself. This is the basic principle of the focusing action of quadrupole magnets. However, a quadrupole magnet with a focusing action in the horizontal direction does not have any focusing action in the vertical direction, but, conversely has a defocusing action as described later. This means that the further an electron moves from the central orbit the stronger the force becomes directing it outwards. Therefore, it is impossible to produce focusing action in both horizontal and vertical directions using a single quadrupole magnet.

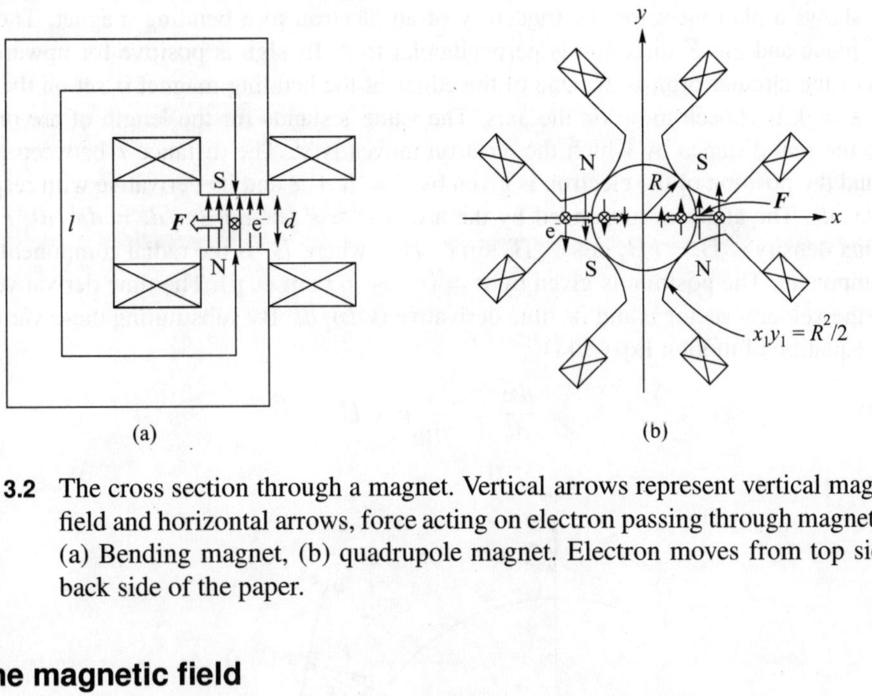

Figure 3.2 The cross section through a magnet. Vertical arrows represent vertical magnetic field and horizontal arrows, force acting on electron passing through magnet gap. (a) Bending magnet, (b) quadrupole magnet. Electron moves from top side to back side of the paper.

3.2.2 The magnetic field

The coordinates used to describe the motion of an electron in orbit are shown in Fig. 1.1. The distance from the origin to any arbitrary point is defined as s. This distance is the linear distance on straight orbits and the length of arc on circular orbits. The deviation of electron position from the central orbit in a horizontal direction is defined as x, which is positive for outward directions, while that in vertical direction, y is positive for upward directions. The magnetic flux density is shown positive for upward directions. Let us consider the magnetic flux density in the gap of a bending magnet whose poles are not necessarily parallel. Its vertical component B_y at a position (x, y, s) is given by

$$B_y(x, y, s) = B_0 + \left(\frac{dB_y}{dx}\right)_{x=0} x \quad = B_0 + B'x, \tag{3.1}$$

where only constant fields and fields linearly dependent on x are taken into account (the linear approximation). B_0 is called the dipole component and $B'x$, the quadrupole component. In the following it is assumed that the magnetic field parallel to the central orbit is uniform within the length of the magnet. The median plane between the two poles is horizontal so that the horizontal component B_x does not have a constant field as B_0 in Eq. (3.1). If $B' = 0$, $B_x = 0$. If $B' \neq 0$, the field proportional to y appears. If there is no electric field in the gap and the magnetic field induced by stored current is neglected, rot $\boldsymbol{B} = 0$. Therefore in the case of $B' \neq 0$,

$$B_x(x, y, s) = \left(\frac{dB_x}{dy}\right)_{y=0} y \quad = \left(\frac{dB_y}{dx}\right)_{x=0} y = B'y. \tag{3.2}$$

3.2.3 The equations of electron motion

Figure 3.3 shows a plan view for the trajectory of an electron in a bending magnet. The orbit plane is the XY-plane and the Z direction is perpendicular to it. Its sign is positive for upward directions. The radius of the circular orbit is R. One of the edges of the bending magnet is set on the X-axis and the origin, $s = 0$, is chosen to be on the axis. The value s stands for the length of arc on the central orbit while the real distance by which the electron moves is s'. The distance r between the center of the circle and the position of the electron is given by $r = R + x$ and its derivative with respect to time, $dr/dt = dx/dt$. The angle ψ suspended by the arc is $\psi \approx s'/r$ and $d\psi/dt = ds'/dt/r \approx c/r$. The magnetic flux density is $\boldsymbol{B} = (B_x \cos\psi, B_x \sin\psi, B_y)$, where B_x is the radial component and B_y, the vertical component. The position is given by $\boldsymbol{r} = (r\cos\psi, r\sin\psi, y)$. The time derivative of position vector $\boldsymbol{r}$ is the velocity vector $\boldsymbol{v}$ and its time derivative is $d\boldsymbol{v}/dt$. By substituting these variables for the relativistic equation of motion Eq. (2.11)

$$\frac{d\boldsymbol{v}}{dt} = \frac{-e}{\gamma m}\boldsymbol{v} \times \boldsymbol{B}. \tag{3.3}$$

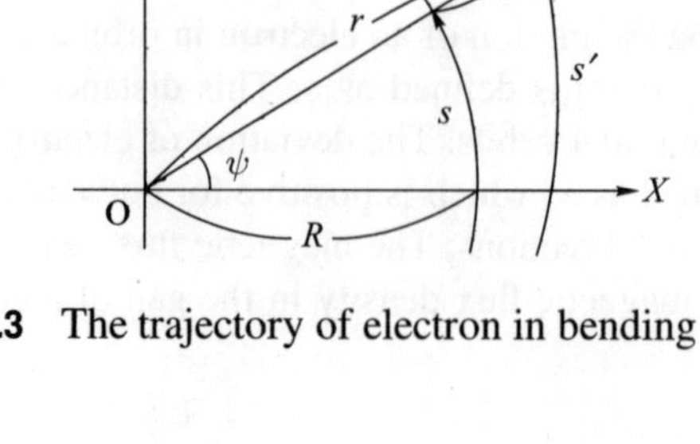

Figure 3.3 The trajectory of electron in bending magnet.

By calculating X and Y components, one can obtain the equation of motion with respect to x as

$$\frac{d^2x}{dt^2} - \frac{c^2}{r} = -\frac{ecB_y}{\gamma m}. \tag{3.4}$$

By using the relations $\gamma mc^2 = E = E_0 + \Delta E$ (where ΔE is the energy deviation from the central energy E_0) and $E_0 = ecB_0R$ (see Sect. **2.1.2**). We then substitute $r = R + x$ and $B_y = B_0 + B'x$ for Eq. (3.4), from which, by neglecting quadratic and higher power terms of small quantities of x/R and $\Delta E/E_0$, and by using $\frac{d^2x}{dt^2} = c^2\frac{d^2x}{ds^2}$, one can finally obtain the differential equation for x in a bending magnet to be

$$\frac{d^2x}{ds^2} + k_x x = \frac{1}{R}\frac{\Delta E}{E_0}, \quad k_x = \frac{1}{R^2} + \frac{B'}{B_0R}. \tag{3.5a}$$

Similarly one can obtain the differential equation of y in bending magnet to be

$$\frac{d^2y}{ds^2} + k_y y = 0, \quad k_y = -\frac{B'}{B_0R}. \tag{3.5b}$$

The corresponding equations for a quadrupole magnet are

$$\frac{d^2x}{ds^2} + k_x x = 0, \quad k_x = \frac{B'}{B_0R}, \tag{3.5c}$$

$$\frac{d^2y}{ds^2} + k_y y = 0, \quad k_y = -\frac{B'}{B_0R}. \tag{3.5d}$$

These equations resemble those for harmonic oscillation. However, in a quadrupole magnet, the signs of k_x and k_y are always opposite, for instance if k_x is positive, k_y is negative. When k is positive, electrons will undergo sinusoidal motion around the central orbit with an increase in s, so that the magnet has a focusing action, while, when k is negative, the electrons will move away from the central orbit. The definition is usually made with respect to x in such a way that when k_x is positive, the magnet is called a focusing magnet and when negative, it is called a defocusing magnet.

3.2.4 The solution to the equation of motion

a. Single electron trajectory: In a bending magnet, the variables expressing motion in horizontal plane are given by $(x, x', \Delta E/E_0)$, where x' stands for dx/ds. The variables in the vertical direction are (y, y'). The variable $\Delta E/E_0$ is absent because accelerators are always set horizontally so that even though the energy deviates, the positions of electrons do not change vertically. When the variables at the initial point s_0 are $(x_0, x'_0, \Delta E/E_0)$ and (y_0, y'_0), those at any arbitrary position s are $(x, x', \Delta E/E_0)$ and (y, y'), with transfer matrices defined as

$$\begin{pmatrix} x \\ x' \\ \Delta E/E_0 \end{pmatrix} = M_x(s \mid s_0)\begin{pmatrix} x_0 \\ x'_0 \\ \Delta E/E_0 \end{pmatrix}, \quad \begin{pmatrix} y \\ y' \end{pmatrix} = M_y(s \mid s_0)\begin{pmatrix} y_0 \\ y'_0 \end{pmatrix}. \tag{3.6}$$

To solve Eq. (3.5a), one must find first the particular solutions for $\Delta E/E_0 = 0$. If $k_x > 0$, the particular solutions of Eq. (3.5a) are $\sin\sqrt{k_x}s$ and $\cos\sqrt{k_x}s$, while if $k_x < 0$, $\sinh\sqrt{-k_x}s$ and $\cosh\sqrt{-k_x}s$. If

$k = 0$, the solution is a quadratic function of s. The general solutions are found for $\Delta E/E_0 \neq 0$ using the particular solutions. When $k_x > 0$, the transfer matrix of a bending magnet in horizontal direction is represented explicitly as

$$M_x(s|s_0) = \begin{pmatrix} \cos\phi & \frac{1}{\sqrt{k}}\sin\phi & \frac{1}{kR}(1-\cos\phi) \\ -\sqrt{k}\sin\phi & \cos\phi & \frac{1}{\sqrt{k}R}\sin\phi \\ 0 & 0 & 1 \end{pmatrix}, \tag{3.7}$$

where $\phi = \sqrt{k_x}\,(s - s_0)$. When $k_x < 0$, $\phi = \sqrt{-k_x}\,(s - s_0)$, sin in Eq. (3.7) is replaced by $-i\sinh$ and cos by cosh. When $k = 0$,

$$M_x(s|s_0) = \begin{pmatrix} 1 & s-s_0 & (s-s_0)^2/2R \\ 0 & 1 & (s-s_0)/R \\ 0 & 0 & 1 \end{pmatrix}. \tag{3.8}$$

In the transfer matrix of the quadrupole magnet, $M_{13} = M_{23} = 0$. In the y direction, similar matrices are obtained except for the fact that the matrices are of 2×2 form. The trajectory of an electron can be determined by matrix calculation according to the configuration of magnets and by choosing the initial values of the variables. However, although the lattice is periodic the trajectory is not periodic. Therefore, it is difficult using this method to understand the collective characteristics of electrons as a beam, defined by the lattice structure. The beam properties are actually characterized not only by the whole lattice, but also by each point of the lattice.

b. The pseudo harmonic solution: The solution enabling us to understand the characteristics of the electron beam as a whole is the pseudo harmonic solution which is a periodic solution with respect to the cells. In the horizontal direction, the deviation x is separated into $x_E(s)$ dependent on $\Delta E/E_0$ and $x_\beta(s)$ which is independent of it. In the vertical direction, the deviation y is only $y_\beta(s)$ independent of $\Delta E/E_0$. They are expressed by

$$x(s) = x_\beta(s) + x_E(s), \tag{3.9}$$

$$x_\beta(s) = a_x\sqrt{\beta_x(s)}\cos\{\phi_x(s) - \phi_{x0}\}, \tag{3.10}$$

$$x_E(s) = \eta(s)\frac{\Delta E}{E_0}, \tag{3.11}$$

$$y(s) = y_\beta(s) = a_y\sqrt{\beta_y(s)}\cos\{\phi_y(s) - \phi_{y0}\}. \tag{3.12}$$

$\beta(s)$ is called the 'betatron function' (not the beta function and v/c) and $\beta(s + L_1) = \beta(s)$ (where L_1 is the length of one cell). The values a_x and a_y are initial ones independent of s. The phase ϕ is given by

$$\phi = \int \frac{ds}{\beta}. \tag{3.13}$$

The initial phases are ϕ_{x0} and ϕ_{y0}. The function $\eta(s)$ is called the 'energy dispersion function'. There are two methods to determine $\beta(s)$ and $\eta(s)$. The first is to use a matrix calculation method. The matrix $M(s + L_1|s) = M_n M_{n-1} \cdots M_2 M_1 = M(s)$ is a transfer matrix for one cell depending on the initial position s. For each value of s, the transfer matrix of one cell is calculated using Eqs. (3.7) and (3.8). On the other hand, the transfer matrix of one cell is given using Eq. (3.9) to Eq. (3.11) as

$$\begin{pmatrix} x(s+L_1) \\ x'(s+L_1) \\ \Delta E/E_0 \end{pmatrix} = \begin{pmatrix} \cos\mu+\alpha\sin\mu & \beta\sin\mu & m_1 \\ -\gamma\sin\mu & \cos\mu-\alpha\sin\mu & m_2 \\ 0 & 0 & 1 \end{pmatrix} \begin{pmatrix} x(s) \\ x'(s) \\ \Delta E/E_0 \end{pmatrix}, \tag{3.14}$$

where μ is the phase difference of one cell, given from Eq. (3.13) as

$$\mu = \int_s^{s+L_1} \frac{ds}{\beta}, \tag{3.15}$$

where $\mathrm{Tr}M = 2\cos\mu + 1$, which is independent of s. From the calculation of the trace, μ is determined and once it is determined, the matrix elements can be obtained by comparison between the calculated transfer matrix at s and Eq. (3.14). The values α, β, γ, m_1 and m_2 are functions of s, and their interrelations are given from Eq. (3.9) to Eq (3.11) by

$$\alpha = -\frac{1}{2}\frac{d\beta}{ds}, \quad \gamma = \frac{1+\alpha^2}{\beta}, \tag{3.16a}$$

$$\eta = \frac{m_1}{2} + \frac{1}{2}(\beta m_2 + \alpha m_1)\cot\frac{\mu}{2}, \tag{3.16b}$$

$$\eta' = \frac{d\eta}{ds} = \frac{m_2}{2} - \frac{1}{2}(\alpha m_2 + \gamma m_1)\cot\frac{\mu}{2}. \tag{3.16c}$$

There is an infinite combination of focusing (and defocusing) forces (ks) of magnets, so that one must calculate $\beta(s)$ using computers. Some excellent programs are available.[4)] Hereafter, the derivative with respect to s is represented by adding superscript $'$.

The second method is to find an analytical solution. First, by substituting Eq. (3.10) into $x''_\beta + k_x x_\beta = 0$, one can obtain

$$\frac{1}{2}\beta\beta'' - \frac{1}{4}\beta'^2 + k\beta^2 - 1. \tag{3.17}$$

Further differentiation of Eq. (3.17) with respect to s yields

$$\beta''' + 4k\beta' = 0. \tag{3.18}$$

Substitution of Eq. (3.11) for Eq. (3.5a) yields

$$\eta'' + k_x\eta = 1/R. \tag{3.19}$$

Using Eqs. (3.14) and (3.16), the values of $\beta(s)$ and $\eta(s)$ and their derivatives can be derived at any arbitrary point. Representing them by β_0, β'_0, η_0 and η'_0 and by using $\alpha'_0 = -\beta'_0/2$, the solutions of Eqs. (3.18) and (3.19) can be obtained.[5)] In the description which follows, the solutions of $\beta(s)$ and $\eta(s)$ at a straight section without focusing action ($k = 0$) are given, for example, by

$$\beta(s) = \beta_0 + \beta'_0 s + \frac{1+{\alpha_0}^2}{\beta_0}s^2, \tag{3.20a}$$

$$\eta(s) = \eta_0 + \eta'_0 s. \tag{3.20b}$$

Figure 3.4(a) shows $\beta(s)$ and $\eta(s)$ for one of the cells of the storage ring illustrated in Fig. 3.1. The values are given in units of m and their signs are positive. Rectangles and squares indicate bending magnets and quadrupole magnets, respectively. Upward and downward squares represent focusing and defocusing, respectively. Figure 3.4(b) illustrates the relation between the electron trajectory and the betatron function. The broken curves give three trajectories for electrons which touch the curves of $\pm a\sqrt{\beta}$. Their amplitudes do not exceed $a\sqrt{\beta}$. The lattice and $\beta(s)$ are symmetric at the center of the long straight section, but the trajectories are not symmetric. This feature is common for all electrons. The maximum amplitude at every point, s, can be expressed using $\sqrt{\beta\,(s)}$. In the following, various parameters concerning the motion of electrons in the ring are derived in terms of $\beta(s)$ and $\eta(s)$.

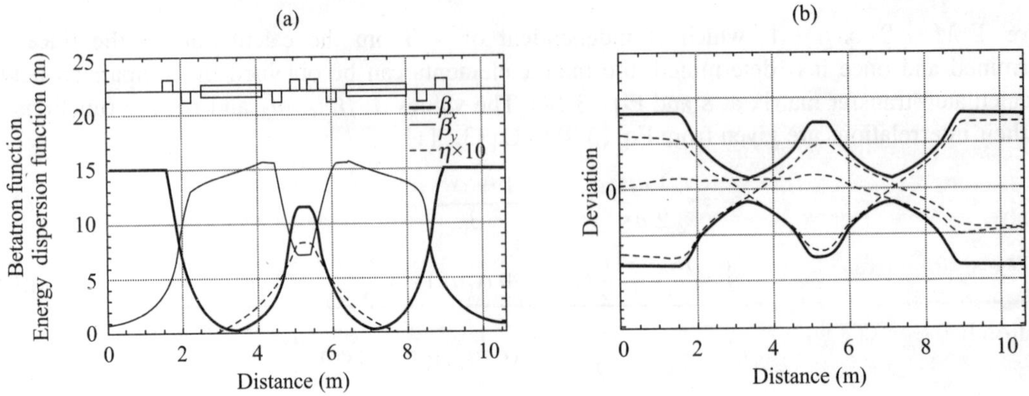

Figure 3.4 (a) The betatron function $\beta_x(s)$, $\beta_y(s)$ and energy dispersion function $\eta(s)$ of synchrotron radiation source shown in Fig. 3.1 and (b) examples of trajectories in the x direction.

3.2.5 The parameters which define the orbit

a. The tune: The tune expresses the number of betatron oscillations (betatron number) made by an electron of $\Delta E/E_0 = 0$ during one turn of the ring. It is given by

$$Q = \frac{1}{2\pi}\oint \frac{ds}{\beta}, \tag{3.21}$$

where Q stands for Q_x or Q_y, the tune in either x or y directions, which are independent. In real accelerators, errors and irregularities of the magnetic field are inevitable. If the tune is integer or half integer, electrons receive an unwanted kick at every turn at the same point wherever errors and irregularities exist. Therefore the amplitude will increase (a resonance) and the electrons cannot circulate. Therefore the value of Q is chosen never to be either integer or half integer.

b. Closed orbit distortion (COD): Machining and other fabrication errors created during magnet manufacture and also those arising from alignment errors during installation will cause the distortion of the equilibrium orbit called COD. Assuming a statistical distribution of the errors, one can estimate the expected maximum COD, $\pm 2Y$ at each point to be

$$Y(s) = \frac{\sqrt{\beta(s)}}{2B_0 R \sin \pi Q}\sqrt{\sum_j (l_j \Delta B_j)^2 \beta_j}, \tag{3.22}$$

where j stands for the j-th magnet, l_j, its length, ΔB_j, its magnetic field error and β_j, average of $\beta(s)$ in the j-th magnet. The COD can be reduced by the use of correcting magnets, but during the commissioning of the ring, it is not easy to establish the conditions for correction unless electrons are circulating in the ring. Therefore, ΔB_j should be small enough for electrons to circulate without correction.

c. The chromaticity: When $\Delta E/E_0 \neq 0$, the tune shifts from that of $\Delta E/E_0 = 0$ by an amount

$$\Delta Q = \xi \frac{\Delta E}{E_0}, \quad \xi = -\frac{1}{4\pi}\int \beta(s)\left\{k - \eta(s)\frac{dk}{dx}\right\} ds, \tag{3.23}$$

where k is given by Eq. (3.5a) to Eq. (3.5d) and subscripts x and y are omitted. The value of ξ for a lattice composed of only bending and quadrupole magnets is called the natural chromaticity, which is usually negative causing beam instability (see Sect. **3.4.3**). The sextupole magnets generate dk/dx.

d. The momentum compaction factor: The circumference of the orbit for electrons with $\Delta E = 0$ is defined to be L. It increases when the electron energy deviates from E_0. This factor expresses the relation between the elongation of the circumference ΔL and the energy deviation ΔE as

$$\frac{\Delta L}{L} = \alpha_p \frac{\Delta E}{E_0}, \quad \alpha_p = \frac{1}{LR}\int_{\text{mag}} \eta ds, \tag{3.24}$$

where the subscript "mag" attached to the integral symbol means integration only in bending magnets.

e. The emittance: The value

$$a_x^2 = \frac{1}{\beta_x}\left\{x_\beta{}^2 + (\beta_x x'_\beta + \alpha_x x_\beta)^2\right\} \tag{3.25}$$

is independent of s as derived from Eq. (3.10) and is a constant. The situation is the same for a_y^2. These values are called the emittance and are expressed in units of m · rad. (In accelerator physics, the emittance was originally defined to be πa^2 but in the case of synchrotron radiation sources, π is often omitted.) The value of emittance is the same at every point of the orbit. Usually the injected beam has an amplitude less than one cm. If there is no mechanism to damp this amplitude, the electron beam will continue to circulate with the same amplitude. That is, neither a nor the emittance will change. However, electrons emit synchrotron radiation after injection, so that the amplitudes of betatron oscillations will decrease. However, at the same time the emission of synchrotron radiation excites these oscillations so that damping and excitation together soon achieve an equilibrium state. This whole phenomenon is called 'radiation damping'. In Eqs. (3.10) and (3.12), a damps within 0.1s for instance, and then reaches equilibrium. The emittance in the x direction is not zero but is finite, and represents the emittance of the light source. The horizontal emittance, obtained by calculation, is called the natural emittance and is given, approximately by

$$\varepsilon_{x0} \approx 2\langle H\rangle_{\text{mag}}\left(\frac{\sigma_E}{E_0}\right)^2, \tag{3.26}$$

$$\langle H\rangle_{\text{mag}} = \frac{1}{2\pi R}\int_{\text{mag}} \frac{1}{\beta_x}\left\{\eta^2 + (\beta_x\eta' + \alpha_x\eta)^2\right\} ds, \tag{3.27}$$

$$\left(\frac{\sigma_E}{E_0}\right)^2 \approx 2\times 10^{-13}\frac{\gamma^2}{R\,(\text{m})}. \tag{3.28}$$

σ_E in Eqs. (3.26) and (3.28) represents the energy spread of the electron beam at equilibrium. The emittance is also mentioned in Sect. **3.3.3**. The emittance in the vertical direction is zero in the ideal case, but in all real cases it is not zero due to the coupling between betatron oscillations in the x and y directions. Using the coupling constant $\kappa\,(0 \le \kappa \le 1)$, the true emittance is given by

$$\varepsilon_x = \frac{\varepsilon_{x0}}{1+\kappa}, \quad \varepsilon_y = \frac{\kappa\varepsilon_{x0}}{1+\kappa}. \tag{3.29}$$

Many low emittance rings make $\eta(s) = 0$ in long straight sections, where insertion devices are situated. These lattices are of the double- or triple-bend achromat type. In such lattices, $\beta_x(s)$ and $\eta(s)$ are small in bending magnets, so that the horizontal emittance can be made small as seen from Eq. (3.27). Moreover the influence of insertion devices on the stored beam can be minimized. A further reduction in emittance can be obtained by making $\eta(s)$ a little larger than 0. The emittance is inversely proportional to the third power of the number of bending magnets in an ideal case. In third generation synchrotron radiation sources the emittances are usually of a few to a few 10s nm · rad and the beam size in the horizontal direction is typically from 0.05 to 0.5 mm. Storage rings with emittances of 10 $\sim$ 100 pm · rad are now being devised which use more than 4 bending magnets to make long straight sections achromatic (multi-bend achromat) and give all bending magnets an appropriate field gradient, in order to achieve X-ray beams close to the diffraction limit.[6)]

f. The cross section and angular spread of the electron beam: The distributions of electrons with respect to position, divergence angle and energy in the beams are Gaussian. The width and angular spread in the horizontal direction are expressed using ε_x, $\beta_x(s)$, σ_E/E_0 and $\eta(s)$. Those due to betatron oscillations are given by

$$\sigma_{x\beta} = \sqrt{\beta_x\varepsilon_x}\,, \quad \sigma'_{x\beta} = \sqrt{\frac{1+\alpha_x^2}{\beta_x}\varepsilon_x}\,. \tag{3.30}$$

Those due to energy spread are given by

$$\sigma_{xE} = \eta\frac{\sigma_E}{E_0}, \quad \sigma'_{xE} = \eta'\frac{\sigma_E}{E_0}. \tag{3.31}$$

The total width and total angular spread in the horizontal direction are given, respectively, by

$$\sigma_x = \sqrt{\sigma_{x\beta}^2 + \sigma_{xE}^2}\,, \quad \sigma'_x = \sqrt{\sigma'^2_{x\beta} + \sigma'^2_{xE}}\,. \tag{3.32}$$

In the vertical direction, $\eta(s)$ does not contribute, so they are given by

$$\sigma_{y\beta} = \sqrt{\beta_y\varepsilon_y}\,, \quad \sigma'_{y\beta} = \sqrt{\frac{1+{\alpha_y}^2}{\beta_y}\varepsilon_y}\,. \tag{3.33}$$

The widths and the angular spreads mentioned in Sect. **1.2** are calculated using these formulae.

3.2.6 The magnets

a. Bending and quadrupole magnets: The magnets of storage rings are electromagnets whose yokes are machined in soft iron, or punched and pressed in silicon steel. As shown in Fig. 3.2(a), d (cm) stands for the gap spacing of a bending magnet and l (cm), the length of its yoke. The ampere · turn (A · T) required for the coil (top and bottom in total) to generate B(T) in the gap is given by

$$NI = \frac{B}{4\pi}\left(d + \frac{l}{\mu}\right) \times 10^5 \text{ (A} \cdot \text{T)}, \tag{3.34}$$

where μ is the relative permeability. The coils are made from hollowed Cu windings which can be cooled by water. The maximum feasible current is about 600 ~ 1000 A/cm^2. In the case of air cooling, the maximum current is about 100 A/cm^2. In a quadrupole magnet, the magnetic field is proportional to x in the horizontal direction (see Fig. 3.2(b)). Its pole shape can be determined by solving the Laplace equation for the magnetic potential Φ given by

$$\frac{\partial^2 \Phi}{\partial x^2} + \frac{\partial^2 \Phi}{\partial y^2} = 0. \tag{3.35}$$

The general solution of Φ is

$$\Phi = a_0 + a_{11}x + a_{12}y + a_{21}x^2 + a_{22}xy + a_{23}y^2 . \tag{3.36}$$

From the symmetry of the quadrupole magnet, $\Phi(x, y) = \Phi(y, x) = \Phi(-x, -y) = -\Phi(x, -y)$. Providing Φ at the surface of the pole is (x_1, y_1) is V_0, $V_0 = a_{22}x_1y_1$. The shape of the pole is hyperbolic. If it is represented as $x_1y_1 = R^2/2$,

$$\Phi = \frac{2V_0}{R^2}xy , \tag{3.37}$$

where R is called bore radius. The magnetic field is given as

$$B_x = -\frac{\partial \Phi}{\partial x} = -\frac{2V_0}{R^2}y , \quad B_y = -\frac{\partial \Phi}{\partial y} = -\frac{2V_0}{R^2}x. \tag{3.38}$$

In real cases, the pole width has to be finite in order to wind the coils. The magnetic field decreases rapidly around the edges of the pole where the coils are wound, so that it is not uniform around the edges. To obtain a uniform field in real magnets, the pole faces are specially shaped by modifying the gaps around the edges (called 'shimming'). The precise shapes are determined by magnetic field calculation.[7)]

b. Insertion devices: Undulator magnets are usually constructed from assemblies of permanent magnets. In the 'Halbach' type, one period of the magnet is composed of 4 magnets (top and bottom in total: 8), which are square pillars magnetized along each side of the square as shown in Fig. 3.5. In this case, the maximum field of the sinusoidal magnetic field in a planar undulator is given by

$$B_0 = 1.8B_r(1 - e^{-\frac{2\pi h}{\lambda_u}})e^{-\frac{\pi g}{\lambda_u}} , \tag{3.39}$$

where B_r is the remnant magnetic flux density of the permanent magnets.[8)] 'Hybrid' undulators are constructed from permanent magnets and soft iron. Many different types of undulator have been

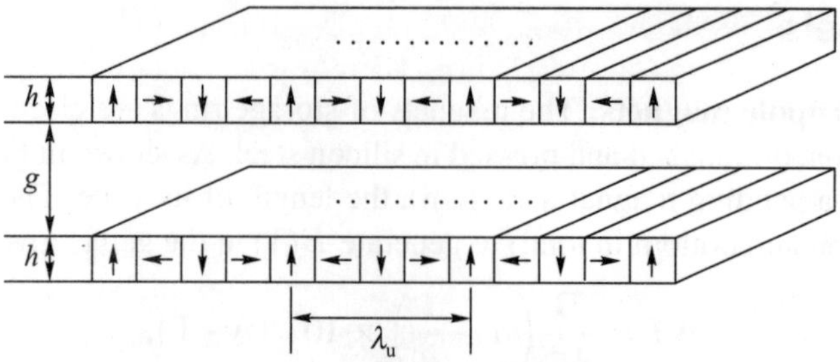

Figure 3.5 Undulator (Halbach type).

described.[9] One outstanding design is the so called in-vacuum undulator. Usually the evacuated beam ducts are situated within the narrow magnet gap, so that it is difficult to obtain a strong magnetic field for a short period undulator. This problem was solved by the in-vacuum undulator design in which the magnets themselves are accommodated within the beam duct, so that the gap can be minimized. The gap is adjustable from outside the duct. Other types are polarization undulators, which can be used to generate right- and left-circularly-polarized light in addition to vertically-, horizontally- and elliptically-polarized light. Recently an undulator has been proposed,[9] in which the orders of higher harmonics are not integral, but are fractional to avoid mixing the higher harmonics (as with higher order light from a monochromator). Multipole wigglers have magnetic structures similar to undulators. The principal difference is that $K \gg 1$ in multipole wigglers. Many superconducting wigglers are in use around the world. There is, as yet, no insertion device for infrared generation.

3.3 RF ACCELERATION

RF cavities are located on the orbit of the electron beam to replace the energy lost by the emission of synchrotron radiation. The average energy lost by each electron per turn is given by $U_0(\text{keV}) = 88.5E_0^4(\text{GeV})/R(\text{m})$ according to Eq. (2.67). Therefore, the required voltage of the RF cavity is to be greater than U_0/e (kV). In real cases, an increased voltage is essential to ensure the long lifetime of the beam. Even those electrons whose energies deviate from the central energy to some extent, can circulate stably because of the action of the RF. There is a phenomenon called 'radiation damping' which affects the motions of all electrons. This is linked to the emission of synchrotron radiation and acceleration and deceleration by the RF. These effects are explained below.

3.3.1 Synchrotron oscillations

Figure 3.6 depicts the RF voltage multiplied by the elementary electric charge in the time domain. The horizontal line indicates the energy loss of an electron, U_0. First, consider an electron of $\Delta E = 0$ circulating on the central orbit, which passes through the RF cavity at a time (phase) when $eV = U_0$. This electron loses U_0 during one turn, but receives an equal amount of energy from the RF cavity, so that it remains on the central orbit after passing through the RF cavity with the same phase of $eV = U_0$. This phase is called the 'synchronous phase'. We describe the motion of electrons from a coordinate moving on the central orbit with the light velocity c at this phase. The abscissa of Fig. 3.6 expresses the phase τ of electrons. If an electron initially located at the synchronous phase has energy deviation $\Delta E_1 > 0$, it circulates on an orbit somewhat greater than the central orbit because of the momentum compaction factor $\alpha_p > 0$. As the result, the time required for one turn increases, so that the phase is delayed from the phase of $eV = U_0$ after 1 turn. Therefore the electron receives an energy less than U_0, that is, the electron is decelerated. When the energy deviation becomes $\Delta E = 0$ after many turns the delay of phase becomes maximum. When $\Delta E < 0$, the electron circulates inside the central orbit and it is continuously decelerated down to $\Delta E = -\Delta E_1$. The phase advances during the circulation until at $\Delta E = -\Delta E_1$, the electron circulates on the innermost orbit and the phase is that of $eV = U_0$. In further circulations, the phase becomes that for acceleration and ΔE gradually reaches 0. At this moment, the electron circulates on the central orbit with the most advanced phase of RF. This motion of the electron can be considered to be an oscillation around the phase of $eV = U_0$. This is the principle of the 'phase stability' and the oscillation is a phase oscillation or energy oscillation (because the energy increases and decreases) and they are called 'synchrotron oscillations'. This behavior is illustrated in Fig. 3.7, where $\Delta E_{\max}$ represents the maximum energy deviation of electrons, which the RF can capture. The deviation in x direction from the central orbit at a certain time at a certain point is given by $\Delta E/E_0$ at that time and $\eta(s)$ at that point.

3.3.2 The equation of motion

The electrons appear to move forward and backward in the direction of the central orbit around the observer at the origin of the chosen coordinate, and inwards and outwards in a horizontal direction perpendicular to the orbit. The deviation of the phase from the synchronous phase $\Delta\tau$ is given by $\Delta\tau = -\Delta L/c = -\alpha_p L \Delta E/E_0 c$. The supplied energy at phase τ is $eV(\tau) = eV_0 \sin\omega(\tau + \tau_0)$ (see

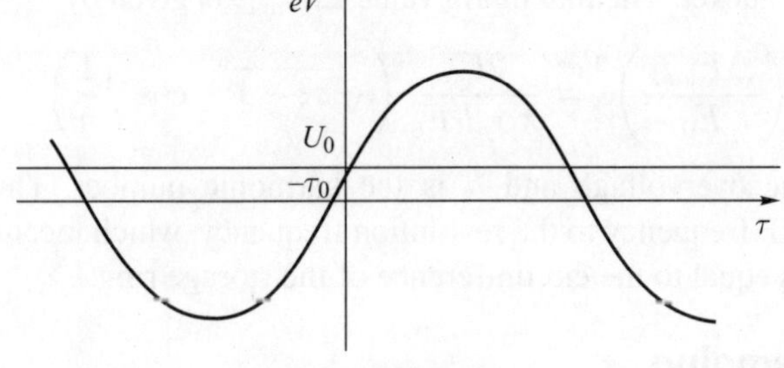

Figure 3.6 RF voltage.

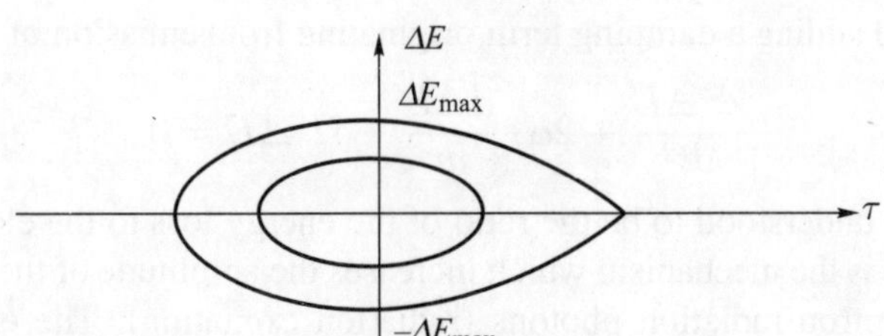

Figure 3.7 Relation between energy deviation and phase of electron around synchronous phase.

Fig. 3.6). The change of phase in time is given by

$$\frac{d\tau}{dt} = \frac{\Delta\tau}{T_0} = -\frac{\Delta L}{cT_0} = -\alpha_p \frac{\Delta E}{E_0}, \tag{3.40}$$

where T_0 is the time required to complete one turn. Since the increment of ΔE is the energy supplied by the RF cavity subtracted by radiation loss U_{rad}, the time dependence is given by

$$\frac{d\Delta E}{dt} = \frac{eV(\tau) - U_{\text{rad}}}{T_0}. \tag{3.41}$$

Using Eqs. (3.40) and (3.41), one can derive

$$\frac{d^2\tau}{dt^2} + \alpha_p \frac{eV(\tau) - U_{\text{rad}}}{E_0 T_0} = 0. \tag{3.42}$$

Providing $U_{\text{rad}} = U_0$, by neglecting the dependence of U_{rad} on ΔE and by using the approximation $eV(\tau) = U_0 + eV'\tau$ ($\omega\tau < 1$, $V' = dV/d\tau$), we finally obtain

$$\frac{d^2\tau}{dt^2} + \alpha_p \frac{eV'}{E_0 T_0}\tau = 0. \tag{3.43}$$

The phases of the electrons execute harmonic oscillations. In other words, the electrons are captured in an 'RF bucket'. The square root of the factor of the second term Ω is the angular frequency of synchrotron oscillation and $\Omega T_0/2\pi$ is the synchrotron number (the number of oscillations in one turn). If ΔE is too large, the oscillation according to Eq. (3.43) cannot be made, so that electrons with such energy must leave the RF bucket. The maximum value $\Delta E_{\max}$ is given by

$$\left(\frac{\Delta E_{\max}}{E_0}\right)^2 = \frac{2U_0}{\pi\alpha_p h E_0}\left(\sqrt{q^2-1} - \cos^{-1}\frac{1}{q}\right), \tag{3.44}$$

where $q = eV_0/U_0$ is the overvoltage and h is the harmonic number. The value h is an integer expressing a ratio of the RF frequency to the revolution frequency, which means that the RF wavelength multiplied by this value is equal to the circumference of the storage ring.

3.3.3 Radiation damping

Radiation damping occurs in both synchrotron and betatron oscillations. Firstly, consider the synchrotron oscillations. Rewriting the equation for phase oscillation Eq. (3.43) using the equation of energy Eq. (3.40) and adding a damping term originating from emission of radiation, we obtain

$$\frac{d^2\Delta E}{dt^2} + 2\alpha_E \frac{d\Delta E}{dt} + \Omega^2 \Delta E = 0. \tag{3.45}$$

$\alpha_E \approx U_0/E_0T_0$ can be understood to be the ratio of the energy loss to the electron energy per second On the other hand, there is the mechanism which increases the amplitude of the electron motion, namely the emission of synchrotron radiation photons (radiation excitation). The emittance ε_{x0} is given at equilibrium between the damping and the excitation processes. With the assumption of $\alpha_E \ll \Omega$, the amplitude of energy oscillation is given by

$$\Delta E(t) = \Delta E_0 e^{i\Omega(t-t_0)} - u e^{i\Omega(t-t_1)}, \tag{3.46}$$

where the energy of an emitted photon is $u = \hbar\omega$. On average, $d\left\langle|\Delta E|^2\right\rangle/dt = N_{\rm p}\left\langle u^2\right\rangle$, where $N_{\rm p}$ stands for the mean number of photons emitted per second (see App. **B**). In this case, the amplitude increases and it is called radiation excitation. On the other hand, for radiation damping, $d\left\langle|\Delta E|^2\right\rangle/dt = -2\alpha_E\left\langle|\Delta E|^2\right\rangle$ according to Eq. (3.45) and the amplitude decreases. At equilibrium, $N_{\rm p}\left\langle u^2\right\rangle = 2\alpha_E\left\langle|\Delta E|^2\right\rangle$. Replacing $\left\langle|\Delta E|^2\right\rangle/2$ by σ_E^2 and by using $N_{\rm p}\left\langle u^2\right\rangle$ in Eq. (B.26), one can obtain $(\sigma_E/E)^2$ similar to Eq. (3.28).

Next, consider the radiation damping of betatron oscillations. An electron will emit light in a tangential direction, but its position and direction of motion do not change at the emitting moment. That is, $\delta x = \delta x_\beta + \delta x_E = 0$ and $\delta x' = \delta x'_\beta + \delta x'_E = 0$. However, the equilibrium orbit will suddenly jump inside at the same time, so that $\delta x_E = -\eta u/E_0$ and $\delta x'_E = -\eta' u/E_0$. By substituting these equations for Eq. (3.25), we obtain

$$a_x^2 = \frac{1}{\beta_x}\left\{\eta^2 + (\beta_x\eta' + \alpha_x\eta)^2\right\}\left(\frac{u}{E_0}\right)^2. \tag{3.47}$$

After detailed calculation and by replacing $(u/E_0)^2$ with $(\sigma_E/E_0)^2$, we obtain Eqs. (3.26) and (3.27). The damping term for betatron oscillations α_β is approximately double that for α_E.

The 'natural electron bunch length' of the beam is defined to be the standard deviation of a Gaussian distribution of electrons around the synchronous phase after radiation damping. This length is equivalent to light pulse durations at experiments which usually extend from a few tens of ps to a few ns. One can minimize the bunch length by 'quasi-isochronous operation', in which α_p is made intentionally small. In real cases, the bunch length is sometimes elongated by instabilities as described later. The interval between bunches is defined by the period of the RF. However, in single bunch operation or operation with bunches separated by equidistant intervals as described in Sect. **1.2.1**, the interval becomes the storage ring revolution time or an integral fraction of it. Irradiation using a femtosecond ($\sim$ 100 fs) laser whose wavelength is the same as the fundamental of undulator radiation, passing along the electron beam which is traversing through an undulator has enabled very short, femtosecond pulsed synchrotron radiation to be obtained from the downstream bending section.[10)]

3.3.4 The RF cavity

The RF frequencies used in storage rings range from a few tens of MHz to around 500 MHz. The frequency is usually chosen by taking both the power required from the ring and the power of available transmitters in MHz region into account. The emitted power from stored current $I(\mathrm{A})$ is $U_0 I/e(\mathrm{kW})$. However in real cases some additional power is always required, because power is also consumed by the RF cavity. For relatively low frequencies, transmitters using triodes or tetrodes had been used but these are now becoming replaced by solid state devices. In the 500 MHz region, the high power klystron is available, so that the frequency 500 MHz has been chosen for use in many rings with large radiation losses. Recently, transmitters using solid state devices have been adopted in this region, though their output power is smaller than that from klystron sources. Once the frequency is chosen, the RF cavity is designed and it will have a resonant frequency equal to the chosen frequency. Figure 3.8 shows a schematic of such a cavity. It is normally made from Al or Cu and is essentially composed of an inner cylinder of radius r_1 and an outer cylinder of radius r_2, with two circular plates separated by h. The inner cylinder is cut into two sections, spaced d apart and the electric field is generated there. (There are other cavities where the inner cylinder is not clearly seen.) Electron beams pass through the inner side of the inner cylinder. The electric force lines of the accelerating field are drawn in the figure with the direction of magnetic force line marked by $\odot$ and $\otimes$. By treating the cavity as a coaxial tube, its inductance will be $L = (\mu_0 h/2\pi)\ln(r_2/r_1)$. The capacitance of two inner cylinders facing each other is $C = \varepsilon_0 \pi {r_1}^2/d$. The resistance of the cavity R is sum of resistances of the inner and the outer cylinders, and two circular plates. The equivalent circuit of this cavity is given in Fig. 3.9. Its resonant frequency is $\omega_0 = 1/\sqrt{LC}$ and the quality factor is $Q = \omega_0 L/R$, which indicates the sharpness of the resonance. In this equivalent circuit, the impedance of the capacitance C, called the shunt impedance is given by $R_s = Q/\omega_0 C$. The frequency $f = \omega_0/2\pi$ and the quality factor Q are given by

$$f = \frac{c}{2\pi}\sqrt{\frac{2d}{h{r_1}^2 \ln(r_2/r_1)}}, \tag{3.48a}$$

$$\frac{1}{Q} = \left(\frac{1/2r_1 + 1/2r_2}{\ln r_2/r_1} + \frac{1}{h}\right)\delta, \tag{3.48b}$$

where δ stands for the skin depth. The higher the frequency, the smaller the cavity becomes. The shape of the cavity is designed to make the resonant frequency equal to the RF frequency by solving Laplace equation.[7] Cavities of 500 MHz are mostly used. The voltage generated on the beam axis is $V = \sqrt{P_c R_s}$, where P_c is the power fed to the cavity.

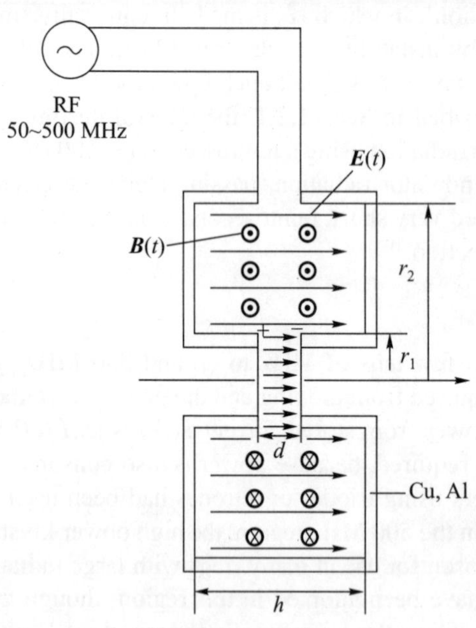

Figure 3.8 Schematic of RF cavity.

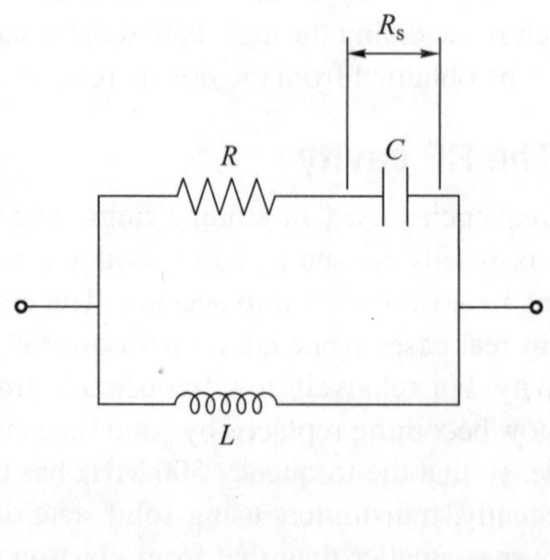

Figure 3.9 Equivalent circuit of RF cavity.

Usually, the cavity is situated on a straight section on the ring. In rare cases, it is situated within the gap of bending magnet.

3.4 THE BEAM LIFETIME AND INSTABILITIES

Electron beams cannot be maintained at a constant circulating current level because the current begins to decay immediately after it has been stored. There are two reasons for this decay. One is due to collisions between the electrons and gas molecules remaining in beam ducts (or orbit chambers) in which the electrons circulate. The second cause is due to electron-electron scattering by Coulomb interaction within bunches, called the 'Touschek effect'. The decay time (the beam lifetime when the current has decayed to the fraction $1/e$) is usually from a few, to several tens of hours for experiments using synchrotron radiation. To obtain such long lifetimes, the vacuum system and the RF system must be extremely carefully designed and fabricated. In some synchrotron radiation facilities, 'top-up injection' is provided to keep the stored current roughly constant by making injection just after current is lost. When the stored current is small, the electron beams circulate stably, but when the current increases, the electron beams become unstable with respect to beam shape for example, due to the electric fields induced by the current itself passing through the beam ducts. Therefore, appropriate counter measures are needed.

3.4.1 The vacuum

a. Collision cross section: There are at least three different loss mechanisms with their associated cross sections which describe the collisions of an electron with molecules of residual gas. The first one is due to elastic scattering with nuclei of the gas molecules, which produce an increase in the amplitude of betatron oscillations, resulting in loss of electrons by collision with the walls of the beam duct. The cross section of the first type is given approximately by

$$\sigma_1 = \frac{4\pi Z^2 e^4}{{E_0}^2 {\theta_c}^2}, \quad \theta_c \approx \frac{A}{\beta_{av}}, \tag{3.49}$$

where Z stands for the atomic number of atoms in the molecule. A is the half height of the beam duct (usually the horizontal width is larger than vertical width) and β_{av}, the average of betatron function in vertical direction. The second type is related to the energy loss by bremsstralung which is emitted by any electron when it passes close to a nucleus in the gas molecule resulting in the escape of the electron from the RF bucket. The cross section for the second kind is given by

$$\sigma_2 = \frac{4}{137} Z^2 {r_0}^2 \ln \frac{183}{\sqrt[3]{Z}} \ln \frac{E_0}{\Delta E_{max}}, \tag{3.50}$$

where r_0 is the classical electron radius and ΔE_{max} is the maximum energy deviation of electrons which can be captured by the RF given by Eq. (3.44). The third mechanism causing energy loss from electrons is associated with residual gas molecules in the beam ducts. The cross section for the third process is given by

$$\sigma_3 = \frac{2\pi {r_0}^2 N_m \times 0.51}{\Delta E_{max}\,(\text{MeV})}, \tag{3.51}$$

where N_m is number of electrons in the gas molecule. The total cross section is $\sigma = \sigma_1 + \sigma_2 + \sigma_3$. The residual gases normally present in beam ducts are CO, initially and H_2 after long periods of storage ring operation.

b. Vacuum pressure and lifetime: We define σ to be the cross section for loss of an electron by collision with a molecule and $\hat{\rho}$, to be the number density of molecules remaining in the vacuum chamber. The probability for loss of an electron by collision per second is given by $c\sigma\hat{\rho}$, because the electron circulates with a velocity close to that of light. Therefore in the case of a stored current with N electrons, the change of electron number is given by $dN/dt = -c\sigma\hat{\rho}N$, for which the solution is $N = N_0 e^{-c\sigma\hat{\rho}t}$. The lifetime dependence on vacuum is given by $\tau_v = 1/c\sigma\hat{\rho}$. Using Boyle-Charles law that $p = \hat{\rho}kT$, where $k = 1.38 \times 10^{-23}$ J/K (Boltzmann constant) and T is absolute temperature, one can obtain $p = kT/c\sigma\tau_v$. If the cross section σ is given, the vacuum pressure p required to guarantee the lifetime τ_v is determined to be

$$p\,(\text{Torr}) = 1 \times 10^{-27}/\sigma\tau_v, \quad p\,(\text{Pa}) = 1.33 \times 10^{-25}/\sigma\tau_v\,, \tag{3.52}$$

where units of σ and τ_v are cm^2 and s, respectively. In any practical case, the vacuum system is designed and fabricated to keep the pressure below 1×10^{-9} Torr (1.33×10^{-7} Pa).

c. Outgassing from the beam ducts (orbit chambers): In the case of ultra-high vacuum, the ultimate pressure is determined by the outgassing rate from the inner surfaces of the beam duct together with the pumping speed.[11] The beam orbit chamber is usually made from stainless steel or Al alloy. When electrons are not stored, outgassing arises only from thermal desorption at room temperature. The amount of outgas per unit time arising from thermal desorption is given by

$$Q_{\text{TH}} = qS\,, \tag{3.53}$$

where S stands for the total inner surface area of the beam duct and q is the outgas rate from unit area of the surface per unit time. The amount of q depends on the kind of material, surface machinery and treatment, and the degree of 'baking' (*i.e.* the heating of the vacuum chambers specifically to stimulate thermal desorption). In the case of Al alloy used in SPring-8, $q = 6.6 \times 10^{-9}$ Pa · m^3/m^2/s.

When electrons are stored in rings, photo-induced desorption also occurs, in which photoelectrons created at the surface by synchrotron radiation hitting the surface, cause desorption of gas molecules (a process called electron-stimulated desorption). The magnitude of this desorption effect is larger than that of thermal desorption. The probability of this desorption decreases with increase in dose of synchrotron radiation (total amount of irradiated synchrotron radiation). The number of desorbed molecules per photon is given by

$$\eta = kD^x \quad \text{(molecules/photon)} \quad (x < 0), \tag{3.54}$$

where dose is expressed by a product of current and time as D (A · hr). A measure of the dose for a new ring, in which the initial outgassing due to synchrotron radiation decreases sufficiently is $D = 100$ A · hr. The design of the vacuum system is made by assuming outgassing at this level. The parameters for oxygen-free Cu in Eq. (3.54) at this dose for H_2 and CO are respectively $k = 7$, $x = -0.67$ and $k = 5$, $x = -6$, for example.[12]. The irradiation wavelengths which generate photoelectrons most effectively are from the VUV to the soft X-ray range. The number of photons which generate photoelectrons is roughly given by integration over the angular flux of synchrotron radiation from a bending section using the long wavelength approximation Eq. (2.74) as

$$N_p = 9.6 \times 10^{17} \left\{ 1 - \left(\frac{u_e}{u_c} \right)^{1/3} \right\} E_0(\text{GeV}) I(\text{mA}) \quad \text{(photon/s)}\,. \tag{3.55}$$

Here u_e represents the threshold energy of photoelectron emission usually chosen to be 10 eV and u_c, the critical energy is chosen as the upper limit of the integration.[13] The outgassing caused by synchrotron radiation is given by

$$Q_{SR} = 0.41 \times 10^{-20} N_p \eta \quad (\text{Pa} \cdot \text{m}^3/\text{s}). \tag{3.56}$$

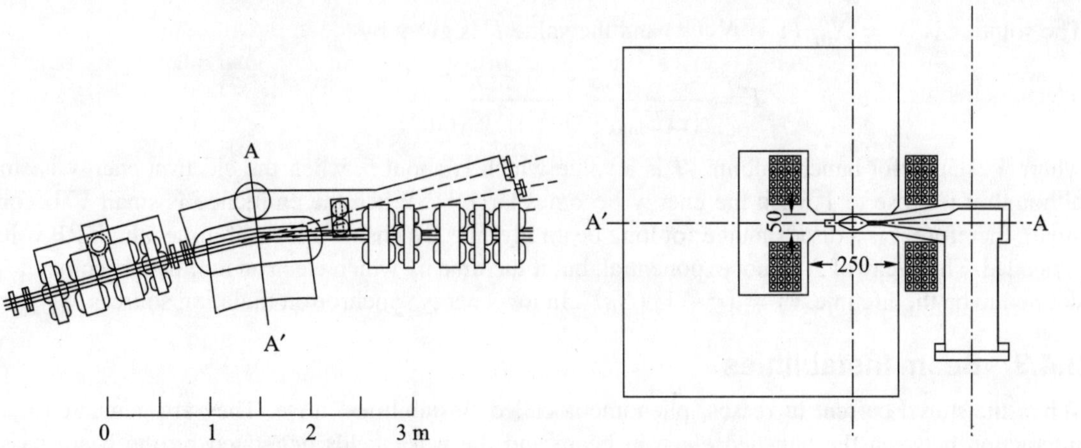

Figure 3.10 Electron beam duct at bending section.

In the case of an orbit chamber having a cross section shown in Fig. 3.10, synchrotron radiation irradiates the wall located outside of the gap of the bending magnet, where the pumps are situated to evacuate the gas desorbed by synchrotron radiation. The wall irradiated by synchrotron radiation is directly cooled by water.

d. The design of pumping system: The ultimate base pressure is given by

$$p = \frac{Q}{S_{\text{eff}}}, \quad \frac{1}{S_{\text{eff}}} = \frac{1}{C} + \frac{1}{S}, \tag{3.57}$$

where Q is total outgas, C, conductance of a pipe connecting a pump and a chamber and S, pumping speed. Units for both C and S are Torr · l/s or Pa · m³/s. The conductance of a circular pipe is given as $C = 12.1\ D^3/L$ (Torr · l/s $= 0.133$ Pa · m³/s), where D and L are respectively its radius and length given in unit of cm. The pipe must not reduce the pumping speed (the effective pumping speed S_{eff} should be as close as possible to the pumping speed S), so that D and L are chosen respectively to be as large and as short as possible. Ion sputter pumps and Ti sublimation pumps are usually employed.

The required pressure is already known, so that the required pumping speed can be calculated. The other method to evacuate the ducts within the bending sections, other than mentioned above (Fig. 3.10), is to distribute small pumps within the orbit chamber of any bending section, usually called an '*in situ* pumping', or an 'integrally pumped' section. At the beginning of the evacuation all chambers are heated to more than 150 degrees (baked!) to promote thermal desorption with rough pumping being achieved by turbo molecular pumps. An important design principle of the beam orbit chamber is described in Sect. **3.4.3**.

3.4.2 The Touschek effect

In a bunch of N electrons, the reduction in the number of electrons per unit time resulting from the Touschek process is given by

$$\frac{dN}{dt} = -\Gamma N^2. \tag{3.58}$$

The solution is $N = N_0/\left(1 + N_0 \Gamma t\right)$ and the value Γ is given by

$$\Gamma = \frac{7.8 \times 10^{-15} J}{\left(\Delta E_{\max}\,(\mathrm{MeV})\right)^2 V\,(\mathrm{cm}^3)} \quad (\mathrm{s}^{-1}), \tag{3.59}$$

where V stands for bunch volume. J is a value which is about 5, when the electron energy is small, although it is close to 1 when the energy becomes large.[14] When the emittance is small V becomes small, therefore it is a disadvantage for long beam lifetime. To ensure a long lifetime a high RF voltage is needed. The decay of N is not exponential, but if the time by which electron numbers becomes $1/e$ is defined to be the lifetime, $\tau_{\mathrm{T}} = (e-1)/N_0\Gamma$. In low energy synchrotron radiation sources, $\tau_{\mathrm{T}} < \tau_{\mathrm{v}}$.

3.4.3 Beam instabilities

When the stored current increases, phenomena called 'instabilities' arise. They are generated by the interaction between the bunched electron beam and the wake fields generated by the beam passing through the metallic beam chamber having uneven or discontinuous parts.[15] Once the phenomenon appears, the beam size and other quantities become unstable. It is damaging to experiments and the stored current can be completely lost in the worst case. The basic counter measure to this instability is to fabricate the interior of the orbit chamber to be as smooth as possible to suppress the generation of any wake fields. For instance, smooth metal pipes are used to cover uneven inner surfaces of bellows and metal strips are spanned across the pumping holes of any beam pipe parallel to the orbit. Assuming that $I(\omega)$ is the ω component of the stored current and $V(\omega)$, that of the voltage induced per turn, the impedance of the beam duct is defined by $V(\omega) = Z_L(\omega)I(\omega)$. The size of the effect of the wake field is expressed by $Z_L(\omega)/n$, which is usually chosen to be about 1 Ω to suppress the instability, where n is the ratio of ω to revolution angular frequency $2\pi c/L$. The calculation of the impedance is omitted here.

A 'head-tail' instability arises from a single bunch in such a way that the front part of the bunch generates a wake field, which influences the rear part. As a result, the length and the cross section of the bunch change with time (they oscillate). In this case, the amplitude of the oscillation increases only when the chromaticity is negative. The natural chromaticity is usually negative, so that sextupole magnets are used to make the chromaticity zero or slightly positive to suppress this instability. However, if the field of sextupole magnets becomes too strong, the dynamic aperture in which the electron beams can stably circulate can become smaller than the physical aperture due to the nonlinear field, so that the lifetime is again influenced. In the case of storage rings with the highest brilliance, this problem becomes more serious, but recently positioning of the sextupole magnets to avoid this problem can be achieved.

In multi-bunch operation, an interaction between bunches can occur which induces an instability in which the bunch interval changes in time. The origin of this instability is linked to the higher harmonic wave generated in the RF cavity by the electron beam. The higher harmonic wave can be suppressed either by choosing the resonance condition by controlling the frequency adjuster of the cavity or the temperature of the cavity. In addition, a novel cavity has been developed, which has parts made using

SiC to absorb the higher harmonic wave. Feedback systems are sometimes employed to suppress other kinds of instability. In some cases, superconducting cavities are used.

When the vacuum pressure is too high, especially at the beginning of beam commissioning in the storage ring, ion trapping occurs, in which the preceding bunch gathers ions which then can capture the following bunch, causing an oscillation which results in an increase of beam size. To suppress ion trapping, the vacuum pressure has to be minimized and ion clearing electrodes are often used to clear these ions. This phenomenon can also be suppressed by a mode of operation in which selected RF buckets are kept empty.

3.5 GENERATION OF COHERENT RADIATION

Usually, the phases of electric fields for light emitted from individual electrons are random. However, the phases can become uniform or correlated under some conditions when the electric fields will superpose, resulting in the coherent emission and extremely brilliant radiation. Infrared coherent radiation and free electron lasers are introduced as examples.

3.5.1 Coherent infrared radiation

Consider the radiation emitted from a bunch of electrons which pass through the field of a bending magnet. The power of the radiation emitted from a bunch of N electrons can be expressed by

$$P(\lambda) = NP_0(\lambda)\{1 + Nf(\lambda)\}, \tag{3.60}$$

where P_0 is the power of synchrotron radiation emitted from an electron, and $f(\lambda)$ is a bunch form factor which can be expressed as

$$f(\lambda) = \left| \int_{-\infty}^{\infty} e^{\frac{2\pi i}{\lambda} z} \rho(z)\, dz \right|. \tag{3.61}$$

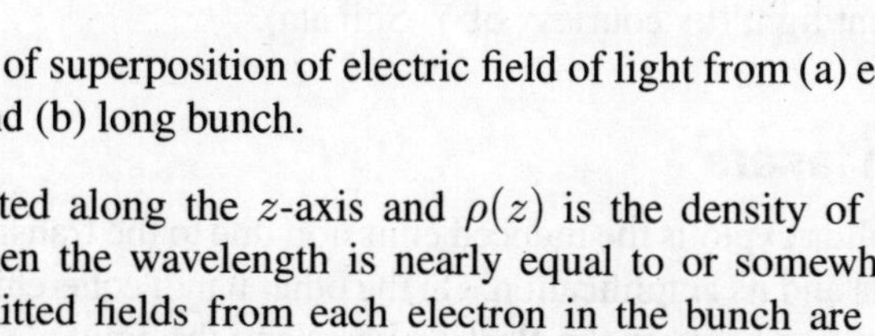

Figure 3.11 Features of superposition of electric field of light from (a) electron beam of short bunch and (b) long bunch.

Here the radiation is emitted along the z-axis and $\rho(z)$ is the density of electrons in a bunch. As shown in Fig. 3.11(a), when the wavelength is nearly equal to or somewhat longer than the bunch length, the phases and emitted fields from each electron in the bunch are uniform even though the light is spontaneous emission. The intensity of the superposed radiation is proportional to the square of sum of the fields emitted from each electron and shows a quadratic dependence on the number of the electrons in the bunch. On the other hand, when the wavelength is shorter than the bunch length as shown in Fig. 3.11(b), the correlation of phases among the light is cancelled. In this case the intensity is proportional to the sum of the intensities emitted from individual electrons and shows a linear dependence. For the radiation from N electron bunches, the intensity is N times stronger in the case of uniform rather than random phases. Such quadratically enhanced radiation is called coherent radiation. Figure 3.12 shows how the observed intensity of the coherent synchrotron radiation depends on the wavelength.[16,17] The measurements were made using a LINAC. The length of the electron bunch is estimated to be several mm. The figure shows an enhancement factor of about 10^6 (roughly equal to the estimated number of electrons in each bunch) compared with the incoherent radiation around the wavelength of 1 mm. Since the emitted radiation from a bunch is coherent with the radiation from the following bunches, further enhancement at specified wavelengths can be made by superposition using an optical resonant cavity. In the case of storage rings, the bunch lengths are typically from several mm to several cm. However recently, coherent infrared radiation has been observed by making the electron bunch shorter using quasi-isochronous operation[18] (see also Ref. 2 in Chap. **13**).

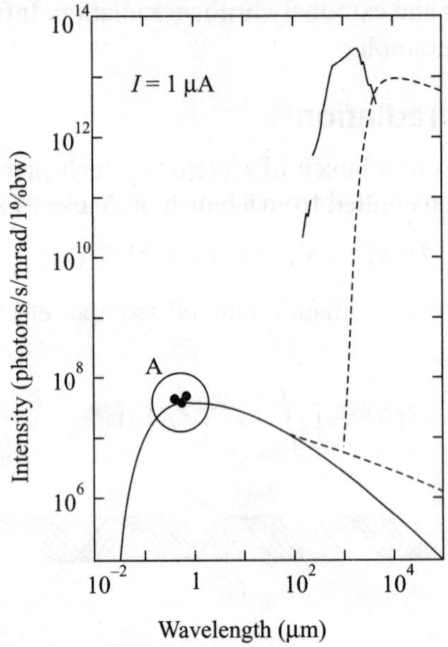

Figure 3.12 Infrared coherent radiation emitted by electron beams from LINAC passing through bending magnet (solid curves in upper part of figure) and A: visible incoherent light (by courtesy of Y. Shibata).[17]

3.5.2 Free electron lasers

Any laser using gases or solids exploits the induced emission due to the transition of bound electrons between discrete energy levels and its amplification. On the other hand, coherent light emission from undulators results from an interaction between the electron beam and the emitted synchrotron radiation itself. This is the basic principle of the free electron laser (FEL). In 1977, the first FEL was operated successfully at wavelength 3.4 μm in the infrared.[19] In this FEL, electron beams from a LINAC were passed through an undulator. The light from the undulator was confined within an optical cavity composed of two mirrors and the light of the wavelength selected from the undulator radiation amplified. The qualitative explanation is as follows. The electron has a velocity component perpendicular to the orbit. Therefore the electron can interact with the photon electric field resulting in two kinds of electrons, which are accelerated and decelerated by the electric field. The accelerated electrons gaining energy and decelerated electrons losing energy gather around the same position. As the result, the electron distribution in the bunch becomes non-uniform in the direction of the orbit, so that microbunches with sizes of the order of the wavelength are formed, thus emitting coherent radiation. (The gain in the FEL under small current density condition is given in App. **E**. Many FELs using optical cavities satisfy this condition.)

Figure 3.13 shows a schematic of an FEL attached to a storage ring using an optical cavity. Figure 3.14 shows examples of emission spectra.[20] In this case, the undulator used is not a simple one, but of the optical klystron type, so that the spectrum from the undulator is not a single peak but has many peaks as shown in the Fig. 3.14. The intensity of the light increases by several orders of magnitude, as

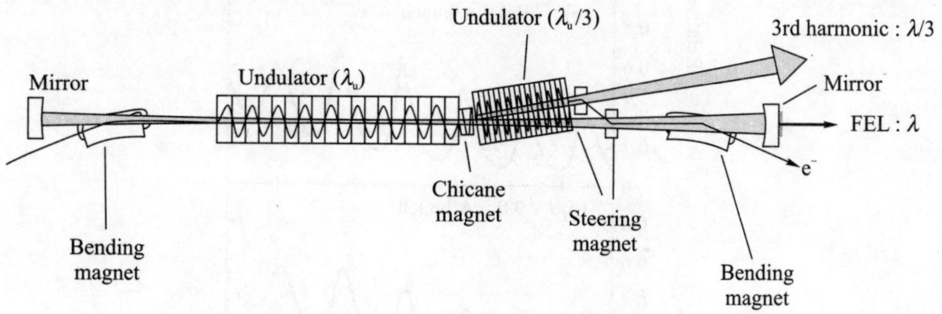

Figure 3.13 Schematic of cavity-FEL.

the stored current is increased from 9.4 mA/bunch to 13.8 mA/bunch and lasing occurs. At present, the shortest wavelength of light obtained from this type of FEL is 180 nm. The reason is that the reflectivity of the mirrors creating the optical cavity is considerably smaller than 100% in VUV region and the gain decreases with decrease in wavelength, so that rather long undulators are required. Third harmonic generation has been proposed to obtain the very shortest wavelengths of light.[21] Such a scheme is shown in Fig. 3.13. The micro-bunched beams generated by the undulator of period λ_u with an optical cavity are allowed to pass through another undulator of period $\lambda_u/3$. Recently cavity-FELs in VUV and hard X-ray region are proposed, which use high reflectance optical elements such as multilayer[22] and monochromatizing crystals,[23] respectively.

There are further types of FEL. One of them is called the SASE-FEL, from which VUV light and X-ray have been obtained.[24,25] In this case, an optical cavity is not used and the electron beams pass through the undulator only once. The light moves faster than the electron. That is, when the light leads by a single period of the undulator, it advances from the electron by one wavelength. Therefore, the light continually interacts with the electrons, which move along in front of the light. As the result, the electron beams receive density modulation and become microbunches. They also emit light and the light interacts with successive microbunches. In hard X-ray region at present, lasing at about 0.12 nm has been achieved (see Refs. 12 and 13 in Chap. **1**).

The other one is laser-seeded FEL which consists of two undulators. When the electron beam passes the first one whose period matches the wavelength of the laser, the laser is irradiated to the electron beam

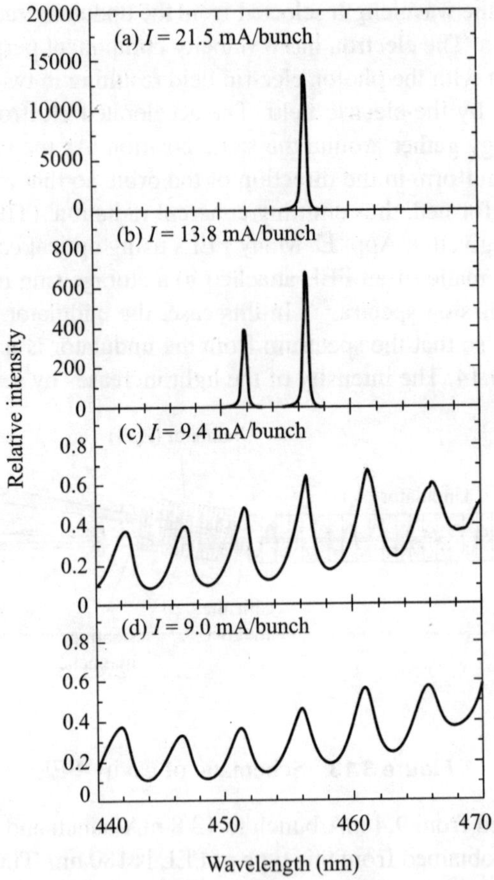

Figure 3.14 Output spectra of cavity-FEL attached to UVSOR storage ring.[20] Undulator is of optical klystron type.

and the electron beam is bunched with the wavelength of the laser. When such bunched electron beam passes the second undulator whose period matches the wavelength of high harmonic of the laser, the electron beam emits the coherent radiation of high harmonic. In the case of storage rings the available space for undulators is limited, but in the case of LINACs the long straight section is available. The radiation intensity of the latter is greater than the former, which is called coherent harmonic generation (CHG)[26] and the latter is called high-gain harmonic generation (HGHG).[27] For example, VUV HGHG of short pulse duration (order of 10 fs) can be obtained using visible laser.

REFERENCES

1. M. Sands: "Physics with Intersecting Storage Rings", Proc. Int. School of Phys. "Enrico Fermi" Course 46. B. Touschek ed. (Academic Press, 1971) p.257.
2. K. Wille: Rep. Prog. Phys. **54** (1991) 1005.

3. H. Winick ed.: "Synchrotron Radiation Sources. A Primer" (World Scientific, 1994).
4. For example, High Energy Accelerator Research Organization: SAD (http://acc-physics.kek.jp/SAD/sad.html).
5. T. Katayama and T. Miyahara: INS-TH-70 (Institute for Nuclear Study, The University of Tokyo, 1971).
6. R. Hettel: Synch. Rad. News **23** (2010) No. 6, 42 and references therein.
7. For example, Los Alamos National Laboratory: POISSON/SUPERFISH (http://laacg1.lanl.gov/laacg/sercices/possup.html).
8. G. Brown, K. Halbach, J. Harris and H. Winick: Nucl. Instr. Meth. **208** (1983) 65.
9. R. P. Walker and B. Diviacco: Synch. Rad. News **13** (2000) No.1, 33 and references therein.
10. B. W. Schoenlein, A. H. Chin, H. H. W. Chong, R. W. Falcone, T. L. Glover, P. A. Heimann, S. L. Johnson, A. M. Lindenberg, C. V. Shank, A. A. Zholents and M. S. Zolotorev: Synch. Rad. News **14** (2001) No.1, 20.
11. A. Roth: "Vacuum Technology" (Elsevier, 1990).
12. SPring-8 Project Part 2, Facility Design (1991).
13. M. Bernardini and L. Malter : J. Vac. Sci. Tech. **2** (1965) 123.
14. U. Völkel : DESY 67/5, March (1967).
15. A. W. Chao: "Physics of Collective Beam Instabilities in High Energy Accelerators" (John Wiley & Sons, 1993).
16. T. Nakazato, M. Oyamada, N. Niimura, S. Urasawa, O. Konno, A. Kagaya, R. Kato, T. Kamiyama, Y. Torizuka, T. Nanba, Y. Kondo, Y. Shibata, K. Ishi, T. Ohsaka and M. Ikezawa : Phys. Rev. Lett. **63** (1989) 1245.
17. Y. Shibata, K. Ishi and M. Oyamada: J. Jpn. Soc. Synch. Rad. Res. **5** (1992) 13, in Japanese.
18. M. Abo-Bakr, J. Feikes, K. Holldack, P. Kuske, W. B. Peatman, U. Schade and G. Wüstefeld : Phys. Rev. Lett. **90** (2003) 94801.
19. D. A. G. Deacon, L. R. Elias, J. M. J. Madey, G. J. Ramian, H. A. Schwettman and T. I. Smith : Phys. Rev. Lett. **38** (1977) 892.
20. S. Takano, H. Hama and G. Isoyama: The result was first reported in UVSOR Activity Report 1992 (ISSN 0911-5730) p.4. It also appeared in Nucl. Instr. Meth. A **331** (1993) 20.
21. M. E. Couprie, D. Nutarelli and M. Billardon : Nucl. Instr. and Meth. B **144** (1998) 66.
22. S. V. Benson, D. R. Douglas, P. Evtushenko, F. E. Hannon, C. Hernandez-Garcia, J. M. Klopf, R. A. Legg, G. R. Neil, M. D. Shinn, C. D. Tennant, S. Zhang and G. P. Williams: Synch. Rad. News **25** (2012) No. 1, 32.
23. K.-J. Kim, Yu. V. Shvyd'ko and R. R. Lindberg: Synch. Rad. News **25** (2012) No. 1, 25.
24. B. Sonntag : Nucl. Instr. Meth. **A467-468** (2001) 8.
25. http://www-ssrl.slac.stanford.edu/lcls/commissioning/documents/th3pbi01.
26. S. Kahn, M. Bakr, M. Höner, H. Huck, R. Molo, A. Nowaczyk, A. Schick, P. Ungelenk and M. Zeinalzadeh: Synch. Rad. News **24** (2011) No. 5, 18.
27. J. B. Murphy and X. Wang: Synch. Rad. News **21** (2008) No. 1, 41.

CHAPTER 4

The Optical Response of Matter

Optical phenomena such as reflection, refraction, absorption, emission and scattering are consequences of the response of matter to light stimulation and are defined by the optical properties of materials. In this chapter the interaction between light and matter is first considered at a macroscopic level, where a knowledge of the dielectric or optical constant is essential in order to understand this interaction. The behavior of electrons in matter is firstly described using classical electron theory and then origin of the dielectric constant and atomic scattering factor are explained. Finally, the absorption and scattering of light are explained using quantum mechanics. Note that it is the response of matter to very short wavelength light and to X-rays in particular that we shall primarily consider.[1,2)]

4.1 ELECTROMAGNETISM, MATTER AND LIGHT

We here describe light propagation using Maxwell's equations and make the assumption that the media are continuous and isotropic. The phenomena of reflection and refraction can be understood through a study of the behavior of light at the interface between two media.

4.1.1 The dielectric and optical constants

First, we derive the relationship between the dielectric and the optical constants. (The dielectric constant is also called dielectric function.) General aspects of the optical properties of matter are described in the next section where we consider isotropic matter in which both constants (dielectric and optical) are scalar. However, for the case of non-isotropic materials, the constants are tensor (see App. **J** describing the magnetooptical effect). Nonlinear optical effects are not included.

Maxwell's equations in a medium are given by Eqs. (2.13) to (2.16). The relationship between the field quantities $\boldsymbol{E}$, $\boldsymbol{H}$, $\boldsymbol{D}$ and $\boldsymbol{B}$, and current density are given by

$$\boldsymbol{D} = \varepsilon_0 \varepsilon \boldsymbol{E},$$

$$\boldsymbol{B} = \mu_0 \mu \boldsymbol{H}, \tag{4.1}$$

$$\boldsymbol{j} = \sigma \boldsymbol{E}, \tag{4.2}$$

where ε and μ are the relative dielectric constant and the permeability of the material, expressed as ratios to ε_0 and μ_0 for vacuum, respectively. In what follows, ε is called the 'dielectric constant' and σ, called the 'electric conductivity'.

Since the medium is usually neutral, separate charge does not exist and one can assume $\rho = 0$. From Eqs. (2.13), (2.14) and (4.1), the following equations are obtained

$$\nabla^2 \boldsymbol{E} = \varepsilon_0 \mu_0 \varepsilon \mu \frac{\partial^2 \boldsymbol{E}}{\partial t^2} + \mu_0 \mu \sigma \frac{\partial \boldsymbol{E}}{\partial t},$$
$$\nabla^2 \boldsymbol{H} = \varepsilon_0 \mu_0 \varepsilon \mu \frac{\partial^2 \boldsymbol{H}}{\partial t^2} + \mu_0 \mu \sigma \frac{\partial \boldsymbol{H}}{\partial t}. \tag{4.3}$$

These are the wave equations for light propagation in any medium with attenuation, which implies absorption.

a. A non-absorbing medium: In the case of dielectric material with $\sigma = 0$, Eq. (4.3) becomes

$$\nabla^2 \boldsymbol{E} = \varepsilon_0 \mu_0 \varepsilon \mu \frac{\partial^2 \boldsymbol{E}}{\partial t^2},$$
$$\nabla^2 \boldsymbol{H} = \varepsilon_0 \mu_0 \varepsilon \mu \frac{\partial^2 \boldsymbol{H}}{\partial t^2}. \tag{4.4}$$

These wave equations represent the situation without attenuation or absorption. One of the solutions of Eq. (4.4) is a plane wave given by a sinusoidal function propagating with a wave vector $\boldsymbol{k}$. In vacuum, the magnitude of $\boldsymbol{k}$, the wave number, is $k = 2\pi/\lambda = \omega/c$. Assuming $\boldsymbol{E}_0$ and $\boldsymbol{H}_0$ to be independent of position, then the electric and magnetic fields can be expressed by

$$\boldsymbol{E} = \boldsymbol{E}_0 \exp\{-i(\omega t - \boldsymbol{k} \cdot \boldsymbol{r} + \phi_0)\},$$
$$\boldsymbol{H} = \boldsymbol{H}_0 \exp\{-i(\omega t - \boldsymbol{k} \cdot \boldsymbol{r} + \phi_0)\}, \tag{4.5}$$

where ϕ_0 is an initial phase. In this formalism, the macroscopic physical quantities which determine the optical response (dielectric constant, optical constant, *etc.*) are included in the wave vectors. This formalism is equivalent to one where light with wave number $\tilde{n}k$ propagates in the medium along the direction satisfying the Snell's law explained later, where $\tilde{n}$ is the complex refractive index given below and k is the wave number in vacuum.[3)] We derive the relationship between the above physical quantities, wave number and angular frequency. Substitution of Eq. (4.5) for Eq. (4.4) yields

$$k^2 = \varepsilon_0 \mu_0 \varepsilon \mu \omega^2 \tag{4.6}$$

which implies that light propagates through the medium with a phase velocity

$$v = \nu\lambda = 1/\sqrt{\varepsilon_0 \mu_0 \varepsilon \mu}. \tag{4.7}$$

The dielectric constant and the permeability in vacuum are $\varepsilon_0 = 8.854 \times 10^{-12}$ C/V/m and $\mu_0 = 1.257 \times 10^{-6}$ Wb/A/m, respectively, and light velocity is $c = 1/\sqrt{\varepsilon_0 \mu_0} = 2.998 \times 10^8$ m/s. In the optical frequency region, generally $\mu = 1$. Note that for radio waves and for a magnetic medium $\mu \gg 1$. The relation

$$(c/\omega)^2 k^2 = \varepsilon \tag{4.8}$$

is then obtained using the expression for the light velocity c given above and in Eq. (4.6). The refractive index n is defined to be the ratio of the propagation velocities of light in vacuum and in the medium, that is, the ratio of wavelengths. In a non-absorbing medium, the refractive index can be expressed by

$$n = c/v = \sqrt{\varepsilon}, \tag{4.9}$$

derived using Eq. (4.7), which defines the very important relationship between the refractive index and the dielectric constant.

Combining Eq. (4.5) with Eqs. (2.13) to (2.16), yields $\boldsymbol{k} \cdot \boldsymbol{E} = \boldsymbol{k} \cdot \boldsymbol{H} = 0$, which means that light is a transverse wave. The relationship between its electric field and magnetic flux density is given by

$$\boldsymbol{k} \times \boldsymbol{E} = \omega \boldsymbol{B}, \tag{4.10}$$

which means that both fields are orthogonal.

b. An absorbing medium: Consider the general case of an absorbing medium. The second order differential equation of Eq. (4.3) includes a first order differential term, so that for k in Eq. (4.5), the solution of Eq. (4.3) is a complex vector, $\tilde{\boldsymbol{k}} = \boldsymbol{k}' + i\boldsymbol{k}''$, where $\boldsymbol{k}'$ and $\boldsymbol{k}''$ are real vectors. The electric and magnetic fields are expressed by

$$\boldsymbol{E} = \boldsymbol{E}_0 \exp\{-i(\omega t - \tilde{\boldsymbol{k}} \cdot \boldsymbol{r})\},$$

$$\boldsymbol{H} = \boldsymbol{H}_0 \exp\{-i(\omega t - \tilde{\boldsymbol{k}} \cdot \boldsymbol{r})\}, \tag{4.11}$$

where the initial phase ϕ_0 is omitted. The electric field is expressed explicitly by

$$\boldsymbol{E} = \boldsymbol{E}_0 \exp\{-i(\omega t - \boldsymbol{k}' \cdot \boldsymbol{r})\} \exp(-\boldsymbol{k}'' \cdot \boldsymbol{r}). \tag{4.12}$$

The magnetic field is expressed similarly. From Eq. (4.12), one can see that the solutions are the product of terms expressing oscillation and attenuation. The condition $\boldsymbol{k}' \cdot \boldsymbol{r} =$ const. for arbitrary $\boldsymbol{r}$ gives a plane orthogonal to $\boldsymbol{k}'$, which is an equi-phase plane. The condition $\boldsymbol{k}'' \cdot \boldsymbol{r} =$ const. for arbitrary $\boldsymbol{r}$ gives a plane orthogonal to $\boldsymbol{k}''$, which is an equi-amplitude plane. In the isotropic medium without boundary, $\boldsymbol{k}' || \boldsymbol{k}''$.

Substitution of Eq. (4.12) for Eq. (4.3) yields

$$\tilde{k}^2 = \varepsilon_0 \mu_0 \varepsilon \mu \omega^2 + i \mu_0 \mu \sigma \omega. \tag{4.13}$$

Let us now introduce the complex dielectric constant, $\tilde{\varepsilon} = \varepsilon_1 + i\varepsilon_2$, whose definition is similar to Eq. (4.8) as

$$(c/\omega)^2 \tilde{k}^2 = \tilde{\varepsilon}. \tag{4.14}$$

The real and imaginary parts are given using Eq. (4.13) by

$$\varepsilon_1 = \varepsilon,$$

$$\varepsilon_2 = \frac{\sigma}{\varepsilon_0 \omega}. \tag{4.15}$$

In the following, we omit ˜ from $\tilde{\varepsilon}$ for the sake of simplicity. The complex refractive index $\tilde{n} = n + i\kappa$ is now introduced and it satisfies the following relation, as in Eq. (4.9) with

$$\tilde{n}^2 = \varepsilon. \tag{4.16}$$

In the field of optics, the complex refractive index is sometimes defined to be $\tilde{n} = n - i\kappa$. In this case, if the magnitude of $\boldsymbol{r}$ is infinite, the wave functions of Eq. (4.11) with a form similar to the wave function in quantum mechanics, will diverge. Therefore to avoid this difficulty, the complex refractive index is defined to be $\tilde{n} = n + i\kappa$. This is solely linked to a choice of sign with no significant change to the theory of optics. From Eq. (4.16)

$$n^2 - \kappa^2 = \varepsilon_1,$$

$$2n\kappa = \varepsilon_2 \tag{4.17}$$

can be derived. The real part n and the imaginary part κ are usually called the 'refractive index' or 'index of refraction' and 'extinction coefficient', respectively. The expression 'optical constants' usually embraces both quantities.

The final exponential term in Eq. (4.12) represents attenuation of the light. From Eqs. (4.14) and (4.16), the following relation

$$k'' = \omega\kappa/c \tag{4.18}$$

is derived between wave vector $\boldsymbol{k}''$ and extinction coefficient κ. In the z direction, when light with intensity I_0 at $z = 0$ passes through the medium, attenuation can be expressed in terms of the absorption coefficient by

$$I(z) = I_0 e^{-\mu z}, \tag{4.19}$$

where $I(z)$ is the intensity at $z = z$. The intensity of light given by Eq. (4.19) is proportional to the square of the absolute value of the electric field Eq. (4.12), or $|\boldsymbol{E}|^2$, thus giving a useful working relationship between the extinction coefficient and the absorption coefficient which is

$$\mu = 2\omega\kappa/c = 4\pi\kappa/\lambda. \tag{4.20}$$

In the case of a light absorbing medium, the optical constant will show a close relationship with the physical quantities described in the following sections which define the interaction between the light and the material. Note that $Im(-1/\varepsilon) = \varepsilon_2/\left(\varepsilon_1^2 + \varepsilon_2^2\right)$ is called 'energy loss function' (see App. **L**).

4.1.2 The Fresnel equation

Consider the phenomena of reflection and refraction when light crosses the boundary between two isotropic media (Fig. 4.1). When a monochromatic plane wave travels from medium 1 to medium 2, part will be reflected and the rest refracted. For simplicity, we assume that medium 1 is non-absorbing (vacuum) and that medium 2 is absorbing. The electric fields of the incident, reflected, and refracted waves can be expressed by

$$\begin{aligned} \boldsymbol{E}_1 &= \boldsymbol{E}_{01} \exp\left\{-i\left(\omega t - \boldsymbol{k}_1 \cdot \boldsymbol{r}\right)\right\}, \\ \boldsymbol{E}_\mathrm{r} &= \boldsymbol{E}_{0\mathrm{r}} \exp\left\{-i\left(\omega t - \boldsymbol{k}_\mathrm{r} \cdot \boldsymbol{r}\right)\right\}, \\ \boldsymbol{E}_2 &= \boldsymbol{E}_{02} \exp\{-i(\omega t - \tilde{\boldsymbol{k}}_2 \cdot \boldsymbol{r})\}. \end{aligned} \tag{4.21}$$

We may also obtain expressions for the magnetic fields by exchanging $\boldsymbol{E}$ for $\boldsymbol{H}$ using Eqs. (4.1) and (4.10). The phase changes at the boundary due to reflection and refraction are included in the amplitudes and the three exponential functions in the above equations are equal at the boundary.

a. Snell's law: For an arbitrary position $\boldsymbol{r}$ in the boundary plane, the dependence of the electric fields on $\boldsymbol{r}$ should be equal. Thus the arguments of the exponentials in Eq. (4.21) are equal, so that we obtain

$$\boldsymbol{k}_1 \cdot \boldsymbol{r} = \boldsymbol{k}_\mathrm{r} \cdot \boldsymbol{r} = \tilde{\boldsymbol{k}}_2 \cdot \boldsymbol{r}. \tag{4.22}$$

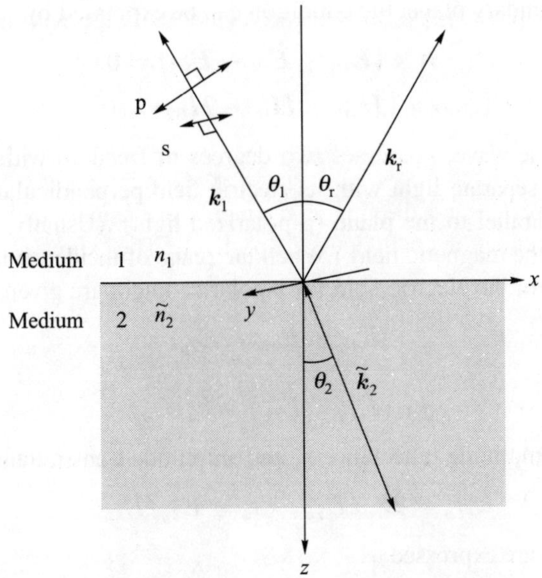

Figure 4.1 Reflection and refraction of light incident at the boundary of different media 1 and 2.

Because the incident and reflected waves travel in the same non-absorbing medium the first and second sides of Eq. (4.22) will be real and the third side will also be real, so that $\boldsymbol{k}_2'' \cdot \boldsymbol{r} = 0$ where $\tilde{\boldsymbol{k}}_2 = \boldsymbol{k}_2' + i\boldsymbol{k}_2''$, meaning that $\boldsymbol{k}_2''$ is perpendicular to the boundary plane, that is the equi-amplitude plane is parallel to the boundary plane. The plane of incidence is defined to be that plane which includes $\boldsymbol{k}_1$ for the incident light and the normal to the boundary plane. If $\boldsymbol{r}$ is perpendicular to the plane of incidence, $\boldsymbol{k}_1 \cdot \boldsymbol{r} = 0$. Therefore from Eq. (4.22) $\boldsymbol{k}_\mathrm{r} \cdot \boldsymbol{r} = \tilde{\boldsymbol{k}}_2 \cdot \boldsymbol{r} = 0$ and this implies that $\boldsymbol{k}_\mathrm{r}$ and $\tilde{\boldsymbol{k}}_2$ are also in the plane of incidence. As shown in Fig. 4.1, the boundary plane is defined to be the xy-plane and the plane of incidence, the xz-plane. If $\boldsymbol{r}$ is directed towards the x direction, then from Eq. (4.22), the relation $k_{1x} = k_{\mathrm{r}x}$ is obtained. Since $k_1 = k_\mathrm{r}$ from Eq. (4.8) and the directions of k_{1z} and $k_{\mathrm{r}z}$ are opposite, $k_{\mathrm{r}z} = -k_{1z}$. This shows that the angle of incidence θ_1 is equal to the angle of reflection θ_r as illustrated in Fig. 4.1 and this is called the 'law of reflection'. Since $\boldsymbol{k}_1 \cdot \boldsymbol{r} = \tilde{\boldsymbol{k}}_2 \cdot \boldsymbol{r}$,

$$k_1 \sin\theta_1 = \tilde{k}_2 \sin\theta_2, \tag{4.23}$$

where θ_2 is the angle of refraction. In this case, θ_2 is complex. Note that θ_2 differs from the angle between $\boldsymbol{k}_2'$ and the normal of the boundary plane. From Eqs. (4.14) and (4.16) this can be written as

$$n_1 \sin\theta_1 = \tilde{n}_2 \sin\theta_2 \tag{4.24}$$

and is called 'Snell's law'. By regarding θ_2 as the light direction in medium 2, one can calculate the reflectance and transmittance.

b. The Fresnel formula: In the boundary plane, the electric and magnetic field components tangential to the boundary plane must be continuous, called the 'continuity condition'. Therefore, using a unit vector $\boldsymbol{n}$ normal to the boundary plane, the condition can be expressed by

$$\boldsymbol{n} \times (\boldsymbol{E}_{01} + \boldsymbol{E}_{0\mathrm{r}} - \boldsymbol{E}_{02}) = 0,$$
$$\boldsymbol{n} \times (\boldsymbol{H}_{01} + \boldsymbol{H}_{0\mathrm{r}} - \boldsymbol{H}_{02}) = 0. \tag{4.25}$$

Light, which is a transverse wave, possesses two degrees of freedom with respect to polarization. It is therefore convenient to separate light with its electric field perpendicular to the plane of incidence (s-polarized light) from parallel to the plane (p-polarized light). (Usually, the 'polarization plane' is defined to be the plane of the magnetic field.) When the plane of incidence is the xz-plane as shown in Fig. 4.1, the relationships for the electric field for s-polarized light are given using Eq. (4.10) by

$$E_{1y} + E_{ry} = E_{2y},$$
$$k_{1z}(E_{1y} - E_{ry}) = \tilde{k}_{2z} E_{2y}. \tag{4.26}$$

For s-polarized light, the amplitude reflectance r_s and amplitude transmittance t_s are defined to be

$$r_\mathrm{s} = E_{\mathrm{r}y}/E_{1y}\,, \quad t_\mathrm{s} = E_{2y}/E_{1y}\,. \tag{4.27}$$

Using wave numbers, they are expressed as

$$r_\mathrm{s} = \frac{k_{1z} - \tilde{k}_{2z}}{k_{1z} + \tilde{k}_{2z}},$$
$$t_\mathrm{s} = \frac{2k_{1z}}{k_{1z} + \tilde{k}_{2z}} = r_\mathrm{s} + 1. \tag{4.28}$$

Using $k_{1z} = k_1 \cos\theta_1$ and $\tilde{k}_{2z} = \tilde{k}_2 \cos\theta_2$, and from Eqs. (4.14) and (4.16), the amplitude reflectance can be expressed in terms of the angles of incidence and the refractive index by

$$r_\mathrm{s} = t_\mathrm{s} - 1 = \frac{n_1 \cos\theta_1 - \tilde{n}_2 \cos\theta_2}{n_1 \cos\theta_1 + \tilde{n}_2 \cos\theta_2}. \tag{4.29}$$

By elimination of the angle of refraction θ_2 using Snell's law Eq. (4.24), Eq. (4.29) can be rewritten as

$$r_\mathrm{s} = t_\mathrm{s} - 1 = \frac{n_1 \cos\theta_1 - \left(\tilde{n}_2{}^2 - n_1^2 \sin^2\theta_1\right)^{1/2}}{n_1 \cos\theta_1 + \left(\tilde{n}_2{}^2 - n_1^2 \sin^2\theta_1\right)^{1/2}}. \tag{4.30}$$

From the continuity condition for the magnetic fields, which are perpendicular to the plane of incidence, the amplitude reflectance for p-polarized light is given by

$$r_\mathrm{p} = t_\mathrm{p} - 1 = \frac{\tilde{n}_2{}^2 \cos\theta_1 - n_1\left(\tilde{n}_2{}^2 - n_1^2 \sin^2\theta_1\right)^{1/2}}{\tilde{n}_2{}^2 \cos\theta_1 + n_1\left(\tilde{n}_2{}^2 - n_1^2 \sin^2\theta_1\right)^{1/2}}. \tag{4.31}$$

These equations are the 'Fresnel formulae' for amplitude reflectance and transmittance. The values are determined solely by the refractive index given in Eqs. (4.30) and (4.31).

The energy reflectance for each polarization is obtained from the squared absolute value of amplitude reflectance, using the complex conjugate by

$$R_\mathrm{s} = r_\mathrm{s}{}^* r_\mathrm{s}\,, \quad R_\mathrm{p} = r_\mathrm{p}{}^* r_\mathrm{p}. \tag{4.32}$$

Figure 4.2 shows calculated s- and p-polarized light reflectance as a function of the angle of incidence for two systems having optical constants of and $n = 0.8$, and $\kappa = 0.3$ and $n = 0.99$ and $\kappa = 0.01$. The former corresponds to those of Au at 40 eV and the latter at 300 eV (see Fig. 5.1). The normal incidence ($\theta = 0°$) reflectance for each polarization is the same and is given in the following equation,

$$R_s = R_p = \frac{(n-1)^2 + \kappa^2}{(n+1)^2 + \kappa^2}. \tag{4.33}$$

When approaching an angle of incidence of $\theta = 90°$ and a grazing angle of $\hat{\theta} = 0°$, the s- and p-reflectances both approach 1. As the angle of incidence is increased from 0°, the reflectance for s-polarized light increases monotonically, while that for p-polarized light decreases and shows a minimum at an angle of incidence, which is called the 'Brewster angle'. It then increases with increase in the angle incidence beyond the Brewster angle. The optical constants $n = 0.99$ and $\kappa = 0.01$ are typical examples of n and κ for materials for the soft X-ray region. In this region, the reflectances at near normal incidence are almost zero and grazing incidence optics is essential because of the total reflection mentioned below. The lower the extinction coefficient κ, the more prominent the total reflection phenomenon becomes, that is, the rise in the reflectance becomes steeper.

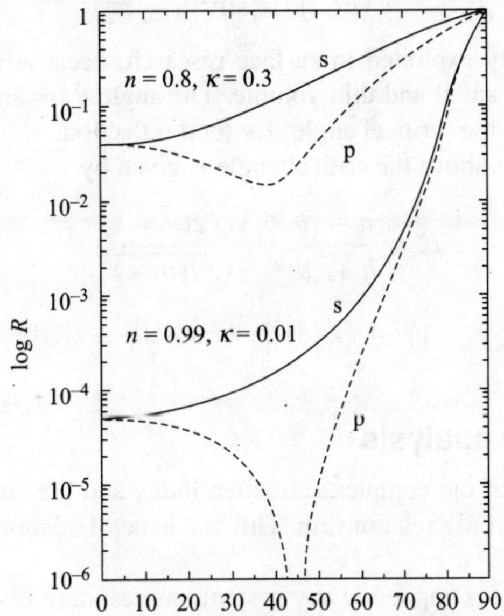

Figure 4.2 s- and p-reflectances as a function of angle of the incidence for two cases $n = 0.8$, $\kappa = 0.3$ and $n = 0.99$, $\kappa = 0.01$.

In all circumstances, the boundary between the media is assumed to be planar. However, reflectance will be diminished due to any surface roughness at the boundary. The scattering effect caused by surface roughness can be represented using a 'Debye-Waller type' factor,

$$F = \exp\left\{ -\left(\frac{4\pi\sigma\cos\theta}{\lambda} \right)^2 \right\}, \tag{4.34}$$

since the roughness dependence of reflectance has an exponential term similar to the Debye-Waller factor representing the temperature dependence of an X-ray diffraction intensity, which will be given in Chap. **9**, where σ is the rms surface roughness. The actual reflectance is given by $R_m = FR$, where R is the energy reflectance Eq. (4.32).

c. Total Reflection: Consider medium 2 to be non-absorbing. In the case of $n_1 > n_2$, if the angle of incidence θ_1 is larger than $\sin^{-1}(n_2/n_1)$, then the real value of the angle of refraction θ_2 satisfying Eq. (4.23) does not exist. Each component of $\boldsymbol{k}_2$ is obtained using Eq. (4.23) as

$$k_{2x} = k_2 \sin\theta_2 = k_1 \sin\theta_1 = (\omega/c)\, n_1 \sin\theta_1,$$

$$k_{2z} = k_2 \cos\theta_2 = (\omega/c)\left(n_2^2 - n_1^2 \sin^2\theta_1\right)^{1/2}. \tag{4.35}$$

In this case, the medium 2 is non-absorbing and k_{2x} is real, but k_{2z} is imaginary, that is, the refracted wave becomes a traveling wave in the x direction and its intensity will decrease exponentially in the z direction. This is called an 'evanescent wave'. Since k_{2z} is imaginary, substitution of the amplitude reflectance Eq. (4.28) for Eq. (4.32) yields an energy reflectance of 1. This phenomenon is called 'total reflection'. The light then enters medium 2 only to a very small depth given by

$$|k_{2z}|^{-1} = (c/\omega)\left(n_1^2 \sin^2\theta_1 - n_2^2\right)^{-1/2}. \tag{4.36}$$

This phenomenon is frequently exploited in surface research, because the light can penetrate but is confined only to an extremely small and thin volume. The angle $\theta_c = \sin^{-1}(n_2/n_1)$ above which the total reflection occurs is called the 'critical angle' for total reflection.

The approximate reflectance above the critical angle is given by

$$R = \frac{h - (\hat{\theta}/\hat{\theta}_c)\sqrt{2(h-1)}}{h + (\hat{\theta}/\hat{\theta}]_c)\sqrt{2(h-1)}}, \tag{4.37}$$

where $h = (\hat{\theta}/\hat{\theta}_c)^2 + [\{(\hat{\theta}/\hat{\theta}_c)^2 - 1\}^2 + (\kappa/\delta)^2]^{1/2}$ and $\hat{\theta}_c = \cos^{-1}(n_2/n_1)$ is the glancing, or grazing, critical angle.

4.1.3 Kramers-Krönig analysis

The real and imaginary parts of the complex refractive index and the complex dielectric constant are related through the Kramers-Krönig relationship. This is a general relationship and is not solely limited to optical properties.

When an external force $F(t)$ is applied to any system, the response $G(t)$ is given by

$$G(t) = \int_0^{\infty} A(\tau) F(t-\tau)\, d\tau, \tag{4.38}$$

where $A(t)$ is the response after time t. According to the causality principle, the response will follow after the applied field. Thus for $t < 0$, $A(t) = 0$. Functions $F(t)$ and $G(t)$ are Fourier-transformed to become

$$F(t) = \int_{-\infty}^{\infty} f(\omega) e^{-i\omega t} d\omega\,, \quad f(\omega) = \frac{1}{2\pi}\int_{-\infty}^{\infty} F(t)\, e^{i\omega t} dt,$$
$$G(t) = \int_{-\infty}^{\infty} g(\omega) e^{-i\omega t} d\omega\,, \quad g(\omega) = \frac{1}{2\pi}\int_{-\infty}^{\infty} G(t)\, e^{i\omega t} dt, \tag{4.39}$$

where $f(\omega)$ and $g(\omega)$ are angular frequency components. Response function $a(\omega)$ is defined to be $g(\omega) = a(\omega) f(\omega)$ and this is obtained by replacing $G(t)$ in fourth integral of Eq. (4.39) by Eq. (4.38). $a(\omega)$ and its Fourier transform can be written as

$$a(\omega) = \int_{0}^{\infty} A(t)\, e^{i\omega t} dt\,, \quad A(t) = \frac{1}{2\pi}\int_{-\infty}^{\infty} a(\omega) e^{-i\omega t} d\omega. \tag{4.40}$$

If we introduce the complex generalized frequency $\omega = \omega_{\rm r} + i\omega_{\rm i}$, then the condition for non-divergence of $a(\omega)$ is given by

$$|a(\omega)|^2 = \left| \int_0^{\infty} A(t)\, e^{i\omega_{\rm r} t - \omega_{\rm i} t} dt \right|^2 \le \int_0^{\infty} |A(t)|^2 e^{-2\omega_{\rm i} t} dt \le \int_0^{\infty} |A(t)|^2 dt. \tag{4.41}$$

This is satisfied when $\omega_{\rm i} \ge 0$. $a(\omega)$ is regular on the real axis and in the upper half plane. $a(\omega)$ in the upper half plane of ω is given from the Cauchy integration formula by

$$a(\omega) = \frac{1}{2\pi i}\int_c \frac{a(z)}{z-\omega} dz. \tag{4.42}$$

The contribution to integration for an infinitely distant arc becomes zero. Thus

$$a(\omega) = \frac{1}{2\pi i}\int_{-\infty}^{\infty} \frac{a(x)}{x-\omega} dx. \tag{4.43}$$

In the limit of $\omega_{\rm i} \to 0$, this yields

$$a(\omega_{\rm r}) = \frac{1}{\pi i}\mathcal{P}\int_{-\infty}^{\infty} \frac{a(x)}{x-\omega_{\rm r}} dx, \tag{4.44}$$

where $\mathcal{P}$ is the 'Cauchy principal integral'. Writing the response functions as $a(\omega_{\rm r}) = a_{\rm r}(\omega_{\rm r}) + i a_{\rm i}(\omega_{\rm r})$ and $a(x) = a_{\rm r}(x) + i a_{\rm i}(x)$, the following equations are obtained

$$a_{\rm r}(\omega_{\rm r}) = \frac{1}{\pi}\mathcal{P}\int_{-\infty}^{\infty} \frac{a_{\rm i}(x)}{x-\omega_{\rm r}} dx,$$
$$a_{\rm i}(\omega_{\rm r}) = -\frac{1}{\pi}\mathcal{P}\int_{-\infty}^{\infty} \frac{a_{\rm r}(x)}{x-\omega_{\rm r}} dx. \tag{4.45}$$

Should $e^{i\omega_{\rm r} t}$ change to $e^{-i\omega_{\rm r} t}$, the response function will not change and this yields the conditions $a_{\rm r}(\omega_{\rm r}) = a_{\rm r}(-\omega_{\rm r})$ and $a_{\rm i}(\omega_{\rm r}) = -a_{\rm i}(-\omega_{\rm r})$. Using these conditions, the integrals Eq. (4.45) can be replaced by integrals with respect to real angular frequency which are

$$a_r(\omega_r) = \frac{2}{\pi}\mathcal{P}\int_0^\infty \frac{x a_i(x)}{x^2 - \omega_r^2} dx,$$

$$a_i(\omega_r) = -\frac{2}{\pi}\mathcal{P}\int_0^\infty \frac{\omega_r a_r(x)}{x^2 - \omega_r^2} dx. \qquad (4.46)$$

These relationships between the real and imaginary parts are called the 'Kramers-Krönig relationships'. Therefore, when either the real part or the imaginary part is given, the other part can be obtained by calculation.

As an example, since the complex refraction index, $\tilde{n}(\omega) = n(\omega) + i\kappa(\omega)$ satisfies the Fourier transform condition $\omega \to \infty$, $\tilde{n}(\omega) - 1 \to 0$ as shown in the next section, the relationship between $n(\omega)$ and $\kappa(\omega)$ can then be shown to be

$$n(\omega) = 1 + \frac{2}{\pi}\mathcal{P}\int_0^\infty \frac{\omega'\kappa(\omega')}{\omega'^2 - \omega^2} d\omega',$$

$$\kappa(\omega) = -\frac{2\omega}{\pi}\mathcal{P}\int_0^\infty \frac{n(\omega') - 1}{\omega'^2 - \omega^2} d\omega'. \qquad (4.47)$$

These expressions are very important, because they enable the consistency of n and κ obtained from independent measurements to be checked and also to obtain n from κ, which is derived using Eq. (4.20) from the measured absorption coefficient. It is also possible to obtain the absorption coefficient from reflectance measurements at near normal incidence and magnetic circular dichroism from the Faraday rotation angle. One can obtain similar relationships for the dielectric constant,

$$\varepsilon_1(\omega) = 1 + \frac{2}{\pi}\mathcal{P}\int_0^\infty \frac{\omega'\varepsilon_2(\omega')}{\omega'^2 - \omega^2} d\omega',$$

$$\varepsilon_2(\omega) = -\frac{2\omega}{\pi}\mathcal{P}\int_0^\infty \frac{\varepsilon_1(\omega') - 1}{\omega'^2 - \omega^2} d\omega'. \qquad (4.48)$$

It is also feasible to derive the dielectric constant from the measured energy loss function.

4.2 THE CLASSICAL THEORY OF OPTICAL RESPONSE

From a macroscopic view point the light-matter interaction can be described using the dielectric constant and the optical constants introduced in the previous section. The interaction of light and matter can also be explained in terms of light-electron and light-ion interactions. In this section we describe the Lorentz and Drude models which are based on classical electron theory. Optical phenomena and constants are always a function of the frequency. The absorption of light by metals in the infrared region originates from free electrons and can be explained by the Drude model. Absorption by metals and insulators observed in the visible and the shorter wavelength regions originates from more tightly bound electrons and is explained by the Lorentz model. The absorption caused by molecular and lattice vibrations with various oscillation modes for the atoms and ions, manifest themselves in the infrared region and such vibrations can also be described using Lorentz model. In those frequency regions higher than that of visible light, the absorption originating from electrons is predominant.

4.2.1 The Lorentz model

External electric fields induce polarization effects in insulators. Polarization is caused by a shift in the center of gravity of the separated positive and negative charges. From a microscopic view point, polarization is due to the movement of positive and negative ions from their equilibrium position or from a change of equilibrium position of bound electrons within the atoms. A pair of such positive and negative ions or such bound electrons can be regarded as a dipole. The situation is similar for the bound electrons in metals.

In the Lorentz model, the electron is considered to be bound to an atom (much heavier than the electron) by a spring. Therefore the motion of the electron can be expressed as that of a damped harmonic oscillator,

$$m\frac{d^2\boldsymbol{x}}{dt^2} + m\Gamma\frac{d\boldsymbol{x}}{dt} + m\omega_0^2\boldsymbol{x} = -e\boldsymbol{E}, \tag{4.49}$$

where m and $-e$ are respectively mass and charge of the electron, and $\boldsymbol{x}$, displacement of the electron from center of the atom (nucleus). The third term on the left side expresses the restoring force (in 'Hooke's law', ω_0 is the resonant frequency) and the second term is the damping force due to various kinds of scattering with a damping coefficient Γ. The right side represents the force on the electron in the external field $\boldsymbol{E}$. The local electric field experienced by the electron is not necessarily equal to the external field. In general, since polarizability in the short wavelength region is rather small we shall assume that the local electric field and the external field are equal. The Lorentz force exerted on the electron with velocity $\boldsymbol{v}$ in the presence of light with magnetic flux density $\boldsymbol{B}$ is neglected for non-relativistic oscillations. Only in the explanation of the magnetooptical effect is the Lorentz force, caused by the applied magnetic field, taken into account. The external field $\boldsymbol{E}$ is the electric field of light which is proportional to $e^{-i\omega t}$. The solution of Eq. (4.49) is

$$\boldsymbol{x} = \frac{-e\boldsymbol{E}/m}{{\omega_0}^2 - \omega^2 - i\Gamma\omega}. \tag{4.50}$$

Since $\boldsymbol{x}$ is the displacement of the electron from the center of the atom, the induced dipole moment is

$$\boldsymbol{p} = -e\boldsymbol{x} = \frac{e^2\boldsymbol{E}}{m}\frac{1}{\omega_0^2 - \omega^2 - i\Gamma\omega}. \tag{4.51}$$

Correspondingly, the polarizability for one electron is

$$\alpha(\omega) = \frac{e^2}{m}\frac{1}{\omega_0^2 - \omega^2 - i\Gamma\omega}. \tag{4.52}$$

Assuming that ρ represents the number of dipoles per unit volume, the macroscopic polarizability $\boldsymbol{P}$ of the material will be given by

$$\boldsymbol{P} = \rho\boldsymbol{p} = \rho\alpha(\omega)\boldsymbol{E}. \tag{4.53}$$

$\boldsymbol{D}$ given by Eq. (4.1), can also be expressed using the following formula

$$\boldsymbol{D} = \varepsilon_0\varepsilon\boldsymbol{E} = \varepsilon_0\boldsymbol{E} + \boldsymbol{P}. \tag{4.54}$$

From Eqs. (4.52), (4.53) and (4.54), the dielectric constant can be represented by

$$\varepsilon_0\varepsilon = \varepsilon_0 + \rho\alpha = \varepsilon_0 + \frac{\rho e^2}{m}\frac{1}{{\omega_0}^2 - \omega^2 - i\Gamma\omega}. \tag{4.55}$$

The real and the imaginary parts of the relative dielectric constant may then be written as

$$\varepsilon_1 = n^2 - \kappa^2 = 1 + \frac{\rho e^2}{\varepsilon_0 m} \frac{\omega_0^2 - \omega^2}{\left(\omega_0{}^2 - \omega^2\right)^2 + \Gamma^2\omega^2}, \tag{4.56a}$$

$$\varepsilon_2 = 2n\kappa = \frac{\rho e^2}{\varepsilon_0 m} \frac{\Gamma\omega}{\left(\omega_0^2 - \omega^2\right)^2 + \Gamma^2\omega^2}. \tag{4.56b}$$

As seen from these equations, the dielectric constant changes markedly around $\omega = \omega_0$ and at $\omega \gg \omega_0$, $\varepsilon_1 \approx 1$ and $\varepsilon_2 \approx 0$. At $\omega \ll \omega_0$, ε_1 becomes a finite value larger than 1 and ε_2 approaches 0. Figure 4.3 shows the angular frequency dependence *i.e.*, the photon energy dependence (spectrum) of ε_1 and ε_2 calculated using Eq. (4.56) with parameters of $\hbar\omega_0 = 4$ eV, $\hbar\Gamma = 1$ eV and $\hbar^2\rho e^2/\varepsilon_0 m = 50$. The change of dielectric constant and refractive index with angular frequency is called 'dispersion'. Except at around the resonant angular frequency ω_0, ε_1 increases with increase in angular frequency and this is called 'normal dispersion'. However, in the region around ω_0, ε_1 decreases with increase in angular frequency and this is called 'anomalous dispersion'. The anomalous dispersion region lies between the frequency ω_1' giving the maximum ε_1 and frequency ω_2' giving the minimum ε_1. The values of ω' are solutions of the following equation derived by equating the derivative of Eq. (4.56a) to 0,

$$\left(\omega_0^2 - \omega^2\right)^2 = \omega_0^2\Gamma^2. \tag{4.57}$$

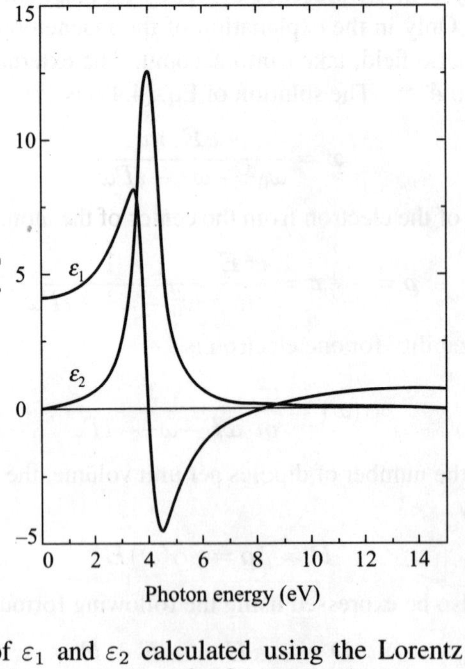

Figure 4.3 The spectra of ε_1 and ε_2 calculated using the Lorentz model for the case of $\hbar\omega_0 = 4$ eV, $\hbar\Gamma = 1$ eV and $\hbar^2\rho e^2/\varepsilon_0 m = 50$.

Since $\omega' \approx \omega_0$, one can obtain the relation

$$\omega_1' = \omega_0 - \Gamma/2, \quad \omega_2' = \omega_0 + \Gamma/2. \tag{4.58}$$

We conclude therefore that the width of the anomalous dispersion region is Γ (see Fig. 4.3). Thus Γ in Eq. (4.49) gives the width of the anomalous dispersion, which corresponds to the peak region of ε_2, that is the resonant oscillation region.

Using Eq. (4.17), the optical constants n and κ can be expressed in terms of the relative dielectric function by

$$n^2 = \frac{({\varepsilon_1}^2 + {\varepsilon_2}^2)^{1/2} + \varepsilon_1}{2},$$

$$\kappa^2 = \frac{({\varepsilon_1}^2 + {\varepsilon_2}^2)^{1/2} - \varepsilon_1}{2} \tag{4.59}$$

Figure 4.4 presents spectra of n and κ showing their dispersion (by solid curves), where the dielectric constants are those given by Fig. 4.3 and the normal incidence reflectance (the broken curve) calculated using Eq. (4.33). As shown in Fig. 4.4, the angular frequencies can be placed into four regions. Region I is that region where $0 < \omega < \omega_0 - \Gamma$ and the dispersion is normal. When $\omega \ll \omega_0$, $\kappa = 0$, which means that the material is transparent and $n^2 = \varepsilon_1 > 1$, so that the normal reflectance $R = (n-1)^2/(n+1)^2$ is not 0. Region II is the region where $\omega_0 - \Gamma < \omega < \omega_0 + \Gamma$ and the dispersion

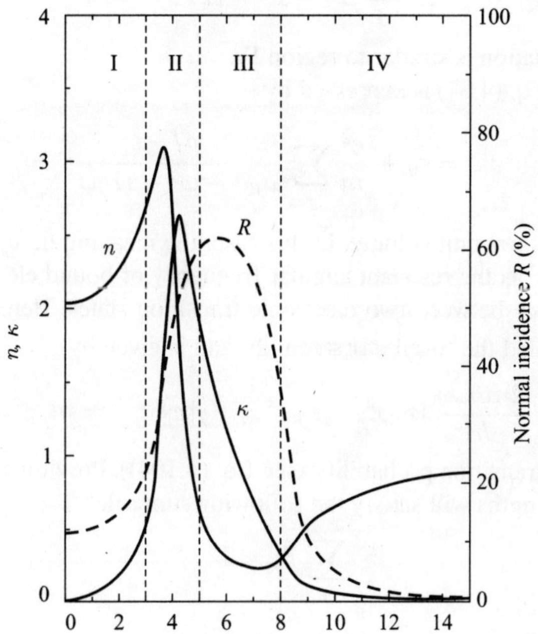

Figure 4.4 The spectra n of κ and (solid curve) calculated using the values of ε_1 and ε_2 given in Fig. 4.3 and normal incidence reflectance (broken curve) using the values of n and κ given in this figure

is anomalous, and n and κ change significantly. In this region κ is large as shown in Fig. 4.4, so that absorption coefficient is large. Region III is the region where $\omega_0 + \Gamma < \omega < \omega_\mathrm{p}$. Here ω_p is the angular plasma frequency. Around $\omega = \omega_\mathrm{p}$, $\varepsilon_1 = 0$ and $Im(-1/\varepsilon)$ reaches a maximum. If $\omega \gg \omega_0 \gg \Gamma$, then ω_p can be obtained from $\varepsilon_1 = 0$ using Eq. (4.56a) as

$$\omega_\mathrm{p}^2 = \frac{\rho e^2}{\varepsilon_0 m}. \tag{4.60}$$

In this region, n is considerably smaller than 1 as shown in Fig. 4.4, so that the normal incidence reflectance is high. This feature can be understood from Eq. (4.33), in which the normal incidence reflectance approaches 1 almost independent of κ as n approaches 0. In fact, the reflectance peak is not in region II, but in region III where κ is larger than n and the material is opaque. Region IV is that region where $\omega_\mathrm{p} < \omega$. In this region κ approaches 0 with increasing ω and the material becomes transparent, while n approaches 1. Thus normal incidence reflectance becomes 0 which differs from the situation of $\kappa = 0$ at $\omega \ll \omega_0$ in region I. This is a very important characteristic of the optical properties of all materials in the X-ray region.

Any discussion of the single resonant frequency system can be extended to a multiple resonant frequency system, where the frequencies of the system are defined to be ω_j and the density of electrons having the resonant frequency ω_j defined to be ρ_j. The dielectric constant is then expressed by

$$\varepsilon_0 \varepsilon = \varepsilon_0 + \frac{e^2}{m} \sum_j \frac{\rho_j}{\omega_j^2 - \omega^2 - i\Gamma_j \omega}, \quad \sum_j \rho_j = \rho. \tag{4.61}$$

If $\omega \gg \omega_j$ for all j, the situation is similar to region IV.

In quantum mechanics, Eq. (4.61) is expressed by

$$\varepsilon_0 \varepsilon = \varepsilon_0 + \frac{e^2}{m} \sum_j \frac{\hat{\rho} \hat{f}_j}{\omega_{j0}{}^2 - \omega^2 - i\Gamma_j \omega}, \tag{4.62}$$

where $\hat{\rho}$ is number of atoms per unit volume. In this case, the meaning of ω_j differs from that given in Eq. (4.61). In Eq. (4.61), ω_j is the resonant angular frequency of bound electrons, while in Eq. (4.62) $\hbar\omega_{j0}$ is the energy difference between two electronic transition states. Here the subscript 0 expresses the initial state and $\hat{f}_j$ is called the 'oscillator strength' and is given by

$$\hat{f}_j = \frac{2m\hbar\omega_{j0}}{\hbar^2} |x_{j0}|^2, \quad |x_{j0}|^2 = |\langle j\,|x|\,0\rangle|^2 = |\boldsymbol{r}_{j0}|^2/3. \tag{4.63}$$

This is proportional to the transition probability (see Eq. (4.104)). Providing that an atom possesses Z electrons, the oscillator strengths will satisfy the following sum rule

$$\sum_j \hat{f}_j = Z \tag{4.64}$$

implying that $\hat{f}_j$ is expressed using the electron number.

4.2.2 The Drude model

In metals when longer wavelength light is applied, ions at lattice points do not react to it, but free electrons do move. They are not bound to the ions, so that a dipole field is not generated. Therefore only the external electric field need be considered for calculation of the polarization of the medium. In the Drude model for metals, the restoring force used in the Lorentz model is not taken into account. Accordingly, the dielectric constants for metals by setting $\omega_0 = 0$ in Eq. (4.56), are given by

$$\varepsilon_1 = n^2 - \kappa^2 = 1 - \frac{\rho e^2}{\varepsilon_0 m} \frac{1}{\omega^2 + \Gamma^2},$$
$$\varepsilon_2 = 2n\kappa = \frac{\rho e^2}{\varepsilon_0 m} \frac{\Gamma}{\omega(\omega^2 + \Gamma^2)}, \tag{4.65}$$

where ρ is regarded as the number of free electrons per unit volume. Bound electrons undergo some resistance caused by collisions. On left side of Eq. (4.49), the second term accounts for this effect. The situation is similar in the case of free electrons. Setting $\Gamma = \tau^{-1}$ and using Eq. (4.60), one can obtain

$$\varepsilon_1 = 1 - \frac{\omega_p^2 \tau^2}{1 + \omega^2 \tau^2},$$
$$\varepsilon_2 = \frac{\omega_p^2 \tau}{\omega(1 + \omega^2 \tau^2)}, \tag{4.66}$$

where τ is called 'mean free time' which means the mean time interval between collisions.

The Drude model resembles the Lorentz model for $\omega \gg \omega_0$ and this can be understood by putting $\omega_0 = 0$ in the Lorentz model. Figure 4.5 presents spectra of the relative dielectric constant, where $\hbar\omega_p = \sqrt{30}$ eV and $\hbar\Gamma = 0.02$ eV are assumed. Figure 4.6 shows plots of the optical constant and the normal incidence reflectance calculated using the dielectric constant given in Fig. 4.5.

In Fig. 4.6 the angular frequencies are placed into three regions similar to Fig. 4.4 for the Lorentz

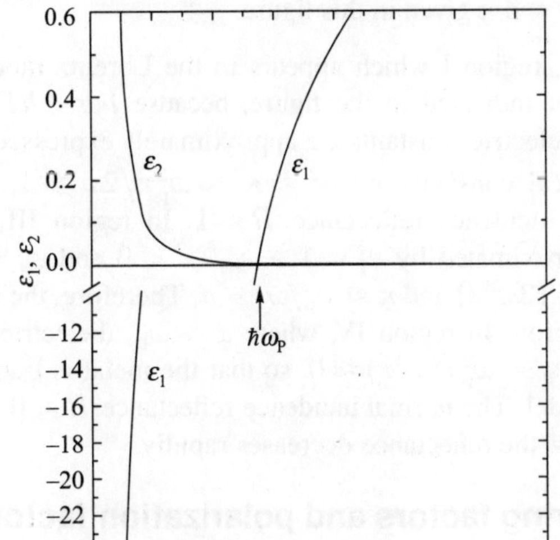

Figure 4.5 The spectra of ε_1 and ε_2 calculated using the Drude model for the case of $\hbar\omega_p = \sqrt{30}$ eV and $\hbar\Gamma = 0.02$ eV. Notice that the scales of positive and negative ε_1 are different.

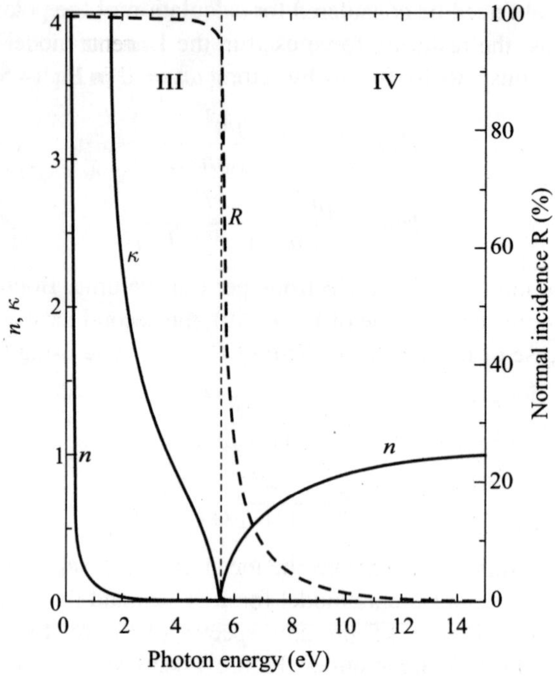

Figure 4.6 The spectra of n and κ (solid curve) calculated using the values of ε_1 and ε_2 given in Fig. 4.5 and normal incidence reflectance (broken curve) using the values of n and κ given in this figure.

model. In the Drude model, region I which appears in the Lorentz model does not appear. Region II is very narrow and is not indicated in the figure, because $\hbar\omega < \hbar\Gamma = 0.02$ eV. In this region, since $\omega\tau < 1$, the relative dielectric constants are approximately expressed by $\varepsilon_1 = 1 - \omega_p^2\tau^2 < 0$ and $\varepsilon_2 = \omega_p^2\tau/\omega > 0$. The optical constants are $n^2 \approx \kappa^2 \approx \omega_p^2\tau/2\omega \gg 1$, so that the metal is opaque and has very high normal incidence reflectance $R \approx 1$. In region III, where $\tau^{-1} < \omega < \omega_p$, the dielectric constants are approximated by $\varepsilon_1 = 1 - \omega_p^2/\omega^2 < 0$ and $\varepsilon_2 = \omega_p^2/(\omega^3\tau) < |\varepsilon_1|$ and the optical constants by $n \approx \omega_p/(2\omega^2\tau)$ and $\kappa \approx \omega_p/\omega \gg n$. Therefore, the normal incidence reflectance is also very high in this region. In region IV, where $\omega > \omega_p$, dielectric constants are expressed by $\varepsilon_1 = 1 - \omega_p^2/\omega^2 \approx 1$ and $\varepsilon_2 = \omega_p^2/(\omega^3\tau) \approx 0$, so that the metal is transparent as for the insulators described in the Lorentz model. The normal incidence reflectance $R \approx 0$ and it is to be noted that just beyond the plasma frequency, the reflectance decreases rapidly.

4.2.3 Atomic scattering factors and polarization factors

a. The atomic scattering factor: Materials have a wide variety of optical properties over the region extending from the visible to the vacuum ultraviolet. These properties will of course reflect the characteristics of bonding states of the materials. For example, graphite and diamond consist of the same element of carbon, but have totally different optical properties. However, this situation changes for the soft and hard X-ray optical regions. When the photon energy $\hbar\omega$ is much larger than the binding energy of the valence electron $\hbar\omega_j$, the optical properties are governed by localized core electrons and can be deduced from the superposition of the optical responses of all constituent atoms. If the photon energy is much larger than binding energy of the core electrons, even the core electrons can be regarded as free electrons. In previous sections, the optical properties of metals and insulators were explained using the classical electron theory. In this section the scattering of the incident light by an electron is considered. Light incident on the electron induces forced electric dipole oscillations generating the scattered wave. In general an accelerating electron will produce an electromagnetic field for which the electric and magnetic fields will be given as in Eq. (2.52). In the case of the harmonic oscillation of an electron, since $\dot{\boldsymbol{x}}$ and $\ddot{\boldsymbol{x}}$ are parallel, the generated electric field at a distance $\hat{r}$ is given by

$$\hat{\boldsymbol{E}} = -\frac{1}{4\pi\varepsilon_0}\frac{e}{c^2\hat{r}}\{\boldsymbol{n}\times(\boldsymbol{n}\times\ddot{\boldsymbol{x}})\}\,. \tag{4.67}$$

The scattering by a bound electron with resonant frequency ω_j in an atom can be treated by the oscillator model Eq. (4.49). The displacement of the electron is given by

$$\boldsymbol{x}_j = \frac{-e\hat{f}_j\boldsymbol{E}_0 e^{-i\omega t}/m}{\omega_j^2-\omega^2-i\Gamma_j\omega}, \tag{4.68}$$

where $\hat{f}_j$ is the oscillator strength. Inserting Eq. (4.68) into Eq. (4.67) and regarding $\boldsymbol{E}_0$ as a scalar, the electric field generated by the total number of electrons will be given by

$$\hat{E} = \frac{e^2}{4\pi\varepsilon_0 mc^2\hat{r}}\sum_j \frac{-\omega^2\hat{f}_j}{\omega_j^2-\omega^2-i\Gamma_j\omega}E_0 e^{-i\omega t} = \frac{r_0}{\hat{r}}fE_0 e^{-i\omega t}. \tag{4.69}$$

Here, $r_0 = e^2/4\pi\varepsilon_0 mc^2 = 2.8\times10^{-13}$ cm is the classical electron radius and f is called the atomic scattering factor, which is a function of photon energy $\hbar\omega$. The electric field of the wave scattered by a free electron can be obtained similarly by setting the physical quantities as $\omega_j = 0$ and $\Gamma_j = 0$ in Eq. (4.69) as

$$\hat{E} = \frac{r_0}{\hat{r}}E_0 e^{-i\omega t}. \tag{4.70}$$

A comparison of Eqs. (4.69) and (4.70) reveals that the scattering intensity of bound electrons is times that of a free electron. This is the definition of the 'atomic scattering factor'. The atomic scattering factor has a complex form $f(\omega) \equiv f_1(\omega) + if_2(\omega)$, which can be explicitly expressed from Eq. (4.66) by

$$f(\omega) = \sum_j \frac{-\omega^2\hat{f}_j}{\omega_j^2-\omega^2-i\Gamma_j\omega}\,. \tag{4.71}$$

From Eq. (4.64), at the limit of ω tending infinity, $f = \sum_j \hat{f}_j = Z$ and f reaches the number of electrons Z in the atom. Using Eq. (4.71), Eq. (4.62) can be written as

$$\varepsilon_0\varepsilon = \varepsilon_0 - \frac{e^2\hat{\rho}}{m\omega^2}f. \tag{4.72}$$

The relative dielectric constant of any system comprised of several kinds of atoms is given by

$$\varepsilon = 1 - \frac{r_0\lambda^2}{\pi}\sum_j \hat{\rho}_j f_j, \tag{4.73}$$

where $\hat{\rho}_j$ is the number of j-th atom with an atomic scattering factor f_j in unit volume. In the soft and hard X-ray regions, in place of the optical constants n and κ,

$$\delta = 1 - n, \quad \beta = \kappa \tag{4.74}$$

are often used. In these regions, $n \approx 1$ and $\kappa \approx 0$ in the limit of $\omega \gg \omega_p$, as seen in Drude and Lorentz models, and both δ and β are very small in comparison with 1. Therefore,

$$\tilde{n}^2 \approx 1 - 2\delta + 2i\beta. \tag{4.75}$$

From this relation and Eq. (4.73), δ and β are expressed by

$$\delta \approx \frac{r_0\lambda^2}{2\pi}\sum_j \hat{\rho}_j f_{1j}, \quad \beta \approx -\frac{r_0\lambda^2}{2\pi}\sum_j \hat{\rho}_j f_{2j}. \tag{4.76}$$

f_2 is proportional to β, that is, to the absorption coefficient. Figure 4.7 shows the atomic scattering factors f_1 and f_2 for Au in the case where the scattering vector introduced later is 0. Prominent structures in f_2 around 2200 and 12000 eV are due to the absorption edges $M_{4,5}$ and $L_{2,3}$ for Au, respectively. Around these edges f_1 changes markedly and this region is therefore called the anomalous dispersion region. In the normal dispersion region, f_1 includes all those electrons, which can be excited by the photon energy.

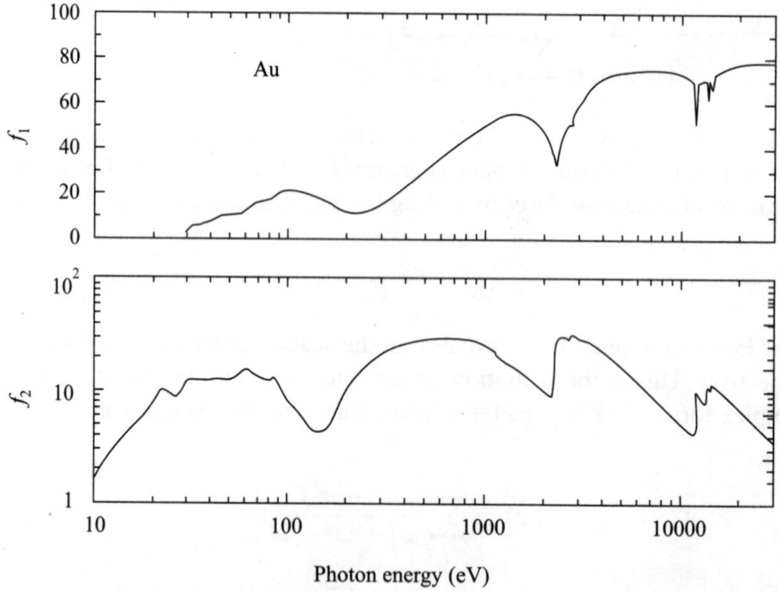

Figure 4.7 The spectra of atomic scattering factors of Au.

There are two choices of sign for f_2. In this book, the electric field of the incident wave is defined to be $\boldsymbol{E}_0 \exp\{-i(\omega t - \boldsymbol{k}\cdot\boldsymbol{r})\}$ with refractive index and relative dielectric constant defined to be $\tilde{n} = n + i\kappa$ and $\varepsilon = \varepsilon_1 + i\varepsilon_2$, respectively. Accordingly, when the atomic scattering factor is defined to be $f \equiv f_1 + if_2$, $f_2 < 0$. On the other hand, when the incident wave defined to be $\boldsymbol{E}_0 \exp\{i(\omega t - \boldsymbol{k}\cdot\boldsymbol{r})\}$ with $\tilde{n} = n - i\kappa$ and $\varepsilon = \varepsilon_1 - i\varepsilon_2$, $f_2 > 0$. In this case, the negative sign is omitted from Eq. (4.76) giving β.

In the soft and hard X-ray regions, one can obtain the approximate optical constants of materials from a summation of the atomic scattering factor f_j using Eq. (4.76) and by neglecting the chemical state of the atoms in a compound. Websites provide complete data on all the atomic scattering factors from 10 eV to 30 keV for the elements up to U with atomic number 92 (see Ref. 6 of Chap. **5**). The data are derived from the experimental results reported so far. Notice that Eq. (4.76) is only approximate. Especially in the XANES region, optical constants depend on bound states and differ markedly from compound to compound, and Eq. (4.76) is not valid. Therefore in order to obtain the most accurate optical constants, new and revised experimental measurements remain as a necessity.

b. The dependence of the atomic scattering factor on the wave number of the scattering vector and anomalous dispersion: The previous description is based on the Lorentz model, which assumes all electrons to be located at the origin. However in the X-ray region, the extent of the electron distribution in the atom is comparable to the wavelength of the actual X-rays and it is therefore necessary to take the actual spatial distribution of the electrons into account. That is, $\boldsymbol{E} = \boldsymbol{E}_0 e^{-i\omega t}$ needs to be replaced by $\boldsymbol{E} = \boldsymbol{E}_0 e^{-i(\omega t - \boldsymbol{k}_1 \cdot \boldsymbol{r}_j)}$ in Eq. (4.49), where $\boldsymbol{r}_j$ is the central position of the electron with resonant frequency ω_j. Accordingly the equation of motion of the electron under forced oscillation becomes

$$m\frac{d^2\boldsymbol{x}_j}{dt^2} + m\Gamma\frac{d\boldsymbol{x}_j}{dt} + m\omega_j^2\boldsymbol{x}_j = -eE_0 e^{-i(\omega t - \boldsymbol{k}_1 \cdot \boldsymbol{r}_j)} \tag{4.77}$$

where the displacement of the electron in the oscillation and wave vector of the incident light are defined to be $\boldsymbol{x}_j$ and $\boldsymbol{k}_1$, respectively. The solution for $\boldsymbol{x}_j$ is given by

$$\boldsymbol{x}_j = \frac{-e\hat{f}_j \boldsymbol{E}_0 e^{-i(\omega t - \boldsymbol{k}_1 \cdot \boldsymbol{r}_j)}/m}{\omega_j^2 - \omega^2 - i\Gamma_j\omega} \tag{4.78}$$

This displacement includes the phase difference of electron oscillations originating from the differing positions of the electron. Assuming that the scattered wave defined to be $\boldsymbol{E}_0' e^{-i(\omega t - \boldsymbol{k}_2 \cdot \boldsymbol{r})}$ has the same amplitude as the incident wave amplitude at $\boldsymbol{r}_j$, the atomic scattering factor taking the electron distribution into account can be expressed by

$$f = \sum_j \frac{-\omega^2 \hat{f}_j e^{i\boldsymbol{q}\cdot\boldsymbol{r}_j}}{\omega_j^2 - \omega^2 - i\Gamma_j\omega}. \tag{4.79}$$

Here the scattering vector is defined to be $\boldsymbol{q} \equiv \boldsymbol{k}_1 - \boldsymbol{k}_2$. Comparing Eq. (4.79) with Eq. (4.71), Eq. (4.79) includes an additional $e^{i\boldsymbol{q}\cdot\boldsymbol{r}_j}$ factor, which incorporates the electronic structure.

At the photon energy $\hbar\omega$, far from the resonant energy as in the case of normal dispersion, $f = \sum_j f_j = \sum_j \hat{f}_j e^{i\boldsymbol{q}\cdot\boldsymbol{r}_j}$, where j indicates the j-th core counted from the valence band. In the case of a free atom, it can be assumed that the electron distribution is spherically symmetric, therefore $\hat{f}_j$ can be replaced by $\rho_j(r)$ and can be rewritten as

$$f_j = 2\pi\int \rho_j(r)\, e^{iqr\cos\alpha} r^2 \sin\alpha dr d\alpha = 4\pi\int \rho_j(r)\, r^2 \frac{\sin qr}{qr} dr\,, \tag{4.80}$$

where α stands for an angle between $\boldsymbol{q}$ and $\boldsymbol{r}$. Therefore,

$$f = \sum_j f_j = 4\pi \int \sum_j \rho_j(r) r^2 \frac{\sin qr}{qr}\, dr. \tag{4.81}$$

f is a function of q and ω. In solids, atoms are not necessarily spherical, but the effect of deviation from spherical symmetry is small (see Ref. 4 in Chap. **9**). When the energy of the incident X-ray is large, all electrons will contribute to scattering. Assuming that all electrons are located at $r = 0$, then $f = Z$. In the case of forward scattering ($q = 0$), $f = Z$ and in general f decreases from the value of Z with q increasing from $q = 0$.[2,4] This feature is schematically illustraited in Fig. 4.8 (the explicit expression of q being given by Eq. (9.1)).

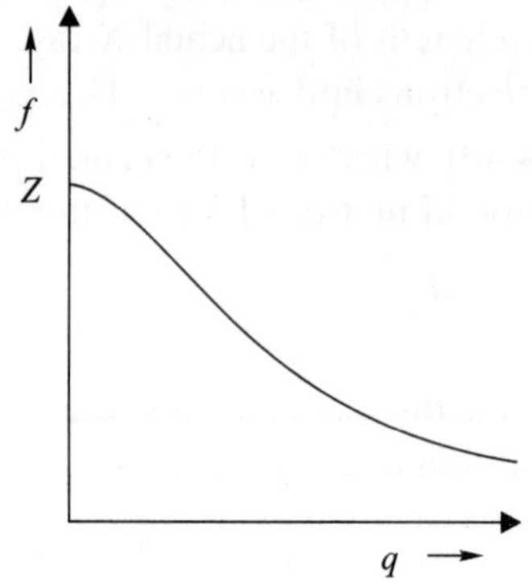

Figure 4.8 The general feature of the dependence of f on q.

By contrast, when $\hbar\omega$ is close to the resonant energy of one of the cores in the atom, the effect of anomalous dispersion of that core becomes prominent. The extent of core electrons distribution is narrow in comparison with the wavelength of the X-ray which excites the core electrons. Thus one can assume that the core electrons are located at $r = 0$. Accordingly the atomic scattering factor including the anomalous dispersion effect of the j-th core can be expressed by

$$\begin{aligned} f(\omega, q) &= \sum_{j'<j} f_{j'} - \frac{\omega^2 \hat{f}_j}{\omega_j^2 - \omega^2 - i\Gamma_j \omega} \\ &= f_0(\omega, q) + f'(\omega) + i f''(\omega). \end{aligned} \tag{4.82}$$

The second and the third terms in Eq. (4.82) are called the anomalous dispersion term. The relationship between the anomalous dispersion term and the atomic scattering factor is $f_1 = f_0(\omega, q) + f'(\omega)$ and $f_2 = f''(\omega)$.

c. The polarization factor: Linearly-polarized light with the electric vector perpendicular to the plane of incidence is called s-polarized light (see Fig. 4.1) and induces dipole motion perpendicular to the plane of incidence. Thus the scattered light intensity emitted from the dipole is constant in the plane of incidence. In other words, the intensity is axially symmetric with respect to the axis of the dipole. On the other hand linearly-polarized light with the electric vector parallel to the plane of incidence, called p-polarized light (see Fig. 4.1), induces dipole motion parallel to the plane of incidence. Thus the scattered light intensity distributes according to a function of $\sin^2(90° - 2\hat{\theta}) = \cos^2 2\hat{\theta}$ as derived from Eq. (2.57). In measurements using linearly-polarized synchrotron radiation when the electric field is parallel to plane of incidence, the measured intensity includes the polarization factor $\cos^2 2\hat{\theta}$. However, when the electric field is perpendicular to the plane of incidence the polarization factor is 1. In addition, for non-polarized light, the polarization factor is the average of the s- and p-components, that is $(1 + \cos^2 2\hat{\theta})/2$.

4.3 THE QUANTUM THEORY OF OPTICAL TRANSITIONS

When an electron encounters an electromagnetic wave, absorption may occur causing changes in electron motion followed by light emission. In the previous section, such an interaction between light and an electron was treated classically. In this section we treat the interaction quantum-mechanically and absorption, emission and scattering are described.[5)] For magnetic scattering, the scattering of the light with magnetic moment see Ref. 6.

4.3.1 The general theory of optical transitions

Using the scalar potential ϕ and vector potential $\boldsymbol{A}$, the electric field $\boldsymbol{E}$ and magnetic field $\boldsymbol{B}$ are expressed by $\boldsymbol{E} = -\nabla\phi - \partial \boldsymbol{A}/\partial t$, Eq. (2.17) and $\boldsymbol{B} = \mathrm{rot}\boldsymbol{A}$, Eq. (2.18), respectively. When these equations are inserted in the Lorentz force equation Eq. (2.10), $\boldsymbol{F} = -e(\boldsymbol{E} + \dot{\boldsymbol{r}} \times \boldsymbol{B})$, which describes the force acting on the moving electron, the non-relativistic equation of motion of the electron is given by

$$m\ddot{\boldsymbol{r}} = e\nabla\phi + e\dot{\boldsymbol{A}} - e(\dot{\boldsymbol{r}} \times \mathrm{rot}\,\boldsymbol{A}), \tag{4.83}$$

where m is electron mass. The equation of motion Eq. (4.83) is derived using the Lagrange equation

$$\frac{d}{dt}\left(\frac{\partial L}{\partial \dot{x}}\right) - \frac{\partial L}{\partial x} = 0, \quad \frac{d}{dt}\left(\frac{\partial L}{\partial \dot{y}}\right) - \frac{\partial L}{\partial y} = 0, \quad \frac{d}{dt}\left(\frac{\partial L}{\partial \dot{z}}\right) - \frac{\partial L}{\partial z} = 0, \tag{4.84}$$

where the Lagrangian L is given by

$$L = \frac{m}{2}(\dot{\boldsymbol{r}})^2 + e\phi - e\dot{\boldsymbol{r}} \cdot \boldsymbol{A}. \tag{4.85}$$

Using L, the momentum $\boldsymbol{p}$ which is the canonical-conjugate of $\boldsymbol{r}$, is obtained as

$$\boldsymbol{p} \equiv \left(\frac{\partial L}{\partial \dot{x}}, \frac{\partial L}{\partial \dot{y}}, \frac{\partial L}{\partial \dot{z}}\right) = m\dot{\boldsymbol{r}} - e\boldsymbol{A}. \tag{4.86}$$

Therefore in the presence of an electromagnetic field, the Hamiltonian for the electron, H can be written as

$$H \equiv \boldsymbol{p} \cdot \boldsymbol{v} - L = \frac{m}{2}\boldsymbol{v}^2 - e\phi = \frac{1}{2m}(\boldsymbol{p} + e\boldsymbol{A})^2 - e\phi. \tag{4.87}$$

When the number of electrons is more than 1, the Hamiltonian of the system is the sum of Hamiltonians of each electron.

Hereafter, the Hamiltonian Eq. (4.87) is written as $H = H_0 + H'$, where

$$H_0 = \frac{1}{2m}\boldsymbol{p}^2 - e\phi,$$
$$H' = \frac{e}{2m}(\boldsymbol{p}\cdot\boldsymbol{A} + \boldsymbol{A}\cdot\boldsymbol{p}) + \frac{e^2}{2m}\boldsymbol{A}^2. \tag{4.88}$$

H_0 expresses the electron's kinetic and potential energies for the case of zero interaction between the electron and electromagnetic field, and H' is the interaction energy.

Equations (4.83) to (4.88) are expressions for the interaction between the electron and electromagnetic field, based on classical theory. In a quantum-mechanical description, $\boldsymbol{p}$ and $\boldsymbol{A}$ should be regarded as operators. Accordingly

$$\boldsymbol{p}\cdot\boldsymbol{A} = \boldsymbol{A}\cdot\boldsymbol{p} - i\hbar\nabla\cdot\boldsymbol{A}, \tag{4.89}$$

so that the interaction Hamiltonian is given using the Coulomb gauge $\nabla\cdot\boldsymbol{A} = 0$ by

$$H' = \frac{e}{m}\boldsymbol{A}\cdot\boldsymbol{p} + \frac{e^2}{2m}\boldsymbol{A}\cdot\boldsymbol{A}. \tag{4.90}$$

When the electromagnetic field is quantized, the operator $\boldsymbol{A}$ is expressed by

$$\boldsymbol{A} = \sqrt{\frac{\hbar}{2\varepsilon_0 V\omega}}\sum_{\boldsymbol{k},\alpha}\left[a_{\boldsymbol{k},\alpha}\boldsymbol{e}_\alpha \exp\{-i(\omega t - \boldsymbol{k}\cdot\boldsymbol{r})\} + a^+_{\boldsymbol{k},\alpha}\boldsymbol{e}_\alpha \exp\{i(\omega t - \boldsymbol{k}\cdot\boldsymbol{r})\}\right], \tag{4.91}$$

where $\boldsymbol{e}_\alpha (\alpha = 1, 2)$ are unit vectors for the two independent polarization directions of the electromagnetic field and a and a^+ are respectively annihilation and creation operators, which satisfy the following commutation relations

$$[a_\mu, a^+_\nu] = \delta_{\mu\nu}, \quad [a_\mu, a_\nu] = [a^+_\mu, a^+_\nu] = 0. \tag{4.92}$$

The eigenvalue and eigenstates of a and a^+ are given by

$$a\,|n\rangle = \sqrt{n}\,|n-1\rangle, \quad a^+\,|n\rangle = \sqrt{n+1}\,|n+1\rangle,$$
$$N\,|n\rangle = a^+a\,|n\rangle = n\,|n\rangle, \tag{4.93}$$

where $|n\rangle$ expresses the state with n phuions and $N \equiv a^+a$ is the number operator for photons.

4.3.2 First order processes

Consider the first order process in the optical transition. The important mechanism is that of light absorption by the atom. Assume that $n_{\boldsymbol{k},\alpha}$ photons of $(\boldsymbol{k}, \alpha)$ (here α stands for $\boldsymbol{e}_1$ or $\boldsymbol{e}_2$) and an atom of state A with an energy E_A exist in the initial state. In this process, one photon is absorbed by the atom resulting in a decrease in the photon number from $n_{\boldsymbol{k},\alpha}$ to $n_{\boldsymbol{k},\alpha} - 1$ and a transition of the atom from the state A with an energy E_A to the state B with an energy E_B in the final state. In general, the transition probability between the initial and the final states is given by a matrix element of the interaction Hamiltonian between the initial and the final states a and a^+ which are included in $\boldsymbol{A}$, Eq. (4.91). In the absorption process the matrix element is non-zero only for the term including a. Therefore, in the case of the first order perturbation, only $\boldsymbol{A}\cdot\boldsymbol{p}$ is taken into account to calculate the interaction in Eq. (4.90). In the first order perturbation of the term $\boldsymbol{A}\cdot\boldsymbol{A}$, the increment or decrement in photon number is 0 or ± 2, so that the term $\boldsymbol{A}\cdot\boldsymbol{A}$ does not contribute to the first order process. For absorption, one can obtain the matrix element using Eqs. (4.91) and (4.93) to be

$$\langle B; n_{\boldsymbol{k},\alpha}-1|\,H'\,|A; n_{\boldsymbol{k},\alpha}\rangle = \frac{e}{m}\langle B; n_{\boldsymbol{k},\alpha}-1|\sqrt{\frac{\hbar}{2\varepsilon_0 V\omega}}\,a_{\boldsymbol{k},\alpha}\exp\{-i(\omega t-\boldsymbol{k}\cdot\boldsymbol{r})\}\boldsymbol{p}\cdot\boldsymbol{e}_\alpha\,|A; n_{\boldsymbol{k},\alpha}\rangle$$

$$= \frac{e}{m}\sqrt{\frac{\hbar n_{\boldsymbol{k},\alpha}}{2\varepsilon_0 V\omega}}\,\langle B|\,e^{-i\boldsymbol{k}\cdot\boldsymbol{r}}\boldsymbol{p}\cdot\boldsymbol{e}_\alpha\,|A\rangle\;e^{-i\omega t}. \tag{4.94}$$

Since the square of absolute value of Eq. (4.94) gives the absorption probability, it is proportional to the photon number in the initial state, namely the intensity of the incident light. Of course, when there is no incident light, that is $n_{\boldsymbol{k},\alpha} = 0$, absorption does not occur.

For the emission process, one can obtain the matrix element in a similar way. Thus,

$$\langle B; n_{\boldsymbol{k},\alpha}+1|\,H'\,|A; n_{\boldsymbol{k},\alpha}\rangle = \frac{e}{m}\langle B; n_{\boldsymbol{k},\alpha}+1|\sqrt{\frac{\hbar}{2\varepsilon_0 V\omega}}\,a^+_{\boldsymbol{k},\alpha}\exp\{i(\omega t-\boldsymbol{k}\cdot\boldsymbol{r})\}\boldsymbol{p}\cdot\boldsymbol{e}_\alpha\,|A; n_{\boldsymbol{k},\alpha}\rangle$$

$$= \frac{e}{m}\sqrt{\frac{\hbar(n_{\boldsymbol{k},\alpha}+1)}{2\varepsilon_0 V\omega}}\,\langle B|\,e^{-i\boldsymbol{k}\cdot\boldsymbol{r}}\boldsymbol{p}\cdot\boldsymbol{e}_\alpha\,|A\rangle\;e^{-i\omega t}. \tag{4.95}$$

In the case of emission, the matrix element indicates that the optical process consists of two parts. One of them is proportional to the photon number $n_{\boldsymbol{k},\alpha}$ and the other arises from 1. The former is called 'induced emission', when the mode of the emitted photons is identical to that of the incident photons. The latter is called 'spontaneous emission' where $n_{\boldsymbol{k},\alpha} = 0$ in the initial state. In this case, an excited atom relaxes, emitting a photon which is uncorrelated with the incident photon, such as a fluorescence photon which can have all possible modes. It should be noticed that only by using quantum mechanics, but not with classical theory, can spontaneous emission be described and the expression is similar to that for induced emission.

The interaction Hamiltonian $H_{\mathrm{I}}(t)$ is dependent on time and is expressed by

$$H_{\mathrm{I}}(t) = H' e^{\mp i\omega t}, \tag{4.96}$$

where H' is the time independent portion of the interaction Hamiltonian. Here, in the case of absorption a negative sign is chosen, while in the case of emission a positive sign is used. The first order expansion coefficient is given by

$$c_B^{(1)}(t) = \frac{1}{i\hbar}\langle B|\,H'\,|A\rangle\int_0^t e^{i(E_B-E_A\mp\hbar\omega)t'/\hbar}dt'. \tag{4.97}$$

Integration of Eq. (4.97) yields

$$|c_B^{(1)}(t)|^2 = \frac{2\pi}{\hbar}|\langle B|\,H'\,|A\rangle|^2 t\,\delta(E_B-E_A\mp\hbar\omega). \tag{4.98}$$

From Eq. (4.98), one can see that $|c_B^{(1)}(t)|^2/t$ is independent of time.

The transition probability, in which photons are emitted in solid angle $d\Omega$ in unit time, is expressed by

$$w_{d\Omega} = \int \{|c_B^{(1)}(t)|^2/t\} \rho_{\hbar\omega,d\Omega} d(\hbar\omega)$$
$$= (2\pi/\hbar) \left|\langle B| H' |A\rangle\right|^2 \rho_{\hbar\omega,d\Omega} , \tag{4.99}$$

which is called 'Fermi's golden rule'. Here the photon energy satisfies the energy conservation relation

$$E_B - E_A + \hbar\omega = 0. \tag{4.100}$$

The density of state $\rho_{\hbar\omega,d\Omega}$ is the number of allowed states for emitted photons in the energy interval between $\hbar\omega$ and $\hbar(\omega + d\omega)$ in a solid angle $d\Omega$. It is given by

$$\rho_{\hbar\omega,d\Omega} d(\hbar\omega) = \frac{V\omega^2 d\Omega}{(2\pi)^3 \hbar c^3} d(\hbar\omega). \tag{4.101}$$

The transition probability for spontaneous emission is obtained by setting $n_{\boldsymbol{k},\alpha} = 0$ in Eq. (4.95) as

$$w_{d\Omega} = \frac{2\pi}{\hbar} \frac{e^2\hbar}{2m^2\varepsilon_0 V\omega} |\langle B| e^{-i\boldsymbol{k}\cdot\boldsymbol{r}} \boldsymbol{p}\cdot\boldsymbol{e}_\alpha |A\rangle|^2 \frac{V\omega^2 d\Omega}{(2\pi)^3 \hbar c^3}. \tag{4.102}$$

Usually the wavelength of the emitted photon is larger than the spatial distribution of electrons participating the transition in the atom and so the following expansion

$$e^{-i\boldsymbol{k}\cdot\boldsymbol{r}} = 1 - i\boldsymbol{k}\cdot\boldsymbol{r} - (\boldsymbol{k}\cdot\boldsymbol{r})^2/2 + \cdots \tag{4.103}$$

is available. By adopting only the first term and by changing the $\boldsymbol{p}$ representation to $\boldsymbol{r}$ representation we can rewrite Eq. (4.102) as

$$w_{d\Omega} = \frac{e^2\omega^3}{8\pi^2\varepsilon_0 m\hbar c^3} |\langle B| \boldsymbol{r} |A\rangle \cdot \boldsymbol{e}_\alpha|^2 d\Omega. \tag{4.104}$$

This is called the 'electric dipole approximation'. The same expression is obtained for absorption.

From Eq. (4.104), the following selection rule for the orbital angular momentum l for the electric dipole transition

$$\Delta l = \pm 1 \tag{4.105}$$

is obtained for a transition between the two states whose angular parts of the wave function are given by spherical harmonic function (see App. **F**).

Inserting the second term on the right side of (4.103) into (4.102), the transition probabilities for the magnetic dipole and the electric quadrupole are obtained. These transition probabilities are much smaller than that those for electric dipole transitions. Therefore an electric dipole transition is the principal type of transition and is often called dipole transition for simplicity.

4.3.3 Second order processes

Let us consider the transition process in which the $(\boldsymbol{k}_1, \boldsymbol{e}_1)$ photon incident on an atom in state A is scattered into $(\boldsymbol{k}_2, \boldsymbol{e}_2)$ by an electron in the atom and its state becomes state B. In this process, it is essential that the photon number does not alter and this differs from the first order process of term $\boldsymbol{A} \cdot \boldsymbol{p}$ where the number of photons does change before and after the transition by ± 1. On the other hand, in the first order perturbation of term $\boldsymbol{A} \cdot \boldsymbol{A}$ calculated using Eq. (4.91), a^+a and aa^+ which lead to satisfy the requirement are included. In the second order perturbation, since $\boldsymbol{A} \cdot \boldsymbol{p}$ appears twice in the matrix element calculation, so that there is a case in which photon number does not change. In the first order perturbation of term $\boldsymbol{A} \cdot \boldsymbol{A}$, the matrix element is given by

$$\langle B; \boldsymbol{k}_2, \boldsymbol{e}_2 | H' | A; \boldsymbol{k}_1, \boldsymbol{e}_1 \rangle = \frac{e^2}{2m} \langle B; \boldsymbol{k}_2, \boldsymbol{e}_2 | \boldsymbol{A} \cdot \boldsymbol{A} | A; \boldsymbol{k}_1, \boldsymbol{e}_1 \rangle$$

$$= \frac{e^2}{2m} \frac{\hbar}{2\varepsilon_0 V \sqrt{\omega_1 \omega_2}} 2\boldsymbol{e}_1 \cdot \boldsymbol{e}_2 \exp\{-i(\omega_1 - \omega_2)t\} \langle B | \exp\{i(\boldsymbol{k}_1 - \boldsymbol{k}_2) \cdot \boldsymbol{r}\} | A \rangle . \quad (4.106)$$

Using $\boldsymbol{k}_1 - \boldsymbol{k}_2 = \boldsymbol{q}$, the matrix element is rewritten as

$$\langle B; \boldsymbol{k}_2, \boldsymbol{e}_2 | H' | A; \boldsymbol{k}_1, \boldsymbol{e}_1 \rangle = \frac{e^2}{2m} \frac{\hbar}{2\varepsilon_0 V \sqrt{\omega_1 \omega_2}} 2\boldsymbol{e}_1 \cdot \boldsymbol{e}_2 \exp\{-i(\omega_1 - \omega_2)t\} \langle B | \exp(i\boldsymbol{q} \cdot \boldsymbol{r}) | A \rangle . \quad (4.107)$$

The first order expansion coefficient is given by

$$c^{(1)}(t) = \frac{1}{i\hbar} \frac{e^2}{2m} \frac{\hbar}{2\varepsilon_0 V \sqrt{\omega_1 \omega_2}} 2 \langle B | \exp(i\boldsymbol{q} \cdot \boldsymbol{r}) | A \rangle$$

$$\times \boldsymbol{e}_1 \cdot \boldsymbol{e}_2 \int_0^t \exp\{i(\hbar\omega_2 + E_B - \hbar\omega_1 - E_A)t'/\hbar\} dt' . \quad (4.108)$$

In the second order perturbation of term $\boldsymbol{A} \cdot \boldsymbol{p}$, by the use of the electric dipole approximation, the expansion coefficient for $\boldsymbol{A} \cdot \boldsymbol{p}$ is given by

$$c^{(2)}(t) = \frac{1}{(i\hbar)^2} \frac{\hbar}{2\varepsilon_0 V \sqrt{\omega_1 \omega_2}} \left(-\frac{e}{m}\right)^2 \int_0^t dt'' \int_0^{t''} dt'$$

$$\times \left[\sum_I \langle B | \boldsymbol{p} \cdot \boldsymbol{e}_2 | I \rangle \exp\{i(E_B - E_I + \hbar\omega_2)t''/\hbar\} \times \langle I | \boldsymbol{p} \cdot \boldsymbol{e}_1 | A \rangle \exp\{i(E_I - E_A - \hbar\omega_1)t'/\hbar\} \right.$$

$$\left. + \sum_I \langle B | \boldsymbol{p} \cdot \boldsymbol{e}_1 | I \rangle \exp\{i(E_B - E_I - \hbar\omega_1)t''/\hbar\} \times \langle I | \boldsymbol{p} \cdot \boldsymbol{e}_2 | A \rangle \exp\{i(E_I - E_A + \hbar\omega_2)t'/\hbar\} \right]$$

$$= -\frac{\hbar}{i\hbar 2\varepsilon_0 V \sqrt{\omega_1 \omega_2}} \left(\frac{e}{m}\right)^2 \sum_I \left(\frac{(\boldsymbol{p} \cdot \boldsymbol{e}_2)_{BI} (\boldsymbol{p} \cdot \boldsymbol{e}_1)_{IA}}{E_I - E_A - \hbar\omega_1} + \frac{(\boldsymbol{p} \cdot \boldsymbol{e}_1)_{BI} (\boldsymbol{p} \cdot \boldsymbol{e}_2)_{IA}}{E_I - E_A + \hbar\omega_2} \right)$$

$$\times \int_0^t dt'' \exp\{i(E_B - E_A + \hbar\omega_2 - \hbar\omega_1)t''/\hbar\} . \quad (4.109)$$

Here we rewrite $\langle B | \boldsymbol{p} \cdot \boldsymbol{e} | I \rangle$ as $(\boldsymbol{p} \cdot \boldsymbol{e})_{BI}$ and so on. The transition $A \rightarrow B$ occurs *via* an intermediate state I. Note that energy conservation is not necessarily satisfied at the intermediate state. The matrix elements are the sum of all possible intermediate states.

The transition probability arising from the first order perturbation of $\boldsymbol{A}\cdot\boldsymbol{A}$ and the second order perturbation of $\boldsymbol{A}\cdot\boldsymbol{p}$ can be obtained using a procedure similar to that used for the derivation of Eq. (4.99) in the form

$$\begin{aligned}\omega_{d\Omega} &= \int \left(|c^{(1)}+c^{(2)}|^2/t\right)\rho_{\hbar\omega_2,d\Omega}d(\hbar\omega_2)\\ &= \frac{2\pi}{\hbar}\left(\frac{\hbar}{2\varepsilon_0 V\sqrt{\omega_1\omega_2}}\right)^2\left(\frac{e^2}{m}\right)^2\frac{V}{(2\pi)^3}\frac{\omega_2^{\ 2}}{\hbar c^3}\,d\Omega\end{aligned}$$

$$\times\left|\langle B|\exp(i\boldsymbol{q}\cdot\boldsymbol{r})\,|A\rangle\,\mathbf{e}_1\cdot\mathbf{e}_2-\frac{1}{m}\sum_I\left(\frac{(\boldsymbol{p}\cdot\mathbf{e}_2)_{BI}(\boldsymbol{p}\cdot\mathbf{e}_1)_{IA}}{E_I-E_A-\hbar\omega_1}+\frac{(\boldsymbol{p}\cdot\mathbf{e}_1)_{BI}(\boldsymbol{p}\cdot\mathbf{e}_2)_{IA}}{E_I-E_A+\hbar\omega_2}\right)\right|^2 \tag{4.110}$$

The differential cross section for scattering is obtained by dividing the above transition probability by photon flux density c/V to be

$$\frac{d^2\sigma}{d\Omega d(\hbar\omega_2)}=r_0^2\left(\frac{\omega_2}{\omega_1}\right)\delta(E_B-E_A+\hbar\omega_2-\hbar\omega_1)$$

$$\times\left|\langle B|\exp(i\boldsymbol{q}\cdot\boldsymbol{r})\,|A\rangle\,\mathbf{e}_1\cdot\mathbf{e}_2-\frac{1}{m}\sum_I\left(\frac{(\boldsymbol{p}\cdot\mathbf{e}_2)_{BI}(\boldsymbol{p}\cdot\mathbf{e}_1)_{IA}}{E_I-E_A-\hbar\omega_1}+\frac{(\boldsymbol{p}\cdot\mathbf{e}_1)_{BI}(\boldsymbol{p}\cdot\mathbf{e}_2)_{IA}}{E_I-E_A+\hbar\omega_2}\right)\right|^2. \tag{4.111}$$

Here, r_0 is the classical electron radius. Equation (4.111) is called the 'Kramers-Heisenberg formula'. Figure 4.9 shows the Feynman diagram for the three scattering processes included in Eq. (4.111). In the figure, the solid lines represent state of the atom and wavy lines represent light. The time is indicated along the direction from the lower to the upper side of the figure. Figure 4.9(a) illustrates the process expressed by the first term of Eq. (4.111), in which the state of the atom undergoes a transition from state A to state B, while light is scattered from $(\boldsymbol{k}_1, \mathbf{e}_1)$ to $(\boldsymbol{k}_2, \mathbf{e}_2)$. Figures 4.9(b) and 4.9(c) illustrate the processes expressed by the second and third terms of Eq. (4.111), respectively. In these processes,

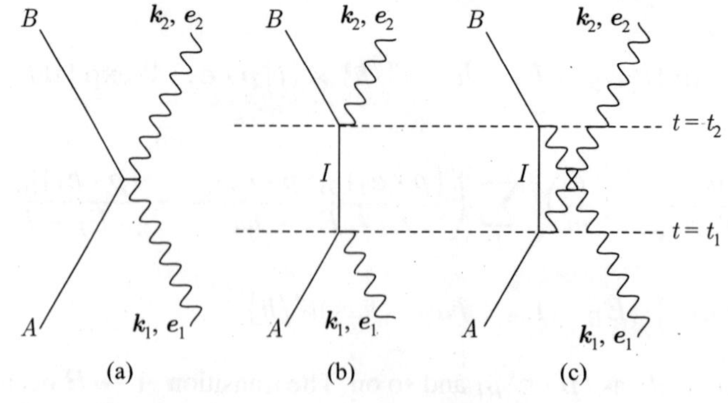

Figure 4.9 Feynman diagrams for light scattering.

absorption and emission are not a succession within the first order process, but are a single process which includes both effects simultaneously.

The above formalism is valid for a one-electron system. For a system with many electrons, the formalism can be extended as follows.

$$\frac{d^2\sigma}{d\Omega d(\hbar\omega_2)} = \sum_B {r_0}^2 \left(\frac{\omega_2}{\omega_1}\right) \delta(E_B - E_A + \hbar\omega_2 - \hbar\omega_1) \times \left| \langle B| \sum_j \exp(i\boldsymbol{q}\cdot\boldsymbol{r}_j) |A\rangle\, \boldsymbol{e}_1 \cdot \boldsymbol{e}_2 \right.$$
$$\left. - \frac{1}{m}\sum_I \left(\frac{(\boldsymbol{p}\cdot\boldsymbol{e}_2)_{BI}(\boldsymbol{p}\cdot\boldsymbol{e}_1)_{IA}}{E_I - E_A - \hbar\omega_1} + \frac{(\boldsymbol{p}\cdot\boldsymbol{e}_1)_{BI}(\boldsymbol{p}\cdot\boldsymbol{e}_2)_{IA}}{E_I - E_A + \hbar\omega_2} \right) \right|^2, \tag{4.112}$$

where $\boldsymbol{p} = \sum_j \boldsymbol{p}_j$. If there are more than two final states, the cross section is given by summation in the above formula with respect to the final state.

a. Non-resonant scattering of hard X-rays: In the hard X-ray region, $\hbar\omega_1$ and $\hbar\omega_2$ are usually much larger than the binding energy of electrons and the scattering is non-resonant. Only the first term in Eq. (4.112) is taken into account and the differential cross section is given by

$$\frac{d^2\sigma}{d\Omega d(\hbar\omega_2)} = \sum_B r_0^2 \left(\frac{\omega_2}{\omega_1}\right) \left|\langle B| \sum_j \exp(i\boldsymbol{q}\cdot\boldsymbol{r}_j) |A\rangle\, \boldsymbol{e}_1 \cdot \boldsymbol{e}_2\right|^2 \delta(E_B - E_A + \hbar\omega_2 - \hbar\omega_1), \tag{4.113}$$

which originates from the first order perturbation of the $\boldsymbol{A}^2$ term. Elastic scattering, when $\hbar\omega_1 = \hbar\omega_2$ and $A = B$ is also called Thomson scattering. Inelastic scattering ($\hbar\omega_1 \neq \hbar\omega_2$), with small $\boldsymbol{q}\cdot\boldsymbol{r}$ ($e^{i\boldsymbol{q}\cdot\boldsymbol{r}} \approx 1 + i\boldsymbol{q}\cdot\boldsymbol{r}$) is called X-ray Raman scattering and inelastic scattering with large $\boldsymbol{q}\cdot\boldsymbol{r}$ ($\boldsymbol{q}\cdot\boldsymbol{r} \gg 1$) is called Compton scattering. Inelastic X-ray scattering will be discussed in Chap. **12**.

b. Resonant inelastic scattering of soft and hard X-rays: Resonant inelastic scattering originates from the second and the third terms of Eq. (4.112) due to second order perturbation of the $\boldsymbol{A}\cdot\boldsymbol{p}$ term. Resonant inelastic scattering becomes prominent when the denominator of these terms tends to 0 and the resonant condition is satisfied. This scattering process is not regarded as successive processes of absorption and emission but as a single process in which absorption and emission interfere with each other. For example, in the case of resonant absorption, the denominator of the second term is 0 and the second term diverges. To avoid this divergent behavior, the lifetime of the intermediate state is taken into account and E_I is replaced by $E_I + i\Gamma_I$. Accordingly, the differential cross section can be expressed by

$$\frac{d^2\sigma}{d\Omega d(\hbar\omega_2)} = \sum_B {r_0}^2 \left(\frac{\omega_2}{\omega_1}\right) \left| \frac{1}{m} \sum_I \frac{(\boldsymbol{p}\cdot\boldsymbol{e}_2)_{BI}(\boldsymbol{p}\cdot\boldsymbol{e}_1)_{IA}}{E_I - E_A - \hbar\omega_1 + i\Gamma_I} \right|^2 \delta(E_B - E_A + \hbar\omega_2 - \hbar\omega_1). \tag{4.114}$$

In the case of resonant absorption, the third term of Eq. (4.112) is omitted, as it is small in comparison with the second term.

4.3.4 The Auger and emission processes

The core hole state generated by absorption of an X-ray photon in the first order process is unstable. The state becomes stable only after the outer electron drops down to the core level and buries the hole.

In this relaxation process, there are two pathways. The first is *via* the emission of a fluorescence photon, where the emitted photon energy is equal to the energy difference between the two states. The second is the Auger process in which another bound electron receives an amount of energy equal to the energy difference between the two states and that electron exits the atom, which is called an Auger electron. These are competing processes. Let p_R and p_A respectively stand for the transition probabilities for the emission and Auger processes in the relaxation of the excited state with a core hole. The mean lifetime of the excited state is given by

$$\tau = 1/(p_R + p_A) \tag{4.115}$$

Using 'uncertainty pinciple', the energy width of the excited state Γ can be written as

$$\Gamma = \hbar/\tau = \hbar\,(p_R + p_A) = \Gamma_R + \Gamma_A, \tag{4.116}$$

where $\Gamma_R = \hbar\, p_R$ and $\Gamma_A = \hbar\, p_A$ originate from the emission and Auger processes, respectively. The experimental and theoretical data for the two processes had been accumulated.[7] The total energy width of the state with an excited core hole is the sum of the energy width originating from both processes. (The energy width Γ in ε_2 given in Eq. (4.56b) introduced in the Lorentz model corresponds to the energy width of resonant absorption.) The photon emission yield ω_R is then given by

$$\omega_R = \frac{p_R}{p_R + p_A} = \frac{\Gamma_R}{\Gamma_R + \Gamma_A}. \tag{4.117}$$

When the hole exists in K core, Γ_R is approximately proportional to Z^4, while the dependence of Γ_A on Z is not significant.[1] Accordingly the above equation can be approximated by

$$\omega_R = \frac{bZ^4}{a + bZ^4} = \frac{1}{1 + a_K Z^{-4}}, \tag{4.118}$$

where $a_K = a/b$ is the adjusting constant which is usually $a_K = 1.12 \times 10^6$. Figure 4.10 shows the transition probabilities for the emission and Auger processes from a K core hole state. Both probabilities are about equal at Z = 31. From this, it is concluded that the Auger process is dominant for light

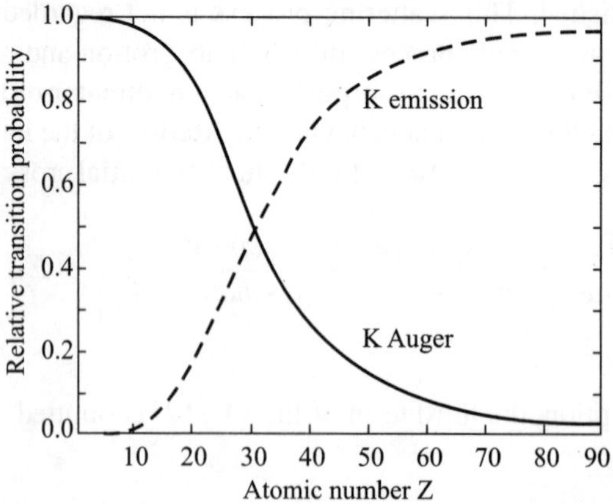

Figure 4.10 Transition probabilities for the emission and Auger processes for the relaxation of a K core hole state.

elements, while the emission process dominates for heavy elements. The Auger process is not an optical process, but a type of autoionization mechanism (see Sect. **6.1.2**).

Auger processes are classified into 'Koster-Krönig' and 'super Koster-Krönig' processes. In the Koster-Krönig process, the principal quantum numbers of the core hole and the electron going out from the atom through Auger process are different, while in the super Koster-Krönig process, they are the same. An Auger process, for instance expressed as LVV, means that the hole exists at L core, the valence electron buries this core hole and another valence electron moves up to a level higher than that of the vacuum level.

REFERENCES

1. B. K. Agarwal: "X-Ray Spectroscopy", 2nd ed. (Springer-Verlag, 1991).
2. A. G. Michette and C. J. Buckley: "X-Ray Science and Technology" (Institute of Physics Publishing, 1993).
3. M. Fox: "Optical Properties of Solids" (Oxford Master Series in Condensed Matter Physics, Oxford University Press, 2010).
4. C. M. Macgillavry and G. D. Rieck eds.: "International Tables for X-Ray Crystallography" Vol. 3 (The Kynoch Press, 1962).
5. J. J. Sakurai: "Advanced Quantum Mechanics" (Addison and Wesley, 1967). eBook is available.
6. J. P. Hannon, G. T. Trammell, M. Blume and D. Gibbs: Phys. Rev. Lett. **61** (1988) 1245.
7. M. O. Krause: J. Phys. Chem. Ref. Data **8** (1979) 307.

CHAPTER

5 Optical Techniques

Optical techniques for measurements in the vacuum ultraviolet, and soft and hard X-ray regions have been made from the beginning of twentieth century but more recently all techniques have continued to evolve alongside the rapid growth of synchrotron radiation research.[1–3] Much of this information is given in Ref. 12 in Chap. **1**. Measuring methods, after monochromatization of radiation from the storage ring are described in the appropriate chapters. Optical techniques in the infrared region are described in Sect. **13.2** and ultra-high vacuum technology is discussed in Sect. **3.4.1**.

5.1 LIGHT ABSORPTION BY MATTER

When light traverses solids, liquids and gases, it is absorbed by interaction with those materials. In the vacuum ultraviolet and soft X-ray regions, light is strongly absorbed by valence and shallow core electrons. In the hard X-ray region however it is absorbed by deep core electrons. In the case when light of wavelength λ with intensity $I_0(\lambda)$ is incident on material of thickness d, the intensity of transmitted light without any reflection effects is expressed by

$$I(\lambda) = I_0(\lambda)\exp\{-\mu(\lambda)d\} \tag{5.1}$$

(see Sect. **4.1.1.b**). Here, $\mu(\lambda)$ is the absorption coefficient in units of inverse thickness. In the VUV and soft X-ray regions the absorption coefficients of most condensed materials lie in the range from 10^4 to 10^6 cm^{-1}, while in the hard X-ray region they lie in the range of from 1 to 10^3 cm^{-1}. The absorption coefficients of heavy elements are about two orders larger than those of light elements. Generally they decrease with reduction in wavelength. Absorption coefficients can be derived by table as described in Sect. **4.2.3**.[4]

The absorption coefficient of air for visible light and for hard X-rays is very small. Therefore they are readily transmitted. However, in regions below the ultraviolet, the absorption by air begins around 200 nm and subsequently increases. For 100 nm light, since the absorption coefficient of air at 1 atm. is 8×10^3 cm^{-1},[5] 90% of intensity of the light is absorbed after passing through only 2.3 µm. The absorption coefficient of any material is roughly proportional to its density and those of gases are proportional to their pressure. When light with a wavelength of 100 nm traverses apparatus with a pass length of a few 10s of cm, a vacuum of a few Pa ($\sim 10^{-5}$ atm.) is needed to reduce air absorption to less than 1% and this is the origin of the term, 'vacuum ultraviolet'. As the wavelength is reduced the absorption coefficient of air also diminishes. At the K absorption edge of N_2 near 32 nm it is 12 cm^{-1} and around 10 nm it is 1 cm^{-1}. The light in this region is called soft X-ray radiation. In the experiments using gaseous samples or samples surrounded by gases in the wavelength region where windows cannot be used, 'differential pumping' systems are employed in which many slits transmitting only monochromatic light are situated on the optical path and each part of the path between two adjacent slits is evacuated. When the wavelength is further reduced, the absorption coefficient of the air becomes extremely small, for instance for the Cu Kα line (0.154 nm, 8.05 keV), it is 2×10^{-2} cm^{-1}, it allows experiments to be carried out in air. X-rays with wavelengths shorter than the Cu Kα line also show high transmissivity for many materials. The W Kα line (0.021 nm, 59.3 keV) and its continuum X-ray are often routinely used for medical and dental X-ray examinations in many differing circumstances (Röntgen photographs).

Any material transparent to the radiation with sufficient mechanical strength to seperate atmosphere and the machine vacuum may be used as a 'window material'. A window material may also used as a support substrate for evaporation of a sample. Short wavelength limits for light transmission for a single crystal or plate are 165 nm for quartz, 135 nm for BaF_2, 123 nm for CaF_2 and 113 nm for MgF_2. LiF provides the shortest wavelength limit of 105 nm.[6] Many lenses, prisms, polarizers *etc.*, are made from these materials and can be used until the wavelength reaches the transmission limit. For even shorter wavelength regions beyond the limit, reflection optical systems must be used except for the special case which involves measurements using transmission gratings, zone plates and 'X-ray refractive lenses' of concave type. The latter two cases are noted in Chap. **14**. In the hard X-ray region, thin foils of light metals are frequently used as windows.

5.2 FOCUSING MIRRORS

Reflection mirrors are used to gather radiation, direct it into monochromator and also to focus the resulting monochromatic light onto the sample. The elements are called the pre- and post-mirrors. The basic features of reflection mirrors will now be considered. Microscopic optical systems can readily generate a spot size with a diameter less than a few μm (a so-called microbeam) or can be used to yield a magnified image as discussed in Chap. **14**.

5.2.1 Reflectances of materials

Since the imaginary part of the indices of refraction of metals are large in the visible and ultraviolet regions, the reflectivity of metal surfaces will be high and therefore metals are usually used as mirror materials in near normal incidence optical systems. By contrast, in the shorter wavelength region of the VUV the refractive indices of almost all materials are close to 1, so that normal incidence reflectances are low. In case of pure Al (not oxidized) for example, the normal incidence reflectivity reaches 93% at 10 eV where $n = 0.044$ and $\kappa = 1.18$, while the reflectivity reduces to 0.03% at 50 eV where $n = 0.97$ and $\kappa = 0.006$. For this reason, it is necessary in the 30 to 40 nm region to change any reflecting optical system from normal incidence to the grazing incidence type. Figure 5.1 shows reflectance spectra for Au which is a typical mirror material in the grazing incidence region. In the figure the reflectivity above 150 eV rapidly diminishes at an angle of incidence of 79°. The decrease of reflectivity around the photon energy of 2 keV originates from M absorption. The decrease around 4 keV at an angle of incidence of 89° and the decrease around 8 keV at 89.5° are due to the effect that the angles of incidence are not sufficiently large for total reflection to occur above those photon energies. In the same energy region, the reflectance spectra of Pt show behavior quite similar to that of Au. In order to obtain high reflectivity in these high energy regions, all focusing mirrors must be installed at grazing incidence angles. In this configuration aberrations in the image become quite large. To minimize this problem aspherical mirrors, or sometimes more than two spherical mirrors, are used. Alternatively, the use of multilayers described below, may help resolve these difficulties to some extent.

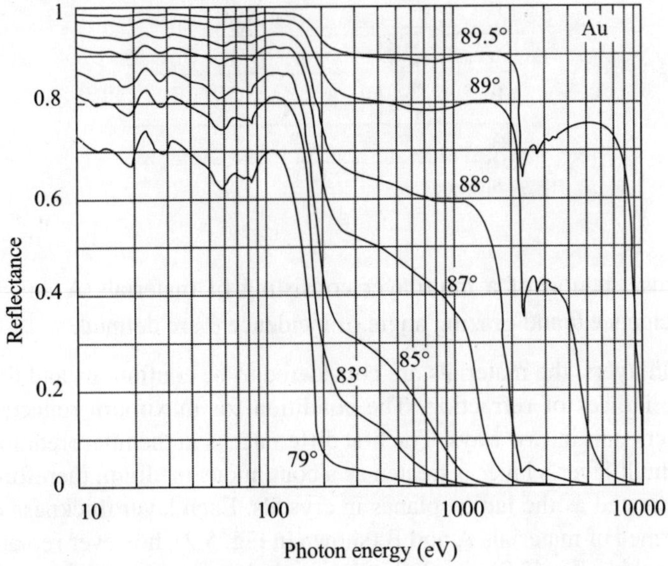

Figure 5.1 Reflectance of Au for s-polarized light. Angles are those between the incident light and the normal to the reflecting surface.

5.2.2 Multilayer devices

When the alternating layers of different materials are deposited periodically to make multilayer reflectors, one can obtain significantly enhanced reflectance for any selected wavelength, resulting from the constructive interference between reflected light from each interface.[7] The relation between layer thickness, angle of incidence and wavelength in the constructive interference in multilayers is analogous to that for Bragg reflection (diffraction) in crystals where the path difference between X-rays reflected from adjacent lattice planes is a multiple of wavelength and is expressed by

$$2d \sin \hat{\theta} = m\lambda. \tag{5.2}$$

For reflections in crystals, d, $\hat{\theta}$, m and λ are the distance between adjacent lattice planes, grazing angle of incidence, order of diffraction and wavelength respectively. Assuming that the indices of refraction of the materials for X-rays to be 1, then the wave vector component vertical to the lattice plane is given by

$$k_z = \frac{2\pi}{\lambda} \sin \hat{\theta}. \tag{5.3}$$

One can also interpret Eq. (5.2) to be the condition in which $k_z d$, the phase difference between two points separated by d in the direction normal to the lattice plane is a multiple of π. In the case of

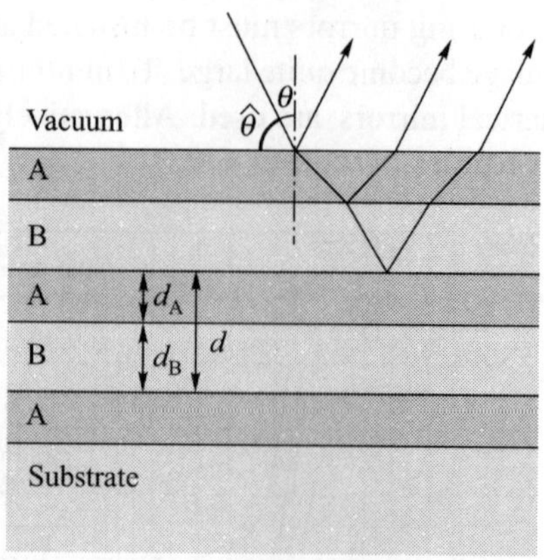

Figure 5.2 Cross section of a multilayer consisting of materials A and B. The angle of incidence θ and grazing angle of incidence $\hat{\theta}$ are defined.

reflections from multilayers, the materials are considered to be continuous and the propagation of light is defined using the indices of refraction. The condition for maximum reflectance from multilayers differs from that for crystals in two ways. The first difference is in the interpretation of distance between lattice planes. Each multilayer's layer is treated as a continuous medium, therefore it is the interfaces in multilayers that are treated as the lattice planes in crystals. Each layer thickness can be represented by d. In a multilayer formed of materials A and B (shown in Fig. 5.2), however replacement of the distance between lattice planes d in Eq. (5.2) simply by the single layer thickness d_A or d_B leads to a difficulty in the condition for the destructive interference because the phase of reflection of light differs from that from A to B by π. The distance d in crystals corresponds to the sum of the two thicknesses $d_A + d_B$ in multilayers. A second difference lies in the effect of index of refraction. In the VUV and soft X-ray regions, δ, the differences between 1 and n, and κ are of order of 10^{-2} and so that are not negligible. In the hard X-ray region, those values are as small as 10^{-6}, but they are not negligible in total reflection. Assuming that light is incident from vacuum to material A which has a complex index of refraction $\tilde{n}_A$ at grazing angle of incidence and then enters material B with complex index of refraction $\tilde{n}_B$ after passing through A. Using Snell's law, the relation between the complex index of refraction and wave number of the light described in Sect. **4.1.1.b** can be written as

$$\tilde{k}_{Az} = \frac{2\pi}{\lambda}\sqrt{\tilde{n}_A^2 - \cos^2\hat{\theta}}, \quad \tilde{k}_{Bz} = \frac{2\pi}{\lambda}\sqrt{\tilde{n}_B^2 - \cos^2\hat{\theta}}, \tag{5.4}$$

where $\tilde{k}_{Az}$ and $\tilde{k}_{Bz}$ are the wave number vector components normal to the layers in materials A and B, respectively, and λ is the wavelength in vacuum. The condition for high reflectance in multilayers is therefore given by

$$2d_A Re\sqrt{\tilde{n}_A^2 - \cos^2\hat{\theta}} + 2d_B Re\sqrt{\tilde{n}_B^2 - \cos^2\hat{\theta}} = m\lambda, \tag{5.5}$$

which is derived by replacing phase difference $k_z d$ in the Bragg reflection, Eq. (5.2) by $Re(\tilde{k}_{Az})d_A + Re(\tilde{k}_{Bz})d_B$. Here the imperfect periodicity associated with interfaces between the vacuum and top layer and between the bottom layer and the substrate is neglected.

The first stage in designing a multilayer mirror is to choose a material pair giving high reflectance. Firstly, two materials which possess quite different terms of indices of refraction are chosen. This can be expressed also as exhibiting a large difference in density in the hard X-ray region. The second important choice is to select two materials with small absorption coefficients to maximize the number of interfaces contributing to constructive interference.[8] Hence, materials whose absorption edges are located just above the design photon energy have great advantage. Additionally, the materials should have chemical properties to enable them to form a smooth interface. In the VUV and soft X-ray regions, normal incidence mirrors selected from material pairs of Mo/Si and Cr/Sc have been specifically developed for the use around 90 eV for EUV lithography and around 390 eV for imaging in the water window,

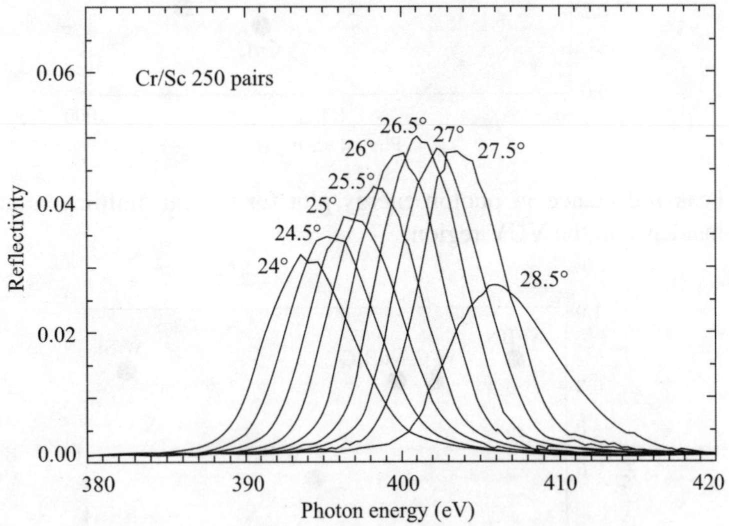

Figure 5.3 Measured reflectance for a Cr/Sc multilayer. The angle of incidence is the angle from the normal to the multilayer ($\theta = 90° - \hat{\theta}$).

respectively. Figure 5.3 shows the measured reflectance spectra of a Cr/Sc multilayer with period length of 1.73 nm. The shift of the energy of the reflectance peak with change in angle of incidence is clearly visible. In the figure, the peak reflectance is about 5%, but the achieved maximum so far is about 30%.[9] Many high reflectance multilayers have been developed and their reflectances reported.[9] In Fig. 5.4 the relation between reflectance at near normal incidence (angle of incidence $\leq 10°$) and photon energy at the peaks for some typical multilayers reported before 2004 is plotted. For hard X-rays, multilayers used at normal incidence cannot be fabricated because the periods (spacings) of the layers should be the same order of the lattice constants of the materials. Multilayers used at grazing incidence angles require periods of the order of nm periods which can be fabricated. Using multilayer mirrors enables the grazing angles to be increased without loss of reflectance. Therefore, for any light beam, the mirror size can be reduced, thus contributing to a decrease in figure error of the mirror. In addition, multilayers are often applied as coating materials to gratings. Figure 5.5 shows the relation between reflectance and grazing angle of incidence for the Cu Kα line for some reported multilayers whose period lengths are from 1 to 10 nm. Hard X-ray multilayers, called 'super mirrors', have been developed to extend the range of high reflectance. A super mirror has an aperiodic structure, in which the period lengths

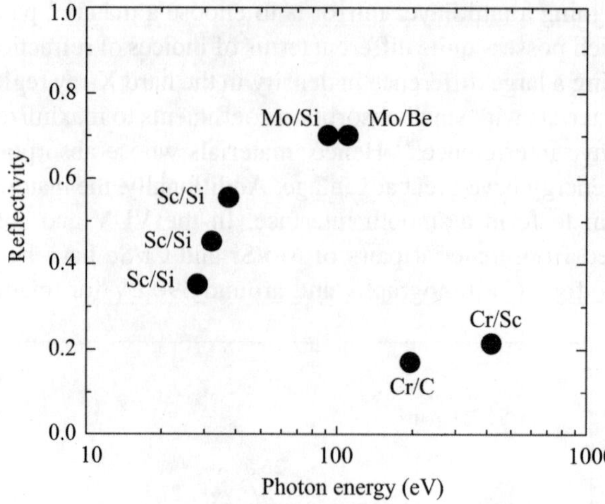

Figure 5.4 Peak reflectance *vs* photon energy plot for typical multilayers at near normal incidence in the VUV region.

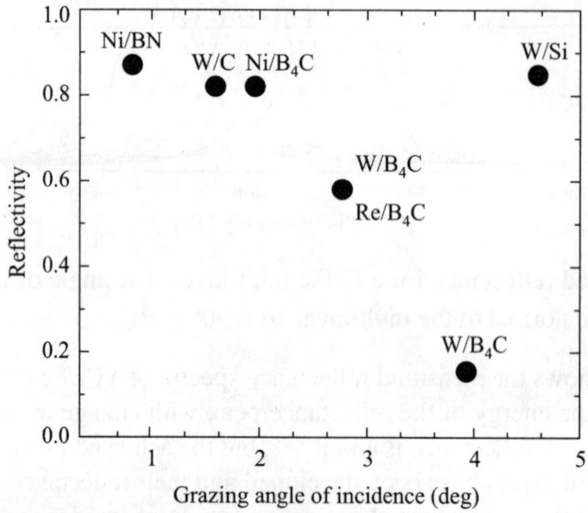

Figure 5.5 Peak reflectance *vs* angle of incidence (grazing angle) plot for typical multilayers for the Cu K line.

change monotonically with depth. Figure 5.6 shows grazing incidence reflectances for a total reflection mirror, a periodic multilayer and a super mirror.[10)] Another type of aperiodic multilayer is one in which distances between the adjacent layers are distributed seemingly at random around an average value but adjusted to make reflectance peak flat-topped and its width wide at the sacrifice of high reflectivity.[11)] Multilayers can also be used as energy selecting and polarizing elements as described later.

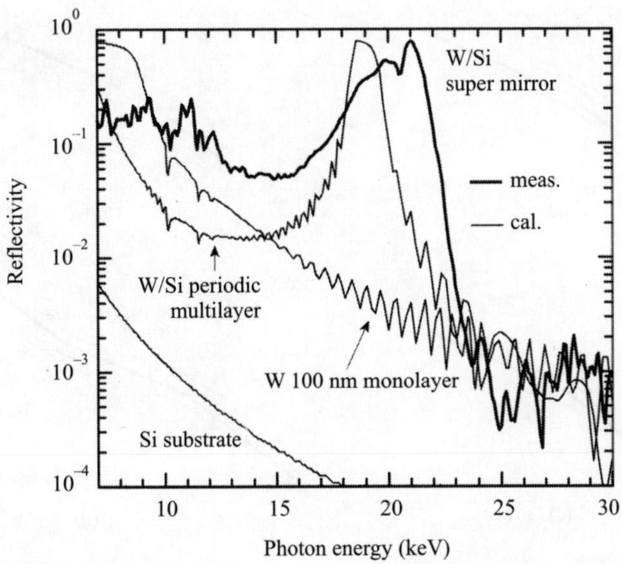

Figure 5.6 Reflectance spectra of a W/Si super mirror, W/Si periodic multilayer, W monolayer and Si substrate at grazing angle of incidence of 0.5°. Thick curves represent measured values and thin curves, calculated values.

5.2.3 Focusing conditions

In optical systems for synchrotron radiation research, grazing incidence optics are often used. Four typical methods for focusing are shown in Fig. 5.7, where the incident ray is drawn on the horizontal plane and the plane of reflection which includes both the incident ray and the normal of the mirror surface is chosen to be on the horizontal plane in Figs. 5.7(a), 5.7(b) and 5.7(c). There are two types of focusing using a single mirror as illustrated in Figs. 5.7(a) and 5.7(b) using a cylindrical mirror. One is 'tangential focusing' in which the mirror has its curvature in a plane of reflection. The other is 'sagittal focusing' in which the mirror has its curvature in a plane vertical to the plane of reflection. In practice, many mirror shapes are used, including cylindrical, spherical, toroidal, ellipsoid of revolution and paraboloid of revolution *etc.*

Let us define r to be the incident arm length (the distance between the emitting point and the mirror), r' to be the exit arm length (the distance between mirror and focusing point) and θ to be an angle of incidence measured from the normal of the mirror surface. For the spherical mirror with a radius of curvature R, the conditions for the tangential focusing and sagittal focusing are given respectively by

$$\frac{1}{r} + \frac{1}{r'} = \frac{2}{R\cos\theta}, \tag{5.6a}$$

$$\frac{1}{r} + \frac{1}{r'} = \frac{2\cos\theta}{R}. \tag{5.6b}$$

The magnification factor of any image is equal to r'/r (see App. **A**). A spherical mirror has the same radius of curvature in the horizontal and vertical directions. In the grazing incidence configuration only the curvature in the plane of reflection plays a role in gathering the light flux, and the curvature vertical

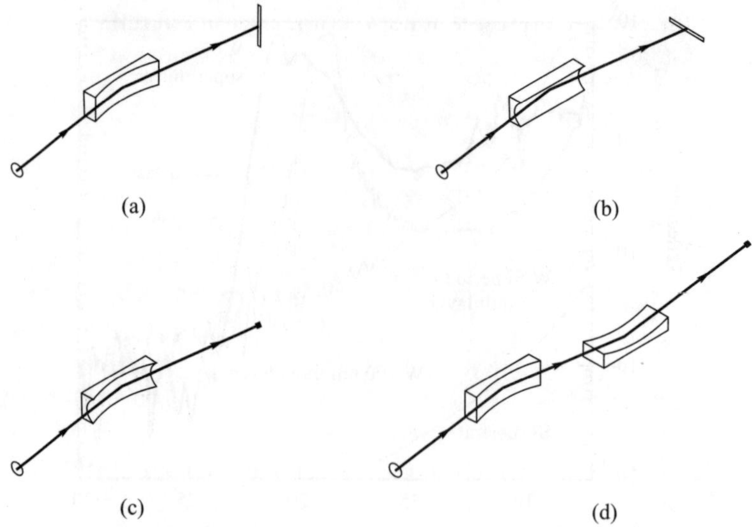

Figure 5.7 Typical focusing. (a) Tangential focusing, (b) sagittal focusing, (c) toroidal mirror and (d) Kirkpatrick-Baez configuration.

to the plane of reflection does not. Therefore, a spherical mirror can be regarded to act as a cylindrical mirror in this case. There are two ways to gather light flux in the horizontal and vertical directions. First is to use a toroidal mirror (as shown in Fig. 5.7(c)) or a mirror of ellipsoid of revolution, whose curvatures differ between the planes of reflection and the plane vertical to it. The second is to use a Kirkpatric-Baez (K-B) configuration, shown in Fig. 5.7(d) in which two spherical mirrors are configured perpendicularly to each other for gathering light flux independently in the horizontal and vertical planes about the light axis.[12)]

For toroidal mirrors, the condition that a tangential focusing point coincides with a horizontal one is derived when r in Eq. (5.6a) is set equal to r in Eq. (5.6b) and r' in Eq. (5.6a) is also set equal to r' in Eq. (5.6b), as

$$\frac{\rho}{R} = \cos^2\theta. \tag{5.7}$$

Here the radii of curvatures for the tangential and the sagittal focusing are respectively defined to be R and ρ. In the case of focusing using a toroidal mirror, a small spot with small aberrations will be obtained only for $r/r' \approx 1$. However, in the case of a light beam with divergence smaller than 1 mrad such as the case in the light beam from undulators, the ray trace calculations reveal that the light beam can be regarded as a paraxial ray resulting in a small spot with a small aberrations even for large deviations of r/r' from 1.

In the K-B configuration, the light beam is gathered in the vertical and in the horizontal direction by two mirrors, so that an image with small aberrations is obtainable. The K-B configuration has the additional merit of being able to set the mirrors independently according to the needs of the optical system. Even when the incident light has a large divergence, this advantage remains. Therefore, the K-B configuration is ideal for the final stage of optical systems where the divergent light must be focused behind the exit slit on the sample. However, notice that the intensity of reflected light is less than that for any single mirror system since two mirrors are used in this configuration.

5.2.4 Mirror profiles and their conservation

A large heat load is imposed on any pre-mirror which is directly exposed to synchrotron radiation. In order to minimize thermal deformation of the mirror special materials to cope with high heat load are chosen, such as SiC,[13] and the mirrors themselves may be directly cooled by water or by other means in order to stabilize the mirror surface shape. Until recently, the curved surface shape has been achieved by mechanical polishing and the surface figures have been restricted to plane, spherical, cylindrical and toroidal. Currently, since the heat load of the post-mirror situated behind the exit slit is small, mechanically bent mirrors can be used in place of conventionally ground and polished mirrors, so that aspherical mirrors such as the ellipsoidal mirror can be selected.[14] When an ellipsoidal mirror is used in the tangential reflection system, the quality of the image is greatly improved by comparison with the use of cylindrical and spherical mirrors since the aberrations are small and it is feasible to obtain a very small spot at the sample position. The technology of shaping and polishing aspherical mirrors has now become highly developed and superpolished mirrors (with rms roughness of only few Å) are required to obtain maximum reflectivity (see Sect. **4.1.2.b**).

In VUV and soft X-ray beamlines, the pre-mirrors and dispersive elements are maintained in ultra-high vacuum conditions but even under such ultra-high vacuum conditions, residual hydrocarbon compounds are decomposed after exposure to synchrotron radiation and the resultant carbon is deposited on the mirrors and the dispersive elements. This causes problems not only arising from the fine spectral structures due to absorption by carbon (C K edge: 283.8 eV) which appears in the reflectance spectrum of the mirror, but also because the reflectance itself gradually decreases. In hard X-ray beamlines, the pre-mirrors and the monochromatizing crystals are also often situated under high vacuum conditions. In these cases too, similar problems occur, resulting in the distortion of the reflected wave front or the appearance of spurious halo patterns.[15] In many cases, cleaning of carbon contamination is achieved using oxygen free radicals obtained by the oxygen discharge method.[16] In addition, the reflected light from mirrors and other optical elements have their own spectral structures, so that we must analyze experimental results with these effects in mind.

5.3 SPECTROSCOPIC SYSTEMS

5.3.1 Grating monochromators and spectrometers

a. Diffraction gratings: In the VUV and soft X-ray regions, diffraction gratings of reflection type are usually used as dispersive elements. The gratings normally have parallel constant spacing grooves ruled on the plane, concave or toroidal substrates. The line spacings ordinarily used are of from 500 to 3000/mm so that groove spacings d are from 0.3 to 2 μm. The equation for diffraction for the light with wavelength λ is given by

$$d(\sin\alpha + \sin\beta) = n\lambda. \tag{5.8}$$

In this equation, α and β are respectively the angles of incidence and diffraction measured from the normal of the grating surface, where the signs of α and β are chosen to be common along a rotational direction and n represents the order of diffraction. When $n = 0$, $\alpha = -\beta$ and the incident light is not dispersed but is reflected as with a mirror, while when n is positive or negative integer, the light is dispersed. 'Varied space gratings' or 'Variable space gratings' have been developed and are often used where the groove spacing varies gradually from one side of the grating to the other. Such 'varied groove spacing' brings its own focussing property, which the 'constant groove spacing' does not possess. Figure 5.8 illustrates the profiles of reflecting surfaces for gratings with (a) the blazed type and (b) the laminar type. In the blazed type, the grating surface is shaped with a saw-tooth-like cross section. Each reflecting facet of the grating between neighboring teeth are inclined at a blaze angle of θ_B, so that

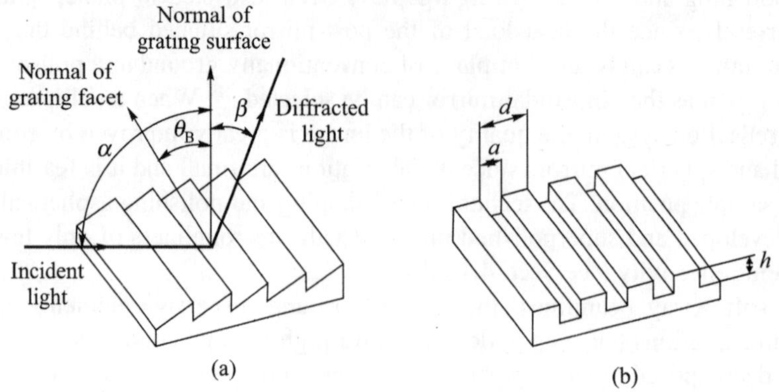

Figure 5.8 Magnified perspective surface view of typical gratings. (a) Blazed type, (b) laminar type.

its normal is also inclined from the normal of the grating substrate surface by the same amount. When an angle of incidence α and an angle of diffraction β satisfy the condition of $-\beta + \theta_B = \alpha - \theta_B$, the incident light is simply reflected. Therefore, this blazed profile considerably enhances the diffraction efficiency around a selected wavelength, which is given by $n\lambda = 2d \sin\theta_B \cos(\alpha - \theta_B)$. The blaze wavelength is defined for a Littrow (by $\alpha = -\beta = \theta_B$ at $n = 1$) mount so that it is given by $\lambda_B = 2d \sin\theta_B$. In a laminar grating surface rulings are rectangular. Light is reflected at the top and the bottom facets. The width a of each groove, the depth h, and the spacing d can be chosen so as to satisfy the interference condition to suppress the intensity of zeroth and even order light, which is derived from the path difference between light reflected by the top and the bottom facets. There are two methods used to create such grooves. One is a mechanical ruling method. The other is a holographic method in which the groove pattern is formed by interference using laser light on a substrate coated with photoresist. The grooves are subsequently created by ion bombardment of the substrate. There are many multilayer-coated gratings enhancing reflectance to meet each demand.[17)] There are also transmission gratings, consisting of the gold thin film with regularly spaced slits made by microfabrication and a Si_3N_4 membrane with an opaque lattice structure of Ta evaporated on it. They are used for special studies.

b. Monochromators: A monochromator is used to extract monochromatic light for experiments from the normally broad range of wavelengths produced by a storage ring. It usually consists of an entrance slit, a grating and an exit slit, and reflection mirrors.[18)] In the region from the visible to 40 nm, optical systems with near normal incidence are available with small aberrations and good focusing properties.

Because the reflectivity is reasonably high, an adequate diffraction efficiency can be obtained. In the shorter wavelength regions however, the reflectivity at near normal incidence is rather low and therefore grazing incidence spectroscopy becomes necessary. In this case, it is essential to install a specifically designed pre-mirror and also other mirrors in the system to diminish the potentially large aberrations.

Firstly, monochromators using spherical gratings with constant spacing grooves will be described. The focusing condition for a spherical grating with a radius of R is given by

$$\frac{\cos^2\alpha}{r} - \frac{\cos\alpha}{R} + \frac{\cos^2\beta}{r'} - \frac{\cos\beta}{R} = 0, \tag{5.9}$$

where r and r' are respectively the incident arm length and the exit arm length defined similarly to the spherical mirror case given in Eq. (5.6). One of solutions is obtained as

$$r = R\cos\alpha, \;\; r' = R\cos\beta. \tag{5.10}$$

This solution is satisfied when both the emitting point and the focusing point locate on a circle of a diameter R which is in contact with the grating surface at its center. This circle is called 'Rowland circle'. In this solution, good focusing can be achieved because the coma aberration is zero and higher order aberrations are small. Accordingly a design incorporating the solution given in Eq. (5.10) called the 'Rowland mounting' is often employed. In the case of near normal incidence when the angles of incidence and diffraction are small and variations of these angles are small for wavelength scanning, then the change of incident and exit arm lengths is small. Under this condition, a mount called 'Eagle mount' was developed and used in the photon energy range below 50 eV. When the Rowland mount is applied in the grazing incidence regime, the angles of incidence or diffraction must be altered to achieve for wavelength scanning in which the incident or the exit arm lengths may vary considerably. In synchrotron radiation facilities where the light source and the experimental apparatus must remain fixed, it is obviously essential to have a well-designed driving mechanism for wavelength scanning. For example, in the 'Grasshopper monochromator', a pre-mirror, with 'Codling type' entrance slit (composed of a single slit blade and a plane mirror situated perpendicular to the slit) and a grating are moved simultaneously, while the exit slit and the direction of exiting beam remain fixed during the wavelength scanning.[19)]

In the case of the on-Rowland mount, the driving mechanism is quite complex and therefore various 'off-Rowland mounts' have been developed. In these mounts, a 'deviation angle' which is the angle between angles of incidence and diffraction ($\alpha - \beta$) is constant, and the position of the entrance slit, the direction of light incident to the grating, the position of the exit slit and the direction of exiting light remain unchanged during wavelength scanning. Using $\hat{r} = r/R$ and $\hat{r}' = r'/R$, which are respectively the incident and the exit arm lengths normalized by R, Eq. (5.9) can be rewritten as

$$\hat{r} = \frac{\cos^2\alpha}{\cos\alpha + \cos\beta - \dfrac{\cos^2\beta}{\hat{r}'}}. \tag{5.11}$$

In Fig. 5.9 each curve represents the relation between $\hat{r}$ and $\hat{r}'$ for light with a wavelength normalised by d, namely $\hat{\lambda} = n\lambda/d$ under constant deviation angle of (a) 152° and (b) 70° and open circles (○) indicate the positions satisfying the Rowland circle condition. For a deviation angle of 152°, the circles lie on a line with a gradient of 0.5 as seen in Fig. 5.9(a). This feature suggests that the Rowland condition

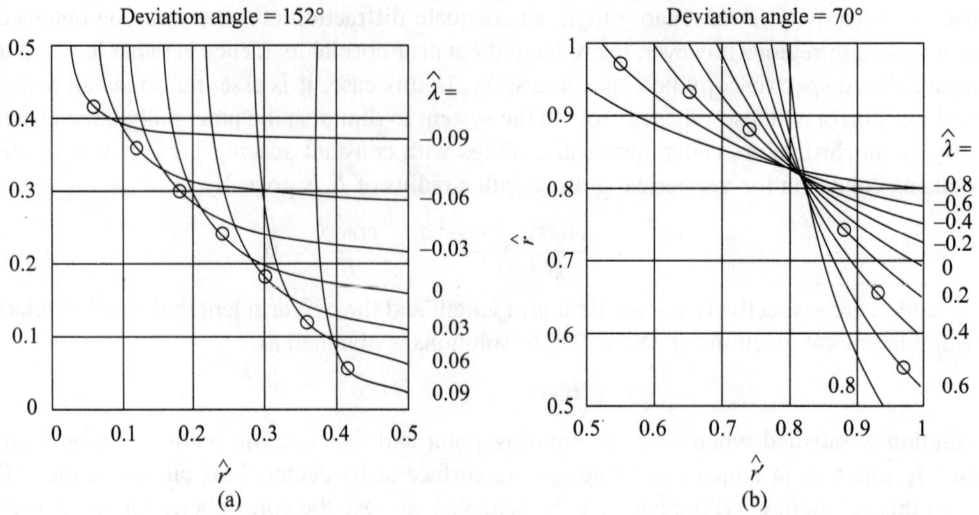

Figure 5.9 Relation between normalized incident arm length and exit arm length at deviation angles of (a) 152° and (b) 70° for some normalized wavelengths in the first order, whose sign corresponds to the sign of the order. ○ indicates positions satisfying the Rowland circle condition.

is almost satisfied if the sum of the incident arm length and the exit arm length is constant.[20] A monochromator utilizing this feature is called a 'constant arm length mount' and has the advantage that focusing is retained across a wide wavelength range. A disadvantage is that a large traverse of the grating is required. For a deviation angle of 70°, each curve passes a point around $\hat{r} = \hat{r}' = 0.82$ as seen in Fig. 5.9(b). A monochromator exploiting this feature is called 'Seya-Namioka mount', which has the advantage that the entrance and exit slits can be fixed, and the wavelength can be scanned by simple rotation of the grating. Therefore the Seya-Namioka mounting is frequently used in VUV region. However this design gradually deviates from the Rowland design as the wavelength increases, resulting in coma aberrations and the diffraction efficiency decreases in the wavelength region below 30 nm due to the decrease in reflectance. In Fig. 5.10, each curve represents the relation between $\hat{r}'$ and $\hat{\lambda}$ for a fixed $\hat{r}$. (In the case for a fixed $\hat{r}'$, the curve also represents $\hat{r}$, if the sign of $\hat{\lambda}$ is inverted.) The mark ○ indicates the position of the exit slit satisfying the Rowland condition, at which each curve shows the minimum and around which the curve is almost flat. This feature means that when $\hat{r}$ is fixed and the grating is rotated for the wavelength scanning, the change of $\hat{r}'$ is small and the good focusing can be achieved with a fixed $\hat{r}'$. Furthermore if $\hat{r}'$ is slightly changed according to the feature of the curve, improved focusing is obtained with small coma aberration, because $\hat{r}'$ is located near the positions satisfying the Rowland condition. This type of monochromator is called the 'Dragon type' and is used at many synchrotron radiation facilities.[21] Those of the inverse type with fixed $\hat{r}'$ are also used.

In the spherical grating monochromator of the on-Rowland type, the wavelength band width of dispersed light is expressed by $\Delta\lambda = wd/nR$, which is obtained by diffentiating Eq. (5.6) with using (5.10), where w stands for the slit widths. Therefore the use of a grating with large radius of curvature is essential to achieve a high resolution or a high resolving power $\lambda/\Delta\lambda$.

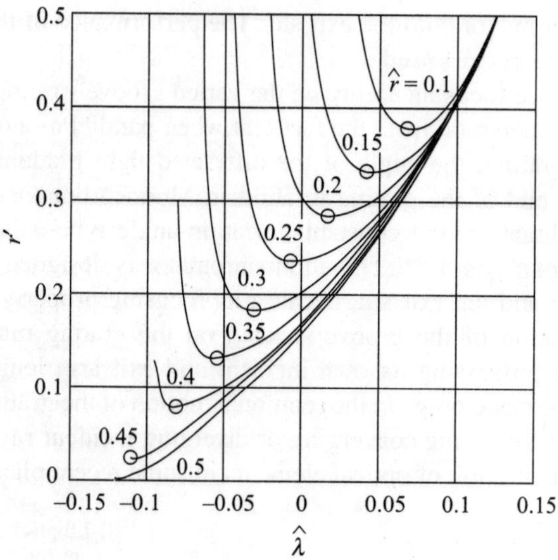

Figure 5.10 Relation between normalized exit arm length and normalized wavelength at deviation angle of 152° for some normalized incidence arm lengths. ○ indicates the positions satisfying the Rowland circle condition.

Monochromators using plane gratings with constant spacing grooves will now be described. In Eq. (5.9), when $R \to \infty$, the exit arm length is given by

$$r' = -r\frac{\cos^2\beta}{\cos^2\alpha}. \tag{5.12}$$

Since $\cos^2\alpha$ and $\cos^2\beta$ are positive quantities, Eq. (5.12) requires that either r or r' is negative. This condition means that an emitting point or focus point is located in the negative region from the grating surface, in other words, behind the grating surface. That is, when r is negative a convergent beam is incident on the grating, while a divergent beam is emitted when r' is negative. Usually the light incident onto the grating is divergent, so that light dispersed by the grating must be focused on to the exit slit using a focusing mirror. In this configuration, mounts with both constant deviation angles and also variable deviation angles are available. In these latter mounts, it is possible to supress the intensity of the higher order light at every wavelength by choosing the angle of incidence and enhancement of the intensity of the diffracted light *via* optimization of the grating parameters and angle of diffraction to satisfy the blaze condition (on-blaze). A typical monochromator of this type is called the 'Petersen monochromator', previously known as the SX-700,[22)] which has a focusing mirror between the grating and the exit slit.

In Eq. (5.12), when $r \to \infty$ and $r' \to \infty$, the value of $\cos^2\beta/\cos^2\alpha$ can be chosen arbitrarily. This condition means when parallel light is incident on the plane grating, parallel diffracted light exits and suggests that one can choose those angles of incidence and diffraction which are optimal to enhance the diffraction efficiency and to supress the higher order light. A monochromator based on this concept is called the 'collimated plane grating monochromator (C-PGM)'.[23)] This type can be used in principle without mirrors due to the small divergence of synchrotron radiation, but in the real case, synchrotron radiation is made parallel by a pre-mirror subsequently incident on the plane grating, and the diffracted light is focused by a further mirror onto the exit slit. The performance of this type of device is mainly dependent on qualities of the mirrors used.

It is also possible to impose focusing ability on the varied groove spacing even though the grating is plane. This ability can be understood from the fact that when parallel monochromatic light illuminates the varied spacing plane grating, the angle of the diffracted light gradually moves according to the illuminated area from one end of the grating to another. Monochromators of this type used at many synchrotron radiation facilities are the constant deviation-angle type using covergent incident light, where $r < 0$, $r' > 0$ and $r/r' \approx -1$.[24] This monochromator is designed to incorporate no focusing mirror between the grating and the exit slit, because its focusing property is better than those of any mirrors. The optimal variation of the groove spacing on the grating must be determined for each individual monochromator possessing its own incident and exit arm lengths and wavelength range. Wavelength scanning can be made only *via* the rotational motion of the grating. Other variable deviation angle types of monochromators using converging or diverging incident rays have been designed.[25,26] Figure 5.11 shows the configuration of optical elements in some recent plane grating monochromators.

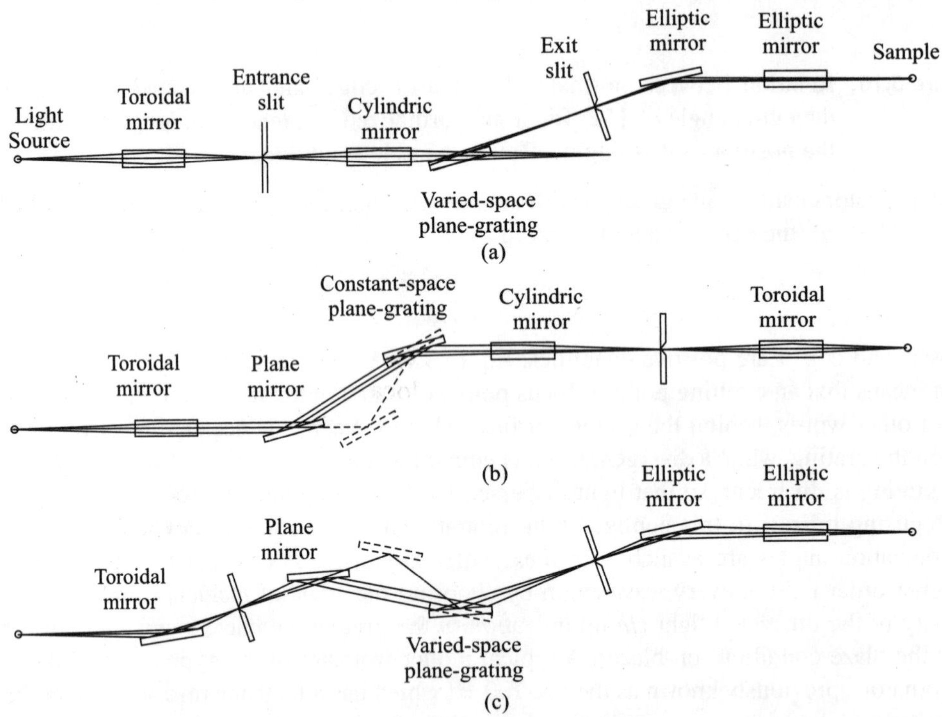

Figure 5.11 Configuration of optical elements in a typical plane grating monochromator. (a) Plane grating monochromator of constant deviation angle using varied space grating, (b) collimated plane grating monochromator with variable deviation angle using a constant space grating and (c) plane grating monochromator with variable deviation angle using a varied space grating.

Figure 5.12 presents an example of a spectrum obtained from a concave grating monochromator of constant deviation type with moving entrance slit situated at a bending magnet beamline of PF.

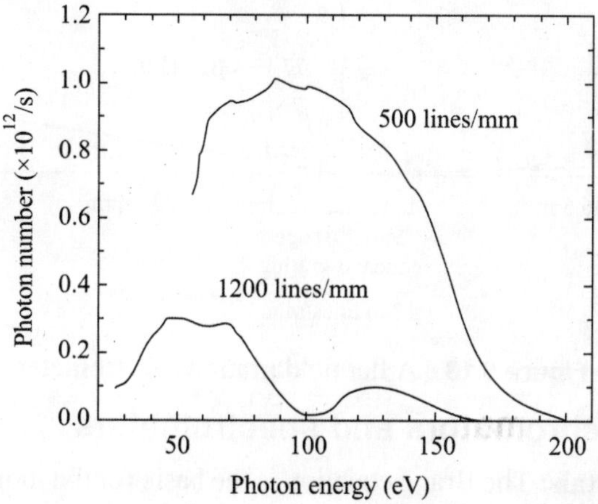

Figure 5.12 An example of the output spectrum from a grating monochromator. The intensity was measured by the TEY (total electron yield) method using Au and was converted to photon number.

Monochromatic light at 70 eV with a band width of 110 meV is obtained with a photon number of 3×10^{11}/s. Diameter of beam spot on the sample is smaller than 1 mm. In experiments described in Sect. **6.1**, monochromatic light of around 860 eV with a band width of 70 meV is used. It is noticed that the polarization chracteristics of synchrotron radiation are conserved at extreme grazing incidence and near normal incidence cases, but otherwise the characteristics are not necessarily conserved except for linear polarization.

c. Spectrometers: The monochromators described here are used to generate monochromatic light for experiments. However, in general spectrometers are used mainly to study and measure emission spectra. There are two kinds of spectrometers for this work. In the first type, a zero-dimensional detector is placed just beyond the exit slit and the dispersed light then measured by wavelength scanning. In the second type, a one- or two-dimensional detector is placed on the focal surface and the dispersed light is measured across a wide wavelength region without scanning. Detectors of the latter type may be photographic plates which were used for the earliest experiments but have now been displaced using CCD and other types of electronic detector (see Sect. **5.6.3**). Usually the focal surface is a part of the cylindrical surface in Rowland type spectrometers equipped with spherical gratings, so that a planar detector such as a CCD is located on the focal surface, which is approximately planar across each narrow wavelength region. If a varied space grating which makes the focal surface plane called 'flat field' is used, one can measure across a rather wide wavelength region using a planar detector. An example is shown in Fig. 5.13. Spectrometers are often used to measure spectra of emitted fluorescence or inelastically scattered light, which are generated by a sample after irradiation by monochromatic light often with similar resolution to the monochromators described above. The spectrometers may also be used to measure absorption spectra when continuum light such as synchrotron radiation illuminates the sample and the light transmitted is subsequently dispersed and detected.

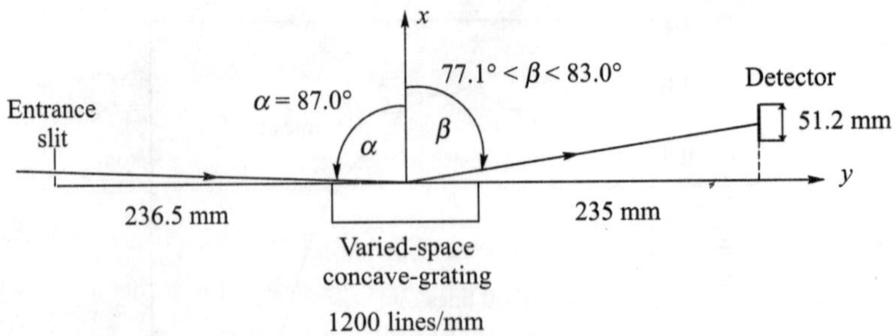

Figure 5.13 A flat field grating spectrometer.

5.3.2 Crystal monochromators and spectrometers

a. Monochromatizing crystals: The Bragg equation is the basis for the monochromatization of X-rays. The wavelength λ of monochromatized light is given by $n\lambda = 2d \sin \hat{\theta}$, where d represents the distance between adjacent lattice planes, $\hat{\theta}$ is the grazing angle of incidence and n is the order of diffraction. The value λ is determined by d and $\hat{\theta}$. The largest wavelength being equal to $2d$. The smaller is d, the smaller λ and also the smaller $\hat{\theta}$, the smaller λ. In the latter case, the less $\hat{\theta}$ becomes, the larger crystal size one needs, so that there is a practical short wavelength limit. Table 5.1 presents $2d$ values for typical monochromatizing crystals and their usable photon energy regions for a range of $\hat{\theta}$ of from 5° to 70° for $n = 1$. Monochromatization of light even above 100 keV is possible, if the grazing angle

Table 5.1 List of monochromatizing crystals and their usable energy region (grazing angle of incidence: 5° ~ 70°)

Crystals	*2d* (Å)	Energy region (keV)
β-Al_2O_3 (002)	22.53	0.586 ~ 6.31
Beryl (1010)	15.965	0.827 ~ 8.91
YB_{66} (400)	11.76	1.12 ~ 12.1
$KTiOPO_4$(KTP) (011)	10.95	1.21 ~ 13.0
SiO_2 (1010)	8.512	1.53 ~ 16.7
InSb (111)	7.4806	1.72 ~ 19.0
Graphite (002)	6.708	1.94 ~ 21.2
Ge (111)	6.532	1.99 ~ 21.7
Si (111)	6.271	2.07 ~ 22.6
LiF (200)	4.027	3.28 ~ 35.3
Si (220)	3.840	3.44 ~ 37.0
Si (311)	3.275	4.03 ~ 43.4
Si (400)	2.715	4.86 ~ 52.4
Si (511)	2.090	6.31 ~ 68.1

of incidence is chosen to be sufficiently small and by using a long crystal. The width of crystals is determined by the spread of incident light on the crystals and is usually a few cm or more. In the case of monochromatization of high power X-rays, the crystals may have to be cooled by water or liquid nitrogen, which flows through channels cut on the rear surfaces of the crystals.

b. Monochromators: The double crystal monochromator is most commonly used which makes use of two crystals situated parallel to each other and rotated by exactly the same angle in the same direction. The light which satisfies the Bragg condition is extracted. The directions of the incident and reflected light remain the same. However with only simple rotation, the position of the light reflected from the first crystal to the second will move and the axis of the reflected light from the second crystal also moves resulting in a problem of beam shift at the sample. To avoid the problem, the crystal must be translated while the crystals are rotated and kept parallel. Figure 5.14 shows three mechanisms in which the incident light illuminates the first crystal and the second crystal is translated. In Figs. 5.14 (a) and 5.14 (b), the first and the second crystals rotate around a center of the first crystal illuminated by the incident light. The translation is achieved using either a computer (Fig. 5.14 (a)) or a mechanical cam (Fig. 5.14 (b)).[27,28] A novel mechanism is shown in Fig. 5.14 (c), in which an L-shaped bar with two orthogonal sides rotates around the crossing point of the two sides. Each crystal slides along each side along rails fixed on the incident and the exiting light axes.[29] When crystals are sufficiently long, the sliding mechanism becomes unnecessary. The rocking curve (angular dependence of reflectance for monochromatic X-ray) for first order light is wider than that for higher order light (band width $\propto 1/n$). By exploiting this fact, one can greatly minimize the higher order light in the double crystal monochromator by shifting the angle of incidence slightly from the Bragg angle (the angle at peak reflectance). This practice is called 'detuning' and the intensity becomes slightly reduced. When the crystals are replaced with synthetic multilayers, one can obtain intense monochromatic light with large band width.[30]

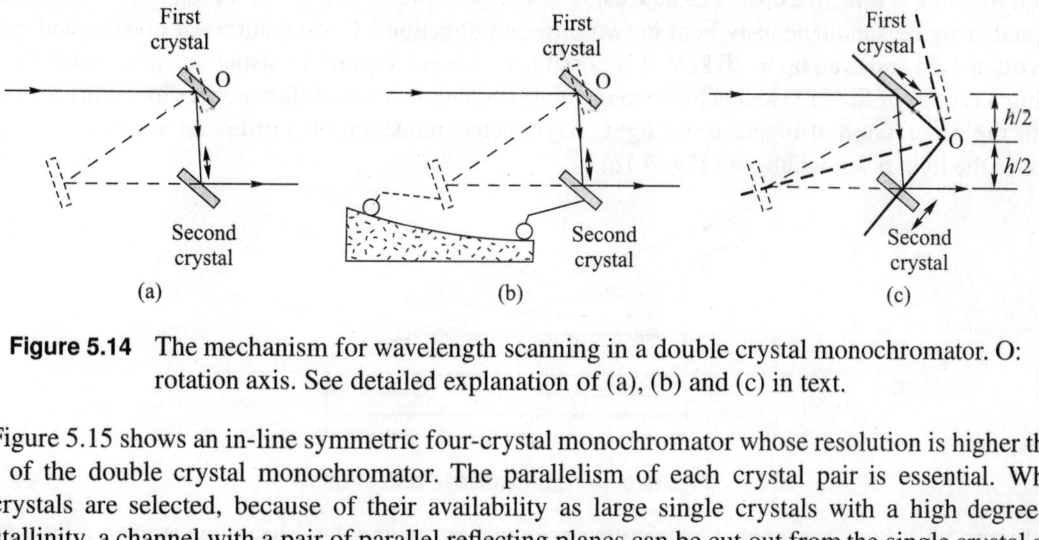

Figure 5.14 The mechanism for wavelength scanning in a double crystal monochromator. O: rotation axis. See detailed explanation of (a), (b) and (c) in text.

Figure 5.15 shows an in-line symmetric four-crystal monochromator whose resolution is higher than that of the double crystal monochromator. The parallelism of each crystal pair is essential. When Si crystals are selected, because of their availability as large single crystals with a high degree of crystallinity, a channel with a pair of parallel reflecting planes can be cut out from the single crystal and therefore the parallelism is automatically guaranteed. When such a 'channel-cut' crystal is incorporated in a double crystal monochromator, the output light axis will shift with wavelength scanning. However, when two channel-cut crystals are employed in the four-crystal monochromator, the exiting light axes remain fixed. That is, one channel-cut crystal replaces the first and the second crystals, and the second replaces the third and the fourth crystals. The channel-cut crystals are rotated in opposite directions by

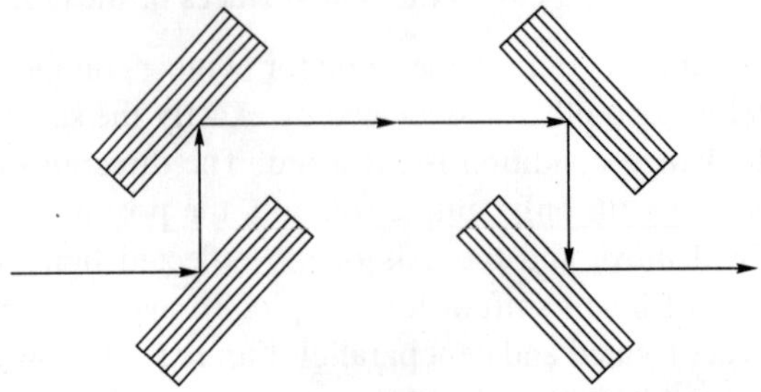

Figure 5.15 Schematic outline of an in-line symmetric four-crystal monochromator.

the same amount. Some monochromators with very high resolution have been developed as follow.[31)] The resolving power of any monochromatizing crystals is derived from differentiation of the Bragg formula and is given by $\lambda/\Delta\lambda = \tan\hat{\theta}/d\hat{\theta}$. The resolving power will become large, when the incident angle approaches the normal, that is $\hat{\theta} \approx 90°$ (back scattering) and the angular spread of the light becomes small. High resolution monochromators are devised to make angle of incidence large and use asymmetric reflection as given in Fig. 5.16. For example, the resolving power becomes 10^8 and the band width, 100 μeV.[32)] There are several ways to focus the light on the sample. One is to make use of post-mirrors which may be either toroidal mirrors or K-B mirrors. An alternative technique is to use shaped monochromatizing crystals.[33)] Sagittal focusing, which is focusing the light in the direction perpendicular to plane of incidence can be obtained simply by bending the crystals in the same direction. Such focusing is effective up to 115 keV using Si (311), Si (400), Si (511) or Si (620). In some cases the crystals may be simultaneously bent in two different directions. Combinations of crystals and mirrors are often used and even up to 30 keV, it is possible to achieve sagittal focusing in a horizontal direction using a crystal of Si (111) and achieve tangential focusing in vertical direction using a mirror together with the suppression of higher order light. Asymmetric reflection also offers an alternative means to reduce the light beam width (see Fig. 5.16).

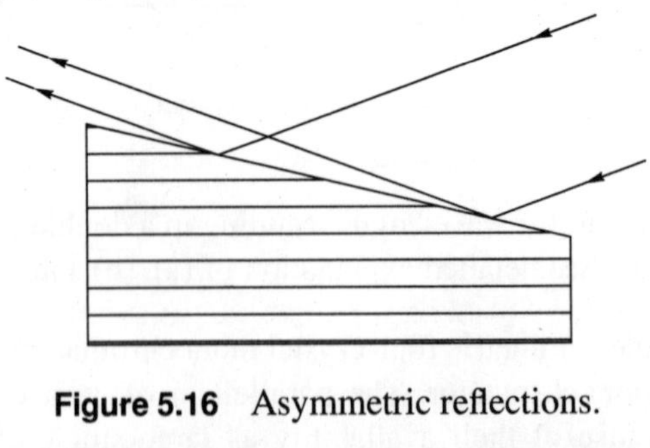

Figure 5.16 Asymmetric reflections.

Figure 5.17 shows the intensity spectrum obtained from an undulator beamline at SPring-8. A double crystal monochromator equipped with Si (111) was used. The light was detected using an ion chamber. The relative band width was 0.02%. The photon number was estimated from the current of the ion chamber to be 4 × 10^{12}/s at 10 keV. The beam size at the sample position was 1.5 mm (W) ×

0.6 mm (H), and was obtained without the use of any post-mirror. It must be noted that polarization characteristics of synchrotron radiation are not necessarily retained when monochromators are used.

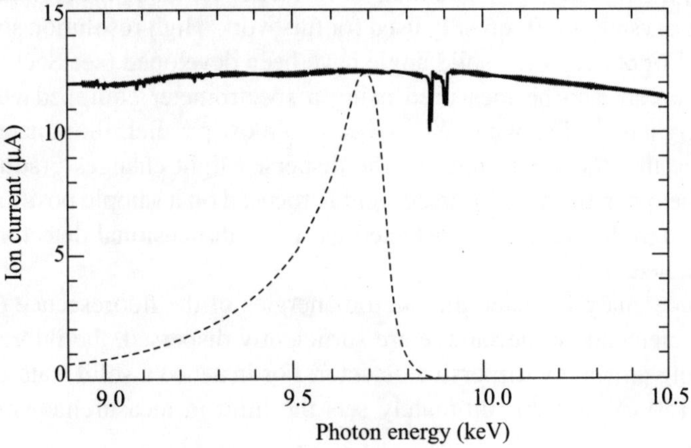

Figure 5.17 The output spectrum from a double crystal monochromator on the undulator beamline at SPring-8. The ordinate represents the ion current. The broken curve represents signals obtained when the undulator gap is fixed, while solid curve, those when the gap is changed so as to obtain maximum intensity with wavelength scanning. The maximum value for the ion current corresponds to a number of photons of 4×10^{12}/s (by courtesy of M. Suzuki).

c. Spectrometers: The spectrum of X-ray fluorescence or of inelastically scattered light from samples irradiated by light or any other excitation sources is possible to measure using spectrometers. Spectrometers may incorporate curved crystals to disperse the light emitted from a large solid angle.

Two types of curved crystal devices are illustrated in Fig. 5.18. The first is of Johann type (Fig. 5.18(a)) and the second is the Johansson type (Fig. 5.18 (b)). In each type, the crystal is bent

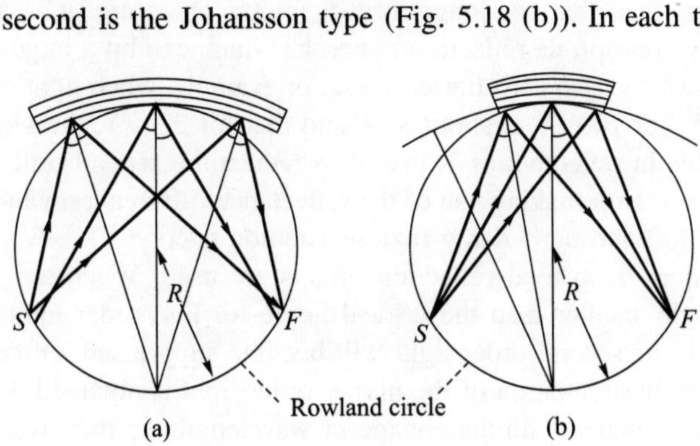

Figure 5.18 Bent crystal. (a) Johann type, (b) Johansson type. Here the radius of curvature in the bent crystal is defined as R according to the definition in ordinary optics, while in X-ray optics it is defined to be $2R$.

into a radius of R and the emitting point is situated on the Rowland circle, and the curved crystal and detector are moved along it. The difference between them is as follows. In the former, the crystal is only bent, but in the latter, the crystal is bent and then its surface is cut along the Rowland circle to reduce the aberrations. Graphite crystals are frequently used for this work. High resolution spectrometers gathering inelastically scattered light over wide solid angle have been developed (see Sect. **12.1**).

Absorption spectra can also be measured using a spectrometer equipped with a specially-curved monochromatizing crystal.[33)] The white X-ray beam, almost parallel, illuminates the crystal which is curved in such a way that the wavelength of the dispersed light changes gradually from one side of illuminated area to the other and the dispersed light is focused on a sample position. The white light and the light passing through the sample are detected by a one-dimensional detector. Then the absorption spectrum can be obtained.

In X-ray florescence analysis, when the photon energies of the fluorescence (characteristic) X-rays from the constituent elements in the sample are sufficiently dispersed, the fluorescence spectra can be obtained only by using an energy dispersive detector. For instance a solid state detector has an energy resolution of about 150 eV and this ultimately sets the limit in measurements of X-ray fluorescence spectra.

5.4 OPTICAL FILTERS

Light from monochromators may contain unwanted components such as higher order light and scattered light originating from the dispersive optical element and elsewhere. The quality of the light reflects the proportion of monochromatic light included in the exiting beam. In order to obtain highest beam quality, many new optical elements and also novel insertion devices have been made or proposed (see Sects. **2.4.2.b**, **3.2.6.b**). Simple filters are also used to minimize the intensity of any unwanted components.

There are two filtering methods. The first exploits the dependence of reflectivity on wavelength and angle of incidence. The second type makes use of the dependence of absorption coefficient on wavelength. As shown in Fig. 5.1 when the angle of incidence is fixed the reflectivity for shorter wavelength light is less than that for longer wavelengths. Therefore, it is quite possible to reduce higher order light using appropriate reflection materials which exhibit a large difference in reflectivity between at the desired wavelength and shorter ones. For example, when light of 100 eV is required, the reflectance of Au at 85° for 100 eV is about 85% and that for 200 eV, the second order light is about 40%, and this difference in reflectivities is useful to reduce the second order light component. Such filtering is used in VUV region making use of the reflectance differences, which do not originate from total reflection. In the soft and hard X-ray regions (photon energy >1 keV), the methods exploiting the reflectance differences from total reflection effects are used. When an X-ray is incident on the mirror at a grazing angle smaller than the critical angle for first order light but larger than that for second order light, then the second order light will become suppressed. Of course, by using multiple reflections, an even greater attenuation of the higher order light is obtained. It is particularly effective to change the angle of incidence with the change of wavelength, so that two or three mirrors may be used to suppress the higher order light, while at the same time, maintaining the exiting light direction and level constant. In the VUV region below 100 eV however, the grazing incidence reflectance does not show any strong dependence on the angle of incidence and it becomes preferable to work at near normal incidence where the reflectivity differs considerably from material to material. The variable deviation-angle monochromators already described so exploit this behavior.

In VUV region down to 105 nm, LiF is used for as a filter to eliminate shorter wavelength light. However in the wavelength region below 100 nm, it becomes necessary to use very thin film filters.[5] The absorption coefficient of In is small in the region from 100 to 80 nm in comparison with in adjacent regions, so that a thin In film can be used as a transmission filter in that region. Similarly, a film of Sn can be used in the region from 70 to 50 nm. For shorter wavelength regions differences between absorption coefficients above and below the core absorption edges are used for filtering. Figure 5.19 presents the energy dependence of the absorption coefficient close to the core absorption edge. In general, the absorption coefficient is small below the core absorption edge, while it is larger above the edge. Therefore, the large difference in absorption between at photon energies of E_1 and E_2 ($E_2 = 2E_1$) can be used to reduce the second order light component. A typical example is an Al thin film with an L edge around 70 eV. In the hard X-ray region, the absorption differences between above and below the absorption edges of many metals (for example, Cu K edge, 8.98 keV) can be used.[4] In the VUV and soft X-ray regions, since the absorption coefficients of materials are large, thin films with the thicknesses of about 100 nm are fabricated and used. In the hard X-ray region, since the absorption coefficient is relatively small, the thickness should be of the order of μm. In this case commercial metal films are often used. Of course, filtration using mirrors and transmission films results in the decrease in the intensities of not only the higher order light, but also the required light.

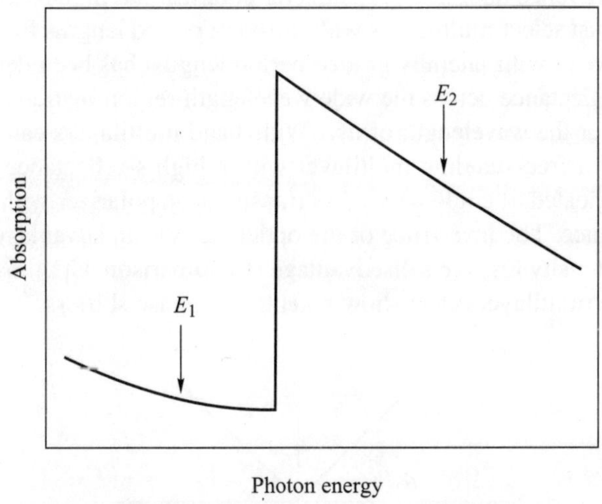

Figure 5.19 A schematic of an absorption spectrum around a core absorption edge.

5.5 POLARIZERS AND PHASE SHIFTERS

Polarizers are used to alter degree of polarization of the light output from monochromators and undertake polarization analysis and phase shifters convert linearly-polarized light to circularly-polarized light for instance.

In the VUV and soft X-ray regions, polarizers using grazing incidence mirrors and multilayers are often chosen. Both polarizers utilize the fact that p-reflectance is smaller than s-reflectance, discussed in Sect. **4.2**. The reflectance of grazing incidence mirrors is high but their degree of polarization is not so high, but one can readily obtain higher polarization by using three or four mirrors in series without a too serious loss of reflectance. Polarizers of this type can be used across a wide wavelength region of particularly relatively longer wavelengths. Figure 5.20 shows a three-mirror polarizer as an example.[34] The polarizers are designed so as to make the optical axis invariant before and after them.

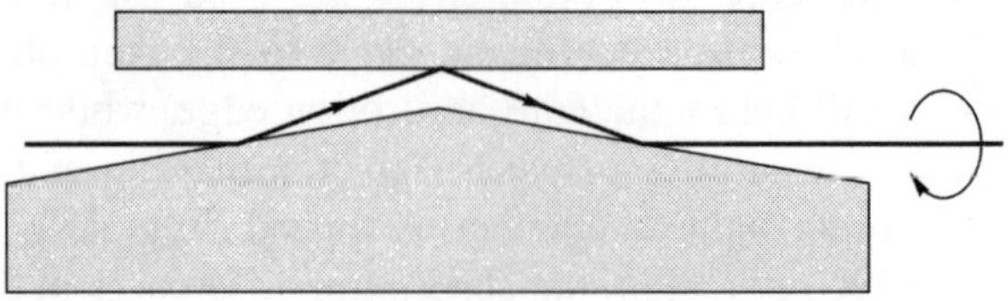

Figure 5.20 Polarizer consisting of three grazing incidence mirrors.

Figure 5.21 shows a cross sectional view of a multilayer polarizer. The angle of incidence is chosen as about 45° to maximize the degree of polarization. That is because the refractive indices of all materials are close to 1 in the VUV and in shorter wavelength regions, where the Brewster angle is about 45°. Because the reflectance of any material at angle of incidence around 45° is small, one can exploit multilayers to increase the reflectances.[35] A multilayer polarizer has high reflectance only across a finite band pass, so that one must select multilayers with different period lengths for use at other wavelengths. A novel multilayer polarizer with laterally graded period lengths has been developed. In this case, one can obtain the highest reflectance across the wide wavelength region by translation of the multilayer to maximize the reflection at the wavelength of use. Wide band multilayers can be also used over a wide wavelength region.[11] If a free-standing multilayer with a high s-reflectance at an angle of incidence around 45° could be fabricated, it could be used as transmission polarizer with an s-transmittance much less than its p-transmittance. The invariance of the optical axis is an advantage while the low degree of polarization and large intensity loss are a disadvantage in comparison with those of reflection multilayer polarizers. Transmission multilayers even show potential as phase shifters.[36]

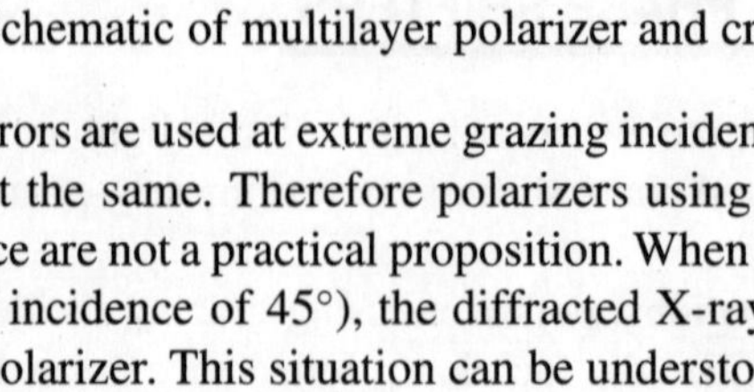

Figure 5.21 Schematic of multilayer polarizer and crystal polarizer.

In the hard X-ray region, mirrors are used at extreme grazing incidence to obtain high reflectance and s- and p-reflectances are almost the same. Therefore polarizers using mirrors in the geometry of Fig. 5.20 at extreme grazing incidence are not a practical proposition. When X-rays are diffracted by a crystal perpendicularly (at an angle of incidence of 45°), the diffracted X-rays are almost s-polarized, which means a crystal functions as a polarizer. This situation can be understood by considering the multilayer in Fig. 5.21 to behave as a crystal. In the hard X-ray region, p- and s-polarizations are sometimes called

π- and σ-polarizations, respectively. They differ from π-component of electric vector of synchrotron radiation normal to the median plane of electron beam and the σ-component parallel to the median plane.

In Sect. **4.1.3**, the real and imaginary parts of a response function are shown to be dependent in general. In reflection, the reflectance and the phase difference of light before and after the reflection have a relation based on causality. Therefore polarizers necessarily act simultaneously as phase shifters. In all these polarizers the phase shift between p- and s-polarized light is large at around angle of incidence where the degree of polarization is maximized. Using this property, some phase shifters have been used in the VUV and soft X-ray regions.[11,34] In the hard X-ray region, quarter wave plates with phase shift of $\pi/2$ are often employed,[37] so that the conversion of linearly-polarized to circularly-polarized light is easily performed.

5.6 DETECTORS

There are many detectors whose specifications are different in different wavelength regions and this includes spatial resolution, sensitivity, dynamic range (the range of signal intensity of the detector proportional to signal size which is often expressed by the ratio of maximum to minimum of the intensity in the appropriate region), response time, energy resolving ability, uniformity of sensitivity in detection area *etc.* In what follows, various detectors are briefly described from zero-dimensional detectors to two-dimensional ones listed from longer wavelengths to shorter ones, that is, in order of VUV, soft X-ray and hard X-ray wavelengths. The detailed descriptions of these detectors are largely left to other texts.[1–3,38]

5.6.1 Zero-dimensional detectors

a. Photoelectron multiplier: In the region from the ultraviolet to the longer wavelength region of the VUV, photoelectron multipliers with windows of quartz (>165 nm) or MgF_2 (>113 nm) are used. They have an advantage of high S/N ratio, because the work functions of their photocathodes are larger than those of ordinary photoelectron multipliers. Furthermore they have another advantage that all second order light is eliminated in the region from the cut-off wavelength of the windows to its double. However, one cannot use them below the cut-off wavelength.

b. Sodium salicylate used with a photoelectron multiplier: The photoelectron multipliers cannot be used in the VUV and soft X-ray regions below 113 nm due to the opacity of their window material. However they can be used in conjunction with a wavelength converter in which all the shorter wavelength absorbed light is re-emitted as fluorescence at a longer wavelength region, which can be readily measured using the photoelectron multiplier. The most often used fluorescent absorbing material is sodium salicylate which emits ultraviolet fluorescence with high conversion efficiency. It has the advantage that the quantum yield of the ultraviolet fluorescence is almost constant across the 200 to 50 nm region. It is often coated on to the front face of the photoelectron multiplier from a solution in ethyl alcohol, and the solution is spread on the glass surface and dried. It does not sublime in vacuum. However, it is not resistant to high temperatures, so that care has to be paid during baking of the chamber. The quantum yield decreases gradually with time, so that one needs to recoat the detector regularly. Since the sensitivity of electron multiplier decreases drastically in the region of wavelength longer than 100 nm and so this method particularly is valuable in this region.

c. Scintillator used with a photoelectron multiplier: In this case, X-rays irradiate the scintillators and the visible or ultraviolet light emitted from them is detected. NaI (Tl) scintillators (Tl-doped NaI single crystal) and plastic scintillators are popular wavelength converters.

d. The electron multiplier: Electron multipliers are similar to photoelectron multipliers except the fact that the former should be used under vacuum, because they have no window. The electron multipliers can be used in the VUV, soft X-ray and hard X-ray regions. Those using Be-Cu (photo) cathodes are the most common. In some cases, the Be-Cu cathodes are demountable, so that alternative cathode materials can be used to measure their total electron yield spectra (see Sect. **7.3**). It is also possible to improve the sensitivity of the cathodes high by coating them with materials of higher electron yield, such as CsI.

e. Channeltrons, ceratrons and microchannel plates: These devices are electron multipliers in which individual dynodes are replaced by a continuous electrode formed from an electrically conducting ceramic with high resistivity. The shape of the conducting ceramics is tubular. Secondary electrons are multiplied in the tube during their transit through it until they reach the terminal anode. The channeltron or ceratron are single pipes with diameter of a few mm, while the microchannel plate (MCP) is a bundle of narrow pipes with diameters of from a few μm to a few 10s of μm, whose lengths (the thickness of the plate) is a few mm. These detectors can detect not only light, but also electrons and ions. When the signal from each channel of a MCP is drawn out using resistance coupling of the anode, it can be used as two-dimensional detector. Furthermore it can also be used as a two-dimensional detector, when the accelerated secondary electrons are made collide with a fluorescent screen situated just behind the MCP. That is, the fluorescent screen emits the visible light which is then detected by a CCD.

f. SSD (Solid state detector): In typical SSD made from Si or Ge, p- and n-type crystals touch and form a P-N junction and a depletion layer will appear under inverse bias voltage. X-rays enter from the surface, reach at the depletion layer, and then excite electrons. Between the sides of the crystal, a voltage is applied, so that electrons generate electrons and holes successively with acceleration. Finally, they reach the electrodes (Au) on each side and are detected as the signal. This detector functions in such a manner that it can be used as an energy dispersive detector simply because the strength of the signal pulse for a single photon is proportional to the energy of light incident upon the detector. When a pulse height analyzer is employed, the energy of the light can be measured with band width of 150 eV. The detectors are usually cooled to liquid nitrogen temperatures.

g. Photodiodes: These devices are similar to those used in the visible region and they are not very effective for weak signals. For use in VUV and soft X-ray regions, the surface coatings of the diodes should be very thin. The output signal from such diodes is often calibrated using an ion chamber and the diode can then be used as a secondary standard in intensity monitoring. Usefully, a PIN photodiode has a very fast response time, so that it is used to observe ns and even ps phenomena.

h. Ion chambers: The ion chamber often used in the VUV and soft X-ray region is a 'double ion chamber', which enables measurements of an absolute number of photons. In this case, the optical path must be evacuated, so that low pressure gas (rare gas) is introduced. When light passes through the gas it generates ions and electrons, which are attracted to the electrodes resulting in the signal current. In the hard X-ray region, it is used at atmospheric pressure. Ion chambers are used at many hard X-ray beamlines as an intensity monitoring device, because monochromatized light will pass through this detector without serious loss of intensity. Its sensitivity is uniform across an area of detection. It is also used to detect the light passing through the sample. However, this detector is not so effective when the intensity of the light is low.

5.6.2 One-dimensional detectors

a. Position sensitive proportional counter: This type of detector exploits discharge in gases. It consists of an anode wire and a cathode and there are a delay line type and a charge division type. In the delay line type, the cathode is divided into many narrow strips being connected successively to part of a delay line. The X-ray will produce a charge to be induced at a cathode strip from which the signal propagates to each end of the delay line. From the time difference between the signals reaching the two ends, one can determine the position of the incident X-ray.

5.6.3 Two-dimensional detectors

The usage of the MCP as a two-dimensional detector is described in Sect. **5.6.1.e**, but there are many other two-dimensional detectors.

a. Photographic films and plates: These detectors have always been used as two-dimensional detectors and are still very useful. They can be used anywhere across in wide wavelength region from visible to hard X-rays. For use in the VUV and soft X-ray regions, the films or the plate are made so as to minimize the amount of gelatin in emulsion to suppress its absorption. The nuclear emulsion plate is specially processed to make its grain sizes small. Especially in experiments to determine light intensities precisely, their limitation is that the dynamic range of photographic density is not wide.

b. Imaging plates: Some ionic crystals doped with impurities of rare-earth atoms, such as BaFBr (Eu) show 'photo-stimulated luminescence'. The mechanism of this emission is as follows. When electrons in those crystals are excited by incident X-rays, they are trapped by the impurities and remain in metastable states until the electrons are re-excited. When they are re-excited, for example by using a He-Ne laser, they will revert again to the ground state and emit light called photo-stimulated luminescence. The lifetime of the metastable state is very long, such that the information following X-ray irradiation is retained or 'memorized' in the material. The imaging plate is a plastic film coated by powders of these materials. It is easily possible to obtain good quality two-dimensional images by detecting the photo-stimulated luminescence following scanning the plate using a He-Ne laser beam.

c. CCD (charge-coupled device): The CCD is a semiconductor device fabricated by lithographic technology, which has pixel size at a minimum value typically of about 5 μm × 5 μm. The grain size of any photographic film or plate is usually of the order of a few microns but CCDs with sub-micron resolution are now being developed.[39] CCDs are used most commonly in visible region. Any CCD used in X-ray region is of the back illumination type where the X-rays penetrate sufficiently deeply to generate the charge to be detected. These detectors have the advantage that the image can be observed in real time, which is not possible with photographic films, plates or with imaging plates. The dynamic range of an imaging plate is larger than that of a CCD.

d. Fluorescence screen with camera: There are several types of fluorescence screens coated by fluorescent materials such as ZnO and films of scintillating materials such as NaI (Tl) and Lu_2SiO_5:Tb.[39] The X-rays are converted to visible light and detected by various camera types, such as a visible CCD camera. Some can have a sub-micron spatial resolution and some exhibit high speed response.[40)]

5.7 TIME-RESOLVED EXPERIMENTS

Synchrotron radiation is emitted as a train of light pulses, where each pulse has a width of 0.01 ~ 1 ns and the pulse interval between them is from ns to the order of μs. This time structure makes a wide range time-resolved spectroscopic experiments possible. For relatively long times (second to minute) synchrotron radiation can be regarded as continuous in time and a light chopper would normally be used to produce pulsed light from the synchrotron radiation. In the first time-resolved experiments using synchrotron radiation, relaxation processes were observed by measuring the decay time of the luminescence emitted from samples excited by the intrinsically pulsed light from the storage ring (see Sect. **7.4**). More recently, many time-resolved experiments, such as in spectroscopy and X-ray diffraction, have been undertaken to understand the change of states as a function of time after application of various kinds of pulsed external fields on samples using synchrotron radiation pulses.

If the sample change is reversible after application of a pulsed field, we can stimulate with excitation pulses and investigate by probe light pulses repeatedly. The change in time of the response to the excitation can be obtained as a function of delay time between excitation and probe pulses. The response time of the detectors is then not so important. However, when the change is irreversible, all measurements must be made with a single excitation pulse and the response time of the detector must be sufficiently fast. In that case, 'single-shot' measurements are required whenever the excitation pulses are pulses of light, electric field, magnetic field, shock wave or strain *etc*. For the case of light pulses, samples are excited electronically or thermally. Time-resolved structure analysis using a pulsed laser is described in Sect. **9.6.4**.

In the 'pump-and-probe' method light pulses are used for the excitation (the pump), and the changes in optical transmittance or reflectance are measured using the probe light. Figure 5.22 illustrates outlines of a pump-and-probe method in two cases using, respectively, so-called DC probe light whose intensity does not change in time and repetitively pulsed probe light. To measure single-shot phenomena, DC probe light and fast optical detectors are essential as shown in Fig. 5.22(a). In the case of reversible phenomena, the probe pulses are applied repeatedly at a certain delay time, Δt, following the pulsed excitation and the intensities of the probe pulses from the sample, such as the transmitted pulses are then accumulated, yielding the average value for a single pulse. After scanning the delay time, the average values are plotted against the delay time and one can then derive the time dependence curve of the transmittance. The intensities of the transmitted pulses are accumulated for each delay time and therefore the response time of the detectors does not need to be fast (see Sect. **13.3.1**). When such information at a single wavelength is insufficient and information across a wide wavelength region is required then a light source with a continuum spectrum such as synchrotron radiation is essential.

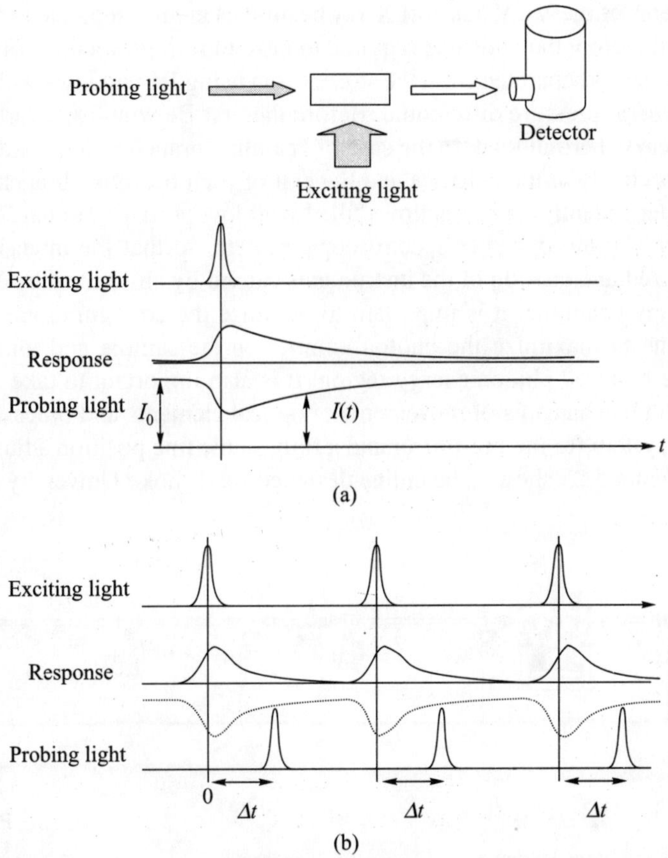

Figure 5.22 Schematic diagram of pump-and-probe system (for transient absorption spectroscopy). (a) Probed by DC light, (b) probed by pulsed light synchronous with a repetitive excitation pulse.

5.8 BEAMLINES

A schematic of beamlines used in conjunction with a storage ring is already shown in Fig. 1.1. The synchrotron radiation is extracted from bending sections or insertion devices and the beamline includes front end, pre-mirror, monochromator, post-mirror and measurement apparatus. The front end consists of the following components, a gate valve to separate the vacuum between the storage ring vacuum chamber (doughnut) and the beamline, an absorber (beam shutter) which protects the gate valve, which is normally closed but is opened when synchrotron radiation is to be used, a radiation shutter (beam stopper) to absorb high energy radiation during injection periods and a gate valve to separate the vacuum between the front end and the downstream experimental beamline. This is essential to protect the storage ring ultra-high vacuum from any accident in the beamline. In order to avoid overheating of optical elements from irradiation by synchrotron radiation, aperture masks are always installed.

The vacuum systems of the VUV and soft X-ray beamlines are not separated from that of the storage ring using windows therefore they are also required to have ultra-high vacuum. However the vacuum in the hard X-ray beamlines is separated from the storage ring using Be windows with sufficient strength to withstand an atmospheric pressure differential. Before the first Be window, a carbon film is sometimes used to absorb the heavy thermal load. In the case of beamline branches, a second and third Be window may be situated respectively at the entrance and the exit of each beamline branch, so that the branches can be evacuated independently and sometimes filled with low pressure He gas. Usually after the third Be window, the X-ray is introduced to spectroscopic system, so that the intensity loss due to the air absorption is minimized and erosion of the instruments caused by air ionized by X-ray is avoided.

In the design of any beamline, it is important to optimize the configuration of all optical elements in the entire beamline to maximize the photon number on the sample and minimize the size of the light spot, across the required photon energy region. It is also important to take account of the size of vacuum chambers and mechanisms of movements of optical elements and other devices and to prepare an effective cooling system for the pre-mirror and a fully stable fine position adjuster for the pre-mirror and other mirrors. Figure 5.23 shows a beamline designed for Tohoku University as an example.

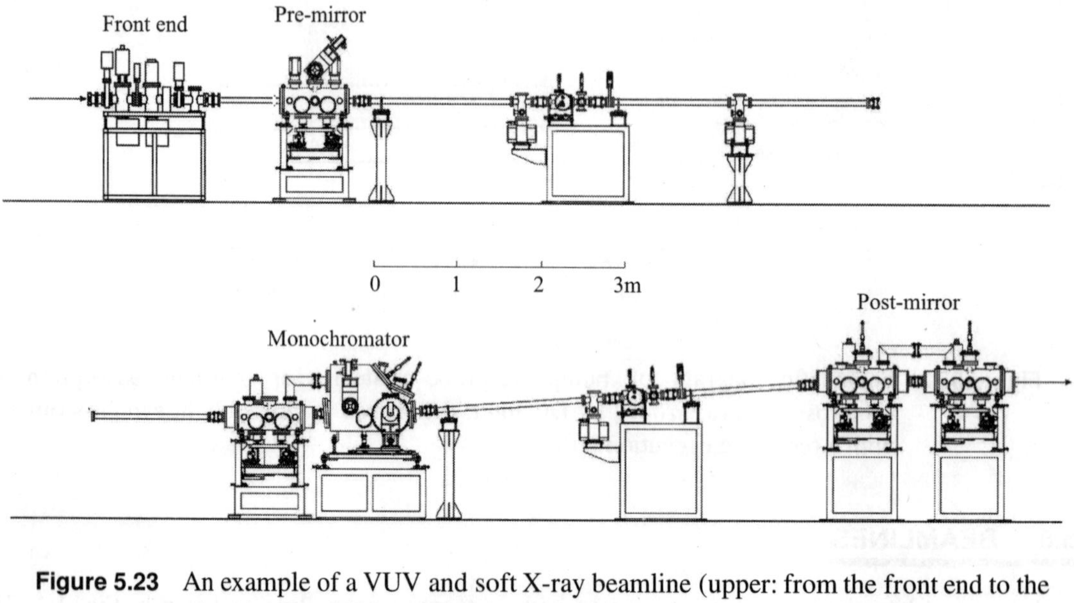

Figure 5.23 An example of a VUV and soft X-ray beamline (upper: from the front end to the pre-mirror and lower: from the monochromator to the post-mirror).

REFERENCES

1. E. Spiller: "Soft X-Ray Optics" (SPIE Press, 1994).
2. J. A. Samson and D. L. Ederer eds.: "Vacuum Ultraviolet Spectroscopy" I, II (Academic Press, 1998).

3. A. Erko, M. Ider, T. Krist and A. G. Michette eds: 'Modern Developments in X-Ray and Neutron Optics" (Springer-Verlag, 2008).

4. B. L. Henke, P. Lee, T. J. Tanaka, R. L. Shimabukuro and B. K. Fujikawa: Atomic Data and Nuclear Data **27** (1982) 1, http://www-cxro.lbl.gov/optical_constants/asf.html.

5. M. Watanabe and G. Isoyama : "Dynamics during Spectroscopic Transition", E. Lippert and J. D. Macomber eds. (Springer-Verlag, 1995) Chap. 13.

6. J. A. Samson: "Techniques of Vacuum Ultraviolet Spectroscopy" (John Wiley & Sons, 1967).

7. For example, T. W. Barbee, Jr.: Opt. Eng. **25** (1986) 989.

8. M. Yamamoto and T. Namioka : Appl. Opt. **31** (1992) 1622.

9. X-ray Optics Center, Lawrence Berkeley Laboratory: http://henke.lbl.gov/multilayer/survey .html.

10. A. Erko, M. Veldkamp,W. Gudat, N. V. Abrosimov, S. N. Rossolenko, V. Shekhtman, S. Khasanov, V. Alex, S. Groth, W. Schröder, B. Vidal and A. Yakshin : J. Synch. Rad. **5** (1998) 239.

11. Z. Wang, H. Wang, J. Zhu, F. Wang, Z. Gu, L. Chen, A. G. Michette, A. K. Powell, S. J. Pfauntsch and F. Schäfers: J. Appl. Phys. **99** (2006) 056108.

12. P. Kirkpatric and A. V. Baez : J. Opt. Soc. Am. **38** (1948) 766.

13. S. Sato, A. Iijima, S. Takeda, M. Yanagihara, T. Miyahara, A. Yagishita, T. Koide and H. Maezawa : Rev. Sci. Instr. **60** (1989) 1479.

14. For example, L. Zhang, R. Hastache, O. Hignette, E. Ziegler and A. Freund: J. Synch. Rad. **5** (1998) 804.

15. Y. Suzuki: private communication.

16. T. Koide, T. Shidara, M. Yanagihara and S. Sato: Appl. Opt. **27** (1988) 4305.

17. M. Koike, M. Ishino, T. Imazono, K. Sano, H. Sasai, M. Hatayama, H. Takenaka, P. A. Heimann and E. M. Gullikson: Spectrochimica Acta B **64** (2009) 756.

18. J. B. West and H. A. Padmore: "Handbook on Synchrotron Radiation" Vol. 2, G. V. Marr ed. (North-Holland, 1987) Chap. 2.

19. F. C. Brown and S. L. Hulbert: Nucl. Instr. Meth. **222** (1984) 42.

20. E. Ishiguro, M. Suzui, J. Yamazaki, E. Nakamura, K. Sakai, O. Matsudo, N. Mizutani, K. Fukui and M. Watanabe: Rev. Sci. Instr. **60** (1989) 2105.

21. C. T. Chen and F. Sette: Rev. Sci. Instr. **60** (1989) 1616.

22. H. Petersen, C. Jung, C. Hellwig, W. B. Peatman and W. Gudat: Rev. Sci. Instr. **66** (1995) 1.

23. R. Follath, F. Senf and W. Gudat: J. Synch. Rad. **5** (1998) 769.

24. M. C. Hettrick, J. H. Underwood, P. J. Batson and M. J. Eckart: Appl. Opt. **27** (1988) 200.

25. M. Koike and T. Namioka: Rev. Sci. Instr. **66** (1995) 2144

26. M. Itou, T. Harada and T. Kita: Appl. Opt. **28** (1989) 146.

27. M. Lemonnier, O. Collet, C. Depautex, J-M. Esteva and D. Raoux: Nucl. Instr. Meth. **152** (1978) 109.

28. T. Matsushita, T. Ishikawa and H. Oyanagi: Nucl. Instr. Meth. A **246** (1986) 377.

29. T. Murata, T. Matsukawa, M. Mori, M. Obashi, S. Naoe, H. Terauchi, Y. Nishihata, O. Matsudo and J. Yamazaki: J. de Phys. **47** (1986) C8, 135.

30. Y. Wang, S. Narayanan, J. Liu, D. Shu, A. Mashayekhi, J. Qian and J. Wang: J. Synch. Rad. **14** (2007) 138.

31. Yu. Shvyd'ko: "X-Ray Optics – High-Energy-Resolution" (Springer, 2004).

32. M. Yabashi: SPring-8 Research Frontiers, 2000/2001, p. 91.

33. T. Matsushita and H. Hashizume: "Handbook on Synchrotron Radiation" Vol. 1, E. E. Koch ed. (North-Holland, 1983) Chap.4.

34. T. Koide, T. Shidara, M. Yuri, N. Kandaka, H. Fukutani and K. Yamaguchi: Rev. Sci. Instr. **63** (1992) 1458.

35. T. Hatano, Y. Kondo, K. Saito T. Ejima, M. Watanabe and M. Takahashi: Surf. Rev. Lett. **9** (2002) 587.

36. J. B. Kortright and J. H. Underwood: Nucl. Instr. Meth. A **291** (1990) 272.

37. K. Hirano, K. Izumi, T. Ishikawa, S. Annaka and S. Kikuta: Jpn. J. Appl. Phys. **30** (1991) L407.

38. G. F. Knoll: "Radiation Detection and Measurement", 4th ed. (John Wiley & Sons, 2010).

39. P.-A. Douissard, A. Cecilia, T. Martin, V. Chevalier, M. Couchaud, T. Baumbach, K. Dupré, M. Kühbacher and A. Rack: J. Synch. Rad. **17** (2010) 571.

40. C. Ponchut, J. M. Rigal, J. Clément, E. Papillon, A. Homs and S. Petitdemange: J. Instrumentation **6** (2011) C01069.

CHAPTER 6

The Spectroscopy of Gases in the Vacuum Ultraviolet and Soft X-Ray Regions

In this chapter, spectroscopic studies on gases (isolated atoms and molecules) in vacuum ultraviolet and soft X-ray regions are introduced. Pioneering work in this field in 1963 obtained the absorption spectrum of the He atom (see Sect. **1.3.1.a**) and revealed for the first time, the Fano effect due to two-electron excitation processes.[1)] Since then many studies have been made in the valence electron excitation region, now augmented by measurements in the shallow-core-electron excitation region.[2–4)] Precise measurements of K(1s) absorption for simple molecules containing a few light elements have been undertaken using high resolution monochromators.[5,6)] What follows is basic research in atomic and molecular science reviewed using experimental results from photoabsorption, photoelectron and photoion studies on rare gases and for some simple molecules using high resolution core excitation. See also Ref. 29 in Chap. **1** and Ref. 7 in this Chapter.

6.1 ATOMIC SPECTROSCOPY

6.1.1 One electron transition spectra

In 1885, Balmer observed emission lines in the visible spectrum of the H atom and noted regularity in their intervals. Lyman also observed three emission lines in the vacuum ultraviolet region in the early stage of 20th century. These series of lines are called the Balmer series and the Lyman series, respectively. The photon energy of the emission line is given as

$$\hbar\omega_n = R\left(\frac{1}{m^2} - \frac{1}{n^2}\right), \tag{6.1}$$

where the Balmer series is represented by $m = 2$ and the Lyman series by $m = 1$ and R is the Rydberg constant equal to 13.6058 eV. R/n^2 corresponds to the binding energy of an electron in the orbital with the principal quantum number n (see App. **F**). A similar series, the Rydberg series, which may be analyzed by a formula like Eq. (6.1) is seen not only in the valence electron absorption spectra but also in the core electron absorption spectra. As an example, the Ne^+ production cross section for 1s electron excitation is shown in Fig. 6.1, which is equivalent to the Ne 1s absorption cross section.[8)] It was obtained using molecular beam apparatus employing a differential pumping system (see Sect. **5.1**) and with a band width below 70 meV using the soft X-ray beamline at SPring-8. The electron configuration of the ground state for the Ne atom is $1s^2 2s^2 2p^6$. Each peak visible in the figure corresponds to excitation of a 1s core electron to one of the np orbitals. The corresponding 1snp excited state energy is given by

$$E_n = I - \frac{R}{(n-\mu)^2}, \tag{6.2}$$

where I is the ionization energy of the 1s electron and $R/(n-\mu)^2$ is the binding energy of the np electron. Here, μ is called the quantum defect. This is the correction term for the hydrogen model derived from many-body effect which occurs when an electron is close to the ion. μ is almost constant and independent of n. The line width of the peak corresponds to the lifetime width of 1snp excited states and is almost constant and independent of n. It is described in the following manner. Following the creation of the 1s hole by photoabsorption, two of the eight electrons in the 2s and 2p orbitals participate in the Auger decay of the 1s hole state, whereas the excited np electron behaves as a spectator during the Auger decay. As a result, the 1s core hole lifetime is independent of the principal number n of the excited np electron. In Fig. 6.1 the peak intensity decreases as n increases. This is explained by the fact that the probability of the 1s → np transition is approximately proportional to and $1/(n-\mu)^3$ hence decreases as n increases.[9] A detailed explanation of peak profiles is given later.

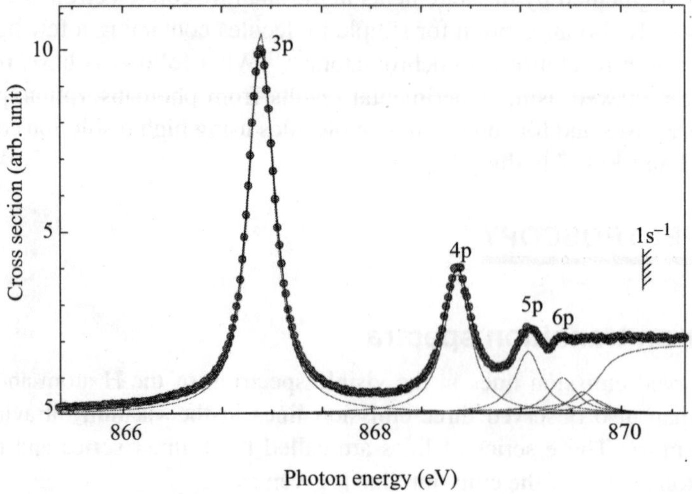

Figure 6.1 1s inner-shell absorption cross section of the Ne atom.[8] The solid line is a fitting curve from the theory. Thin solid lines reveal components.

6.1.2 Two-electron excitation and autoionization

The ground state of the neutral He atom is $1s^2\,{}^1S_0$. Photoabsorption mostly leads to the excitation of electrons to the discrete $1sn\mathrm{p}\,{}^1P_1$ and continuum $1s\varepsilon\mathrm{p}\,{}^1P_1$ states. However, owing to the interaction between two electrons (an electron correlation effect) both two 1s electrons in the ground state can be excited to two unoccupied states, either the discrete doubly excited state 2snp or the $2\mathrm{p}n\mathrm{s}\,{}^1P_1$. As these doubly excited states have higher energy than the threshold energy of ionization of the 1s electron, one electron in the doubly excited state is emitted outside the atom. In other words, the doubly excited state is automatically ionized to the continuum state which is composed of the ion of the 1s $^2S_{1/2}$ ground state and an electron with positive energy.

This phenomenon is called 'autoionization' and the process is essentially the same as the Auger decay following core electron excitation as described in the previous section. Autoionization is the result of electron correlations, or 'configuration interaction' between the doubly excited state and the singly ionized continuum state. Of course the He atom can also be directly ionized to the singly ionized continuum state $1s\varepsilon p\,^1P_1$ by absorption of a photon with the same energy as the autoionization resonance energy. Because these two processes, namely, resonant photoionization and direct photoionization, have common initial and final states they interfere quantum-mechanically with one another. As a result, asymmetric absorption peaks appear in the region of doubly excited states and these are shown in Fig. 6.2. This feature is called the Fano effect and the peak profile of the absorption cross section σ is expressed by

$$\sigma = \sigma_a \frac{(\varepsilon + q)^2}{\varepsilon^2 + 1}. \tag{6.3}$$

$\varepsilon = (\omega - \omega')/(\Gamma/2)$ is normalized and relates the difference between the photon energy $\hbar\omega$ and the resonance energy $\hbar\omega'$ to the half width of the resonance $\Gamma/2$. σ_a represents the absorption cross section of the continuum state coupling to the discrete state and q is a profile index defined by Fano (App. **G**).

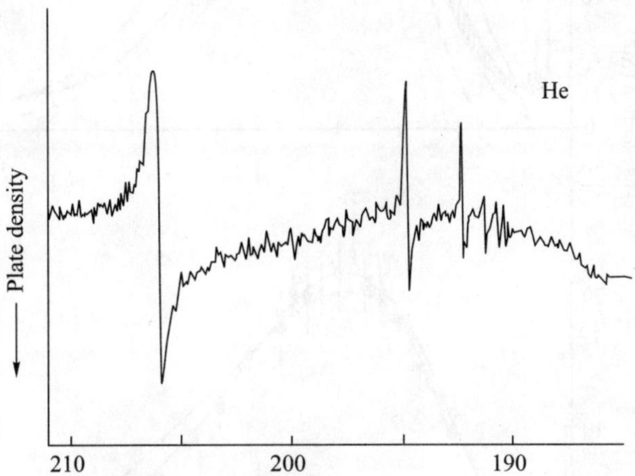

Figure 6.2 The absorption spectrum of the two-electron excitation in the He atom.

One might expect the Fano effect to show in the core electron excitation spectrum because the one-electron core excited states are also subject to autoionization, or Auger decay and could thus interfere with the continuum states. However, as shown in the previous section, peak profiles are symmetric in the 1s excitation spectrum of Ne. This symmetric feature arises from cancellations of interference effects among Auger decay channels to many different continuum states.[8)]

6.1.3 Giant resonances

The electron configuration of the ground state of Xe is $1s^22s^22p^63s^23d^{10}4s^24p^64d^{10}5s^25p^6\ ^1S_0$. We discuss the 4d core electron ionization of this atom. Figure 6.3 shows measured 4d partial ionization cross sections and calculated ones.[10)] In the lower part of the figure the cross sections measured by different groups are plotted using different marks. In the upper part the hatched area represents the fluctuation of the measured cross sections given in the lower part. σ_{abs} is the total absorption cross section. The binding energy of the 4d electron in Xe is 68.3 eV. The 4d photoionization cross section rises from this threshold energy to attain a maximum value around 100 eV. This maximum of the cross section is called a 'giant resonance' or 'shape resonance'.

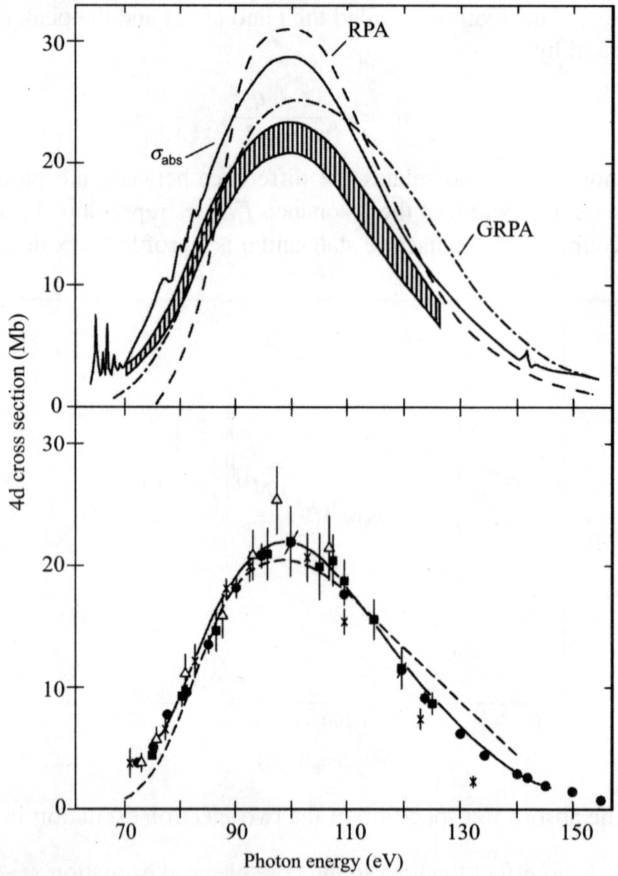

Figure 6.3 The 4d ionization cross section in a Xe atom (by courtesy of V. Schmidt, © IOP Publishing 1992).[10)] The details are in the text.

In general, any electron ionized from the atom may interact with the Coulomb potential $-e^2/4\pi\varepsilon_0 r$, when it is far away from the remaining ion and any electron near the nucleus, $-Ze^2/4\pi\varepsilon_0 r$. If the electron is in between however, it experiences the potential $V(r)$ which is different from the Coulomb potential. Every atom shows variations of $V(r)$. According to the independent particle approximation, the wave function of an electron is determined by solving the wave equation with an effective potential, which is the sum of the atomic potential $V(r)$ and the centrifugal potential $\hbar^2 l(l+1)/2mr^2$ (see App. F).[9] Sometimes a potential barrier appears in the effective potential owing to the contribution of the centrifugal potential. Figure 6.4 shows a schematic view of the effective potential with this potential barrier. When the electron energy is low, the εf (l = 4) continuum wave function cannot penetrate into the ion core but when the electron energy is close to the barrier height, the εf wave function has a large amplitude in the region of the 4d orbital. Then, the 4d → εf photoionization cross section is resonantly amplified. This resonant structure feature is the above-mentioned giant resonance. With further increase in the energy, the εf wave function oscillates sinusoidally. Then cancellations among the positive and negative contributions of the transition matrix element take place, resulting in a decrease in the photoionization cross section. When the 4d photoionization cross section is calculated on the basis of the independent particle approximation, a shape resonance appears, although the calculated result is a few times larger than experimental data. We cannot include it in Fig. 6.3, because it is out of scale. Moreover the resonant width is too narrow to be reproduced clearly. Such discrepancies were of long-standing before they could be resolved theoretically.

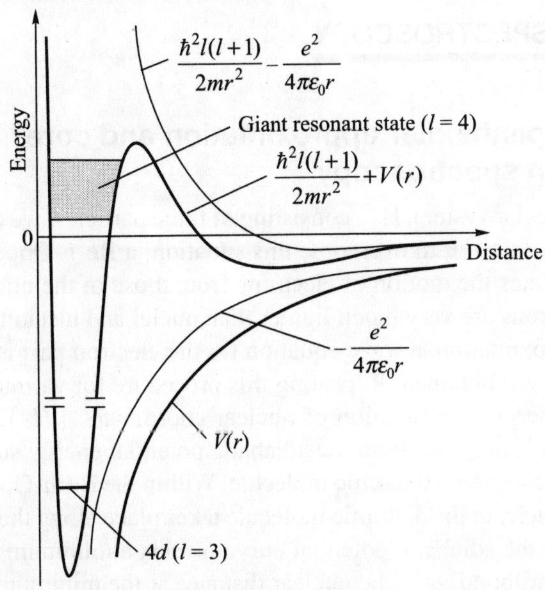

Figure 6.4 Schematic view of the effective potential in an atom.

To eliminate the discrepancies, electron correlations were theoretically taken into consideration. The most important among them is that of polarization of an ion caused by a generated core hole. The broken line in the upper part of Fig. 6.3 shows a result calculated by the RPA (random phase approximation), which is the self-consistent field calculation combined with the polarization effect. The theoretical curve is presented in the same figure showing the experimental data. The next significant electron correlation effect arises from the differences between electron orbitals of neutral atoms and ions with a 4d hole. GRPA is RPA theory renormalized to deal with this correlation. The result calculated by GRPA is also shown by the one-dotted chain-line in the upper part of Fig. 6.3. The GRPA result is closer to the experimental result than RPA, but some differences are still observed in the higher energy region.

To achieve even better agreement with the experimental results it was necessary to take into account the coupling between the 4d and other orbitals. A broken line in the lower part of Fig. 6.3 shows the result obtained by taking this coupling (4d + 5s + 5p) into account. Since the intensity transfers to the ionization of other orbitals through this coupling, the 4d cross section decreases. The resulting absolute value almost coincides with the actual measurements. However some slight differences between the profiles remain. To overcome this problem the inelastic scattering effect originating from the excitation of the 5s or 5p electron by the 4d photoelectron is introduced. The solid line in the lower part of Fig. 6.3 gives the final result. From the figure we see that the experimental results can be almost perfectly reproduced once all the above theoretical considerations are fully taken into account.

It was in 1964 that the first giant resonance was observed in photographic measurements.[11] However, precise measurements were not feasible until the late 1980s, using photoelectron spectroscopy.[12,13] In 1990, the theoretical debate was finally resolved.[14] Thus, it took 26 years to reach a quantitative understanding of the giant resonance from the date of its first observation.

6.2 MOLECULAR SPECTROSCOPY

6.2.1 The Born-Oppenheimer approximation and core photoelectron spectroscopy

Even for the simplest molecular system, H_2^+, consisting of three particles, we cannot solve Schrödinger's wave equation analytically. In order to overcome this situation, a Born-Oppenheimer approximation is used (App. **H**) which separates the motion of electrons from those of the nuclei. This approximation is based on the fact that electrons are very much lighter than nuclei and instantly respond to movement of the nuclei. Under this approximation, a wave equation for the electron part is solved for a fixed nuclear position and the total energy is obtained. Repeating this procedure for various configurations of nuclei, one can obtain the total energy as a function of nuclear coordinates $\{\boldsymbol{R}_k\}$. When the total energy is drawn as a function of $\{\boldsymbol{R}_k\}$, one can attain the adiabatic potential energy surface.

For simplification let us consider a diatomic molecule. Within the Born-Oppenheimer approximation, the relative motion of the nuclei in the diatomic molecule takes place along the adiabatic potential energy curve (see Fig. 6.6). When the adiabatic potential curve reveals a minimum, it is called 'bonding' and for no minimum, called 'anti-bonding'. The nuclear distance at the minimum energy is the equilibrium nuclear distance R_e. In the case of bonding, discrete vibrational levels appear in the potential well. The maximum and minimum internuclear distances of the vibrations are given by the outer and inner intersections of the potential curve with the vibrational level, respectively. The dissociation energy is defined to be the difference between the energy of the infinitely separated atoms (nuclei) and the minimum energy.

Let us now consider the photoabsorption of diatomic molecules. Because the electronic transition occurs instantaneously we can assume that the internuclear distance is unchanged before and after the transition. This assumption is called Franck-Condon principle. The energy separation of the vibrational components in the spectrum provides the coefficient of the quadratic term of the potential curve. Usually, in the case of the transitions from the ground state (both electronic and vibrational) to the electronically-excited state with some vibrational components, their relative intensities will depend on the square of overlapping of the vibrational wave functions of the initial and final states. That is, the relative intensities are approximately described by the Franck-Condon factors. Furthermore, when the adiabatic potential curves are approximated only by the quadratic term, the intensity distribution of vibrational components follows a Poison distribution (App. **H**).

Taking the CO molecule as an example, we briefly explain the concepts of molecular orbitals. When the C and O atoms are remote from each other, the electron configuration of each atom is $1s^2 2s^2 2p^2$ and $1s^2 2s^2 2p^4$, respectively. When the two atoms approach, the 2s and 2p atomic orbitals configure the molecular orbitals surrounding the two atoms, although the 1s atomic orbitals are scarcely influenced. These molecular orbitals are calculated under the assumption of self-consistent field approximations, similar to the case of atoms and are defined as one-electron orbital functions. In CO, molecular orbitals composed of atomic 2s and $2p_z$ orbitals are symmetrical on the rotation around z-axis defined to be the molecular axis. In other words, these are the σ orbitals having no nodal plane that includes the molecular axis. In contrast, $2p_x$ and $2p_y$ states form molecular orbitals having one nodal plane that includes the molecular axis. They are called π orbitals. The electronic configuration of the CO molecule is represented as $1\sigma^2\ 2\sigma^2\ 3\sigma^2\ 4\sigma^2\ 1\pi^4\ 5\sigma^2$, where 1σ and 2σ are composed of O 1s and C 1s core orbitals, respectively. 3σ and 4σ are mainly composed of O 2s and C 2s, respectively. 5σ is mainly composed of O $2p_z$ and C $2p_z$, and 1π, O $2p_{x,y}$ and C $2p_{x,y}$. In addition, the highest energy occupied orbital 5σ of the ground state is called HOMO (highest occupied molecular orbital) and the lowest energy unoccupied orbital 2π, LUMO (lowest unoccupied molecular orbital).

Molecular photoelectron spectroscopy can be viewed as an experimental method to determine the potential energy curves of the molecular ion which is formed by removing one electron from one of the molecular orbitals. For electron energy analyzer, see Sect. **7.3.2.b**. In photoelectron spectroscopy, monochromatic light with photon energy of $\hbar\omega$ irradiates the sample. When the kinetic energy of the emitted photoelectron is $E_{\rm kin}$, the binding (potential) energy of the ionized state, $E_{\rm b} = \hbar\omega - E_{\rm kin}$ (of which meaning is different from that in solids used in Chap. **7**) is obtained. Figure 6.5 shows O 1s photoelectron spectrum for CO.[15] In this spectrum, the molecular vibrations of the O 1s ionized state

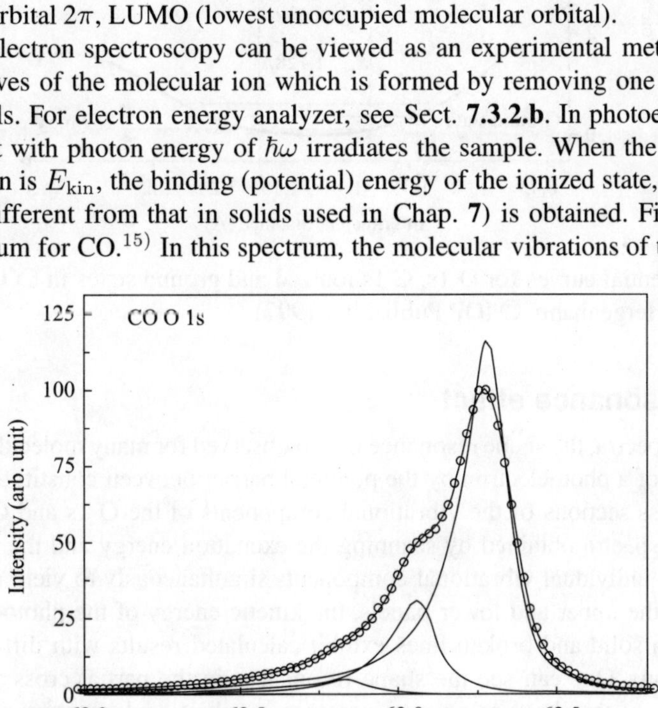

Figure 6.5 The O 1s inner shell photoelectron spectrum in CO.[15]

are observed. The width of each vibrational component comes from the short lifetime (about 4 fs) of the O 1s hole state. From the analysis of vibrational energy, the shape of the potential curve is derived. Intensity ratios among the vibrational components are analyzed according to the Poisson distribution and the differences of the equilibrium nuclear distances between initial and final states are acquired.[16)] Figure 6.6 shows potential curves of the O 1s and C 1s ionized states obtained from such analyses[17)] together with the ground state potential which is well known from investigations in the infrared region. The equilibrium nuclear distance of the O 1s ionized state increases by 0.039 Å from the ground state, while that of the C 1s ionized state decreases by 0.049 Å.

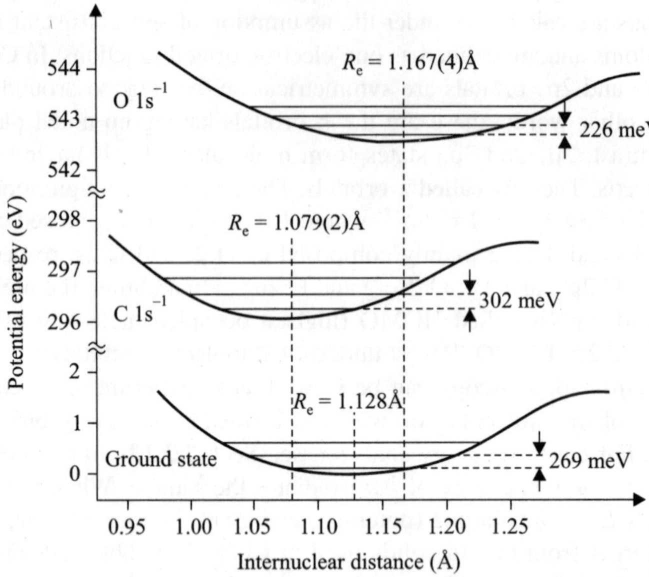

Figure 6.6 Potential curves for O 1s, C 1s ionized and ground states in CO (by courtesy of U. Hergenhahn, © IOP Publishing 1997).[17)]

6.2.2 Shape resonance effect

As is seen in atomic spectra, the shape resonance is also observed for many molecules, which arises from the temporal capture of a photoelectron by the potential barrier between constituting atoms. Figure 6.7 shows the partial cross sections of the vibrational components of the O 1s and C 1s ionized states in CO.[15)] They are the spectra obtained by scanning the excitation energy and the kinetic energy of the photoelectrons of the individual vibrational components simultaneously to yield a 'constant final state (CFS)' spectrum. In the upper and lower panels, the kinetic energy of the photoelectrons is indicated on the abscissa. Both solid and broken lines exhibit calculated results with different approximations (see the figure caption). One can see the shape resonances in the partial cross sections. In the O 1s ionization, the energy of the shape resonance increases as vibrational quantum number of the excited state, v' increases. On the other hand, in the case of C 1s ionization, the energy decreases as v' increases, as shown in the figure. Let us consider the reason why different vibrational components labeled by v'

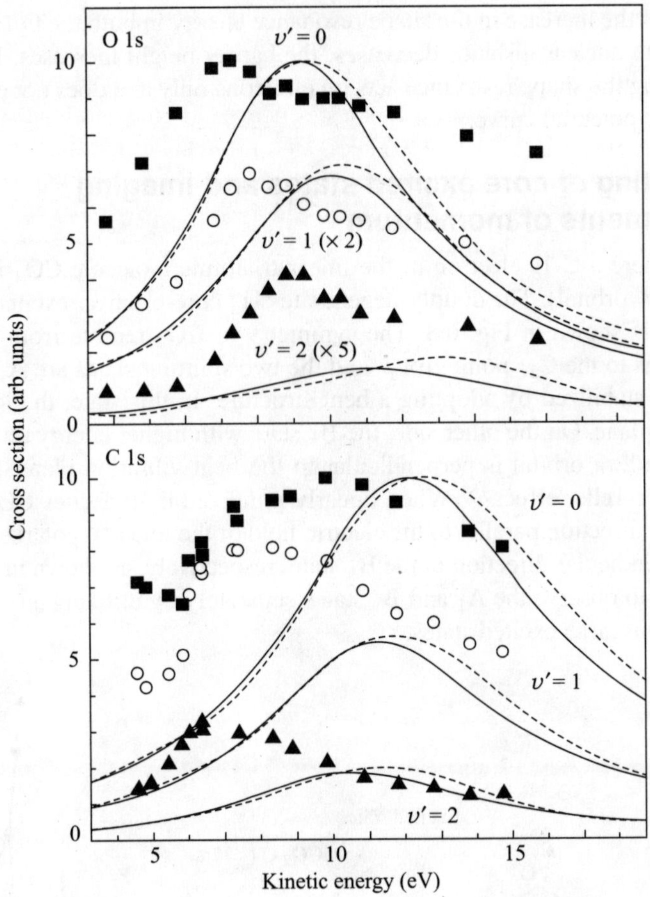

Figure 6.7 The partial cross section for the ionized O 1s and C 1s vibration states ($v' = 0, 1, 2$).[15] ■, ○ and ▲ show experimental results. Solid and broken lines indicate calculated results by taking into consideration the dependence of the transition dipole moment on nuclear distance and the dependence on effective nuclear distance estimated from vibration state v', respectively.

affect the increase or the decrease of the resonant energy. For $v, v' = 0$, the probability distribution of the internuclear distance follows a Gaussian distribution around the equilibrium internuclear distance. For $v, v' = 1, 2, 3, ...$, the internuclear distance probability is distributed around the inner and outer walls of the potential well. The transition intensity depends on the magnitude of the overlap between initial and final vibrational wave functions. With this in mind and by examining the potential curves in Fig. 6.6, we consider the following situation. In the O 1s ionization, the transition to $v' = 0$ occurs at an internuclear distance of C–O which is slightly shorter than its equilibrium distance. With increasing v', the transitions will occur effectively at shorter nuclear distances. In the case of C 1s ionization, the process proceeds inversely, suggesting that the resonance energy depends strongly on the internuclear distance. In other words, the intensity distributions among the vibrational components of the photoelectron spectra in the shape resonance region cannot be explained by the Frank-Condon factors. Therefore, the decrease in the nuclear distance gives the increase in the shape resonance energy in both the O 1s and C 1s ionizations. This means that as the nuclear distance decreases, the barrier height increases. It is also clear that the potential barrier giving the shape resonance acts on electrons only and does not explicitly appear in the shape of the adiabatic potential curve.

6.2.3 The splitting of core excited states and imaging measurements of momentum

Consider the case where a C 1s electron of the linear triatomic molecule CO_2 in $D_{\infty h}$ point group is excited to LUMO ($2\pi_u$ orbital). The doubly-degenerate $^1\Pi_u$ core-electron-excited state is split into two states. The situation is shown in Fig. 6.8. The symmetry of CO_2 remote from the origin of the bent coordinate Q_2 belongs to the C_{2v} point group, and the two splitting states are A_1 and B_1. The A_1 state with lower energy is stabilized by adopting a bent structure. In this state, the lobe of the $2\pi_u$ orbital lies in the molecular plane. On the other side, the B_1 state with higher energy retains a linear structure. Then, the lobe of the $2\pi_u$ orbital is perpendicular to the bent-vibration plane. The splitting of these states is called Renner-Teller effect.[18] When linearly-polarized light excites CO a centered C atom is displaced towards the direction parallel to the electric field of the linearly-polarized light in the A_1 state and towards the perpendicular direction in the B_1 state, respectively, as shown in Fig. 6.8. We will now describe experiments to observe the A_1 and B_1 states separately by utilizing anisotropic displacements of the central C atom in these excited states.

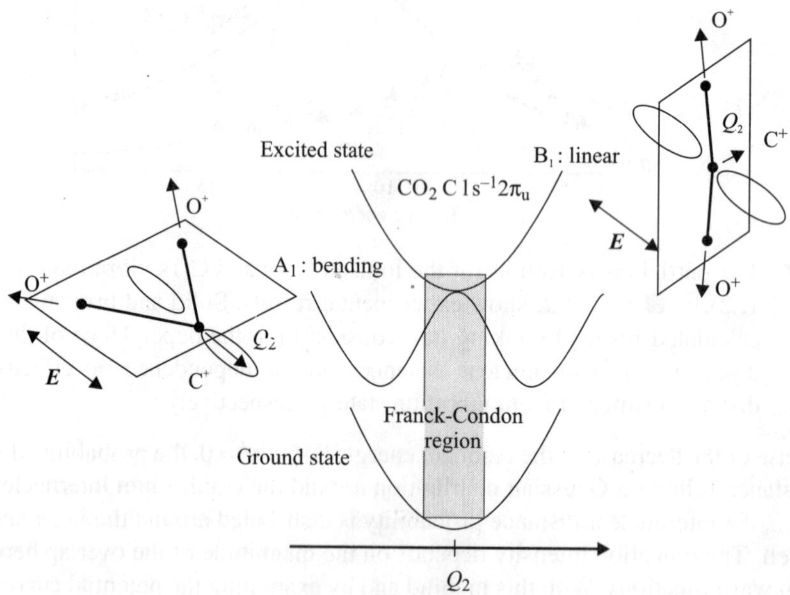

Figure 6.8 A conceptual figure for the Renner-Teller state splitting from the $1s^{-1}2\pi_u\ {}^1\Pi_u$ state in CO_2 and $2\pi_u$ orbitals with bent planes for the A_1 and B_1 states.

The triply ionized mother ion $CO_2{}^{3+}$, produced by the Auger process from the core excited state, bursts due to strong Coulomb repulsion, and instantly dissociates into three atomic ions, C^+, O^+ and

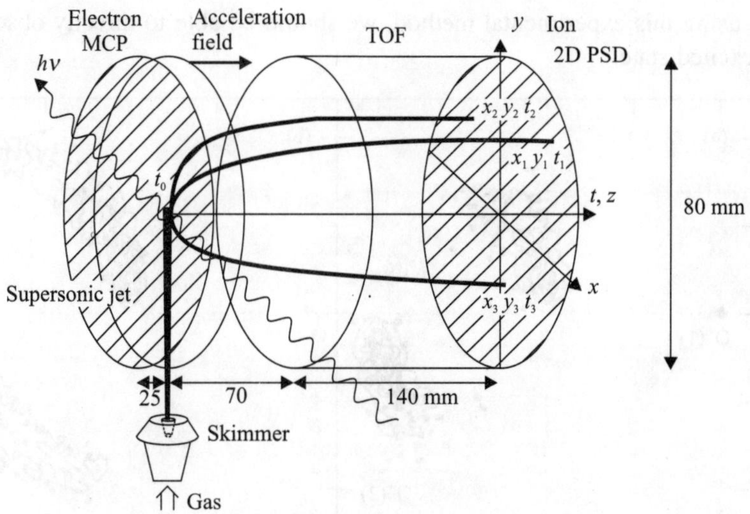

Figure 6.9 An outline of the experimental apparatus for the momentum-imaging measurement. MCP: microchannel plate, TOF: time-of-flight tube, PSD: position sensitive detector.

O^+. The momenta of these three ions reflect the shape and spatial arrangement of the mother ion at the time at which the Coulomb explosion occurs. These phenomena were observed by the apparatus schematically shown in Fig. 6.9 using linearly-polarized light. After each ion is accelerated and has moved by some distance, its position and time of flight, namely (x, y, t) are detected. The values of (x, y, t) of three kinds of ions seem to distribute randomly on the detector plane at first glance. However one can find sets of three fragment ions, C^+, O^+ and O^+ correlated in time by coincidence measurements and also determine the positions of these ions. Then, (x, y, t) obtained for each ion is converted to (p_x, p_y, p_z). In examining the momentum correlations of these three ions it can be checked that all three ions obey momentum conservation laws. Also to be taken into account is whether the direction of the displacement of the C atom is parallel or perpendicular to the electric vector of the incident light and whether the events originating from the A_1 and B_1 states are separated. Figure 6.10 shows the measured momentum correlation among the three ions. Here, for each event, we normalize the magnitude of the momentum of the early arrival O^+ (1) to be unity, define its direction to be x direction and identify the coordinates as (–1, 0). Then, the momentum of the C^+ ion is plotted on the upper half of xy-plane in which its components perpendicular to x direction is chosen to be on the positive y direction. Consequently that of the later arrival O^+ (2) is plotted on lower half of the plane. The intensity distribution is given by contour lines. Figures 6.10(a) and 6.10(b) correspond to the molecular break-up originating from the B_1 and A_1 states, respectively.[19)] One can see that the break-up originating from the B_1 state results in a small contour island of C^+ as shown in Fig. 6.10(a), while the break-up from the A_1 state yields a contour island of with a large-elongated shape as shown in Fig. 6.10(b). This illustrates that the C^+ ion originating from the A_1 state emitted towards the perpendicular direction of molecular axis gains a momentum larger than that from the B_1 state. These experimental results imply that $CO_2{}^{3+}$ is stabilized by taking a linear structure in the B_1 state while a bent structure in the A_1 state. In addition, the C^+ ion originating from the linear B_1 state gets also a small momentum to the perpendicular direction of the molecular axis. This is due to zero-point bending oscillation of the molecule. By using this experimental method, we should be able to directly observe the shape of molecules in the excited state.

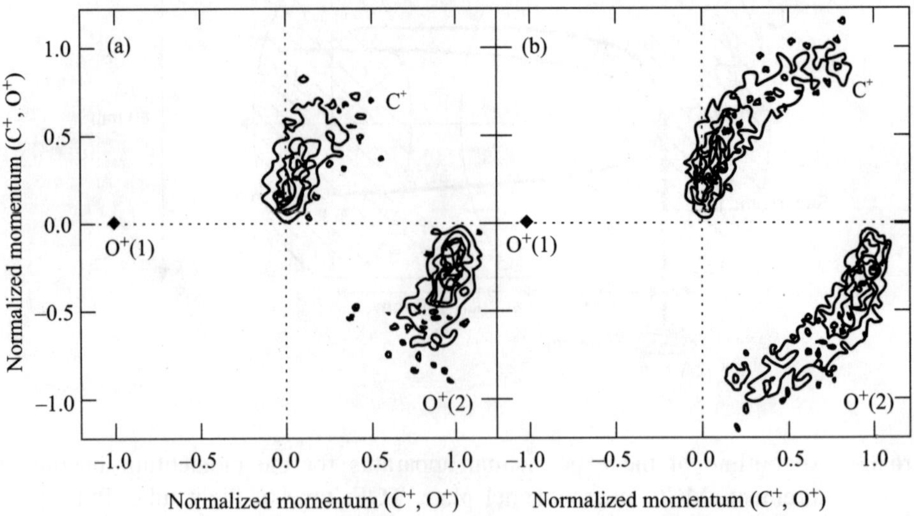

Figure 6.10 Momentum correlation between three ions C^+, O^+ and O^+ produced by the break-up of triply charged parent ion $CO_2{}^{3+}$. (a) Break-up from the B_1 state (linear), (b) break-up from the A_1 state (bending).

Studies on interatomic or intermolecular electronic relaxation processes in the loosely bonded clusters, called interatomic or intermolecular Coulombic decay, in which Coulombic explosion takes place after the electronic decay, have been made intensively using coincidence measurements of photoions and also of emitted electrons.[20]

References

1. R. P. Madden and K. Codling: Phys. Rev. Lett. **10** (1963) 516.
2. U. Becker and D. A. Shirley eds.: "VUV and Soft X-Ray Photoionization" (Plenum Press, 1996).
3. V. Schmidt: "Electron Spectroscopy of Atoms using Synchrotron Radiation" (Cambridge University Press, 1997).
4. Y. Hatano: "Chemical Applications of Synchrotron Radiation" Part I, T.-K. Sham ed. (World Scientific, 2002) Chap. 2.
5. N. Kosugi: "Chemical Applications of Synchrotron Radiation" Part I, T.-K. Sham ed. (World Scientific, 2002) Chap. 5.
6. K. Ueda: J. Phys. B: At. Mol. Opt. Phys. **36** (2003) R1.
7. M. N. Piancastelli, M. Simon and K. Ueda: J. Electr. Spectr. Rel. Phenom. **181** (2010) 98.

8. A. De Fanis, N. Saito, H. Yoshida, Y. Senba, Y. Tamenori, H. Ohashi, H. Tanaka and K. Ueda: Phys. Rev. Lett. **89** (2002) 243001.
9. U. Fano and J. W. Cooper: Rev. Mod. Phys. **40** (1968) 441.
10. V. Schmidt: "Photoionization of atoms using synchrotron radiation", Rep. Prog. Phys. **55** (1992) 1483.
11. D. L. Ederer: Phys. Rev. Lett. **13** (1964) 760.
12. B. Kämmering, H. Kossmann and V. Schmidt: J. Phys. B: At. Mol. Opt. Phys. **22** (1989) 841.
13. U. Becker, D. Szostak, H. G. Kerkhoff, M. Kupsch, B. Langer, R. Wehlitz, A. Yagishita and T. Hayaishi: Phys. Rev. A **39** (1989) 3902.
14. M. Ya Amusia, L. V. Chernysheva, G. F. Gribakin and K. L. Tsemekhman: J. Phys. B: At. Mol. Opt. Phys. **23** (1990) 393.
15. D. A. Mistrov, A. De Fanis, M. Kitajima, M. Hoshino, H. Shindo, Y. Tamenori, H. Tanaka, A. A. Pavlychev and K. Ueda: Phys. Rev. A **68** (2003) 022508.
16. L. S. Cederbaum and W. Domcke: J. Chem. Phys. **64** (1976) 603.
17. B. Kempgens, K. Maier, A. Kivimaki, H. M. Köppe, M. Neeb, M. N. Piancastelli, U. Hergenhahn and A. M. Bradshaw: "Vibrational excitation in C 1s and O 1s photoionization of CO", J. Phys. B: At. Mol. Opt. Phys. **30** (1997) L741.
18. J. Adachi, N. Kosugi, E. Shigemasa and A. Yagishita: J. Chem. Phys. **107** (1997) 4919.
19. Y. Muramatsu, K. Ueda, N. Saito, H. Chiba, M. Lavoliēc, A. Czasch, T. Weber, O. Jagutzki, H. Schmidt-Böcking, R. Moshammer, U. Becker, K. Kubozuka and I. Koyano: Phys. Rev. Lett. **88** (2002) 133002.
20. U. Hergenhahn: J. Electr. Spectr. Rel. Phenom. **184** (2011) 78.

CHAPTER 7

The Spectroscopy of Solids in the Vacuum Ultraviolet and Soft X-Ray Regions

Spectroscopic studies of solids and solid surfaces using synchrotron radiation began with simple absorption spectroscopy in the VUV and soft X-ray regions,[1] but now photoelectron spectroscopy (PES) is routinely used as the technique of choice.[2,3] The measurement of optical spectra is important not only to characterize materials, but also to elucidate their electronic structures, and sometimes to test theoretical predictions of electronic structures. Such studies embrace not only fundamental research but also many practical applications linked to the study of electronic and magnetic devices. When any new material is discovered, such as graphene,[4,5] for example, its electronic structure and physical properties have to be totally understood before its commercial potential can be fully realized and effectively exploited. This chapter provides an overview of the electronic structure of crystals such as metals, semiconductors, ionic crystals, organic crystals and strongly correlated electronic systems. This is followed by a description and explanation of optical transitions in such crystals. The techniques both for absorption and photoelectron spectroscopies are fully explained, and finally studies on various types of materials, including mesoscopic systems and solid surfaces are introduced, which also include some results using laboratory light sources. For the most recent developments, see Ref. 29 in Chap. **1**.

7.1 THE ELECTRONIC STRUCTURE OF CRYSTALS

Every atom has discrete energy levels below the ionization level. When atoms are condensed the electron wave functions which belong to each atom will overlap with each other. As a result of this overlapping the discrete energy levels broaden, resulting in an energy band. The distribution of electrons within each energy band is not uniform. The number of states accessible for occupation by electrons per unit volume and unit energy is called the 'density of states (DOS)'. The schematic energy diagram for a crystal is shown in Fig. 7.1, where the energy of each band is shown on the ordinate and the density of states is shown on the abscissa. The figure gives the energy of an electron in the average field originating from the charges on nuclei and from other electrons. This one-electron approximation is called 'band theory'. In semiconductors and insulators, there is a band gap between the valence band, which is normally filled with electrons and the unoccupied conduction band. In normal metals a valence electron moves freely and is not bound to the original atom, although the ions of the original atoms remain fixed at each lattice point. 'Metallic bonding' is created by cooperation between these ions and the free electrons. The electrons do not fill the valence band completely but only up to an energy which is called 'Fermi level' E_F. When an electric field is applied, these electrons conduct the current. In addition, there exist core levels with binding energies larger than those of valence bands. However, since a core state does

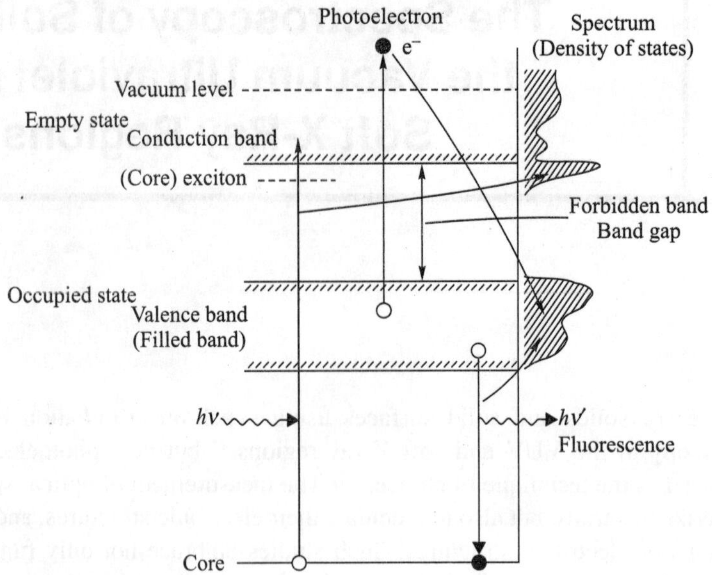

Figure 7.1 A schematic view of the electronic structure of a solid (semiconductor or insulator) and associated optical processes.

not take part in the bonding, that is the core wave functions of the atom do not overlap with those of other atoms, the width of energy level is not broadened. Even if the core level is described by the same principal quantum number, its energy shows a quite different value which depends on the particular element. This fact was discovered by Moseley who first proposed the dependence of characteristic X-ray energy on the atomic number.

The wave number and the wave number space (called the Brillouin zone) are defined by the translational and rotational symmetries of the atomic arrangements in crystals. For each point in wave number space the energy is determined and then the relationship between $\boldsymbol{k}$ and E can be established. This relationship is called the 'band dispersion'. In many cases, the band dispersion, sometimes called the 'band structure', is drawn along directions of symmetry in the crystal, sometimes with the density of states. The density of states corresponds to the number of states which have the same energy in $\boldsymbol{k}$-space. Band theory provides a good explanation for electronic structure and for the optical properties of semiconductors and ionic crystals, although the effects of excitons (a bound pair of an electron and a hole) are prominent in the latter. Band theory is also applicable to metals including alkali metals, alkali earth metals and noble metals. In many molecular crystals the molecules are bound to each other by van der Waals forces and the specific characteristics of the molecules are retained. Accordingly the absorption spectra may show the characteristics of the individual molecules and in this case, the spectra are interpreted in terms of the molecular orbital theory described in Chap. **6**.

The spectral profiles for the density of states specific to the crystal dimensions are as follow (see Fig. 7.34 and App. **I**). In the case of a three-dimensional crystal, when the free electron approximation holds and E can be expressed by a quadratic function of $\boldsymbol{k}$ where E is chosen to be zero at $\boldsymbol{k} = 0$, the density of states $J(E)$ is proportional to $\sqrt{E}$. When the crystal is regarded as a system in two-dimensions (a quantum well), in one-dimension (a quantum wire) and of zero-dimension (a quantum dot), then the density of states will show steps, asymmetric peaks and sharp lines, respectively. These features characterize nano-structured materials which are described later.

There are also many 'strongly correlated materials', in which Coulomb repulsion among the valence electrons (correlation) has a very large influence on the electronic states and whose electronic structure is difficult to explain using band theory. When such correlations are strong, then these electrons have a tendency to become stabilized at each lattice point. In many cases the constituent atoms of the materials involved have d or f valence electrons (*e.g.* in the case of transition elements or rare-earth elements). These electrons have a large angular momentum, being originally localized in nature originating from centrifugal potentials (see Sect. **6.1.3**). When the number of valence electrons is equal to that of sites, up-spin and down-spin electrons are configured to be antiferromagnetic. When the correlation is weak, it is possible for both up-spin and down-spin electrons to exist at the same lattice point. An insulating stabilized material due to localization is called a 'Mott insulator'. Even though correlation may be strong, the material will not necessarily be an insulator. There are many metallic strongly correlated materials which include f electrons and which exhibit the so-called 'Kondo effect'. This effect is manifest as an anomaly in the electric resistivity which exhibits a minimum in its temperature dependence, originating from the scattering of conduction electrons by localized electron spins. In those cases where the localized spin originates from the f electrons, hybridization between the wave functions of the f and conduction electrons near to the Fermi level results in 'valence fluctuation'. In the photoelectron spectra of such valence fluctuation materials, one can derive experimental evidence showing that the materials have more than two valencies, which then results in the mixed valence state. When the f electrons act in the role of the conduction electrons, the effective mass of the conduction electrons may increase by up to a thousand times and become much heavier than in normal metals. These materials are called 'heavy electron systems'. In this case however, band theory is not the most appropriate model. To calculate electronic structures for strongly correlated materials, the Anderson or Hubbard models are applied. Since the photoelectron spectra of strongly correlated insulators do not reflect a one electron state, but rather a correlated state between the excited electron and the other electrons, we cannot simply compare the photoelectron spectrum with the results of the band calculation.[6)] The band model alone is not able to explain superconductivity but it does however provide a useful starting point.

Since there are no atoms above a clean solid surface, the atomic bonds at the surface must differ from those within the bulk material, yielding completely different characteristics. Clean surfaces are usually chemically very active and atoms or molecules are frequently adsorbed on them. Sometimes chemical reactions can be induced at the surface generating new compounds. The study of such clean and adsorbed surfaces is now of vital concern in materials science research.

The basic processes associated with optical transitions in solids and their methods of measurement are now discussed. The spectra associated with many types of solids and surfaces will be introduced and their electronic structures, described mainly using the one-electron model.

7.2 OPTICAL TRANSITIONS IN CRYSTALS

a. Intraband transitions: When free electrons in a metal absorb light, in terms of the band model an electron will make a transition from a lower occupied state below the Fermi level to an upper unoccupied state within the same energy band. Absorption is very strong in the infrared region, but diminishes toward the visible region (see Sect. **4.2.2**). In this case, optical transitions are allowed within the uncertainties of energy and momentum originating from the uncertainty of τ in Eq. (4.66). Interband transitions will begin in the high energy regions of visible light as described below. The behavior in the infrared region is considered separately in Chap. **13**.

b. Interband transition: Absorption by valence electrons will occur whenever the sample is exposed to the light with energy higher than the band gap energy. The transition probability is proportional to $\omega J_{mn}(\hbar\omega)\ |r_{mn}|^2$. The shape of the measured spectrum will reflect the joint density of states $J_{mn}(\hbar\omega)$ between the valence and conduction bands. Since the energy of the core level has no wave number dependence, we can deduce only the density of states in the conduction band. This 'partial density of states' is determined by the selection rules for angular momentum for those transitions from the core level to the conduction band, where the angular momentum of the core level is clearly defined. The transition occurs at the same $\boldsymbol{k}$ point in Brilluion zone, because the momentum of light is negligible for those wavelength regions greater than soft X-rays.

c. Excitonic absorption: The above processes are considered in terms of the one-electron approximation. In semiconductors or ionic crystals a new bound state called an 'exciton state' may be generated by the Coulombic attraction between an excited electron and the positive hole left behind. In many semiconductors the excitonic absorption is weaker than that from absorption due to the interband transition, which occurs between the infrared, visible and ultraviolet regions. In ionic crystals, exciton absorption is of the same order of magnitude or often greater than that due to the interband transition. In both materials, exciton absorption is observed at the low energy edge of the interband transition. As the result of the Coulombic attractive force between the electron-hole pair, atoms between the pair are polarized and the Coulombic attraction can be expressed by $V(r) = -e^2/4\pi\varepsilon_0\varepsilon r$. In this case the dielectric constant ε is not as that for the static case but applies at the frequency at which the electron rotates around the hole. When the interband transition is allowed and direct, a series of absorption spectra similar to those for a hydrogen atom is observed. The separations of spectral lines are however very narrow due to the effect of dielectric constant mentioned above. The exciton will ionize to become a free electron and a hole when the energy of excitation light is larger than the band gap energy. This excitonic effect also appears in core absorption. A valence band exciton can move about because the valence positive hole can move, but a core exciton cannot move but remains fixed because the core hole cannot move.

d. Photoelectron emission: When an electron is excited to a state above the vacuum level by light absorption, the electron can escape from the crystal and is then called a 'photoelectron'. For measurements in the core energy region, Auger electrons, also associated with core excitation, may be measured. Even when the energy of the exciting light is constant, the energy of all photoelectrons escaping from the crystal will depend on the energy of their initial state in the optical transition. In photoelectron spectroscopy, the number of photoelectrons is counted for each kinetic energy E_{kin}. The electron energy in the initial state is defined to be the binding energy E_{b}, and E_{b} can be determined when the equation $E_{\text{b}} = \hbar\omega - E_{\text{kin}}$ applies. (*n.b.* In many cases, E_{b} is chosen to be positive.) In the technique of angle-integrated photoelectron spectroscopy, the total density of states for valence bands can be obtained. By angle-resolved photoelectron spectroscopy, the relationship between E_{b} and $\boldsymbol{k}$, that is, the band dispersion can be experimentally determined along symmetry directions in the crystal. Photoelectron diffraction experiments are also feasible using similar techniques.

e. Light emission: For those cases in which electrons absorb light and remain in an excited state for a short time, they will subsequently relax into the ground state by the emission of light or by exchanging energy with other electrons as well as to the crystal lattice. In these cases the emitted light is called fluorescence or luminescence. The transition by which the excited electrons change their state instantaneously is known as 'scattering' (see Sect. **4.3.3**).[7] When any ionic crystal is stimulated with ultraviolet light, the difference between the photon energy of the excitation light and the fluorescence is large and is absorbed *via* crystal lattice distortions. For core excitations, the electron in the valence band is transferred to the core hole vacancy and the fluorescence subsequently emitted. The excitation photon energy differs from the photon energy of the fluorescence. The core fluorescence spectrum gives a measure of the partial density of states in the valence band. In cases where the width of the fluorescence spectrum is comparable to its central energy, the intensity must be modified using the factor ω^3 to obtain the true density of states. This can be understood by the quantum-mechanical formulation of the optical transition Eq. (4.104). The classical explanation is as follows. The total emitted power all over solid angles from an oscillating electron is given by the formula for Larmor radiation Eq. (2.57). The motion of the oscillator is expressed by $x = x_0 e^{-i\omega t}$. After calculation of $\ddot{x}$ from this equation, the time-averaged energy of the oscillator $\overline{P}$ is given by $\overline{P} \propto \omega^4 x_0^2$. Since the emitted photon number I is given by $I = \overline{P}/\hbar\omega$, I is proportional to $\omega^3 x_0^2$. In the case in which the photon energy $\hbar\omega$ is not a single value, but is distributed over the width of the valence band, and $I(E)$ is given per unit energy, we can derive $I(E) \propto \omega^3 J(E) |\boldsymbol{r}_{mn}|^2$, where x_0^2 is replaced by $|\boldsymbol{r}_{mn}|^2$ for the quantum mechanical expression.

7.3 SPECTROSCOPIC TECHNIQUES

7.3.1 Absorption and emission spectroscopy

In order to measure the fundamental optical properties of solids, we can study their absorption spectra using monochromators. The absorption coefficients of materials in the VUV and soft X-ray regions are often very large and of the order of from 10^4 cm^{-1} to 10^6 cm^{-1}. It is therefore essential to be able to prepare quite thin films with thicknesses of from a few tens to a few hundred nanometers. In some cases, samples can be produced either by evaporation or sputtering and therefore appropriate transparent substrates for use in the VUV and soft X-ray regions with sufficient strength such as Si_3N_4 membranes, polymer films are needed. Samples are often damaged or destroyed during evaporation. In these, or in cases where we wish to obtain the spectra of single crystals, the near normal incidence reflectance spectrum can be measured, followed by a Kramers-Krönig analysis (Sect. **4.1.3**) in order to obtain the absorption spectrum. In the soft X-ray region such measurements may be extremely difficult, because the normal incidence reflectance is so low. However there is a convenient alternative method called the 'total electron yield (TEY) spectroscopy'.[8] In general when any sample is irradiated by light, photoelectrons, secondary electrons and Auger electrons are emitted from its surface. When the sample is isolated electrically from the measurement chamber and is connected to the measurement chamber *via* a current meter, then the current generated by these emitted electrons can be measured and the TEY spectrum subsequently derived from the measurement of this current. The TEY spectrum where the intensity is normalized for the incident light intensity (by using, for example, the quantum efficiency spectrum of Au) has a spectral profile quite similar to that of the absorption spectrum.[8] This

TEY method can also be exploited in the hard X-ray region. When the absorption coefficient is very large however we must be aware of possible sources of distortion to the spectral profile obtained.[9] We cannot derive the absorption coefficient directly from the simple TEY measurement. For measurements on insulators, methods must be employed to avoid charge-up of the sample. When a sample is liquid, a narrow cell (called 'wet cell') consisting of Si_3N_4 membranes or polymer films is used.[10]

In emission measurements, the sample can be irradiated using monochromatic light and the emission measured using a spectrograph or spectrometer. The emission should include the fluorescence as well as scattering light in inelastic scattering. An optical spectrometer is frequently used for the measurement of the visible-ultraviolet emission. A vacuum spectrometer must be used for the observation of core emission in the VUV and soft X-ray regions, a crystal spectrometer is used in the hard X-ray region. The emission or fluorescence yield spectrum is often called the 'excitation spectrum'. In the case of core emission, the profile of the emission yield spectrum is also very similar to that of the core absorption spectrum. However a close correspondence between the spectral profiles of the emission yield spectrum and the absorption spectrum is not always the case in the visible-ultraviolet emission region.

7.3.2 Photoelectron spectroscopy

a. The escape depth of an electron: Any electron excited in a solid will usually be scattered before it reaches the surface. The number of photoelectrons reaching the surface without scattering is designated as I and expressed as $I \propto I_0 \exp(-d/l)$, where the number of electrons generated at the depth d is I_0 and l is called 'electron escape depth'. Figure 7.2 shows the dependence of escape depth on the kinetic energy of the electron. The dependence is not expressed by a simple curve, but by a band. None the less, it is called the 'universal curve'.[3] The maximum value of l is a few times larger than the minimum value at any kinetic energy. The value of l for most materials lies within this band. Prior to reaching the surface, electrons are scattered by electron-electron scattering. Since the cross section for electron-electron scattering is given by $Im(-1/\varepsilon)$, the escape depth is determined in terms of dielectric function ε for every material. Excited electrons in solids with a kinetic energy above 10 eV may be regarded as free electrons. In this case the dielectric function is determined by the mean electron density. The mean electron density is similar for most materials so that the universal curve can be obtained by calculation,[11] and it can also be determined from experiments.[12] In the case of electrons with kinetic energies below 10 eV, the escape depth is significantly different for different materials.

The escape depth achieves a minimum value of about 0.3 nm in the region with kinetic energy of about 50 eV. Almost all electrons with this kinetic energy may traverse about only one atomic layer before scattering. Therefore, it is clear that the observed photoelectron spectrum will provide information almost exclusively about the surface layer. In Fig. 7.2, photon energies generated by a He discharge lamp and an X-ray tube are indicated by vertical lines. When the He-II resonance line (40.8 eV) is used to investigate the electronic state near to the Fermi level E_F, that of the surface is observed. By contrast, when a low energy photon from a He-I resonance line (21.2 eV) source or a high energy photon from a Mg Kα (1.25 keV) line is used, the escape depth reaches a few 10s of nm, so that the electronic state of the bulk is observed. In some materials however within the universal band, the escape depth is small in the low kinetic energy region, but the escape depth becomes large in the high kinetic energy region at around a few 100 eV. In these materials high energy excitation is needed to study the bulk properties.[13] There are two ways to distinguish the surface spectrum from the bulk by exploiting the differences in the escape depth. The first is the manner in which electrons with a

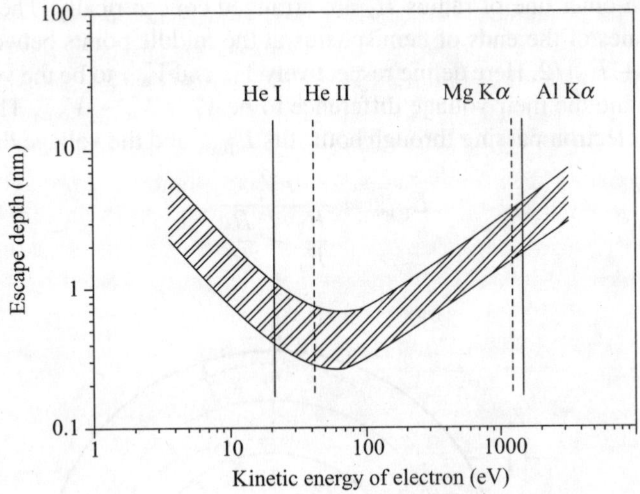

Figure 7.2 Dependence of escape depth on kinetic energy for an electron traversing a solid.

certain binding energy are excited by different photon energies, yielding differing kinetic energies and thus making it feasible to distinguish between them by the dependence of photoelectron intensity on the kinetic energy. The second is the technique in which the electrons are excited by the same photon energy, while the dependence of photoelectron emission on the observation angle is measured. There is even a third possibility in which the photoelectron yields are measured before and after the deposition of some monolayer onto the sample surface to distinguish between the surface contribution (the spectrum changes) from the bulk contribution (the spectrum does not change).

b. Energy analysis: Photoelectrons possess three physical quantities, energy, momentum and spin. First we consider only the analysis of the photoelectron energy. In order to analyze the energy of a photoelectron, we must use an electric or magnetic field or perhaps use the dependence of electron velocity on the kinetic energy. This technique is called the 'time-of-flight method' in which synchrotron radiation pulsed structure is exploited using an RF signal from the storage ring itself as a timing reference signal. A magnetic field may be used to act as an analyzer for very high energy photoelectrons or used in a time-of-flight analyzer in the so called 'magnetic bottle' where photoelectrons are gathered over a very large solid angle.[14)] There are two methods of analysis which exploit electric fields. The first is a dispersive type analyzer in which only those electrons with a certain energy can traverse through the exit slit plane. This is also called an electrostatic energy analyzer. Another is a non-dispersive analyzer resembling a triode vacuum tube in which electrons with energies less than the stopping voltage of the grid are blocked, while those with larger energies can reach the collector through the grid. This is called a retarding-field type analyzer. Energy analyzers of the electrostatic type are widely used in photoelectron spectroscopic measurements. They are classified into four types, that is concentric sphere,[15)] parallel plate, cylindrical,[16)] and cylindrical mirror analyzers,[17)] according to the shapes of the electrodes. The concentric sphere types also include hemispherical types with electrodes which may be complete half spheres and also the sector types with the electrode of about 160°-cut sphere and the partial sphere type with the electrodes shaped as part of a sphere. Among them, the hemispherical types are perhaps the most popular.[15)] Shown in Fig. 7.3, an inner electrode with radius R_1 and an outer one of radius R_2 are arranged concentrically. The entrance and exit slits are located on the planes of the ends of hemispheres at the middle points between the outer and inner electrodes, $R = (R_1 + R_2)/2$. Here define respectively V_{in} and V_{out} to be the voltages of the inner and outer spheres, and define the their voltage difference to be $V_{\text{f}} = V_{\text{in}} - V_{\text{out}}$. The relation between the kinetic energy of any electron passing through both slits E_{pass} and the voltage difference V_{f} is given by

$$E_{\text{pass}} = \frac{eV_{\text{f}}}{\dfrac{R_2}{R_1} - \dfrac{R_1}{R_2}} . \tag{7.1}$$

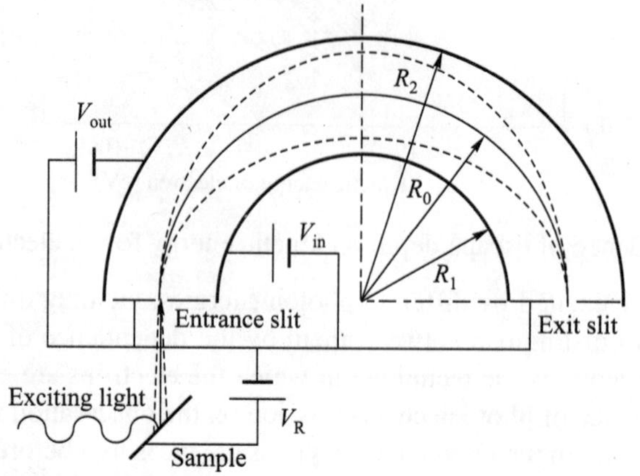

Figure 7.3 Structure of the hemispherical electron energy analyzer.

The relation between the voltages applied to the electrodes measured from the potential of the slits are given by

$$V_{\text{in}} = \frac{E_{\text{pass}}}{e}\left(\frac{R_2}{R_1} - 1\right), \quad V_{\text{out}} = \frac{E_{\text{pass}}}{e}\left(\frac{R_1}{R_2} - 1\right). \tag{7.2}$$

Since $R_2 > R_1$, V_{in} and V_{out} are positive and negative, respectively. Notice that the actual relation between V_{f} and E_{pass} may deviate slightly from the formula Eq. (7.1) due to possible errors in machining and in assembling the electrodes.

The energy resolving power of the hemispherical analyzer with both entrance and exit slit of width W is given by

$$\frac{E_{\text{pass}}}{\Delta E} = \frac{2R_0}{W}, \tag{7.3}$$

where ΔE stands for the half width of the energy resolution. Since ΔE is proportional to E_{pass}, we must set E_{pass} to be constant in order to measure the spectra with constant resolution. For this purpose photoelectrons emitted from the sample are accelerated or decelerated so as to have a constant kinetic energy at the entrance slit. The resolving power is determined by R_0, W and E_{pass}. In order to obtain a high resolving power, a spherical analyzer with large R_0 should be used. The use of narrow slits, *i.e.*, small W slits also leads to high resolving power, but has the disadvantage that the photoelectron current passing through the slits becomes very small. To attain good energy resolution ΔE, it is necessary to make E_{pass} smaller, that is the photoelectrons should be decelerated. This method is called the retardation method in which an additional high precision voltage source is necessary to supply retarding voltage V_R and an electric lens of excellent quality is needed to gather enough electrons in the hemispherical analyzer after deceleration.

The cylindrical mirror analyzer is also used frequently.[17)] Shown in Fig. 7.4 is the inner cylinder with radius of R_{in} and an outer one of R_{out}, both coaxially installed, and two apertures with a width of W cut on the inner cylinder at optimal positions. A negative voltage is applied to the outer cylinder. Photoelectrons emitted from the sample which is located on the central axis pass through the first aperture on the inner cylinder and are deflected by the field between two cylindrical electrodes. They pass through the second aperture and are focused at the exit slit on the central axis for detection. A calculation shows that the best focusing condition is given when the angle of incidence from the central axis to the first aperture is chosen to be 42.308° and the distance between entrance and exit slits is $L_0 = 6.188R_{in}$. The relation between the applied voltage V and the kinetic energy of a passing electron is given by

$$E_{pass} = \frac{1.3098\,eV}{\ln(R_{out}/R_{in})}. \tag{7.4}$$

Although the entrance slit defines the source size, it is not effectively used in the cylindrical mirror analyzer and only the sample is placed at the slit position. If electrons emitted from an electron source with the axial length s and the radius ρ pass through the first aperture with divergence angle of 2α, the resolving power can be expressed by

$$\frac{E_{pass}}{\Delta E} = \frac{1}{\dfrac{s + W + 1.1\rho}{11.2R_{in}} + 3.2\alpha^3}. \tag{7.5}$$

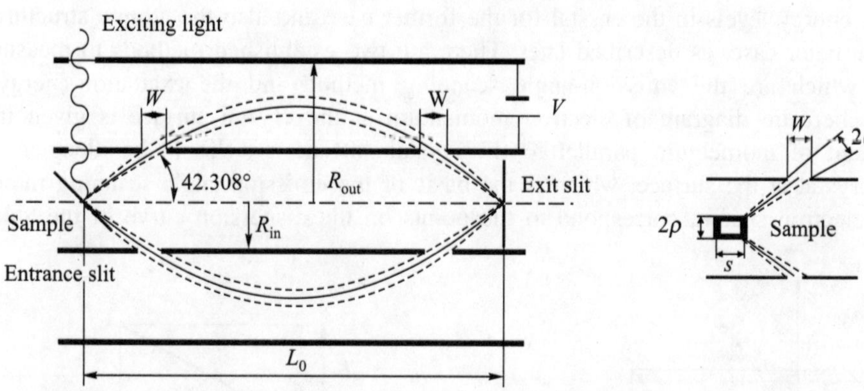

Figure 7.4 Structure of the cylindrical mirror electron energy analyzer.

In many cases, the entrance sides of both cylinders are shortened to make sample exchange easier, resulting in some disturbance of the electric fields, which then must be corrected using a correction electrode. To make the field around the sample uniform, a spherical mesh electrode is used. For achievement of even better resolution, a serial connection of two cylindrical mirrors is used, in which the position of exit slit in the first stage analyzer coincides with the entrance slit of the second analyzer. This is called the 'double cylindrical type'.

Since the orbits of electrons with kinetic energy lower than a few tens of eV are easily perturbed by the electric and magnetic fields surrounding them, accurate measurements cannot be made without counter measures. For example, electric fields may be caused by charge-up of insulators around the sample, so that insulators must be used which are not easily charged up and these insulators should be protected from the excitation light. The fields caused by the insulators must be well shielded at the sample position. Some magnetic fields originate from geomagnetism with strength of about 30 μT and there will be flux leakage from magnets used by the ion sputtering pump. In order to reduce these fields, the energy analyzer has to be magnetically shielded and there are cases where photoelectron measurement chamber in ultra-high vacuum including the energy analyzer is entirely made from magnetic shielding metal. As a consequence the magnetic field on electron orbits from the sample to the detector may be reduced to 1/100 that of geomagnetism, that is to about a few 100 nT. In many cases, the sample preparation chamber is connected to the measurement chamber and preparation of samples by evaporation, cleaving, cleaning *etc.*, is made in this adjacent chamber and then transferred to the measurement chamber thus retaining the ultra-high vacuum condition.

The binding energy origin for the obtained spectrum of any sample is determined by comparison with the energy of the measured Fermi level edge for Au evaporated on the sample. For measurements of insulators, a thin film sample is necessary to avoid charge-up. Sometimes, a special electron gun called flood gun, is used to feed electrons to the positively charged sample to achieve charge compensation.

c. Angle-resolved photoelectron measurements: Figure 7.5 shows the definition of angles for both the incident light and the emitted photoelectrons. The intensity of photoelectrons depends on the polar angle θ and azimuthal angle ϕ. The angular dependence of photoelectron emission originates from the electron momentum of $\hbar\boldsymbol{k}$ inside or surface of a crystal, or from the scattering by surrounding atoms before leaving the crystal resulting in photoelectron diffraction. The angular dependence measurement, that is, the angle-resolved measurement, enables us to determine the band dispersion of energy levels in the crystal for the former case and also the atomic structures near the surface in the latter case, as described later. There are two established methods to measure electron momentum, which are the emission-angle scanning method and the excitation-energy scanning method. A schematic diagram of electron momentum at the crystal surface is given in Fig. 7.6. The component of momentum parallel to the crystal surface $k_{||}$, should be identical inside and outside the crystal at the surface, which is the basis of the emission-angle scanning method. Peaks in the photoelectron spectra correspond to the points on the dispersion curve of the valence band.

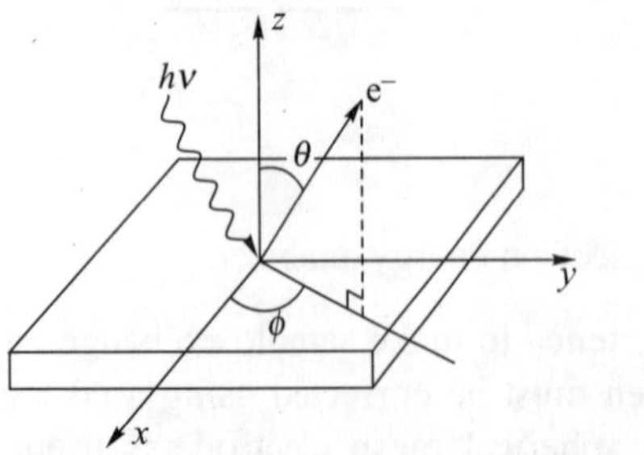

Figure 7.5 Definitions of the emitting angles of photoelectrons.

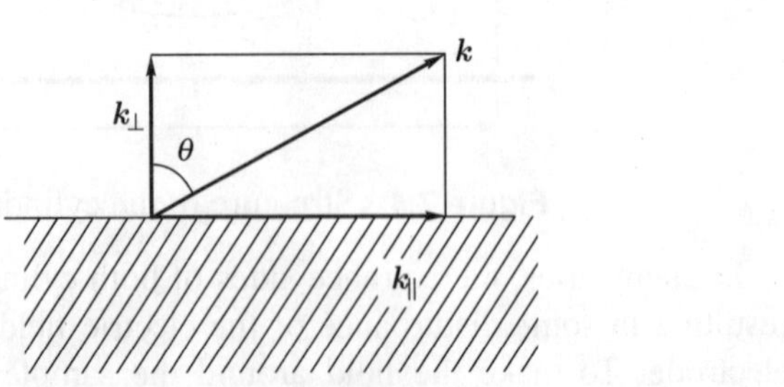

Figure 7.6 The horizontal and vertical components of electron momentum at the crystal surface.

The kinetic energy of the peak gives the binding energy at the point on the dispersion curve. Then $k_{||}$ is expressed by

$$k_{||} = \sqrt{2mE_{\text{kin}}}\sin\theta/\hbar, \;\; E_{\text{kin}} = \hbar\omega - E_{\text{b}}(k_{||}). \tag{7.6}$$

In Eq. (7.6), the valence band dispersion $E_{\text{b}}(k_{||})$ is taken to be positive. The θ dependence of the photoelectron peaks is measured under constant excitation energy, by rotation of energy analyzer around the sample or by rotation of the sample under the condition in which the energy analyzer situated at a fixed position. $k_{||}$ is obtained from substitution of E_{kin} for Eq. (7.6). The band dispersion curve is experimentally determined from the relationship between the binding energy and the momentum derived from the peak position. In the case of the excitation-energy scanning method, only those photoelectrons emitted vertically are measured. The vertical component of the momentum $k_{\perp}$ is then different for the inside and the outside of the crystal. If we regard the emitted photoelectron to be a free electron, $k_{\perp}$ is obtained from,

$$\frac{\hbar^2 k_{\perp}^2}{2m} = \hbar\omega + E_0 - E_{\text{b}}(k_{\perp}), \tag{7.7}$$

where E_0 is chosen to be the most probable positive value (see Fig. 7.7). The momentum can be determined by scanning the excitation energy and using Eq. (7.7). To derive the band dispersion in the reduced zone, in both methods, $k_{||}$ or $k_{\perp}$ should be determined taking the reciprocal lattice vector into account.

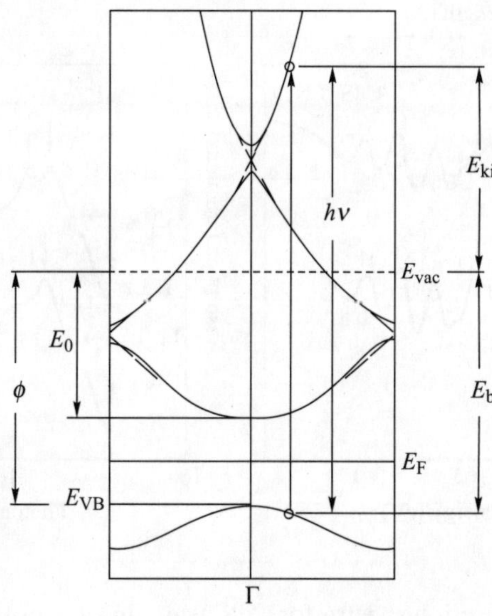

Figure 7.7 Schematic of dispersion curves for the valence and conduction bands. The absorption process is indicated by a vertical arrow. ϕ and E_{VB} represent respectively work function and energy at the top of the valence band (see parameters in text).

d. Spin analysis: The spin polarization is measured by connecting a spin detector to the exit of the electron energy analyzer.[18,19] There are two designs of spin detector. The first type is the 'Mott detector' in which spin-polarized electrons are made to collide with a thin gold (Au) foil after acceleration to an energy of the order of from a few 10s to 100 keV. The spin polarization is then estimated from the difference in intensities scattered by Au between the right and left sides. The second method makes use of a spin LEED detector, in which low energy electrons with about 100 eV collide with a clean W single crystal. The spin polarization can be estimated from the diffraction intensity difference between the right and left sides in the LEED pattern.

7.4 SEMICONDUCTORS AND IONIC CRYSTALS

We have already noted that the core absorption spectrum reflects the partial density of states in the conduction band. Figure 7.8(a) shows the S 2p core absorption spectrum for a CdS thin single crystal (flake) where the *c*-axis is parallel to the plane of the flake surface illuminated by linearly-polarized light with the electric vector parallel and also perpendicular to the *c*-axis.[20] The S 2p core level splits into L_3 and L_2 sub-levels with 1.1 eV separation and with an intensity ratio of 2 to 1. Taking this into consideration, the L_3 spectra can be extracted as shown in Fig. 7.8(b). In the figure, Γ and U represent good symmetry points in the Brillouin zone for a hexagonal crystal structure. The spectral shape reflects van Hove singularities such as of the M_0 type (see App. **I**) Making a comparison of the onset energies of the singularities with the calculated dispersion curve of the conduction band and by taking the polarization effect into account, the validity of the band calculation can be confirmed.

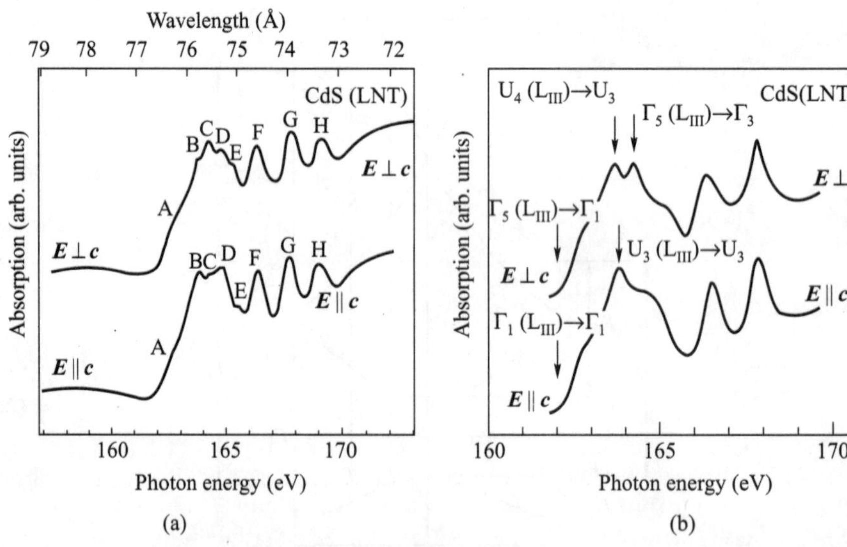

Figure 7.8 S $L_{II,III}$ absorption spectra for CdS using linearly-polarized light. (a) Raw data, (b) L_{III} spectrum derived. Vertical arrows indicate the assignments of transitions obtained by a band calculation.

Figure 7.9 shows the angle-integrated photoelectron spectra for the valence band of GeTe, which reflect the density of states of the valence band.[21] When GeTe is evaporated on to a cooled substrate, the evaporated sample is amorphous. This sample subsequently becomes irreversibly crystalline after annealing. In Fig. 7.9 experimental results are presented both for amorphous and for crystalline samples together with the density of states for the crystalline sample calculated theoretically.[22] From a comparison between them, their spectral features are seen to be rather similar. Peaks I, II and III correspond to a bonding p-state composed from Ge 4p and Te 5p states, the Ge 4s state and the Te 5s state, respectively. When the structure of the sample is changed from amorphous to crystalline, the valence band shifts towards the low binding energy side. This result suggests that the band gap in the amorphous sample is larger than that in the crystalline sample. In fact, this tendency is also observed in infrared absorption measurements. The energy separation between peaks I and II in the amorphous phase is smaller than that in the crystalline sample, suggesting that s-p hybridization is larger in the amorphous sample than in the crystalline sample. In other words, the degree of covalency is larger in the amorphous sample.

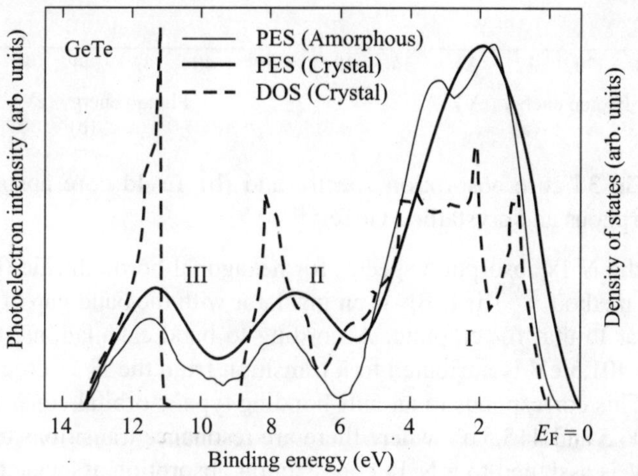

Figure 7.9 Valence band photoelectron spectra for GeTe (by courtesy of K. Fukui).[21] Thin curve: amorphous, thick curve: crystal, broken curve: calculated density of states by M. Haung *et al.*

The Ge 3d core absorption spectra for GeTe are shown in Fig. 7.10(a) for both amorphous and crystalline samples.[22] In this figure, the effects of structural changes are clearly seen in the spectra. The spectrum of the amorphous sample shows a broad band, which appears to reveal the density of states for the conduction band. However, the spectrum of the crystalline sample shows sharp core exciton absorption peaks, which are considered to correspond to intraatomic transitions within the Ge atom. These facts all suggest that the bottom of the conduction band in the crystalline sample is predominantly composed of the Ge 4p state, while the Te 5p states are dominant in the amorphous sample. This means that covalency is greater in the amorphous sample than in the crystalline sample for GeTe. Figure 7.10 (b) shows the Te 4d core absorption spectra for both amorphous and crystalline samples of GeTe. The spectrum for the crystalline sample shows broad bands while that for the amorphous sample reveals sharp peaks. These features in the Te 4d absorption spectra appear to be contrary to those in the Ge 4d absorption spectra. However, the results for Te 4d absorption also suggest that covalence is more significant in the amorphous sample than in the crystalline sample as the results for Ge 3d reveals. These conclusions are supported by the photoelectron measurements.

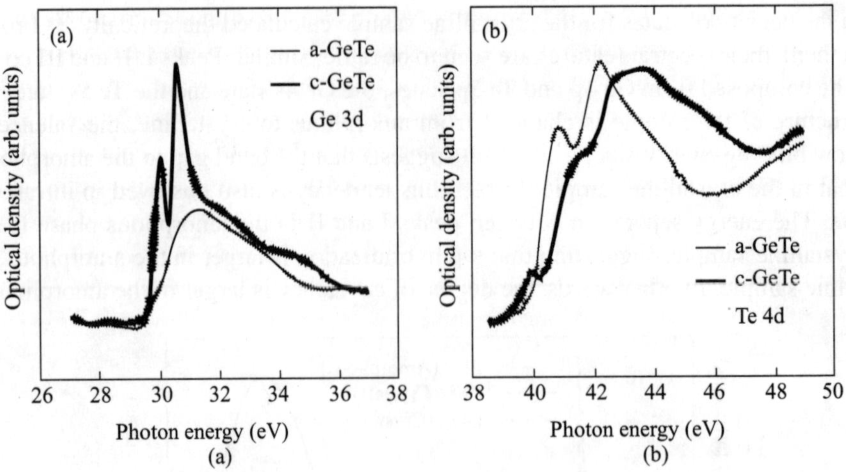

Figure 7.10 (a) Ge 3d core absorption spectra and (b) Te 4d core absorption spectra for amorphous and crystalline GeTe.[22].

Figure 7.11 shows the N 1s absorption spectra for hexagonal boron nitride (h-BN) measured using the total electron yield method.[23] The h-BN is an insulator with the band gap of 5.6 eV and possessing a layer structure similar to that for graphite. According to band calculations, the sharp peak near to the absorption edge at 401.5 eV is attributed to a transition from the N 1s core level to the bottom of the conduction band. This corresponds to an anti-bonding type π orbital consisting of N $2p_z$, and has broad peaks around 406.5 and 415.5 eV where there are resonance transitions to N $2p_{x,y}$ orbitals. The small peak at 400.3 eV is assigned to a N 1s core exciton absorption associated with a transition to a N $2p_z$ orbital. The vertical arrows labeled with alphabetical letters indicate the energies of the emission excitation.

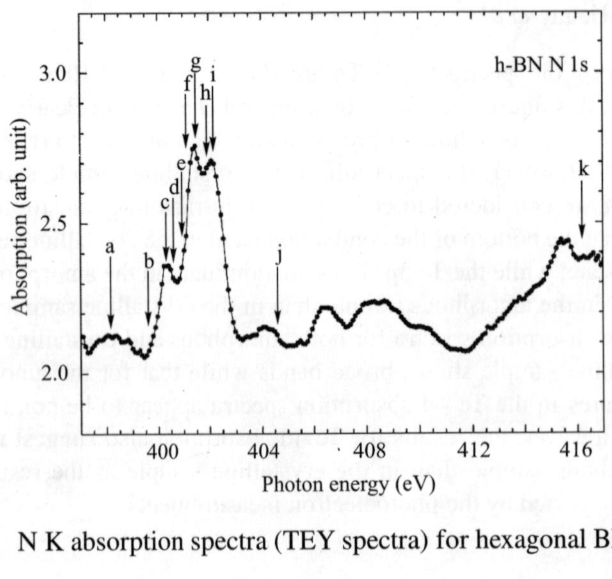

Figure 7.11 N K absorption spectra (TEY spectra) for hexagonal BN (h-BN).[23]

Figure 7.12 shows the N 1s soft X-ray emission spectra, where the letter and the number attached to each spectrum respectively correspond to the letter and the excitation energy given in Fig. 7.11. Emission excited using photons with energies greater than 401 eV (in other words around the energies of interband transitions) is called 'normal emission'. For this region the spectral shape scarcely changes with change of excitation energy. Within it, the sharp peak at 394.6 eV originates from the N $2p_z$ orbital and is called 'π emission'. The small peaks at 385, 390 and 393 eV originate from the N $2p_{x,y}$ orbital and are called 'σ emission'. In normal emission, both π and σ emissions are observed. Soft X-ray emission therefore can directly provide the partial density of states for the valence band. This corresponds to the initial state of the allowed dipole transition to the core level, which is flat and therefore dispersionless. However, the spectral features of those emissions excited by the exciton absorption energies of 400.3 and 400.6 eV differ substantially from those for normal emission. In this case the π emission becomes very weak and the energy position for the σ emission shifts towards the low energy side. Such emission is called 'spectator emission'. This anomalous behavior of the π emission is also observed in the B 1s emission for this material.[24] Even though spectator emission depends on the excitation energy, its origin cannot be explained by Raman scattering. Spectator emission must be explained by considering the N

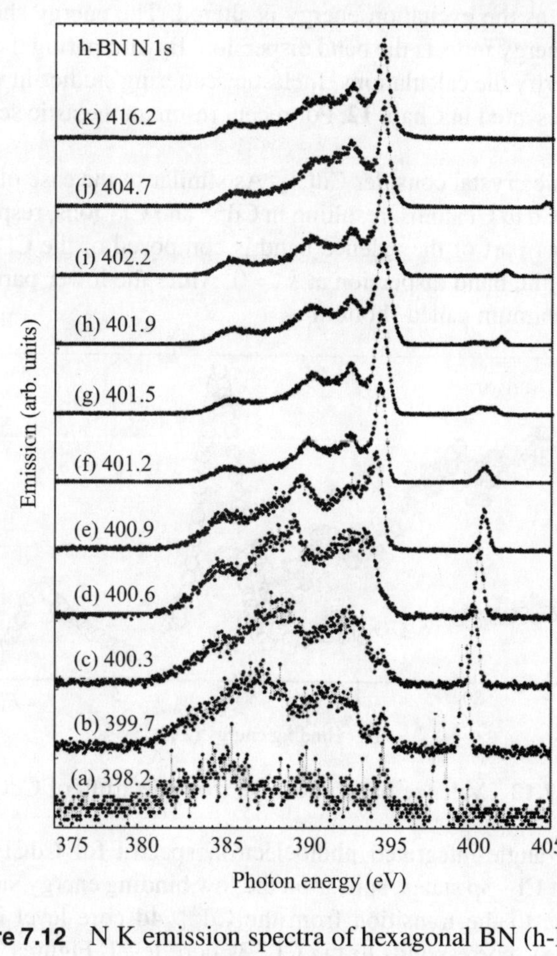

Figure 7.12 N K emission spectra of hexagonal BN (h-BN).[23]

$2p_z$ excitonic electron to be localized and combined with the reconfiguration of the valence electrons arising from the Coulomb potential of the core hole. In other words, in spectator emission, the N 1s core hole potential is screened by the localized N $2p_z$ electron and the π electrons are repelled from the N atom site which then results in a lowering of the π emission intensity. Some spectral features for the transition metal compounds and rare-earth metal compounds can be explained by an effect in which the core hole potential is screened by the hybridization of d or f electrons with the valence electron. In this case the d or f electrons which are localized play an essential role. This type of effect is observed in even the s-p electron insulator.

In resonant inelastic scattering processes (Sect. **4.3.3.b**), absorption occurs from a core level to a conduction band and simultaneously a resonant emission occurs from the valence band to a core level. The conservation law for momentum can be derived using the one-electron approximation, because the momentum of the photon in the VUV region is negligible in comparison with that of the electron.[7] Therefore the absorption and emission occur at the same crystal momentum ($\hbar\boldsymbol{k}$) in the Brillouin zone. When the excitation energy is chosen to be around the absorption edge, the resonant emission (resonant inelastic scattering light) appears, whose profile differs from that of normal emission and which changes remarkably as the excitation energy is altered. The energy shift of the emission peaks with change of excitation energy reflects the band dispersion. By comparing the shift with the calculated band dispersion, one can verify the calculations. Inelastic scattering studies in which photon momentum is taken into account are presented in Chap. **12**. For recent resonant inelastic scattering experiments, see Ref. 11 in Chap. **12**.

As an example of an ionic crystal consider $CdCl_2$. As similar to the case of alkali halides, in $CdCl_2$, electrons transfer from the Cd to Cl atoms, resulting in Cd^{2+} and Cl^- ions, respectively which are linked by ionic bonding. The upper part of the valence band is composed of the Cl^- 3p state which exhibits a maximum called 'top' in the band dispersion at $k = 0$, while the lower part of the conduction band, Cd^{2+} 5s state exhibits a minimum called 'bottom'.

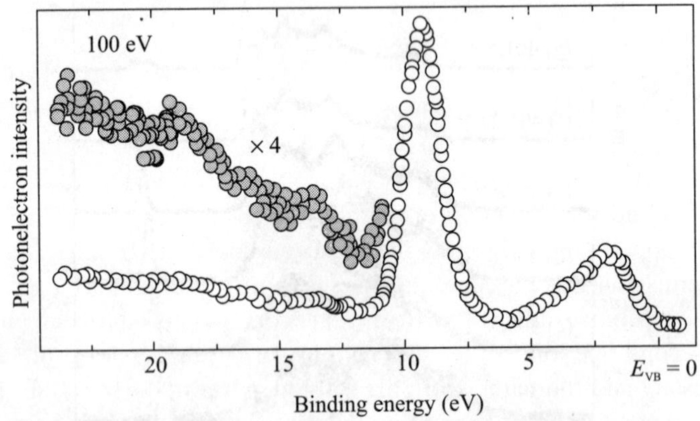

Figure 7.13 Valence band photoelectron spectrum of $CdCl_2$.[25]

Figure 7.13 shows the angle-integrated photoelectron spectra for $CdCl_2$.[25] The spectra for the valence band, composed of Cl^- 3p states, appear on the low binding energy side. As the binding energy increases, a sharp line due to the transition from the Cd^{2+} 4d core level is observed at around 10 eV. The small peak at 14 eV corresponds to the Cl^- 3s core level. Figures 7.14(a) and 7.14(b) show the near normal incidence reflectance spectrum of $CdCl_2$ measured at 35K, and the imaginary part of the dielectric constant together with energy loss function $Im(-1/\varepsilon)$ which was derived from the reflectance spectrum using a Kramers-Krönig analysis. The excitation at the lowest photon energy, the fundamental absorption edge energy, corresponds to the return of the electron from the negative to the positive ion. The sharp peak at 6 eV corresponds to exciton absorption and the step-like structure around 6.7 eV to the onset of an interband transition. Absorption lines originating from the Cd^{2+} 4d core from 15.5 to 17 eV can be seen. However no absorption structure originating from the Cl^- 3s core excitation is visible and this could be due to the forbidden transition from Cl^- 3s to Cd^{2+} 5s or due to line broadening by the Fano effect.

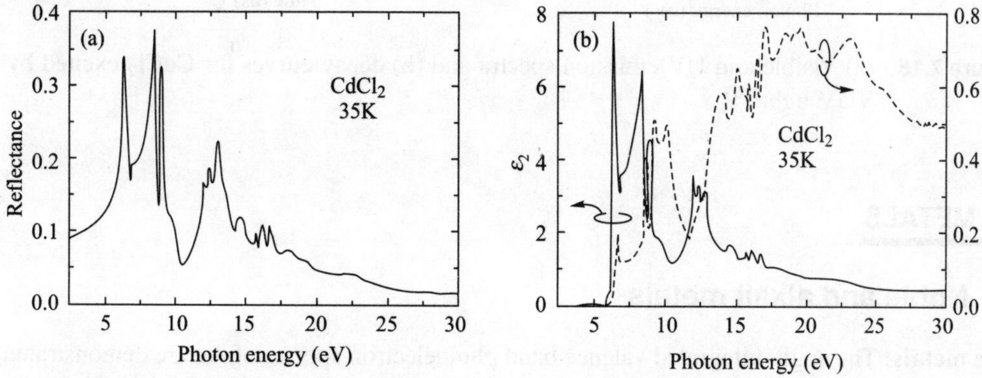

Figure 7.14 (a) Near normal incidence reflection spectrum for $CdCl_2$ at 35K and (b) its ε_2 and $Im(-1/\varepsilon)$ spectra obtained from a Kramers-Krönig analysis.[25]

The emission spectra for $CdCl_2$ measured at 75 and 11K are exhibited in Fig. 7.15.[26] The excitation energy is 6.2 eV, which corresponds to the peak energy of the exciton absorption. Two broad emission bands are observed at 3.76 and 2.3 eV. The band at 3.76 eV is interpreted to be the direct transition from the exciton in the self-trapped state having the lowest energy. The band at 2.3 eV originates from the transition which should be forbidden, but has become allowed due to lattice relaxation. The intensity and the life time are both stronger and increased in the 2.3 eV emission compared to the 3.76 eV emission. In general, when any atom with spherical symmetry enters a non-spherical field within a crystal its degree of symmetry is lowered and the transition originally forbidden becomes allowed. The degree of the symmetry of the self-trapped state exhibiting the localized nature is considerably reduced. The decay time for the 3.76 eV emission was measured using synchrotron radiation as a pulsed excitation source and to provide the time reference signal. Two decay time components have been observed, a shorter life time of 2.2 ns and a second one longer than 600 ns. The former originates from the allowed transition from the spin singlet state and the latter, from the forbidden transition from the spin triplet state. The spins of the electron and positive hole are anti-parallel in the singlet state and parallel in the triplet state. The reason why the forbidden transition from the spin triplet state becomes allowed is attributed to a mixing of spin singlet and spin triplet states by spin-orbit interaction in Cl^- 3p state.

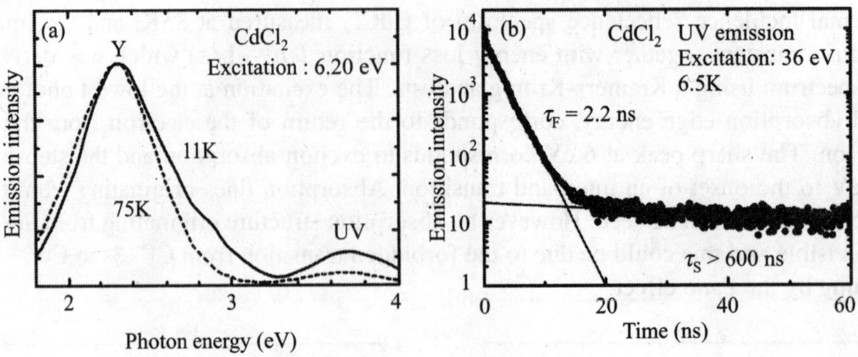

Figure 7.15 (a) Visible and UV emission spectra and (b) decay curves for $CdCl_2$ excited by VUV light.[26]

7.5 METALS

7.5.1 Noble and alkali metals

a. Noble metals: The angle-integrated valence band photoelectron spectra of Au are demonstrated as a typical example of photoelectron measurements for metals and are shown in Fig. 7.16.[27] The excitation source used was a He-I resonance line. Au is usually used for determination of the Fermi level of a sample because of its stability arising from a high resistance to oxidation. In the figure, the binding energy origin is chosen to coincide with the Fermi level. In Fig. 7.16(a) a clear step structure in the

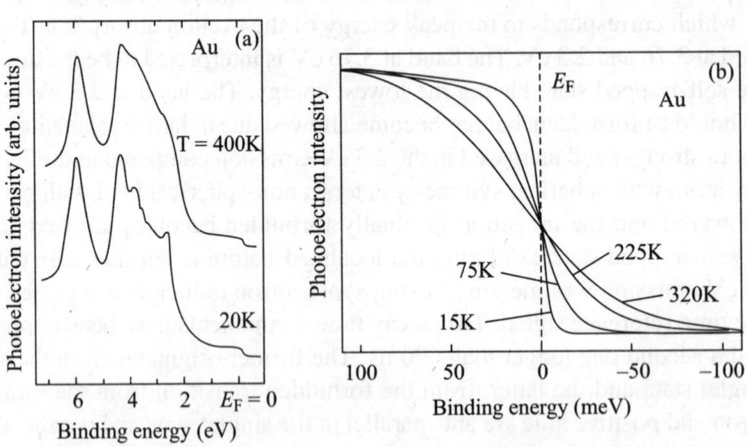

Figure 7.16 A valence band photoelectron spectra for Au ((by coutesy of T. Takahashi) (a) In the binding energy region of 0~7 eV (© AGNE Gijutsu Center 1995)[27] and (b) around Fermi level obtained with an excitation energy of 21.2 eV (© World Scientific 2002).[28]

photoelectron intensity is observed at the Fermi level. This structure is ascribed to the 6s electrons in the valence band. Peaks at the binding energies of 4 and 6 eV respectively, are ascribed to states of total angular momentum with $j = 5/2$ and $j = 3/2$ originating from the spin-orbit splitting of the Au 5d level. Figure 7.16(b) also shows the temperature dependence for the spectral profile of the Au valence band close to the Fermi edge, measured by high resolution photoelectron spectroscopy.[28] When the temperature is 225K, the step structure around the Fermi edge exhibits a gentle slope whereas when the temperature becomes 15K, it shows a steep rise. This result is an experimental demonstration of the temperature dependence of the Fermi-Dirac distribution function.

b. Alkali metals: Figure 7.17 shows some absorption spectra of outermost core states in alkali metals, namely for Na 2p, K 3p, Rb 4p and Cs 5p states, respectively.[29] If the absorption spectrum from the core state reflects only the unoccupied density of states above Fermi level which are mainly composed of s states, then the spectrum would show a gradual increase toward high energy side following a steep rise at the absorption edge. In fact, the measured spectrum reveals an anomalous spike structure around the edge and the steep rise at the edge is followed by a rapid decrease toward the high energy side, all within an energy range of only 1 eV. This phenomenon is called the 'Fermi edge anomaly', and can be understood in terms of many-body effects where free electrons aggregate in order to shield the remaining positive hole. This behavior is also called the excitonic effect in metals.[30,31] According to theory, all absorption coefficients at the higher energy side of the edge should be expected to be proportional to $\omega^{-\alpha}$, where $\alpha > 0$ for a p $\rightarrow$ s transition and $\alpha < 0$ for an s $\rightarrow$ p transition. In fact, no spike structures are observed in s $\rightarrow$ p absorption spectra, therefore the theories are supported by the experiment.

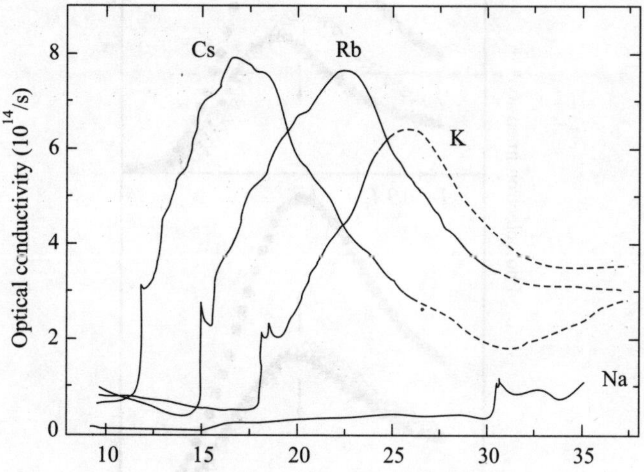

Figure 7.17 Shallow core absorption spectra for alkali metals (Na 2p, K 3p, Rb 4p, Cs 5p).[29]

7.5.2 Transition metals

a. Spin-resolved photoelectron spectroscopy: In the early 1980's, spin-resolved photoelectron spectroscopy was first carried out on single crystals of Fe and Ni and the results were contrasted with band calculations.[32,33] Figure 7.18 shows the angle- and spin-resolved photoelectron spectra for a Ni single crystal.[34] The structure of Ni is fcc and the (110) plane is chosen to be the emission plane of the photoelectron. The sample is shaped into the window frame wound with the coil for magnetization. The magnetization is in the (110) plane and forms the closed loop along the frame. Measurements are using the excitation-energy scanning method, in which photoelectrons are detected in the [011] direction perpendicular to the emission plane (the sample surface). Only those results using excitation light of 21.2 eV (obtained using undulator radiation) are shown in Fig. 7.18. The upper and lower parts of the figure give the experimental data for $T/T_c = 0.5$ and 0.9, respectively, where T and T_c represent the measurement and Curie temperatures. At $T = 0.5T_c$, the total density of states curve (●) splits with a majority (up, △) and a minority (down, ▲) of spin sub-bands with different populations. This implies that the ferromagnetism of Ni originates from the differences between these two spin sub-bands. When the temperature rises to $0.9T_c$, only very little difference is observed between these sub-bands, which implies that Ni is almost paramagnetic. Clearly, spin-resolved photoelectron spectroscopy offers a powerful tool for the study of magnetism and magnetic effects. Once the excitation energy reaches 21.2 eV, a state of binding energy of 0.55 eV corresponds to the valence band at the X point in the Brillouin zone.

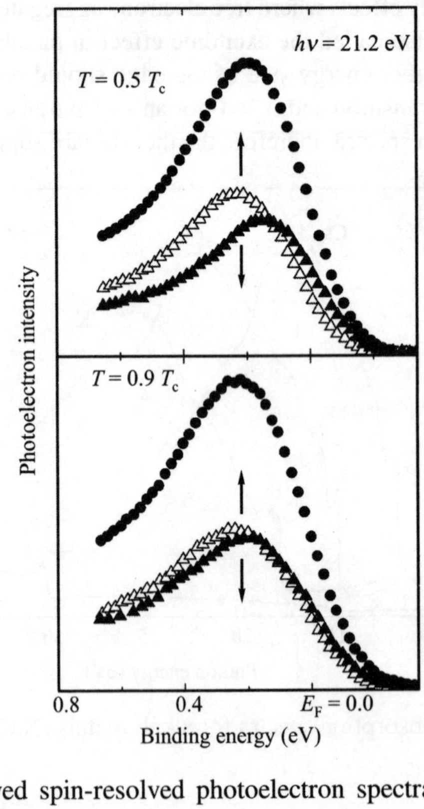

Figure 7.18 Angle-resolved spin-resolved photoelectron spectra of Ni (by courtesy of A. Kakizaki).

b. Magnetooptical spectroscopy: Magnetism in transition metals originates from the existence of incomplete d shells. The core absorption corresponds to an electronic transition to the incomplete shell from a core level whose orbital and spin angular momenta are well defined. The magnetic moments are obtained from spectral analysis. In many cases, the magnetic circular dichroism (or MCD), which represents the difference between absorption coefficients under a magnetic field for right and left circularly-polarized light ($\mu^+ - \mu^-$) can be measured. In the case of the L absorption spectra for transition metals such as Ni, the initial state of the 2p level ($L_{2,3}$ states) may split by more than 10 eV due to the spin-orbit interaction. The majority spin band in 3d final states is almost occupied, while the minority spin band remains unoccupied. Define the light propagation direction to be along the z-axis. The magnetic field is then applied along the z-axis and the direction of the rotation of circularly-polarized is also defined with respect to the z-axis (App. **J**). The transition probability from the L_2 level for left circularly-polarized light in which L_z becomes $L_z + 1$ (where L_z expresses the quantum number of the orbital magnetic moment), differs from that for right circularly-polarized light in which L_z becomes $L_z - 1$. The transition probability from the L_3 level for left circularly-polarized

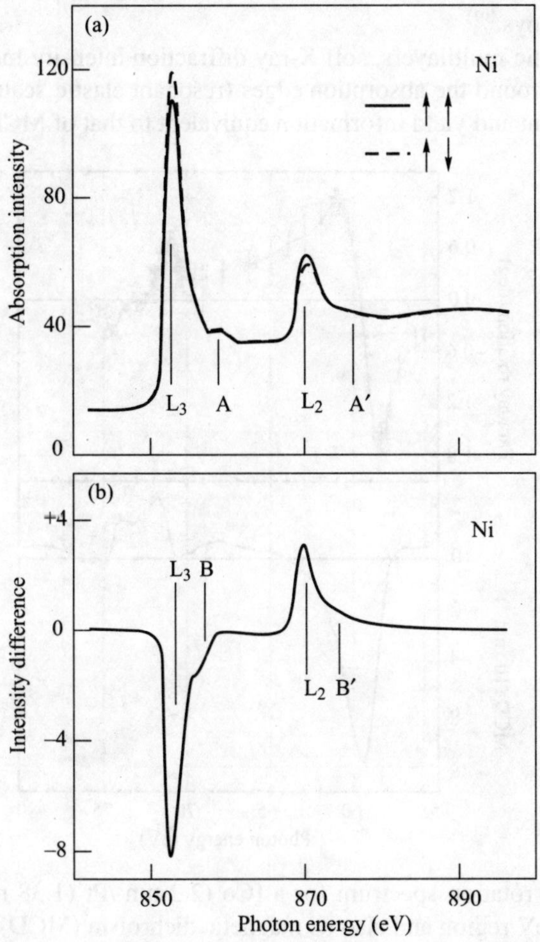

Figure 7.19 (a) The magnetoabsorption spectra for Ni at its $L_{2,3}$ edge and (b) its MCD (by courtesy of C. T. Chen, © The American Physical Society 1990).[35]

light also differs from that for right circularly-polarized light. Note that the signs of both differences are opposite. It is then possible to obtain the magnetic moment for the 3d final state using the difference between these transition probabilities (MCD) (see App. **K**). Figure 7.19 shows (a) the magnetoabsorption spectra at the Ni $L_{2,3}$ edge and (b) its MCD, respectively.[35] A large MCD effect is observed. Since core absorption is element-specific, this method is very useful to study magnetic alloys and multilayer films.

Magnetooptical rotation spectroscopy is also a useful technique in studies of magnetism (see App. **J**). In this case it is the linear polarization property of synchrotron radiation which is exploited. Both three- and four-mirror reflectors or multilayer films have been used as rotation analyzers for linearly-polarized light (see Sect. **5.5**).[36] Figure 7.20 shows the Faraday rotation spectrum for a Co/Pt magnetic multilayer film. Faraday rotation is observed not only at the $M_{2,3}$ edge of magnetic Co but also at the $N_{6,7}$ edge of (usually) non-magnetic Pt.[37] These facts indicate that Pt films are probably magnetized in Co/Pt multilayers, a result which would support the results from MCD experiments.[38] Similar features are observed in bulk $CoPt_3$ alloys.[39]

For the study of magnetic multilayers, soft X-ray diffraction intensity measurements with scanning the incident light energy around the absorption edges (resonant elastic scattering) in the presence of a magnetic field are important and yield information equivalent to that of MCD.[40]

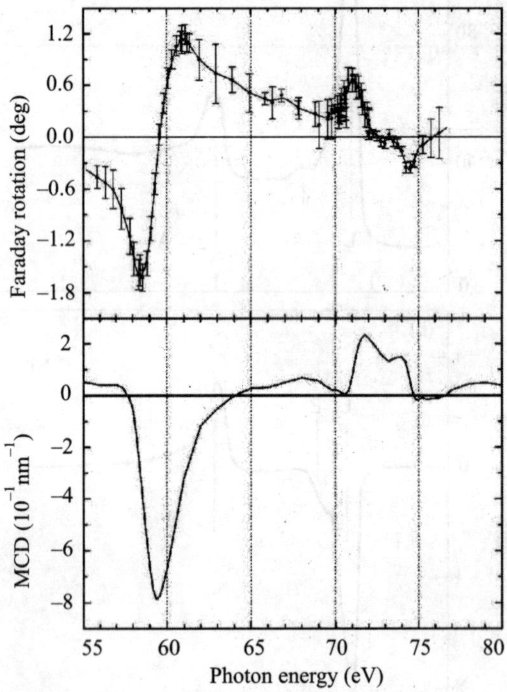

Figure 7.20 Faraday rotation spectrum for a [Co (2.2 nm)/Pt (1.38 nm)] multilayer in the 55∼80 eV region and also its magnetic dichroism (MCD) spectrum obtained by Kramers-Krönig analysis.[37]

7.5.3 Rare-earth alloys

a. Resonant photoelectron spectroscopy: When valence electrons are excited by photons with an energy close to a core absorption threshold, some interference may occur between the valence state with one positive hole and the valence state with two positive holes generated by the Auger process of 'super Koster-Krönig' type. By scanning the excitation energy around the threshold, some peaks in the photoelectron spectrum are enhanced, giving 'resonant photoelectron emission'.[41] Using this effect, we can assign the origin of some wave functions in the valence band. Figure 7.21(a) gives photoelectron spectra for the f electron system of UPd_2Al_3 with different excitation energies around the U 5d core.[42] When the excitation energies lie close to the U 5d threshold, that is from about 97 to 100 eV, the spectral intensity close to E_F is remarkably enhanced. Figure 7.21(b) shows the 'constant initial state (CIS)' spectrum which represents the excitation energy dependence of the intensity of photoelectrons emitted from the identical initial state. The solid line shows experimental results taken with the binding energy of the valence band $E_b = 0.5$ eV. Two peaks are found at 98 and 107 eV corresponding to the spin-orbit splitting pairs of the U 5d level, namely, $5d_{5/2}$ and $5d_{3/2}$, respectively. Clearly this intensity enhancement reflects the U 5d-5f resonance and indicates that the valence band in the vicinity of the Fermi level originates mainly from a U 5f state. The asymmetric profiles of the two peaks are interpreted in terms of the Fano effect (see Sect. **6.1.2**). The broken line shows the result at $E_b = 4.0$ eV, which shows a monotonic decrease in intensity. It is proposed that the structure of the valence band in the region with binding energy from 3 to 5 eV does not originate from the U 5f state, but may be from the Pd 4d state.

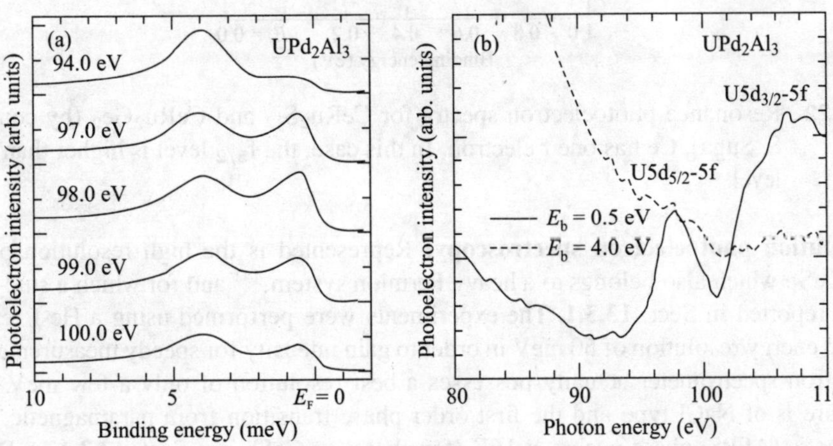

Figure 7.21 (a) Resonance photoelectron spectrum and (b) constant initial state (CIS) spectrum for UPd_2Al_3.

Figure 7.22 gives the resonant photoelectron spectra for $CeRu_2Si_2$ and $CeRu_2Ge_2$. This is the valence band photoelectron spectrum derived using 882.6 eV photons, which excite Ce 3d electrons.[43] In $CeRu_2Si_2$ for instance, the effective mass obtained from electric conductivity and specific heat measurements is about 1000 times larger than that for a normal electron. Thus, this material is one in the heavy Fermion category and it has a Kondo temperature of 20K. In the figure, the 4f component in the valence band is exaggerated due to the 3d-4f resonance. The peak close to the Fermi level is not found in the spectrum obtained from 4d-4f excitation but it is observed in spectra obtained by high energy (882.6 eV) excitation at 20K, which of course reflects the bulk rather than the surface properties. The intensity of the peak diminishes at higher binding energies, perhaps corresponding to the tail of the Kondo peak.

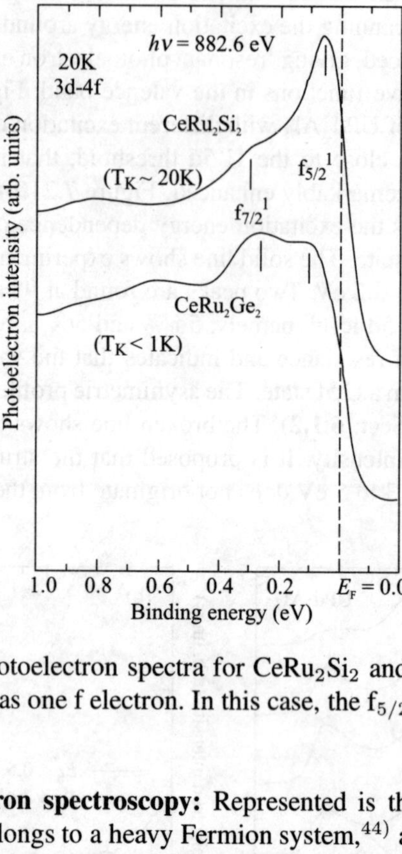

Figure 7.22 Resonance photoelectron spectra for $CeRu_2Si_2$ and $CeRu_2Ge_2$ (by courtesy of S. Suga). Ce has one f electron. In this case, the $f_{5/2}$ level is higher than the $f_{7/2}$ level.

b. High resolution photoelectron spectroscopy: Represented is the high resolution photoelectron spectrum of CeSb which also belongs to a heavy Fermion system,[44] and for which a study by infrared reflectance is reported in Sect. **13.3.1**. The experiments were performed using a He-I resonance line source with an energy resolution of 50 meV in order to gain intensity for speedy measurements although the photoelectron spectrometer actually possesses a best resolution of only a few meV. The sample crystal structure is of NaCl type and the first order phase transition from paramagnetic (P) phase to antiferromagnetic (AFP5) phase occurs at 16K (for phases of CeSb, see Sect. **13.3.1.a**). Photoelectron measurements were made both for the P phase at 30K and the AFP5 phase at 10K. Figure 7.23 shows the angle-resolved photoelectron spectra for both phases. The spectra exhibit complicated variations to their profiles as the incident angle of excitation light is changed. The second order differentiation of these spectra with respect to the binding energy enhances the spectral features and the spectral structures become easier to observe. In Fig. 7.24, the intensities of the structures of the second derivative are represented by the black-white gradation (gray scale), which are plotted on the plane of the binding energy (ordinate) and the $k_{||}$ (abscissa). Since the observation directions are [001] for the upper and [011] for the lower figures, the figures represent the band dispersion experimentally determined along Γ–Δ–X and Γ–Σ–M directions respectively in the Brillouin zone. At around the Γ point, the shape of

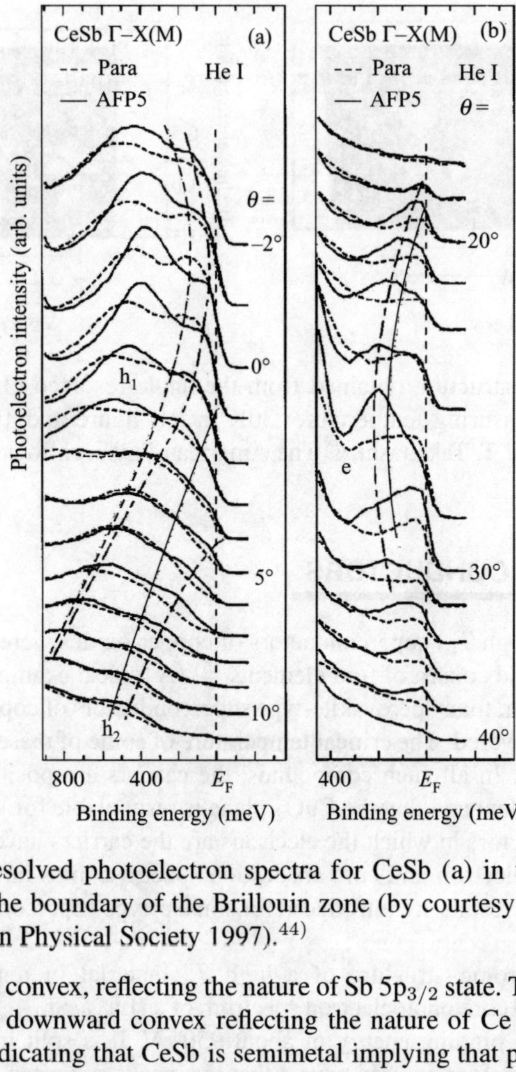

Figure 7.23 Angle-resolved photoelectron spectra for CeSb (a) in the central part and (b) around the boundary of the Brillouin zone (by courtesy of T. Takahashi, © The American Physical Society 1997).[44]

the valence band is upward convex, reflecting the nature of Sb $5p_{3/2}$ state. The bottom of the conduction band at around X point is downward convex reflecting the nature of Ce 5d state. The Fermi surface crosses these two bands indicating that CeSb is semimetal implying that positive holes around Γ point and electrons around X point are carrying the electric charges. The spread of the Fermi surface at around Γ and X points are called 'pockets'. From a comparison between the left and the right figures, it can be seen that pockets both for electrons and positive holes enlarge when the phase changes from paramagnetic (P, 30K) to antiferromagnetic (AFP, 10K). In AFP phase, Ce $4f_{5/2}$ state is supposed to locate near to the Fermi level and to hybridize with the Sb $5p_{3/2}$ state. This hybridization will ensure that the energy of Sb $5p_{3/2}$ ascends and that of Ce level descends. Therefore, high resolution angle-resolved photoelectron spectroscopy can provide detailed information close to the Fermi surface of not only semimetals as mentioned above, but also for metals.

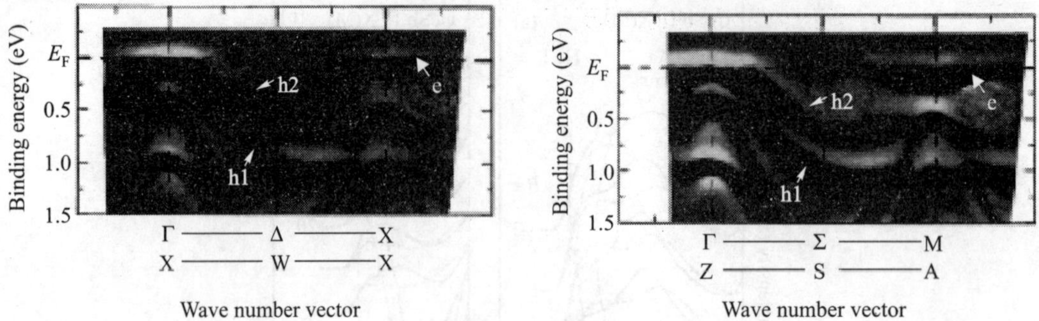

Figure 7.24 The band structure obtained from the angle-resolved photoelectron spectra of CeSb. Measuring temperature: 30K in left figure and 10K in right figure (by courtesy of T. Takahashi, © The American Physical Society 1997).[44]

7.6 HIGH T_c SUPERCONDUCTORS

High critical-temperature (high T_c) superconductors of copper oxides were discovered in 1986, which have perovskite structures and consist of four elements.[45] (A typical example of the crystal structure is shown in Fig. 9.8.) Thereafter, many perovskite-type superconductors of copper oxide composed of four or more elements were discovered. The critical temperature of some of these compounds exceeds liquid nitrogen temperature (77K). In all such compounds, the carriers are positive holes (p-type) and it is supposed that copper is divalent and that the CuO_2 plane is responsible for the superconductivity. More recently, n-type superconductors in which the electrons are the carriers have been discovered.[46] Many experiments on their photoelectron and core absorption spectroscopies have been performed.[47-49] In the following, spectroscopic results are introduced for some p-type superconductors and their electronic structure described.

First, consider the electronic structure of a high T_c material in a non-superconducting state. Figure 7.25 shows the Cu 2p core photoelectron spectrum of a $(La_{0.96}Sr_{0.04})_2\,CuO_{4-y}$ single crystal.[50] The Cu 2p core level has a binding energy of about 950 eV. It is split into $j = 3/2$ and 1/2 levels by spin-orbit interaction, therefore one can expect that two peaks appear in the photoelectron spectrum with an intensity ratio of 2 to 1. However, four peaks are found in this material. Similar phenomena have been observed for many materials with strongly-correlated electron systems, so that this material most probably falls into the same category. According to theory,[51] the appearance of four peaks originates from two different rearrangement processes for the electrostatic shielding of one Cu^{2+} 2p positive hole by valence electrons. It is explained as follows. The valence state is a hybridized between the Cu 3d and O 2p states. O is the ligand and its 2p state is represented by L. One of two processes is described as a $\underline{2p}3d^9$ state having two positive holes both in Cu 2p and Cu 3d orbitals, where __ means a state with one hole. The other is a $\underline{2p}3d^{10}\underline{L}$ state in which one d positive hole in Cu is filled with an electron, which is supplied from the valence state L thus resulting in the valence state with one positive hole. Since this process does not occur when Cu is monovalent, the experimental result strongly suggests

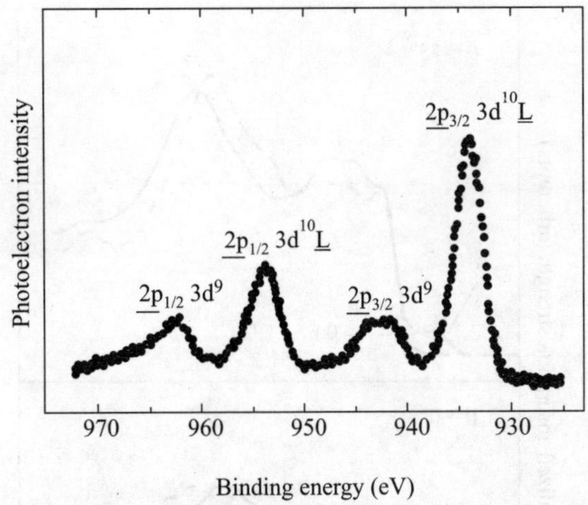

Figure 7.25 Cu 2p core photoelectron spectrum of $(La_{0.96}Sr_{0.04})_2CuO_{4-y}$ single crystal (by courtesy of T. Takahashi, © The American Physical Society 1988).[50)]

that Cu is divalent in this material in the non-superconducting state. The energy difference between the split peaks of $2p_{3/2}$ or those of $2p_{1/2}$ yields a Coulomb repulsion energy U_{dd} (d : d electron). U_{dd} is from 7 to 8 eV and this suggests that it is stable for an electron to remain close to the Cu atom. U_{dd} is of almost the same magnitude of the valence band width W. Therefore, the electronic structure of the high T_c superconductor is best described using the Hubbard model or the charge transfer model. We can next see how this picture changes when a non-superconducting state (an insulator) changes into a superconducting state.

Figure 7.26 shows the linear polarization dependence of the O K absorption spectra in $Bi_2Sr_2Ca_{1-x}Y_xCu_2O_8$ $(x = 0 \sim 0.6)$.[52)] For $x = 0$, this material is a superconductor with $T_c = 81$K, while for $x = 0.6$, it is a non-superconductor even at 4K. The total electron yield (TEY) measurements were made at room temperature at the PF. In Fig. 7.26, the spectra for $\theta = 0°$ were taken when the light was incident parallel to the c-axis of the crystal $i.e.$, perpendicular to the ab-plane. In this case the electric vector $\boldsymbol{E}$ is perpendicular to the c-axis $(\boldsymbol{E} \perp \boldsymbol{c})$, and in other words, $\boldsymbol{E}$ is parallel to the ab-plane. The spectra for $\theta = 75°$ correspond to those when the light was incident with an angle of incidence of 75° to the normal of the ab-plane (grazing incidence to the ab-plane). In this case, $\boldsymbol{E}$ is approximately parallel to the c-axis. In the case of $\theta = 75°$, the absorption spectra near the threshold of O K absorption for both $x = 0$ and $x = 0.6$ samples rise steeply around 530 eV and resemble each other, except for a small shoulder to be seen at 528.5 eV for $x = 0$. These results suggest that the electronic state along the direction parallel to c-axis does not differ very much between the superconducting and non-superconducting samples. In the case of $\theta = 0°$, however, the absorption structures at the O K edge for $x = 0$ and $x = 0.6$ samples differ quite strikingly from each other, that is, the sharp peak at 528.5 eV for $x = 0$ is not seen for $x = 0.6$. Since Ca is divalent and Y is trivalent, the decrease in x implies that there is an increase in the number of the positive holes in this sample. Correspondingly, the sharp peak arises from the positive hole doping and is a characteristic of the superconducting sample. The photon energy of this peak corresponds to the lowest excitation energy for 1s electrons in the O atom. In photoelectron experiments,[52)] the Fermi edge can be observed for $x = 0$, but not for $x = 0.6$,

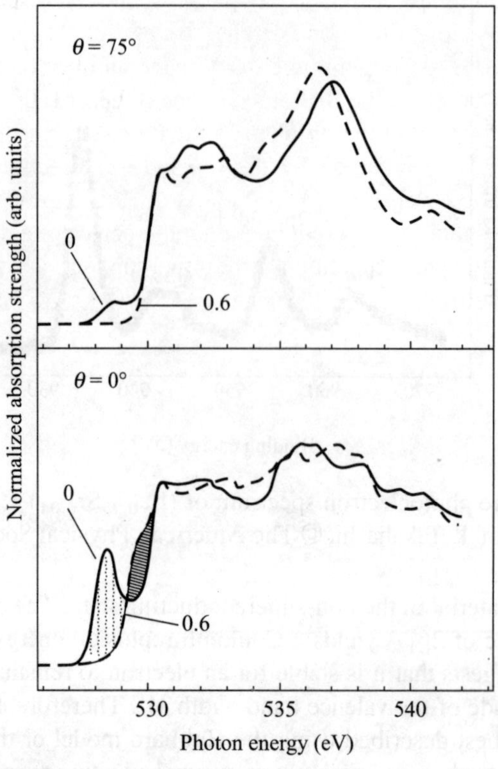

Figure 7.26 O K absorption (TEY) spectra of $Bi_2Sr_2Ca_{1-x}Y_xCu_2O_8$ ($x = 0 \sim 0.6$) at room temperature. θ is the angle of incidence of the light measured from a normal to the *ab*-plane (by courtesy of T. Takahashi, © World Scientific 1993).[52]

and therefore it can be said that the insulator changes to metal due to doping by the positive hole creating a new state near the Fermi edge. Since this phenomenon is observed only for $\theta = 0°$, it suggests that superconduction occurs in the ab-plane, that is, CuO_2 plane. Moreover, it is suggested that the O $2p_{x,y}$ positive hole is the actual carrier in superconduction, because the sharp peak in the O K (1s) absorption indicates the existence of the O $2p_{x,y}$ positive hole. In the case of $\theta = 0°$, if the integrated absorption intensity is assumed to be equal for the samples of $x = 0.6$ and 0, when x changes from 0.6 to 0, the intensity of the shoulder structure at 529.5 eV for the $x = 0.6$ sample seems to move to the sharp peak at 528.5 eV for the $x = 0$ sample. These results perhaps indicate that the structure at 529.5 eV in the insulator for the $x = 0.6$ sample corresponds to the upper Hubbard band while the new peak at 528.5 eV in the superconductor for $x = 0$ corresponds to the new electronic state generated within the electron transfer gap. In conclusion, this material is thought to change from a Mott insulator to a metal which can be a superconductor.

One of the direct observations of a superconducting energy gap is given here for $Bi_2Sr_2CaCu_2O_8$. Figure 7.27 shows the high resolution angle-resolved photoelectron spectra of this material in its superconducting state (at 25K) and in its non-superconducting state (normal state at 100K).[53] Measurements were made using the emitting-angle scanning method combined with an excitation energy of 21.2 eV in the ab-plane. Coordinate points (0, 0), (0, π), (π, 0) and (π/2, π/2) in the Brillouin zone of this perovskite structure are designated as Γ, X, Y and $\overline{\text{M}}$, respectively. Measurements were made on the emitted electrons along directions parallel to the *a*-axis (ΓX) and the *b*-axis (ΓY), as well as the diagonal direction (Γ$\overline{\text{M}}$). As shown in the inset, wave number vectors for the emitted electrons correspond to the A, B and C points in the Brillouin zone and their magnitudes can be determined from emission angle θ against the normal of the *ab*-plane. When the state changes from its 'normal' state to a 'superconducting' state, the Fermi energies for the ΓX and ΓY directions locate at the same position to those in the normal state, however, for (Γ$\overline{\text{M}}$), it shifts about 10 meV to the higher binding energy side. This result suggests to us that the superconducting gap opens along the (Γ$\overline{\text{M}}$) direction. From this conclusion, the symmetry of the wave function for the superconducting state is seen not to be a normal isotropic s state but is an anisotropic $d_{x^2-y^2}$-like state.

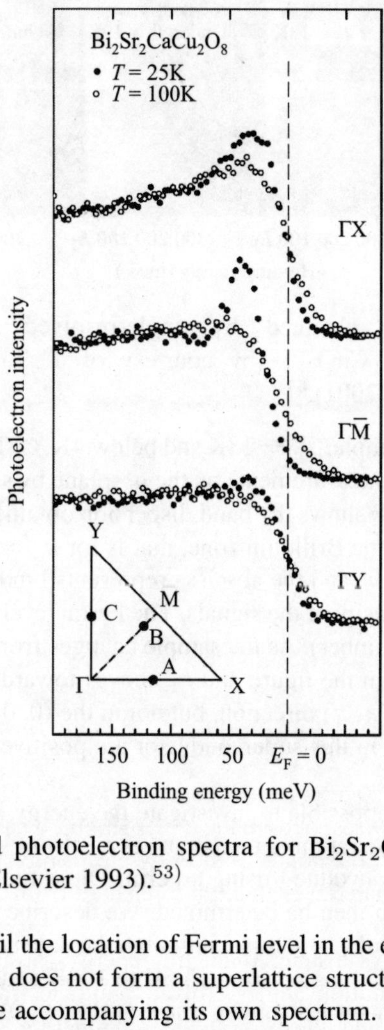

Figure 7.27 Angle-resolved photoelectron spectra for $Bi_2Sr_2CaCu_2O_8$ (by courtesy of T. Takahashi, © Elsevier 1993).[53)]

We can also investigate in detail the location of Fermi level in the energy band. The example given is $Bi_{1.80}Pb_{0.38}Sr_{2.01}CuO_{6-\delta}$, which does not form a superlattice structure, in contrast to $Bi_2Sr_2CuO_{6-\delta}$, which has a superlattice structure accompanying its own spectrum. The measurements were made on under-doped and also on over-doped samples, where the number of positive holes is fewer and also larger than the optimal value giving the highest T_c, respectively. For the under-doped sample, $T_c = 18\text{K}$

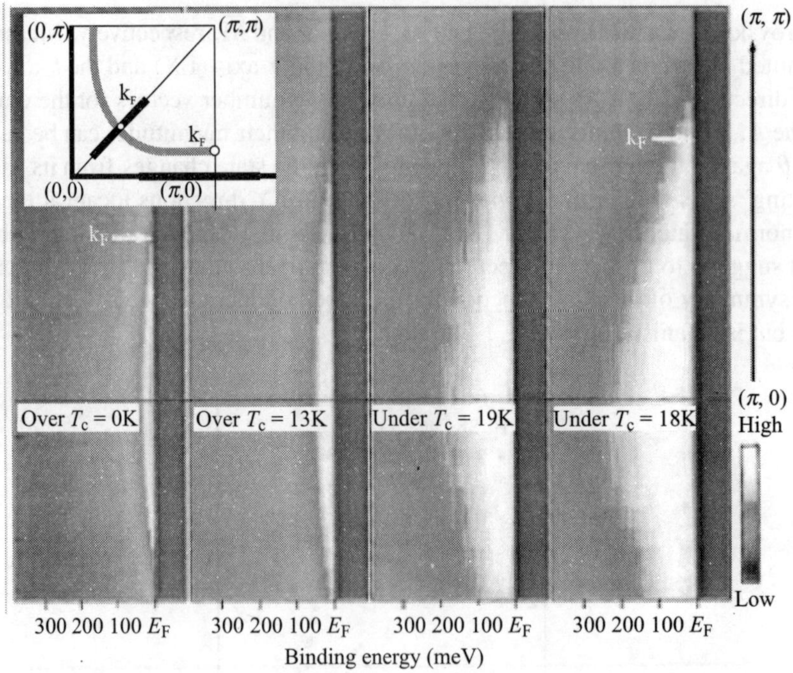

Figure 7.28 Band structure obtained by angle-resolved photoelectron spectra of $Bi_{1.80}Pb_{0.38}Sr_{2.01}CuO_{6-\delta}$ (by courtesy of T. Takahashi, © The American Physical Society 2001).[54]

and 19K, and for the over-doped sample, $T_c = 13$K and below 4K.[54] The band dispersion was obtained from angle-resolved photoelectron measurements on the ab-plane by scanning both the azimuthal angle ϕ and the polar angle θ. Figure 7.28 shows the band dispersion obtained along the direction connecting the $(\pi, 0)$ and the (π, π) points in the Brillouin zone, that is for $k_x = \pi$ and $-0.13\pi \leq k_y \leq 0.25\pi$. In this figure, the ordinate represents k_y and the abscissa represents binding energy, and the brightness of the plotted points indicates the intensity of the signals. The Fermi level crosses the energy band at points indicated by k_F (the Fermi wave number). As the sample changes from over-doped to under-doped and the T_c changes in the order given in the figure and k_F moves towards the (π, π) point. Similarly, the crossing was found in the (0, 0) – (π, π) direction, but not in the (0, 0) – $(\pi, 0)$ direction. These results lead to the conclusion that carriers in this superconductor are positive holes and that the Fermi surface locates at around the (π, π) point.

When a sample is at 0K, it is impossible to investigate the energy band above the Fermi level using photoelectron spectroscopy. However at finite temperatures electrons spreading over the Fermi level are detectable. Measured data are deconvoluted using a Fermi-Dirac function to remove any temperature effects and the band dispersion can then be determined. We describe here the experimental results for the over-doped $Bi_{1.80}Pb_{0.38}Sr_{2.01}CuO_{6-\delta}$ sample, although the data are not given.[55] The measurement was made at a temperature above T_c ranging from 13.5K to 140K and it was found that the Fermi level crosses the energy band with a downward convex shape at the $(\pi, 0)$ point in the Brillouin zone. The overall summary is that electron-like and hole-like carriers locate at around the $(\pi, 0)$ and the (π, π) points, respectively.

Finally, it is to be noted that iron-based superconductors have been found[56] and that many efforts have been made to achieve high T_c.[57]

7.7 ORGANIC MATERIALS

7.7.1 Soft X-ray absorption by organic materials

Organic materials are now being widely used as electronic materials, *e.g.*, in devices such as organic transistors, solar cells and electroluminescent elements.[58–61] Soft X-ray absorption spectroscopy and photoelectron spectroscopy are ideal tools for investigating and understanding their behavior and characteristics.[62] As an example, we consider the study of molecular orientation applying soft X-ray absorption spectroscopy in a study of the absorption fine structure near the C K edge. Measurements were made on PTFE (poly(tetrafluoroethylene)) thin films using linearly-polarized light.[63] This polymer is commercially well-known as Teflon. The molecular structure of PTFE is $(CF_2)_n$, where C atoms are bound to each other with a single bond forming a molecular chain in which the F atoms are configured around the chain. In a PTFE thin film evaporated on a Cu substrate in vacuum, it is supposed that the direction of the molecular chain is not exactly linear, but that its orientation changes along the length of the chain as illustrated in the inset of Fig. 7.29(a). Figure 7.29(a) shows absorption spectra obtained from TEY measurements taken near to the C K edge for the thin film after evaporation. Two peaks designated as σ^* (C–F) and σ^* (C–C) appear near to the absorption edge. The first one originates from the transition from the C 1s level to the σ^* orbital, which is a σ type empty orbital around the C–F bond which is perpendicular to the C–C bond. The second originates from the transition to the other σ^* orbital around the C–C bond, which is parallel to the C–C bond. According to selection rules, the transitions are allowed when the electric vector of the incident light is parallel to the bond orientation of the σ^* orbital but is forbidden when perpendicular. That is, for the C–C bond, the transition is allowed, when the electric vector is parallel to C–C bond, while for C–F bond, it is allowed, when perpendicular to the C–C bond. From this, we are able to assess the molecular orientation of PTFE. In reality, the PTFE molecule possesses a spiral structure and the relationships in direction between the C–C and C–F bonds are quite complex, but in spite of this, the following investigation yields useful information. The intensity of the σ^* (C–C) peak for normal incidence ($\hat{\theta} = 90°$) is larger than that for grazing incidence ($\hat{\theta} = 15°$). This leads to the conclusion that the molecular chain must be roughly parallel to the substrate. If the distribution of the molecular orientations is to be Gaussian then the half width of the distribution is estimated to be 68°, which means that the sample has a significantly random structure. Figure 7.29(b) shows the spectra from the sample after a rubbing treatment, in which the surface of the thin film was abraded by a long-haired felt sheet. It is generally assumed that molecular chains in polymers become aligned along rubbing direction. Following this abrasion, the spectral features changed remarkably. Since the intensity is stronger for the σ^* (C–C) peak than for the σ^* (C–F) peak in the normal incidence case, it is concluded that the direction of the molecular chain is essentially parallel to the substrate. The orientation distribution width became narrower and was estimated to be 34°, implying that the degree of molecular orientation is altered and increased by rubbing.

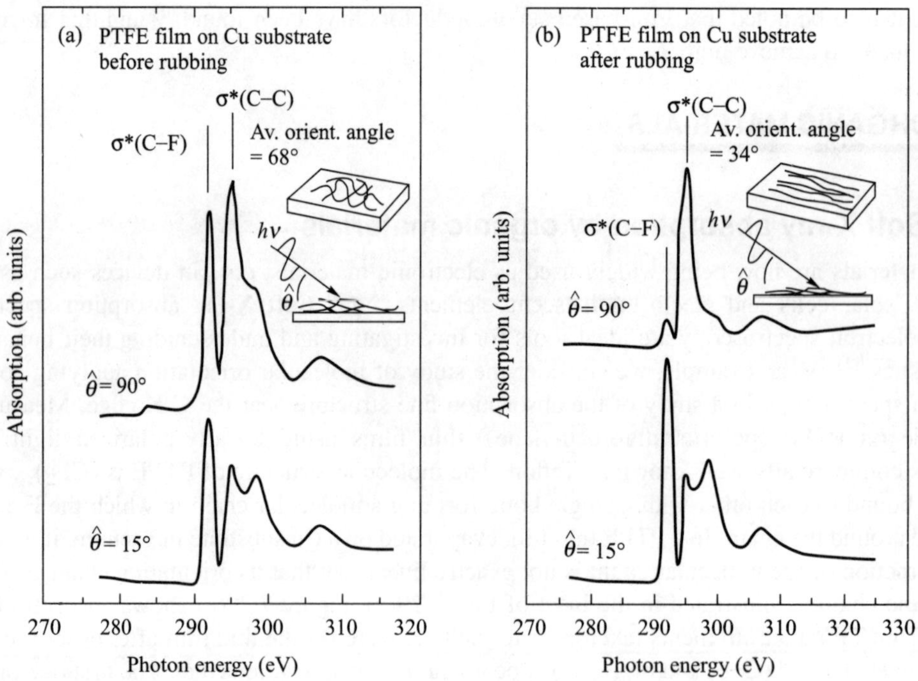

Figure 7.29 C K absorption spectra obtained using the TEY method for a Teflon (PTFE) evaporated thin film on a Cu substrate. (a) Before (as evaporated) and (b) after rubbing treatment.

7.7.2 The photoelectron spectroscopy of organic materials

Photoelectron measurements on organic materials have sometimes been unsatisfactory in the past because sample preparation, including evaporation was undertaken in the same ultra-high vacuum chamber accommodating the energy analyzer, so that the inside of the chamber often became seriously contaminated. However this problem has been largely overcome by separating the preparation chamber and the analyzer chamber. The prepared sample is simply transferred from a preparation chamber directly to an analyzer chamber under ultra-high vacuum conditions. There may also be problems of sample charge-up and of radiation damage caused by the incident VUV light.[62)] In many cases, organic materials are insulators and when they are in the form of powder or crystal, they are easily charged up during photoelectron measurements because any positive holes created cannot be properly neutralized, thus causing deformation and a shift of the spectrum. To overcome this, it is important to prepare thin samples with thicknesses of only a few tens of nm by vacuum evaporation. Otherwise the sample must be prepared in the atmosphere by a spin coating method, in which a solution of the organic material is spread on a circular plate being rotated at high speed yielding a uniform thin film. Since synchrotron radiation is exceptionally brilliant, the exit light from a monochromator can be focused sharply on the sample. Sometimes a mesh-screen has to be inserted into the light path to reduce the light intensity to avoid charging effects. This method is also useful to decrease radiation damage to the sample, which may result in photodissociation or in changes to molecular orientation. It is very important during measurement of organic materials to check reproducibility of the data by measuring several times under the same conditions.

a. The valence band of ionic liquids: The density of states of the valence band obtained from angle-integrated photoelectron spectra can provide important information useful for the selection of organic materials in applications to electronic devices. The upper part of the valence band, in particular, will establish characteristics such as their electrical and optical properties. Knowledge of the energy of the top of the valence band, corresponding to the ionization potential, allows us to determine the kind of carriers needed for charge conduction. If the top of the valence band is shallow then the material will be an electron-donor and if the top is deep the material will be an electron-acceptor. In the former a positive hole can be injected and transported more easily than an electron but *vice versa* in the latter. The magnitude of the density of states at the top of the band will have a strong influence on the electrical conductivity. Since it is possible to measure the dependence of intensity of spectral peaks on the excitation energy using synchrotron radiation, the peaks can be assigned in terms of atomic orbitals and then a more detailed analysis on the density of states become possible. As examples we give some results for one of the molecular ionic liquids, melting salts at room temperature, which have an extremely low vapor pressure and attract attention from their use as stable electrolytic liquids and as solvents for clean chemistry. Figure 7.30 shows the photoelectron spectrum of $Bmim^+PF_6^-$ containing C, N, F as well as P (see the chemical structure in the inset). This is presented together with the results

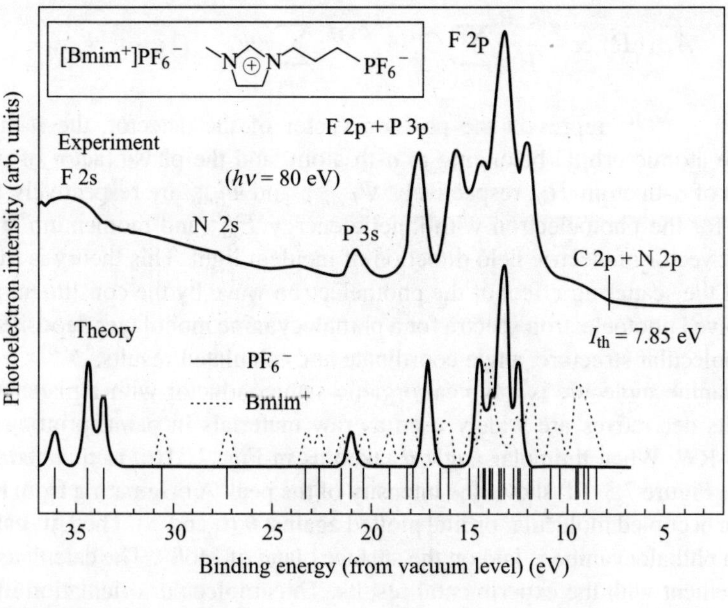

Figure 7.30 Photoelectron spectrum and molecular orbital calculation results for the ionic liquid $Bmim^+PF_6^-$.

from molecular orbital calculations.[64] However in the figure, only the photoelectron spectrum with an excitation energy of 80 eV is given, although the spectra were taken using various excitation energies. By comparing these spectra with each other and with the calculated results, the valence electronic structure could be analyzed. In the figure, the structures are assigned and the measured spectrum is well reproduced by the intensity sum of the theoretical curves for positive $Bmim^+$ ion (dotted curve) and for negative PF_6^- ion (solid curve). At the right-hand side in the spectrum, only the contribution of the $Bmim^+$ ion is found, which means that the top of the valence band, giving the particular properties to this liquid originates from the valence state of the positive $Bmim^+$ ion. This is in marked contrast to the fact that the top of the valence band in alkali halides such as Na^+Cl^- originates from the valence states of the negative ion. In the ionic liquid, the positive and negative ions are surrounded by each other and the ionic liquid becomes stabilized. The stabilization energy can be estimated using the difference in binding energy of both ions given by the theoretical curves.

b. The structure determination of thin films from the angular dependence of their photoelectron emission: When organic molecules align in thin films, the angle-resolved photoelectron spectra will show a remarkable angular dependence reflecting the structure and orientation of the molecule. The orientation of the molecules in the film can be qualitatively determined by taking symmetry into consideration and using selection rules. The calculation methods for photoelectron intensity have been newly developed using the independent atomic center (IAC) approximation on the basis of molecular orbital theory, which give satisfactory results.[65] In this theory, any photoelectron wave function, that is the wave function of the final state, is approximated by a sum of waves of free electrons emitted from each atomic orbital in every molecular orbital. Therefore, it includes interference effects which can be exploited to analyze the molecular structure and orientation. The photoelectron intensity from n-th molecular orbital is expressed by $|A_{\text{tot}}(\boldsymbol{R})|^2$, where $A_{\text{tot}}(\boldsymbol{R})$ is given by

$$A_{\text{tot}}(\boldsymbol{R}) \propto \frac{e^{i\boldsymbol{k}\cdot\boldsymbol{R}}}{R} \sum_{a} C_a^n e^{-i\boldsymbol{k}\cdot\boldsymbol{R}_a} \sum_{l,m} \langle \Psi_{E_{\text{kin}},k} | \boldsymbol{e}\cdot\boldsymbol{r} | \Psi_{lmn} \rangle. \tag{7.8}$$

Here, R, C_a^n and $e^{-i\boldsymbol{k}\cdot\boldsymbol{R}_a}$ represent the position vector of the detector, the n-th molecular orbital coefficient for the atomic orbital belonging to a-th atom, and the phase factor of photoelectron wave from the position of a-th atom $\boldsymbol{R}_a$, respectively. $\Psi_{E_{\text{kin}},\boldsymbol{k}}$ and Ψ_{lmn} are respectively the wave functions of the final state for the photoelectron with kinetic energy E_{kin} and momentum $\boldsymbol{k}$ and of the atomic orbital. $\boldsymbol{e}$ is a unit vector of electric field direction of incident light. This theory is further developed by taking account of the scattering effect of the photoelectron wave by the constituent atoms. Figure 7.31 shows angle-resolved photoelectron spectra for a phthalocyanine monolayer deposited on a MoS_2 single crystal with its molecular structure, angle coordinate and calculated results.[65,66]

The phthalocyanine molecule is a typical organic semiconductor with a planar structure as shown in Fig. 7.31(a). Its derivatives are widely used as raw materials in paint, printing ink and recording materials for CD-RW. When the polar angle θ taken as in Fig. 7.31(c) is changed, the spectral shape alters remarkably. Figure 7.31(d) shows the intensity of the peak A originating from HOMO, the highest energy state in the occupied molecular orbital plotted against θ (○ and △). The IAC calculation was made for the case when phthalocyanine is laid on the surface plane of MoS_2. The calculated result (solid line) shows good agreement with the experimental results. Thus molecular orientation in thin organic films can be basically determined by the combination of the angle-resolved photoelectron data and the results from IAC calculations. In the case of thin films having only short range order or for ultra-thin films of monolayers, it is however difficult to use X-ray diffraction to determine the molecular orientation. Therefore in such cases the angle-resolved photoelectron measurements are very important.

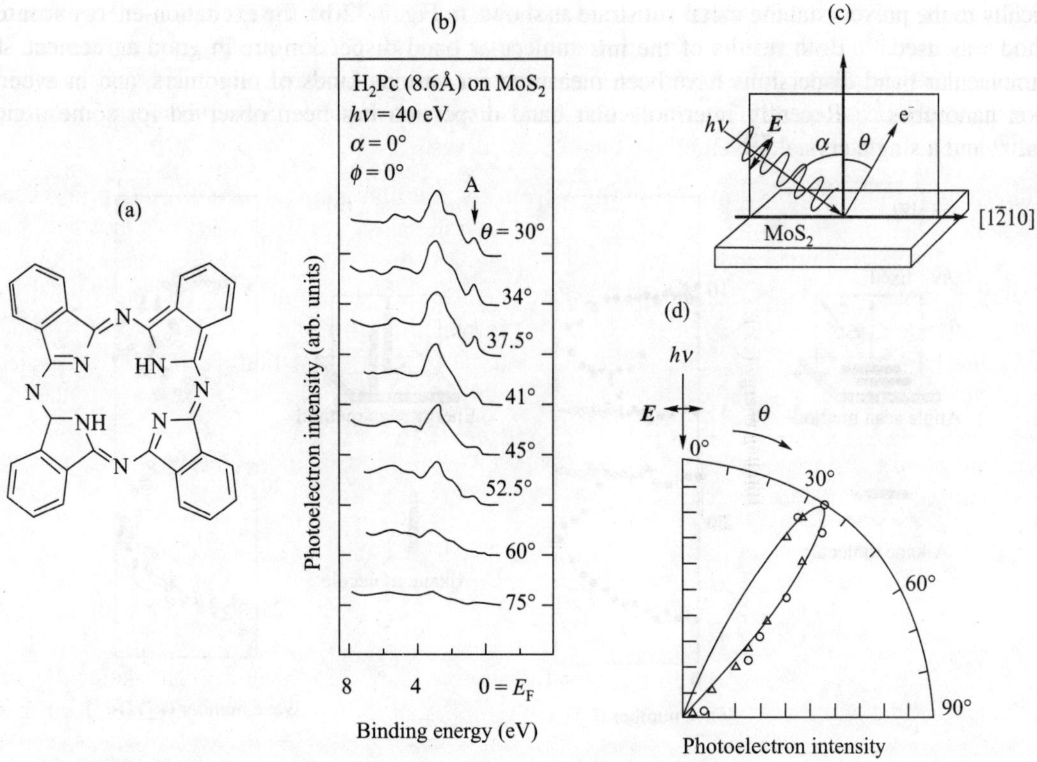

Figure 7.31 Angle-resolved photoelectron spectra and angular dependence of the intensity of the HOMO peak (A) for a hydrogenated phthalocyanine film on a MoS_2 single crystal (by courtesy of N. Ueno). See details in text.

c. Intramolecular and intermolecular band dispersion: The energy band dispersion in crystals can be directly determined from angle-resolved photoelectron experiments. For many inorganic crystals where periodic lattice is formed by strong interaction among atoms or ions, the band dispersion has been determined by such experiments. However for organic crystals where the periodic lattice is formed by molecules, the experiments to determine the band dispersion, in this case called the intermolecular band dispersion, has proved to be difficult since the band dispersion is small resulting from weak intermolecular interaction due to the small overlapping of wave functions of neighboring molecules. In the case of polymers however, where molecules align periodically forming a normal chain, the overlapping of wave functions of neighboring atoms is quite large and results in observable dispersion called intramolecular band dispersion. The intramolecular band dispersions of polymers have been studied using oriented films of their oligomers in which the degree of polymerization ranges from 2 to about 20, as model compounds. Figure 7.32 shows the intramolecular band dispersion determined by angle-resolved photoelectron measurements for one of the alkane molecules having a normal chain and is a model compound for polyethylene.[67] The alkane molecule is a stick-like structure in which C atoms form a zig-zag chain as shown in the figure. When the sample was an epitaxial film in which alkane molecules lie on Cu (100) surface as shown in Fig. 7.32(a), the emission-angle scanning method was used.[66] When the sample was the evaporated film in which the alkane molecules stand vertically to the polycrystalline metal substrate as shown in Fig. 7.32(b), the excitation-energy scanning method was used.[68] Both results of the intramolecular band dispersion are in good agreement. The intramolecular band dispersions have been measured for various kinds of oligomers, and in even for carbon nanotubes.[69] Recently intermolecular band dispersion has been observed for some organic films,[70] and a single crystal.[71]

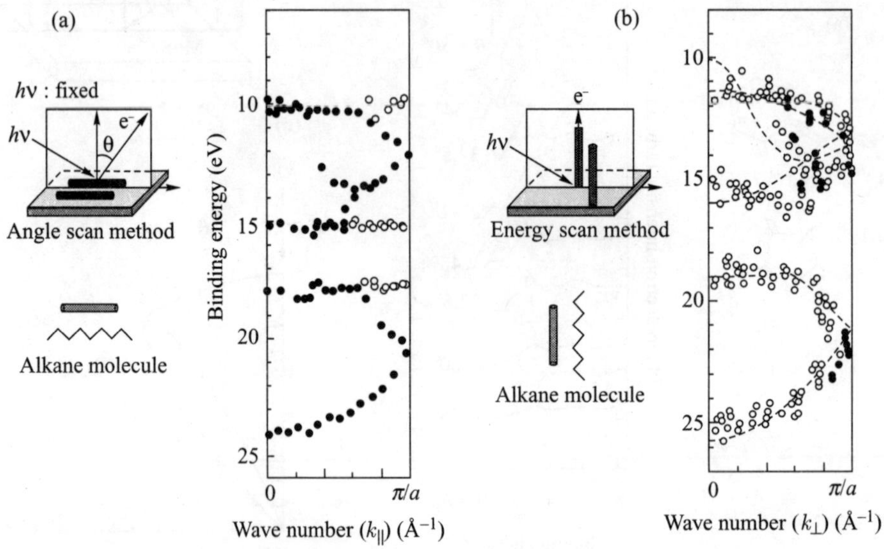

Figure 7.32 Intraband dispersion in a normal chain alkane molecule measured (a) using the emission-angle şcanning method for molecules lying parallel to the substrate and (b) using the excitation-energy scanning method for molecules standing vertically on the substrate. The broken curves represent band calculation results by Karpfen (the binding energy in the calculation is multiplied by 0.8). $E_{\rm vac} = 0$.

d. The valence band of metal fullerene: Fullerene represented by the molecular formula C_{60} appeared in 1985 as a new material composed of only C atoms.[72] C_{60} is known to be a molecule having a shape like a soccer ball. In the solid state C_{60}, the soccer balls are configured at each lattice point of an fcc structure. When C_{60} combines with metal, two different kinds of compound are formed. One is type in which metal atoms are configured interstitially between lattice points of the balls and another, included inside the balls. Superconductivity occurs in both types of compounds. Alkali-metal-doped C_{60} belongs to the former type. Valence band photoelectron spectra of Rb_xC_{60} are shown in Fig. 7.33.[73] This sample was prepared by evaporation of Rb on the fullerence film which was evaporated on the Mo substrate coated with an Au film. Measurements of electrical resistance were made on those evaporated on the quartz substrate which is situated near the Mo substrate. In the figure, the number attached to each spectrum represents integrated evaporation times in minutes. The energy position of the Fermi level is calibrated with that of the evaporated Au film. Measurements were made at room temperature. Initially, the pure C_{60} sample shows no density of states near Fermi level. However, a finite density of states is observed near the Fermi level after evaporation times of from 15 to 45 minutes, showing the metallic nature. However when the evaporation was increased, the density of states at the Fermi edge weakened and eventually disappeared. From electric resistance measurements on Rb_xC_{60}

films on the quartz substrate, it was found that the resistance diminished gradually with changing the evaporation time from 0 to 45 minutes and at 45 minutes, it reached a minimum value. For times over 45 minutes, the resistance began to increase. From another experiment, it is known that Rb_3C_{60} exhibits superconductivity.[74] Therefore, the sample obtained by evaporation for 45 minutes can be considered to correspond to Rb_3C_{60}. As seen from the experiment, the change of the film from insulator to metal is clearly reflected in the change of magnitudes of density of states at the Fermi level.

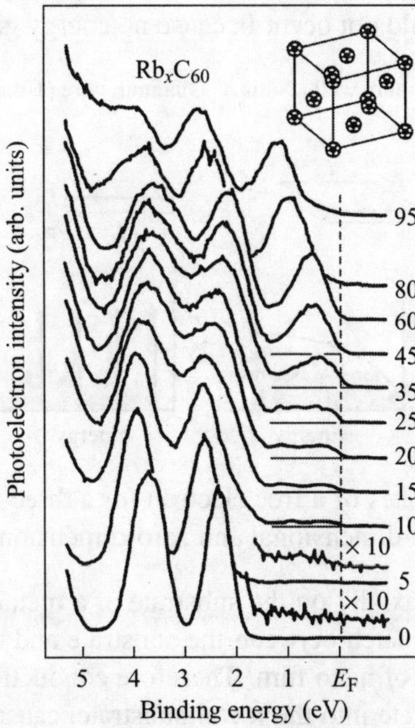

Figure 7.33 Photoelectron spectra for Rb_xC_{60}. The numbers attached to the curves indicate the integrated evaporation time in minute for Rb on a C_{60} film on an Au-coated Mo substrate (by courtesy of T. Takahashi).[73]

7.8 THE PHOTOELECTRON SPECTRA OF MESOSCOPIC METAL SYSTEMS

A mesoscopic system is a group of materials, in which more than one of their size parameters of thickness, width and length is of the order of nm. Such fine materials exhibit intermediate characteristics which lie between atoms, molecules and bulk crystals.[75] Figure 7.34 shows schematic density of states in three-, two-, one- and zero-dimensional materials. Here the energy band dispersion is assumed to be $E = a(k_x^2 + k_y^2 + k_z^2)$, $E = a(k_x^2 + k_y^2)$ and $E = ak_x^2$ for three-, two- and one-dimensional cases, respectively (see App. **I**). In the zero-dimensional case, the energy band has no wave number dependence. In the low-dimensional structures as shown in the figure, the motions of electron in the direction in which the structure size is narrowed, is confined by potential barriers at the structure surface. This effect quantizes, that is makes electronic states discrete, so that it is called the 'quantum size effect'. The shape of density of states depends on both the dimension and the size in the confinement direction. A two-dimensional thin film with nm order thickness creates a 'quantum well', where the freedoms of motion of the electron remains in *xy*-plane parallel to the thin film, while movement in z direction, (the thickness direction) is restricted. This confinement of motion causes a change in the shape of the density of states from the smooth function of $\sqrt{E}$ in the three-dimensional case to a step function. In so-called 'semiconductor superlattices' in which wide gap and narrow gap semiconductors are alternately layered, the confinement potential for electron and positive hole arises from the difference between the energy gaps of both kinds of semiconductors resulting in the quantum size effect. Such semiconductor superlattices have been already applied to various kinds of electronic devices. In metal mesoscopic films, hereafter called nano films, quantum size effects have been observed by photoelectron spectroscopy, although it is considered that such effects should not occur because no energy gap exists above the Fermi level.

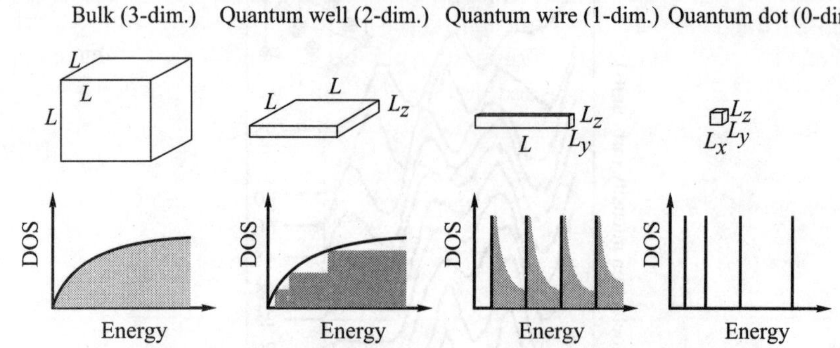

Figure 7.34 The density of states of a free electron for a three-dimensional crystal, and two-dimensional, one-dimensional and zero-dimensional mesoscopic materials.

In a metal nano film grown epitaxially on the substrate of a metal single crystal, the relative energy gap originating from the band mismatch between the substrate and the nano film exists in z direction, the direction normal to the surface of nano film. Therefore conduction electrons traveling in a vertical direction in the film cannot propagate into the metal substrate, causing confinement of the conduction electrons. This phenomenon cannot be observed by electric conductivity measurements, which include information over wide $\boldsymbol{k}$-space. Of course, angle-resolved photoelectron measurements can specify the $\boldsymbol{k}$ of emitted photoelectrons and when the photoelectrons are measured in the vertical direction for the nano film, the quantum well states resulting from the quantization of $\boldsymbol{k}$ vector in z direction can be observed. The upper part of Fig. 7.35 shows the angle-resolved photoelectron spectrum in the z direction for an Ag (111) nano film with the thickness of 5 nm grown on the Cu (111) seed layer on Si (111) 7×7 clean surface.[76,77] In it, a prominent peak is found just below the Fermi level, which corresponds to the surface state (called a 'Shockley state') for a Ag (111) clean surface. On its high binding energy side, weak peaks marked by n, whose energy positions depend on the film thickness are found, originating from quantum well states and are not found in bulk Ag. These weak peaks are only observed in the binding energy range from 0.3 to 0.9 eV corresponding to the energy difference of the tops of filled valence bands in the z direction for bulk Ag and Cu, as illustrated by gray range in the lower part of Fig. 7.35. Since this energy range corresponds to the s-p hybridization gap for bulk Cu, the s-p valence electrons in the Ag nano film are reflected at the boundary of Ag and Cu, when the electrons satisfy the phase condition of $2k(E)d + \Phi(E) = 2n\pi$ and wave number in z direction is quantized, resulting in quantum well state. Here, $k(E)$ is the wave number expressed as a function of binding energy, the inverse function of the band dispersion for the valence band to be quantized, $\Phi(E)$ is the summation of phase shift caused by the reflection at the boundaries of the nano film and the Cu seed layer, as well as the nano film and the vacuum, and n is quantum number (integer). The band dispersion of the quantum well state within the xy-plane where freedoms of motion for electrons still remain can be also observed using angle-resolved photoelectron spectroscopy.[76)]

In addition to these studies on two-dimensional metal nano films, studies on zero-dimensional metal nano particles is progressing using photoelectron spectroscopy, which is a very useful probe for the investigation of other quantum properties such as single electron tunneling and single electron charging phenomena. For example, in the photoelectron spectrum of metal nano particles decorated by organic molecules held on the HOPG (highly-oriented pyrolitic graphite) substrate, the spectral profile is affected by the tunneling effect in which positive holes created in the final states of photoelectron emission propagate from nano particles to the substrate *via* surface decorating molecule.[78)]

Recently novel electronic states depending on spin have been studied, which appear in some particular low-dimensional systems formed at surfaces or interfaces.[79,80)] The low-dimensional electronic state specific to graphene has also been investigated.[5)]

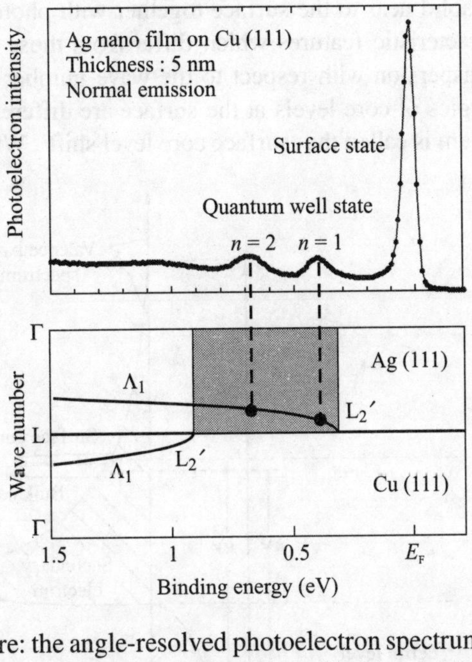

Figure 7.35 Upper figure: the angle-resolved photoelectron spectrum of a 5 nm thick Ag nano film grown on Cu (111) measured in a direction vertical to the film surface.[77)] Lower figure: dispersion of the s-p valence band around the Fermi level in the vertical direction (ΓL direction in Briliouin zone) for bulk Ag and Cu. n is a quantum number.

7.9 SOLID SURFACES

Atomic bonds and structures at the solid surface are quite different from those inside the solid because of the absence of matter above the surface. The surfaces of semiconductors and metals are rather active sites where physical or chemical adsorption phenomena occur and where chemical reactions can be stimulated, sometimes generating new compounds. Studies on atomic structures and electronic states as well as of surface photochemical reactions on clean and adsorbed surfaces are fundamentally important in the development of semiconductor devices and catalyzers *etc*. There are very many different kinds of surface studies using synchrotron radiation.[81] Here we describe structure studies on clean and adsorbed surfaces employing photoelectron spectroscopy. Studies on adsorbed surfaces using fluorescence X-ray analysis and infrared spectroscopy are described respectively in Sects. **11.1** and **13.2**. Photochemical reactions at adsorbed surfaces are given in Sect. **8.1**. Studies of chemical reactions at solid surfaces such as catalyzers under ambient pressures of reaction gases using photoelectron spectroscopy in conjunction with differential pumping systems are given elsewhere.[82]

7.9.1 Photoelectron spectra of solid surfaces

As described in Sect. **7.3.2**, the escape depth of a photoelectron in any solid is estimated to be a few or a few ten times of the lattice constant and therefore photoelectron spectra respond very sensitively to the electronic state at a solid surface. Figure 7.36 shows a schematic picture of the electronic states at a surface and of the bulk solid near to the surface together with photoelectron spectra. The valence state of the surface has characteristic features which differ from those of the bulk, because the state has two-dimensional band dispersion with respect to the wave number vector parallel to the surface. In addition, the binding energies of core levels at the surface are different from those in the bulk. The energy difference between them is called the 'surface core level shift'. When valence and core electrons

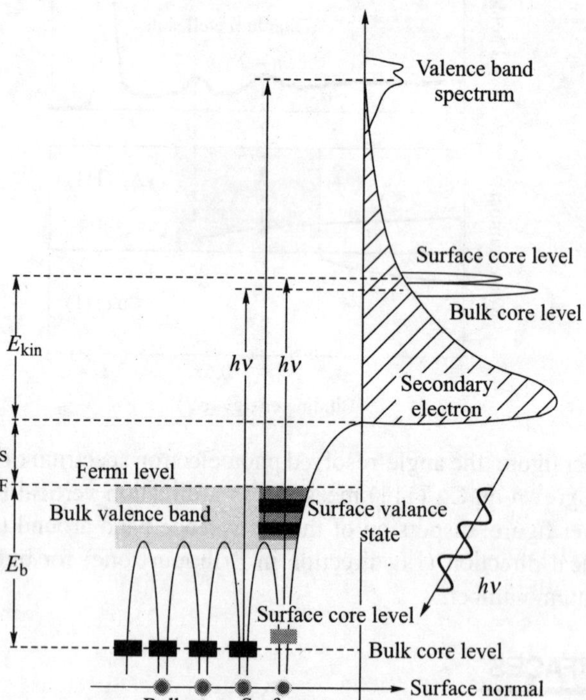

Figure 7.36 Schematic views of the electronic state and the photoelectron spectrum of a solid surface and of the neighboring bulk solid.

are excited by VUV or soft X-rays, a photoelectron spectrum is obtained as shown in Fig. 7.36. In the valence photoelectron spectrum, the peaks originating from the surface and bulk states overlap, some of which are distinguishable where the relative intensities between those peaks depend on the nature of the materials. Similarly in the core photoelectron spectrum, the peaks originating from the surface, whose binding energy is core-level-shifted and those from the bulk will overlap and some of them are also distiguishable.[83] As illustration, a surface study using core photoelectron spectroscopy will be introduced with specific reference to a Si (001) 2×1 surface, which is equivalent to a Si (100) 2×1 surface.

Figure 7.37 shows a schematic drawing for the Si (001) 2×1 surface. If the Si (001) surface is cut ideally as shown in Fig. 7.37(a), there remain two dangling bonds per Si atom, giving rise to an unstable state. To diminish a number of dangling bonds, neighboring atoms combine each other, forming a symmetric dimer as shown in Fig. 7.37(b). However, it is known that asymmetric dimers are formed eventually as seen in Fig. 7.37(c) by the partial transfer of valence electrons in one Si atom to the other resulting in a lowering of the surface energy. The regular alignment of asymmetric dimers with a c (4×2) structure is observed when the Si (001) surface is cooled to low temperature.[83] However, at room temperature, since the alignment of asymmetric dimers is random, the observed structure is not the c (4×2) structure, but a (2×1) structure.[83]

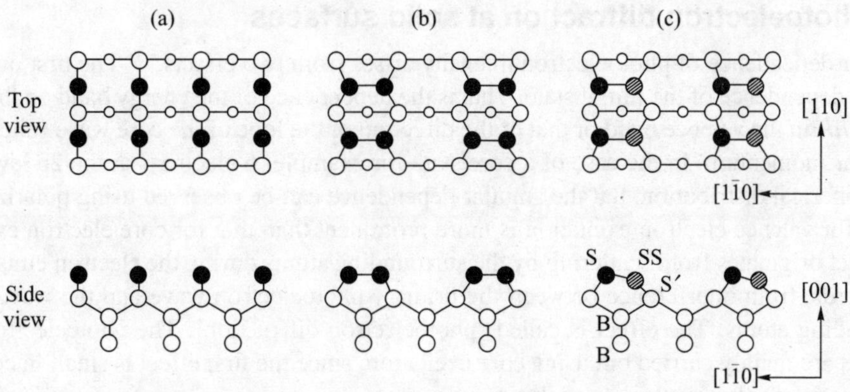

Figure 7.37 Structures for Si (001) surfaces (a) cut ideally, (b) formed by symmetry dimers and (c) formed by asymmetry dimers.

Figure 7.38 shows a Si 2p core photoelectron spectrum using an excitation energy of 130 eV, for a Si (001) 2×1 surface. The Si 2p level splits into $2p_{1/2}$ and $2p_{3/2}$ sub-levels with a separation of 0.6 eV and an intensity ratio of 1 to 2 due to spin-orbit splitting. Unexpectedly, the observed spectrum is composed of three peaks as seen in the figure. These peaks can be described in terms of five components, namely, S, C, B, SS and S′, which correspond to the atoms emitting photoelectrons indicated by the same letters as given in Fig. 7.37(c). In the figure, S and SS are ascribed to the Si atoms forming the asymmetric dimer, in which S lies above SS. S′ is the Si atom located in the second layer, which is directly connected to the dimer. B is the Si atom lying deeper than the second layer and can be regarded as a bulk Si atom. The origin of C is not confirmed, but it may be attributed to the Si atoms near defects on the surface caused by deficiency of dimers. This result was first reported in 1992.[84]

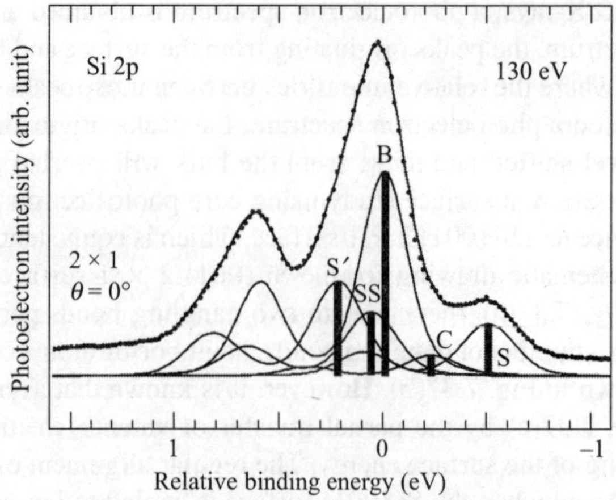

Figure 7.38 A Si 2p photoelectron spectrum for a Si (001) 2 × 1 surface (by courtesy of R. I. G. Uhrburg, © The American Physical Society 1992).[84)]

7.9.2 Photoelectron diffraction at solid surfaces

The angular dependence of photoelectron intensity arises from two effects.[85)] The first originates from the angular dependence of the initial state. That is the dependence of the energy band on the momentum of electron $\hbar k$ in the valence band or that of the direction of the lobe of the core wave function specified with angular momentum. In the case of a core state for example, 6 electrons at the 2p level of Si atom exhibit a spherical distribution, but the angular dependence can be observed using polarized light. The first effect for valence electron excitation is more prominent than that for core electron excitation. The second effect originates from scattering by the surrounding atoms during the electron emission process, in other words, from interference between the primary photoelectron wave and the wave scattered by the surrounding atoms. The effect is called 'photoelectron diffraction'. The photoelectron diffraction experiments are mainly carried out using core excitation, since the first effect is small in core excitation resulting in a relatively simple data analysis.

Figure 7.39 illustrates the principle of photoelectron diffraction. The angular dependence of the peak intensity, for example, the intensity of the Si $2p_{3/2}$ electron, on the emission angle gives rise to the photoelectron diffraction pattern. In the figure, ψ_0 identifies the primary photoelectron wave from the emitter atom in which the core electron is excited, while ψ_1 is the scattered wave from the neighboring

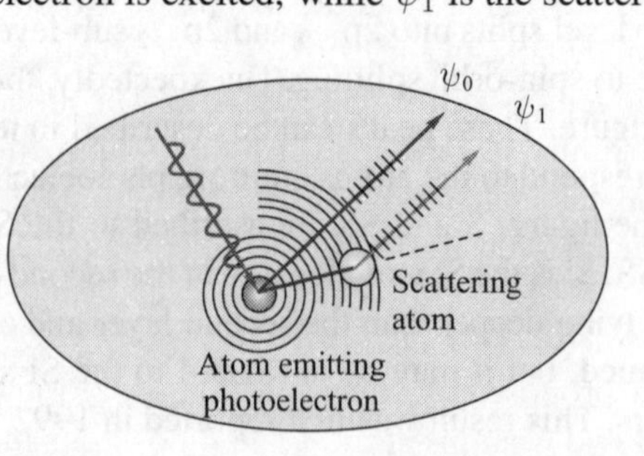

Figure 7.39 Schematics showing photoelectron diffraction.

atom (the scatterer). The photoelectron diffraction pattern originates from mutual interference between ψ_0 and ψ_1.

The features of waves scattered by atoms are classified by two groups which depend on the kinetic energies of the photoelectrons.[86] Figure 7.40 illustrates this classification, where in (a) the kinetic energy is about 100 eV and in (b) above 400 eV. In the case of (a), the scattered waves are nearly isotropic resulting in successive scatterings by surrounding atoms, called 'multiple scattering'. However, in the case of (b), the amplitude of the scattered wave is large in the forward direction because of the small scattering angle, so that a 'single scattering' analysis is possible if there exists no atom along the forward direction. In that case, since the intense FF (forward focusing) peak is generated along the forward direction, a qualitative analysis of the diffraction becomes possible if any atoms lie in that direction. However as the scattering intensity from atoms located in the substrate is weak, there is the disadvantage of uncertainty about the relation between the locations of the emitter and the substrate. In the case of (a), there is no such disadvantage, but much work is necessary for the multiple scattering analysis. Examples of both cases will be introduced.

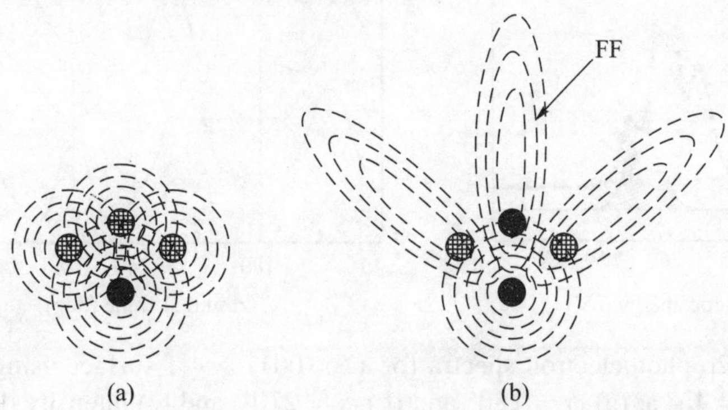

Figure 7.40 Two different types of photoelectron scattering (see details in text).

Firstly, in case (a), a study using a multiple scattering analysis is presented. Figure 7.41 shows the angle-resolved photoelectron spectra and photoelectron diffraction patterns of Si 2p core electrons for a Si (001) 2 × 1 clean surface measured at room temperature and using an excitation energy of 130 eV.[87] The spectra shown in Figs. 7.41(a) and 7.41(b) are results measured at the azimuth angles ϕ of 180° and 270°, respectively. It is evident that the intensity of the component S originating from the upper Si atom in the asymmetric dimer is much larger at 180° than at 270°. This feature originates from the photoelectron diffraction phenomenon. In Fig. 7.41(c) the intensities of the S components are plotted against the azimuthal angle together with calculated results from a multiple scattering analysis for several inclining angles ω of dimer in the range from 0° (symmetric dimer) to 23°. By comparing the measured pattern with the calculated results, the inclining angle of the asymmetric Si dimer is determined to be 19°. This conclusion is also supported by other studies.[88]

Finally, as an example of case (b), a study derived using the single scattering analysis is given. The sample was the benzene molecule (C_6H_6) adsorbed on Si (001) 2 × 1 surface,[89] for which three kinds of structure model have been proposed as shown in Fig. 7.42. Figure 7.42(a) exhibits the standard butterfly (SB) model in which two C atoms at the 1 and 4 positions in the benzene molecule are bound by σ-bonding to two dangling bonds of a Si dimer in the substrate. Figure 7.42(b) shows the tight bridge (TiB) model in which four C atoms at the 1, 2, 3 and 4 positions are bound *via* σ bonding to

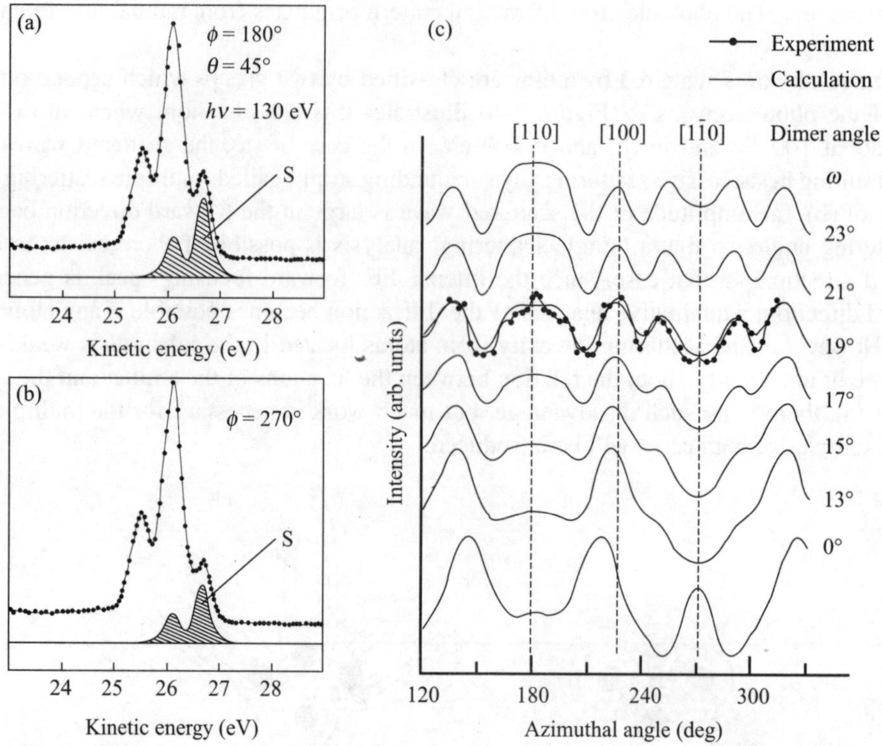

Figure 7.41 Si 2p photoelectron spectra for a Si (001) 2 × 1 surface using 130 eV light at $\theta = 45°$ at (a) $\phi = 180°$ and (b) $\phi = 270°$, and (c) intensity dependence of the peak S on the azimuthal angle measured and calculated with changing inclining angles ω.[87]

Figure 7.42 Three adsorption models for benzene on a Si (001) 2 × 1 surface (© World Scientific 2003).[89] See details in text.

four dangling bonds of two Si dimers. Figure 7.42(c) shows the twisted bridge (TwB) model, which is similar to the TiB model but it differs because the benzene molecule is rotated by 90° from the TiB case. In all the above three models, the dimer in the Si substrate became symmetric ($\omega = 0$) as a result of benzene adsorption. Since the binding energies of C 1s level are close to each other in all three models, it has been difficult to choose the most probable model from the theoretical analysis of photoelectron spectra.[88] Figure 7.43 shows photoelectron diffraction patterns of the benzene-adsorbed Si surface obtained by detecting C 1s electrons of E_{kin} = 968 eV using the Mg Kα emission line (1253.6 eV) together with calculated results. The relative intensities of the C 1s photoelectrons are represented by the gray scale, in which brighter represents the stronger. The detection angles for the photoelectrons lie on the plane of the page. Figure 7.43(d) shows the experimental results. Figures 7.43(a), 7.43(b) and 7.43(c) show calculated results using the single scattering analysis for the three models respectively shown in Figs. 7.42(a), 7.42(b) and 7.42(c). In the calculation the diffraction intensities originate from the forward focusing (FF) peak which lies on C–C bond direction. The FF peak is not very strong because all the C atoms in benzene are photoelectron sources. As seen in Fig. 7.43, the order of good reproducibility of the measured pattern by calculation is the order of SB, TiB and TwB. In this study, similar measurements using two excitation energies of 346 and 486 eV were made and analyzed, supporting the validity of the SB model.[89] For recent progress in photoelectron diffraction research, see Ref. 90.

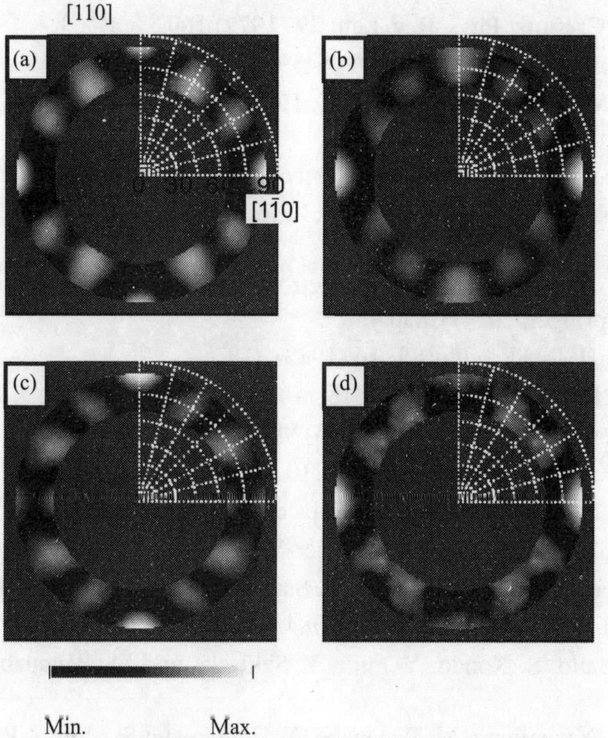

Figure 7.43 C 1s photoelectron diffraction patterns for benzene adsorbed on Si (001) 2 × 1 surface (© World Scientific 2003).[89] See details in text.

REFERENCES

1. F. C. Brown: "Solid State Physics" Vol. 29, H. Ehrenreich, F. Seitz and D. Turnbull eds. (Academic Press, 1974) p.1.
2. M. Cardona and L. Ley eds.: "Photoemission in Solids" Vol. 1, 2 (Springer-Verlag, 1978, 1979).
3. S. Hüfner: "Photoelectron Spectroscopy: Principles and Applications", 2nd ed. (Springer-Verlag, 1996).
4. A. K. Geim and K. S. Novoselov: Nature Materials **6** (2007) 183.
5. A. Bostwick, J. McChesney, T. Ohta, E. Rotenberg, T. Seyller and K. Horn: Prog. Surf. Sci. **84** (2009) 380.
6. N. Tomita and K. Nasu: Phys. Rev. B **60** (1999) 8602.
7. A. Kotani and S. Shin: Rev. Mod. Phys. **73** (2001) 203.
8. W. Gudat and C. Kunz: Phys. Rev. Lett. **29** (1972) 169.
9. T. Ejima: Jpn. J. Appl. Phys. **42** (2003) 6459.
10. For instance, M. Nagasaka, T. Hatsui, T. Horigome, Y. Hamamura and N. Kosugi: J. Electr. Spectr. Rel. Phenom. **177** (2010) 130.
11. H. Ibach: "Electron Spectroscopy for Surface Analysis" (Springer-Verlag, 1977).
12. V. V. Nemoshkalenko and V. G. Aleshin: "Electron Spectroscopy of Crystals" (Plenum Press, 1979) p.36.
13. S. Suga and A. Sekiyama: "Photoelectron Spectroscopy –Bulk and Surface Electronic Structures" (Springer, 2014) Chap 4.
14. P. Kruit and F. H. Read: J. Phys. E **16** (1983) 313.
15. E. M. Purcell: Phys. Rev. **54** (1938) 818.
16. G. A. Harrower: Rev. Sci. Instr. **26** (1955) 850.
17. H. Z. Sar-el: Rev. Sci. Instr. **38** (1967) 1210.
18. J. Kirschner: "Polarized Electrons at Surfaces" (Springer-Verlag, 1985).
19. J. Kessler: "Polarized Electrons" (Springer-Verlag, 1985).
20. M. Watanabe, H. Yamashita, Y. Nakai, S. Sato and S. Onari: Phys. Stat. Sol. (b) **43** (1971) 631.
21. K. Fukui: J. Phys. Soc. Jpn. **61** (1992) 1084.
22. K. Fukui, T. Saito, S. Kondo, Y. Fujii, Y. Sakisaka and M. Watanabe: J. Phys. Soc. Jpn. **59** (1992) 4161.
23. N. Miyata, M. Yanagihara, M. Watanabe, Y. Harada and S. Shin: J. Phys. Soc. Jpn. **71** (2002) 1761.
24. M. Yanagihara, N. Miyata, T. Ejima, M. Watanabe and Y. Kayanuma: J. Phys. Soc. Jpn. **66** (1997) 1626.
25. M. Fujita, H. Nakagawa, H. Matsumoto, T. Miyanaga, M. Watanabe, K. Fukui, E. Ishiguro, Y. Fujii and Y. Sakisaka: J. Phys. Soc. Jpn. **59** (1990) 338.
26. M. Kitaura, H. Nakagawa, K. Fukui, T. Miyanaga and M. Watanabe: J. Electr. Spectr. Rel. Phenom. **79** (1996) 175.

27. T. Takahashi: Solid State Physics **30** (1995) 102, in Japanese.

28. T. Takahashi: "Progress of high-resolution photoemission spectroscopy in strongly correlated electron systems", Surf. Rev. Lett. **9** (2002) 995.

29. S. Sato, T. Miyahara, T. Hanyu, S. Yamaguchi and T. Ishii: J. Phys. Soc. Jpn. **47** (1979) 836.

30. G. D. Mahan: Phys. Rev. **163** (1967) 612.

31. P. Nozieres and C. T. De Dominics: Phys. Rev. **178** (1969) 1097.

32. E. Kisker, K. Schroder, W. Gudat and M. Campagna: Phys. Rev. B **31** (1985) 329.

33. H. Hopster, T. Raue, G. Guntherodt, E. Kisker, R. Clauberg and M. Campagna: Phys. Rev. Lett. **51** (1983) 829.

34. J. Fujii, A. Kakizaki, K. Shimada, K. Ono, T. Kinoshita and H. Fukutani: Solid State Commun. **94** (1995) 391.

35. C. T. Chen, F. Sette, Y. Ma and S. Modesti: Phys. Rev. B **42** (1990) 7262.

36. J. B. Kortright, M. Rice and R. Carr: Phys. Rev. B **51** (1995) 10240.

37. K. Saito, M. Igeta, T. Ejima T. Hatano, A. Arai and M. Watanabe: J. Electr. Spectr. Rel. Phenom. **144-147** (2005) 757.

38. N. Nakajima, T. Koide, T. Shidara, H. Miyauchi, H. Fukutani, K. Iio, T. Katayama, M. Nyvlt and Y. Suzuki: Phys. Rev. Lett. **81** (1998) 5229.

39. T. Shishidou, S. Imada, T. Muro, F. Oda, A. Kimura, S. Suga, T. Miyahara, T. Kanomata and T. Kaneko: Phys. Rev. B **55** (1997) 3749.

40. G. van der Laan: C. R. Physique **9** (2008) 570.

41. C. Guillot, Y. Ballu, J. Paigne, J. Lecante, K. P. Jain, P. Thiry, R. Pinchaux, Y. Pétroff and L. M. Falicov: Phys. Rev. Lett. **39** (1977) 1632.

42. T. Ejima, S. Suzuki, S. Sato, N. Sato, S. Fujimori, M. Yamada, K. Sato, T. Komatsubara, T. Kasuya, Y. Tezuka, S. Shin and T. Ishii: J. Phys. Soc. Jpn. **63** (1994) 2428.

43. A. Sekiyama, K. Kodama, K. Matsuda, T. Iwasaki, S. Ueda, S. Imada, S. Suga, R. Settai, H. Azuma, Y. Ōnuki and Y. Saitoh: J. Phys. Soc. Jpn. **69** (2000) 2771.

44. H. Kumigashira, H.-D. Kim, A. Ashihara, A. Chainani, T. Yokoya, T. Takahashi, A. Uesawa and T. Suzuki: Phys. Rev. B **56** (1997) 13654.

45. J. G. Bednorz and K. A. Müller: Z. Phys. B **64** (1986) 189.

46. Y. Tokura, H. Takagi and S. Uchida: Nature **337** (1989) 345.

47. Z.-X. Shen, D. S. Dessau, B. O. Wells, D. M. King, W. E. Spicer, A. J. Arko, D. Marshall, L. W. Lombordo, A. Kapitulnik, P. Dickinson, S. Doniach, J. DiCarlo, T. Loeser and C. H. Park: Phys. Rev. Lett. **70** (1993) 1553.

48. D. S. Marshall, D. S. Dessau, A. G. Loesor, C.-H. Park, A. Y. Matsuura, J. N. Eckstein, I. Bozovic, P. Fournier, A. Kapitulnik, W. E. Spicer and Z.-X. Shen: Phys. Rev. Lett. **76** (1996) 4841.

49. N. Nücker, H. Romberg, X. X. Xi and J. Fink: Phys. Rev. B **39** (1989) 6619.

50. T. Takahashi, J. Maeda, H. Katayama-Yoshida, Y. Okabe, T. Suzuki, A. Fujimori, S. Hosoya, S. Shamoto and M. Sato: Phys. Rev. B **37** (1988) 9788.

51. A. Kotani and Y. Toyozawa: J. Phys. Soc. Jpn. **37** (1974) 912.

52. T. Takahashi, S. Suzuki, T. Kusunoki and H. Katayama-Yoshida: "Collapse of Mott-Hubbard framework by hole doping in $Bi_2Sr_2CaCu_2O_8$" in Proc. of 10th Int. Conf. on Vacuum Ultraviolet Radiation Physics, eds. F. J. Wuilleumier, Y. Petroff and I. Nenner (World Scientific, 1993) p.330.

53. T. Yokoya, T. Takahashi, A. Fujiwara and Y. Koike: Solid State Commun. **87** (1993) 553.

54. T. Sato, T. Kamiyama, Y. Naitoh, T. Takahashi, I. Chong, T. Terashima and T. Takano: Phys. Rev. B **63** (2001) 132502.

55. T. Sato, T. Kamiyama, T. Takahashi, J. Moset, A. Kaminski, J. C. Campuzano, H. M. Fretwell, T. Kakeuchi, H. Ding, I. Chong, T. Terashima and M. Takano: Phys. Rev. B **64** (2001) 54502.

56. Y. Kamihara, H. Hiramatsu, M. Hirano, R. Kawamura, H. Yanagi, T. Kamiya and H. Hosono: J. Am. Chem. Soc. **128** (2006) 10012.

57. J. Padlione and R. L. Greene: Nature Physics **6** (2010) 645.

58. J. D. Wright: "Molecular Crystals" (Cambridge University Press, 1987).

59. M. Pope and C. E. Swenberg: "Electronic Process in Organic Crystals and Polymers" 2nd ed. (Oxford University Press, 1999).

60. N. Karl: "Organic Electronic Materials", R. Farchioni and G. Grosso eds. (Springer-Verlag, 2001) Chap. 6.

61. In the field of conducting and semiconducting organic materials, Int. Conf. on Science and Technology of Synthetic Metals is held every 2 years. The proceedings are published in Synthetic Metals **153** (2005) 1, Synthetic Metals **159** (2009) 2157 *et al.*

62. K. Seki, H. Ishii and Y. Ouchi: "Chemical Applications of Synchrotron Radiation" Part I, T.-K. Sham ed. (World Scientific, 2002) Chap. 8.

63. K. Nakagawa, M. Sei, R. Matsumoto, E. Ito, T. Araki, H. Ishii, Y. Ouchi, K. Seki and K. Kondo: J. Electr. Spectr. Rel. Phenom. **78** (1996) 375.

64. D. Yoshimura, T. Yokoyama, T. Nishi, H. Ishii, R. Ozawa, H. Hamaguchi and K. Seki: J. Electr. Spectr. Rel. Phenom. **144** (2005) 319.

65. S. Hasegawa, S. Tanaka, Y. Yamashita, H. Inokuchi, H. Fujimoto, K. Kamiya, K. Seki and N. Ueno: Phys. Rev. B **48** (1993) 2596.

66. N. Ueno, K. Suzuki, S. Hasegawa, K. Kamiya, K. Seki and H. Inokuchi: J. Chem. Phys. **99** (1993) 7169.

67. D. Yoshimura, H. Ishii, Y. Ouchi, E. Ito, T. Miyamae, S. Hasegawa, K. K. Okudaira, N. Ueno and K. Seki: Phys. Rev. B **60** (1999) 9046.

68. K. Seki, U. Karlsson, R. Engelhardt and E. E. Koch: Chem. Phys. Lett. **103** (1984) 343.

69. N. Ueno and S. Kera: Prog. Surf. Sci. **83** (2008) 490.

70. J. Choi, I. A. Samayoa, S.-C. Lim, C. Jo, Y. C. Choi, Y. H. Lee and P. A. Dowben: Phys. Lett. A **299** (2002) 601.

71. S. Machida, Y. Nakayama, S. Duhm, Q. Xin, A. Funakoshi, N. Ogawa, S. Kera, N. Ueno and H. Ishii: Phys. Rev. Lett. **104** (2010) 156401.

72. H. W. Kroto, J. R. Heath, S. C. O'Brien, R. F. Curl and R. E. Smalley: Nature **318** (1985) 162.

73. The result was first reported in T. Takahashi, T. Morikawa, S. Sato, H. Katayama-Yoshida, A. Yuyama, K. Seki, H. Fujimoto, S. Hino, S. Hasegawa, K. Kamiya, H. Inokuchi, K. Kikuchi, S.

Suzuki, K. Ikemoto and Y. Achiba: UVSOR Activity Report 1991 (ISSN 0911-5730) p.38. It also appeared in Physica C **185-189** (1991) 417.

74. R. C. Haddon, A. F. Hebard, M. J. Rosseinsky, D. W. Murphy, S. J. Duclos, K. B. Lyons, B. Miller, J. M. Rosamilia, R. M. Fleming, A. R. Kortan, S. H. Glarum, A. V. Makhija, A. J. Muller, R. H. Eick, S. M. Zahurak, R. Tycko, G. Dabbag and F. A. Thiel: Nature **350** (1991) 320.

75. John H. Davies: "The Physics of Low-Dimensional Semiconductors: An Introduction" (Cambridge University Press, 1998).

76. K. Takahashi, A. Tanaka, M. Hatano, H. Sasaki, S. Suzuki and S. Sato: J. Electr. Spectr. Rel. Phenom. **88-91** (1998) 347.

77. K. Takahashi, A. Tanaka, H. Sasaki, W. Gondo, S. Suzuki and S. Sato: Phys. Rev. B **60** (1999) 8748.

78. A. Tanaka, Y. Takeda, T. Nagasawa and S. Sato: Phys. Rev. B **67** (2003) 033101.

79. K. He, Y. Takeichi, M. Ogawa, T. Okuda, P. Moras, D. Topwal, A. Harasawa, T. Hirahara, C. Carbone, A. Kakizaki and I. Matsuda: Phys. Rev. Lett. **104** (2010) 156805.

80. D. Hsieh, Y. Xia, L. Wray, D. Qian, A. Pal, J. H. Dil, J. Osterwalder, F. Meier, G. Bihlmayer, C. L. Kane, Y. S. Hor, R. J. Cava and M. Z. Hasan: Science **323** (2009) 919.

81. R. Z. Bachrach ed.: "Synchrotron Radiation Research. Advances in Surface and Interface Science" Vol.1, 2 (Springer, 1992).

82. M. Salmeron and R. Schlögl: Surf. Sci. Rep. **63** (2008) 169.

83. For example, W. F. Egelhoff, Jr.: Surf. Sci. Report **6** (1987) 253.

84. E. Landemark, C. J. Karlsson, Y.-C. Chao and R. I. G. Uhrberg: Phys. Rev. Lett. **69** (1992) 1588.

85. For example, S. A. Chambers: Adv. Phys. **40** (1991) 357.

86. For example, S. Kono: J. Electr. Spectr. Rel. Phenom. **126** (2002) 43.

87. E. L. Bullock, R. Gunnella, L. Patthy, T. Abukawa, S. Kono, C. R. Natoli and L. S. O. Johansson: Phys. Rev. Lett. **74** (1995) 2756.

88. T. Abukawa, C. M. Wei, K. Yoshimura and S. Kono: Phys. Rev. B **62** (2000) 16069.

89. M. Shimomura, M. Munakata, K. Honma, S. M. Widstrand, L. Johansson, T. Abukawa and S. Kono: "Structural study of benzene adsorbed on Si (001) surface by photoelectron diffraction", Surf. Rev. Lett. **10** (2003) 499.

90. D. P. Woodruff: J. Electr. Spectr. Rel. Phenom. **178-179** (2010) 186.

CHAPTER 8 Applications of Synchrotron Radiation Irradiation in Manufacturing

Synchrotron radiation is extensively used with many applications in the study of materials in order to evaluate their electronic properties and atomic structures. However, in this chapter our focus will be shifted from analytical studies to those manufacturing processes which make use of the unique properties of synchrotron radiation. Two important examples are given. The first is in the cleaning of semiconductor surfaces, new material deposition and in surface etching. Both examples exploit chemically stimulated reactions on surfaces *via* photoexcitation, the process which generated the original idea for such surface treatments and which are normally undertaken using gas plasma discharges. The second is in methods of microfabrication and lithography, which are the manufacture of microstructures, sometimes called micromachining for the production of sub-micron scale devices either in metals or ceramics on a commercial basis. These processes exploit the optical response of materials to very short wavelength light where diffraction effects are minimal.

8.1 SURFACE PHOTOCHEMISTRY

8.1.1 The basic reaction mechanisms for photoexcitation at solid surfaces

Studies of surfaces using photoelectron spectroscopy were described in Sect. **7.9** and those using infrared spectroscopy in Sect. **13.3**. Consider what happens when solid surfaces are irradiated by high energy light such as synchrotron radiation. Non-thermal chemical reactions involving movement or desorption of atoms from the solid surface occur in the first instance.[1,2)] This is due to electronic transitions from the ground to excited states by photoabsorption, which may induce local instabilities in the crystal lattice, such as the movement of atoms from lattice points[3)] or the breaking of chemical bonds. The phenomenon in which atoms, ions *etc*., are emitted from the surface after the break of a chemical bond is called photo-stimulated desorption (PSD). The mechanisms of PSD can be interpreted using the following models. First is the Menzel-Gomer-Redhead (MGR) model, based on valence electron excitation to anti-bonding orbitals (see Sect. **6.2.1**), in which the PSD yield is unusually large for a specific electronic transition.[4)] Second is the Knotek-Feibelman (KF) model based on Coulomb repulsion between positive holes generated *via* the Auger process after core electron excitation (see Sect. **4.3.4**).[5)] In this case the PSD yield spectrum resembles the absorption spectrum.[6)] However, many different models have been suggested, [7,8)] since there are a number of complex PSD features which depend on the nature of the bonding between the adsorbed atoms or molecules and the substrate which is a metal, semiconductor or insulator. Figure 8.1 presents a schematic view of a substrate with an adsorbed surface layer situated in the gaseous atmosphere. We shall interpret the photochemical reactions at the solid surface using this figure.

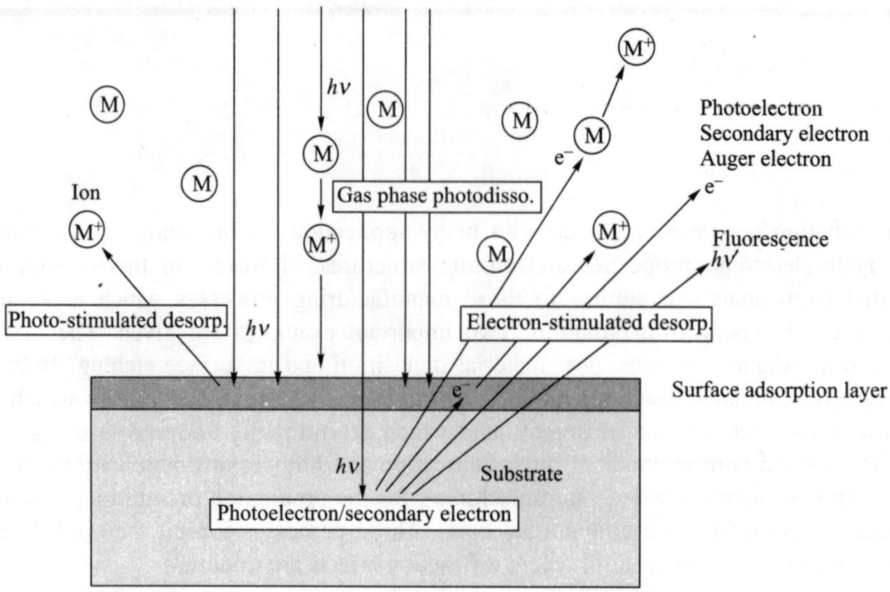

Figure 8.1 A schematic view of the basic processes of photoexcitation reactions at a solid surface under a gaseous atmosphere.

As shown in Fig. 8.1, light is absorbed by the gas, the adsorbed surface layer and the substrate, where each absorption induces a different kind of surface or chemical reaction.[9,10] When gas absorbs the light, the gas is excited and active species such as positive ions and neutral radicals are generated. They arrive at the surface creating a chemical reaction on it which is primarily adsorption. Hereafter this process is called 'gas excitation'. In some cases, thin films may accumulate under low temperature conditions.[11] Absorption at the adsorbed surface layer will excite the layer and may promote a chemical reaction such as PSD and this process is called 'surface excitation', which can be used very effectively for surface cleaning and etching.[10] Absorption in the substrate generates primarily excited electrons and some of them will generate secondary electrons by inelastic scattering. In that case, the electrons will escape from the surface as photoelectrons which can excite the adsorbed surface layer perhaps causing electron-stimulated desorption (ESD).[13]

After escaping from the surface, they may subsequently collide with atoms or molecules and cause their ionization or dissociation.[14] The cross section for ionization by an electron is larger than that for a photon and many secondary electrons may be generated in solids, so that surface chemical reactions such as ESD easily occurs. When synchrotron radiation is incident parallel to the surface, only the gas will be excited. Otherwise, both the gas and the surface layer and also the substrate are excited. Irradiation with light therefore may induce surface chemical reactions accompanied by the emission of electrons and desorption of atoms, ions *etc*. If the electron current emitted from any sample is measured while scanning the photon energy of the incident light, a total electron yield (TEY) spectrum can be obtained giving specific information on absorption at the surface. When photoelectron spectroscopy or Auger electron spectroscopy are undertaken, the bonding state of the surface can be investigated. When the ion desorption yield is measured, it will directly provide information on the efficiency of the photoexcitation reaction.[11,12)] Clearly, the incident light can be used both for the excitation of surface reactions and also for the study of the resulting products. Therefore it is an excellent measurement probe for the reaction. When the photoexcitation and observation are made using the same light there is the significant advantage that the excitation and observation areas on the surface are identical without any interference which might be introduced by extra probes. Therefore this method using the same source for inducing the reaction and also for probing the electrons and ions released from the surface can be crucial in studying basic processes in surface photochemical reactions.[9,15)]

8.1.2 Cleaning of semiconductor surfaces using photoexcitation

A good illustration of a surface photochemical reaction induced by valence electron excitation is in the removal of hydrogen from a hydrogen-terminated (passivated) Si (111) surface.[12,15)] The H-terminated Si (111) surface is produced by soaking Si with HF solution. The surface is flat and almost covered with mono hydride (Si–H). When VUV light with energy from 10 to 40 eV is used to irradiate the surface, the desorption of H^+ is observed. As shown in lower part of Fig. 8.2, the PSD (H^+ desorption) yield reveals a strong dependence on photon energy and the maximum peak is observed at 23.3 eV. Weaker structures in the region from 10 to 15 eV arise from higher order light from the monochromator itself. The TEY spectrum measured on the same surface (the solid line in the upper part of Fig. 8.2) shows peaks at around 14 eV, 17 eV and 23 eV. Because H and Si have no core levels in this energy region, all

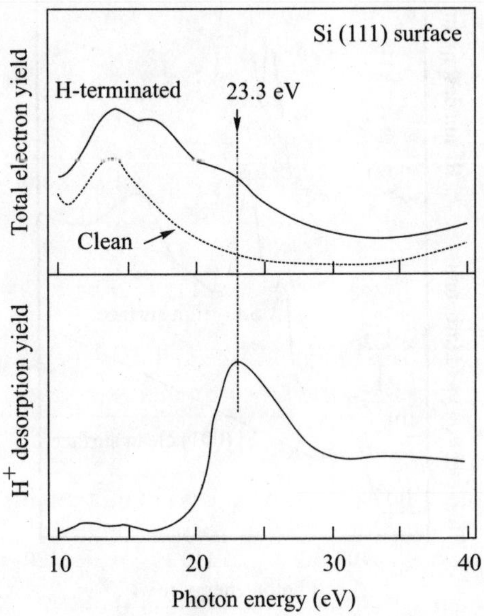

Figure 8.2 Spectra of H^+ ion desorption yield (upper) and the total electron yield (lower) from H-terminated Si (111) surface.[15)]

these structures must come from absorption by valence electrons. In the TEY spectrum on the Si (111) 7×7 clean surface (the dotted curve in upper part of Fig. 8.2), only the peak around 14 eV appears, therefore the peaks at around 17 and 23 eV can be attributed to contributions from adsorbed hydrogen. Thus, the TEY is very sensitive to the adsorption species on the surface. Comparison of the PSD and the TEY spectra reveals that only the TEY peak around 23 eV corresponds to the PSD of H^+, which means that the absorption around 23 eV causes noticeable PSD of H^+. It is concluded therefore that absorption in the Si substrate does not contribute to the PSD of H^+ ions. In other words, ESD of H^+ ions by photoelectrons including secondary electrons from the substrate is negligibly small. As shown in Fig. 8.2 there is no PSD peak for the H^+ ion corresponding to the TEY peak around 17 eV and it is clear that every absorption due to adsorbed hydrogen does not contribute to the PSD of H^+ ions, but only the absorption leading to excitation of valence electrons to anti-bonding orbitals contributes. In this case the phenomena are well explained using the MGR model.

We now describe the surface photochemical reactions at a SiO_2/Si (001) surface by core electron excitation using the PSD of positive ions (H^+ ions with a small percentage of OH^+ ions).[16] The sample is a 30 Å thick SiO_2 film made from the thermal oxidization of a Si (001) surface. As seen in Fig. 8.3, the TEY of clean Si (001) 2×1 surface (curve (c)) rises at 99 eV, the absorption edge of the Si 2p core level shows a peak at 100 eV and from the higher energy side of the peak monotonically increases. In addition to the TEY structure of the clean Si (001) surface, the TEY of the SiO_2/Si (001) surface (curve (b)) shows three peaks A, B and C between 105 and 115 eV. These structures are attributed to absorption

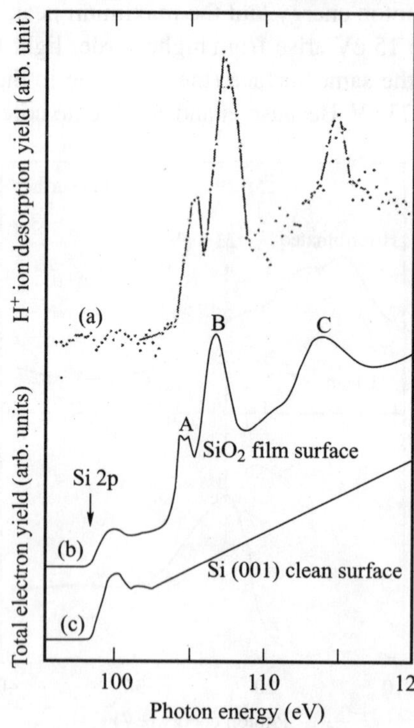

Figure 8.3 Spectra of (a) H^+ ion desorption yield and (b) total electron yield from SiO_2/Si (001) surface, and (c) total electron yield from Si (001) surface.[16]

from the chemically shifted Si 2p level of the oxidized surface whose binding energy E_b lies between 104 and 105 eV. The PSD yield spectrum of the H^+ ion from the SiO_2/Si (001) surface (curve (a)) does not show any structure at 100 eV, although other features appear which correspond to the peaks A, B and C in the TEY spectrum. H^+ ions in the PSD spectrum are generated by the desorption of $O_3 \equiv$ Si–H or $O_3 \equiv$ Si–OH bondings in structural defects near to the surface of the SiO_2 film. Good correspondence between the spectral structures from the TEY and PSD suggests that PSD can be described in terms of the KF model which is based on the electronic transition from the chemically shifted Si 2p core of the Si atom located at the defects. There is no PSD peak corresponding to the peak at 100 eV in TEY which suggests that the ESD due to the excitation of the substrate does not contribute to the PSD. This is because the kinetic energies of the Si 2p photoelectron and the Si LVV Auger electron are insufficient to excite PSD. This would require kinetic energies of Si 2p excited electrons to be greater than 104 ~ 105 eV. That is, the required photon energy should be larger than 200 eV. Thus we can cut a specific chemical bond of any adsorbed species by adjusting the photon energy since the core electron excitation has such a high selectivity for surface photochemical reactions. Synchrotron radiation is the ideal source for fundamental research on surface photochemistry because we can select any photon energy.

8.1.3 Material deposition and surface etching using photoexcitation reactions

Synchrotron radiation across the region from the VUV to soft X-rays of about 1000 eV is sometimes used as a white light source without any monochromator simply to fully exploit its very high intensity. Semiconductor processing using synchrotron radiation excitation in this way has been studied.[17,18] The deposition of thin films on Si surfaces and the etching of Si surfaces are presented as examples which are important in semiconductor device fabrication. Figure 8.4 shows the temperature dependence of the Si growth rate during chemical vapor deposition (CVD) using Si_2H_6 gas with a pressure of 1.3×10^{-3} Torr (0.17 Pa) on a H-terminated Si surface.[11] In the case of thermal CVD (□), very little growth is recorded below 500°C. This is because H at the surface disturbs the dissociative adsorption of Si_2H_6. The rapid increase in growth rate above 500°C is due to the increase in active Si dangling bonds resulting from H desorption from the surface due to thermal excitation. When the synchrotron radiation is incident parallel to the surface of the substrate (●), corresponding primarily to gas excitation, a remarkable increase in growth rate is observed even below 500°C because the Si_2H_6 is photo-dissociated and the resultant SiH_x ($x = 0 \sim 3$) radical is adsorbed onto the substrate, in spite of the presence of H at the surface. Since the growth rate of the Si thin film shows little dependence on temperature below 450°C, the growth below 450°C is attributed to radical production in the gas. Above 550°C however, there is little difference in growth rate between gas excitation made at any stored current and thermal CVD. This is because the removal of H from the surface is dominant in the Si film growth process above 550°C, in contrast to radical production. For the case in which synchrotron radiation illuminates the surface of the substrate at normal incidence, both gas excitation and surface excitation will occur and will contribute to the growth. Figure 8.4 shows that when synchrotron radiation illuminates the surface normally, the growth rate (○) increases by about 5 times that for gas excitation (●) and that by thermal CVD above 550° C. This increase in growth rate is attributed to surface excitation caused by the removal of surface hydrogen *via* PSD as given in Fig. 8.2. At low temperatures, below 300°C, the thin film growth of Si is impossible *via* the thermal CVD, but is feasible using synchrotron radiation excitation CVD.

There are some trials to fabricate metal films from organometallic molecules adsorbed on substrates cooled to liquid nitrogen temperatures using synchrotron radiation irradiation.[19] Figure 8.5 shows photoelectron spectra from $W(CO)_6$ adsorbed on a Si (111) surface and cooled to 100K before and after

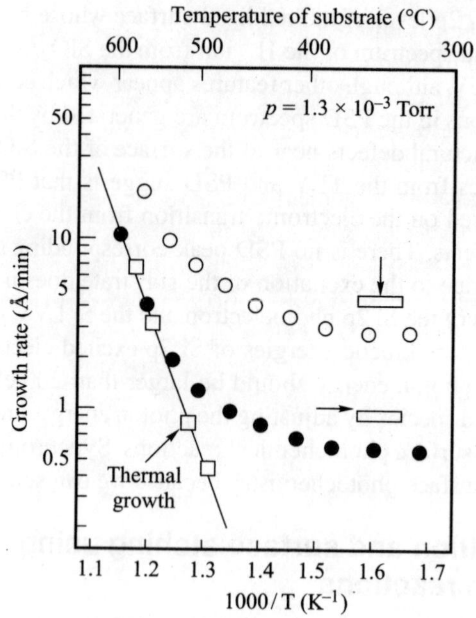

Figure 8.4 The temperature dependence of the rate of Si growth on H-terminated Si by CVD using Si_2H_6 gas and synchrotron radiation ($I = 300$ mA), and by thermal growth (by courtesy of H. Akazawa, © American Institute of Physics 1995).[11)]

synchrotron radiation irradiation. In the figure, the ordinate and the abscissa indicate the intensity and kinetic energy of the photoelectrons, respectively. Before irradiation, the spectrum reflects the molecular spectrum, but following synchrotron radiation irradiation the spectral shape changes strikingly and the Fermi level appears giving evidence of its metallic nature. From this result, it is clear therefore that

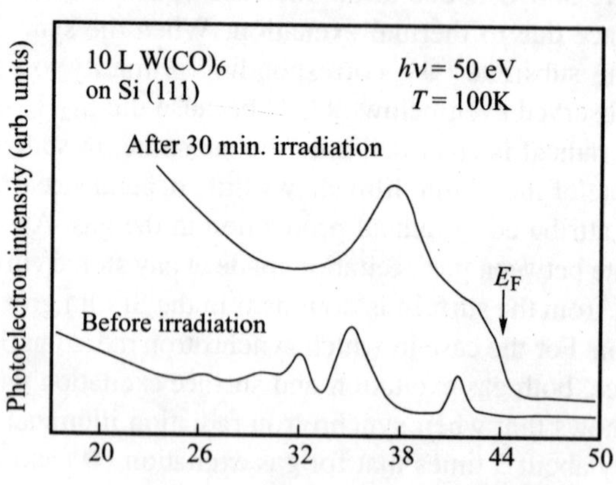

Figure 8.5 Photoelectron spectra from $W(CO)_6$ deposited on Si (111) at 100K before (lower) and after (upper) the synchrotron radiation irradiation (L: Langmuir (see Sect. **13.3.3.b**)) (by courtesy of G. Margaritondo, © Elsevier 1990).[19)]

W metal is produced at low temperatures. The technique of MOCVD (metal organic chemical vapor deposition) has been used as a method to coat semiconductor films on substrates at room temperature using synchrotron radiation irradiation on mixtures of two gaseous metal organic molecules.[20]

The etching of Si and SiO_2 surfaces has been achieved using synchrotron radiation through a mask with an aperture of 2 mm radius with the etching gas SF_6 at 0.1 Torr (13.3 Pa) at room temperature.[21] In this case, the synchrotron radiation is incident normal to the surface and the photoexcitation includes both gas and surface excitations. Figure 8.6 shows the etching profiles. Around the central region of the hole, which corresponds to the synchrotron radiation irradiation point, the depths of both Si and SiO_2 holes are almost identical. Around the boundary region of the hole, the depth of the Si hole gradually diminishes and the etched region is extended even beyond the irradiated region. However, the SiO_2 hole shows a steep rise, indicating that etching is restricted to within the irradiated region. These results illustrate that the synchrotron radiation etching processes differ for SiO_2 and for Si surfaces. Neither surface can be directly etched by neutral SF_6 molecule but both can be adsorbed by free fluorine radicals generated *via* excitation of SF_6 gas. The fluorine radicals can be supplied both inside, and also close to the irradiated region. Since the Si surface is etched outside the irradiated region it is proposed that the Si surface is directly etched by adsorbed fluorine radicals. The desorption species is SF_4. In contrast to this, the steep rise in the etching profile of SiO_2 hole suggests that surface excitation is also

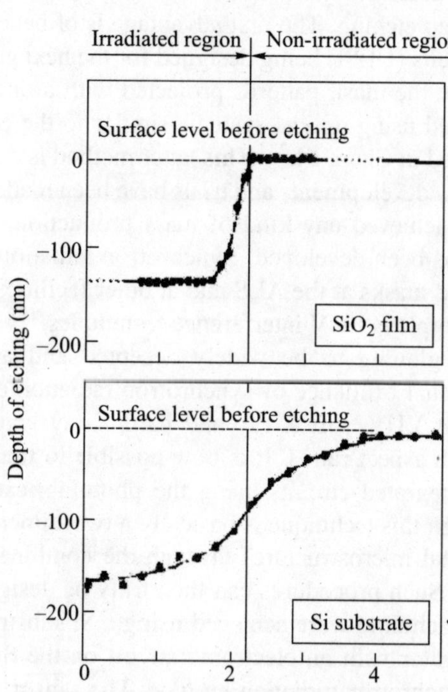

Figure 8.6 The cross sections of Si (lower) and SiO_2 film (upper) near to surfaces etched by synchrotron radiation irradiation using SF_6 (by courtesy of T. Urisu, © American Vacuum Society 1987).[21]

essential to generate the desorption species. In summary, synchrotron radiation etching mechanisms are rather different for Si and SiO_2 surfaces and this may offer useful selectivity in the etching of materials consisting of Si and SiO_2.[21)] Several trials are active to fabricate electronic devices *via* etching and deposition techniques using synchrotron radiation. For example, a SiO_2 film deposited on InP single crystal was perforated with many small holes and semiconductor quantum dots were grown in them as luminescent elements.[22)] Photonic crystals have also been fabricated. Experiments using halogens have also been reported.[23)] Teflon is a good insulator chemically stable withstanding high temperatures, is easily evaporated and can be etched only with irradiation by X-rays. These properties can be exploited for the microfabrication of this material as discussed in the next section.

8.2 MICROFABRICATION AND LITHOGRAPHY

There are at least two advantages in microfabrication using synchrotron radiation irradiation with mask patterns. The first is that good spatial resolution is obtained. This is because the diffuseness around edges of mask pattern images caused by diffraction of light is minimized through the use of short wavelength light. The second advantage is that microfabrication with high aspect ratios (the depth to width ratio) is so easily achievable. This is because the short wavelength light can penetrate deeply into the photoresist yielding deep etching. The first advantage is of benefit mainly in the production of ultra-large-scale integrated circuits (ULSI) being designed for the next generation of electronic devices. To fabricate ULSI, not only are the mask patterns projected with a one to one ratio,[24)] but also their reduced images can be projected using optical systems similar to the Schwarzschild objective coated with Mo/Si multilayers described in Chap. **14**.[25)] This latter method is called EUV (extreme ultraviolet) lithography. Although numerous developments and trials have been made, the fabrication of ULSI using synchrotron radiation has not achieved any kind of mass production, while EUV lithography using laboratory sources at 13 nm has been developed. Synchrotron radiation has been used to characterize the reducing mirror systems and masks at the ALS and at other facilities,[26,27)] as well as being used to explore nm-patterning which employs EUV interference techniques.[28)]

Micromachining is now beginning to be widely accepted and applied in industry because it exploits to the full the exceptional brilliance of synchrotron radiation combined with full wavelength selectivity, even at the shortest VUV, soft X-ray and hard X-ray wavelengths, leading to excellent spatial resolution and very high aspect ratios. It is now possible to make nanoscale complex systems in assembled form such as integrated circuits using the photofabrication method which transcribes mask patterns by light. Although this technique is basically a two-dimensional technology, we can also produce quasi three-dimensional microstructures through the combination of deeply etched patterns and two-dimensional patterns. Such procedures can then truly be designated as micromachining. The first examples of such micromachining were achieved using a Si substrate. We can now make devices combining a sensor or an actuator with an electronic circuit on the Si substrate from the projection of their mask patterns with synchrotron radiation *en bloc*. The sensor or the actuator is fabricated by deep etching. As a result, it is possible to down-size a system having quite complicated and advanced functions. The second is that of 'two-point-five-dimensional fabrication', where very deep etching on a flat surface keeping a two-dimensional mask pattern is made. There are two possible routes to fabricate a structure with very high aspect ratio. The first is by metal-plating the mold of the photoresist made by irradiation with light. The second is *via* the deep reactive ion etching (RIE) method, in which substrate material is reactively evaporated by accelerated ions. The former is a technique using the synchrotron radiation (λ = 0.2 ~ 0.5 nm) developed at Forschungszentrum Karulsruhe in Germany in 1986, which has been called LIGA (LIthographie Galvanoformung Abformung).[29] 'Galvanoformung' and 'Abformung' mean electrical plating and molding, respectively. By this LIGA process, structures with the width of a few μm and depths of up to a few hundred μm can be manufactured. Figure 8.7 shows the fabrication procedure for electrical plating in the LIGA process. The photoresist used is normally PMMA (poly (methylmethacrylate)). Once the PMMA has been exposed to light, the exposed area of the PMMA plate is dissolved away, so that the PMMA can then be employed as a positive resist.

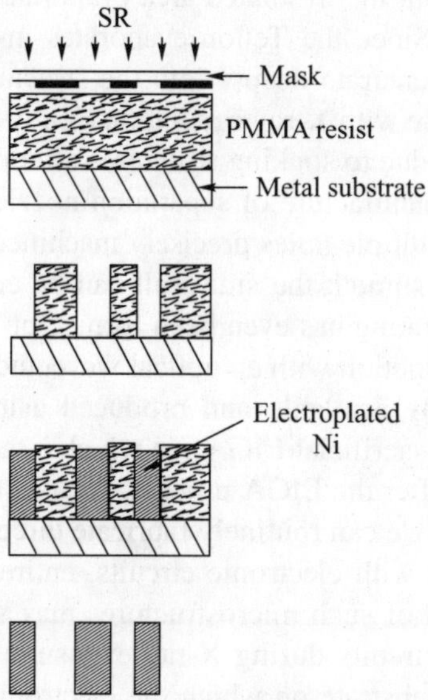

Figure 8.7 A schematic of the LIGA process (details in text).

Figure 8.8 A high aspect-ratio structure in negative resist fabricated by X-ray exposure.

Since X-rays with shorter wavelengths than those of visible light are used, the PMMA is therefore exposed much more deeply and with minimal diffuseness due to diffraction. Using the holed microstructure of the developed resist as the mold, we can complete the microstructure of metals, for example Ni. In this case, the metals are electrically plated on the mold of the resist and subsequently the resist is removed. Finally, using the metal structure as a mold we can fabricate fine structures in plastics *via* polymerization of the raw material in the mold. In this method the metal mold can be repeatedly used and it is not necessary to use synchrotron radiation more than once, so that many products could, in principle be produced quite cheaply. For most metal structures, the aspect ratio is about 100, but for plastic structures it is about 10. Ceramic structures can also be made using a procedure in which the mold of the plastic structure is filled with raw material, and then baked and sintered.

The shorter the wavelength of X-rays which are used, the more deeply the photoresist will become exposed. Using X-rays with λ = 0.06 nm at the synchrotron radiation facility in Wisconsin University, USA, structure with a depth of 1 cm was fabricated.[30] When X-rays with high transmittance are used, it is necessary to use masks with thick metal. Figure 8.8 gives an example of a structure with high aspect ratio made on a negative resist (THB-NI, Japan Synthetic Rubber Co. Ltd.) using SORTECH at

Tsukuba. Electric plating of metal was achieved using this structure as the mold.[31] When the negative resist is irradiated by X-rays, the exposed area polymerizes and remains even after development. Since X-rays are attenuated by absorption during penetration into negative resist, the dose of X-rays at the bottom of the structure becomes rather small causing some weakness in the deeper parts of the structure. As a result, the structures may collapse and stick to each other by surface tension during the drying process after development. This difficulty has been resolved by drying the structure after substitution of developing solvent using a liquid with lower surface tension (such as Fluorinate from Sumitomo 3M Ltd.). When Teflon is exposed to synchrotron radiation, the irradiated area evaporates giving self-development.[32] This is etching without gas excitation. Since the Teflon evaporates instantaneously during irradiation of synchrotron radiation, X-rays can penetrate deeply into the structure. Although microfabrication with Teflon is usually difficult, it is feasible with X-ray irradiation (X-ray exposure).[33] When we use the Teflon as a resist, there are no problems due to sticking structures after development.

The LIGA process was originally developed for the manufacture of separation nozzles to produce enriched uranium.[29] The nozzle resembles a filter with multiple holes precisely machined. As the side wall of the structure made by the LIGA process is very smooth the side wall can be employed as a mirror which is required in some devices. A diffraction grating has even been formed on the side wall and a small integrated spectrometer made using it in conjunction with an optical waveguide. Figure 8.9 shows such a micro spectrometer which is commercially available and produced using ANKA in Germany.[34] There is a further technology called SLIGA (sacrificial LIGA), in which a sacrificial layer is formed beforehand in the substrate and later removed after the LIGA process. As a result a movable structure released from the substrate can be formed. Once we can routinely fabricate three-dimensional microstructures using the LIGA process on Si substrates with electronic circuits, entirely new fields of applications will emerge. However, the direct creation of such microstructures may sometimes be difficult due to the damage sustained by the electronic circuits during X-ray exposure. In that case, a method to paste the fabricated microstructures to the substrate on which the electronic circuits are formed can be employed. The LIGA process is being developed also at Ritsumeikan University[33] as well as University of Hyogo in Japan.[35] Recent progress in the world is summarized in Ref. 36.

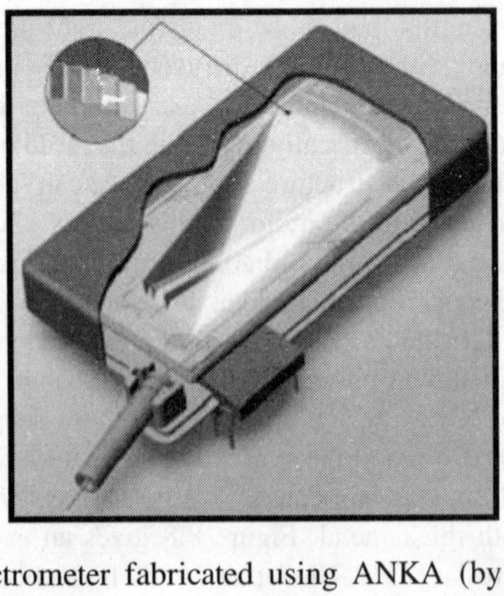

Figure 8.9 A micro spectrometer fabricated using ANKA (by courtesy of V. Saile and INSION company).

In the LIGA process using synchrotron radiation, it is possible to create microstructure with very large aspect ratios from several tens to several hundreds. Recently, structures with aspect ratio of 10 (width 20 μm, height 200 μm) have been fabricated using ordinary UV light. In this process, SU-8 resist based on an epoxy resin developed at IBM, has been used, which can be also used for LIGA as an X-ray negative resist. Its sensitivity is about hundred times larger than that of PMMA. This is good news for the LIGA process which requires rather long exposure times.[37)]

There is also an alternative technology in which an acryl plate is exposed to a proton beam. The acryl plate can be developed up to the penetration depth of the proton beam which is determined by its kinetic energy.[38)] The area irradiated by the proton beam can even be further chemically expanded after exposure to a solvent vapor, so that one could make a small lens using this technique.

References

1. Int. Workshop on Desorption Induced by Electronic Transitions (DIET) has been held every 3 years. The proceedings are published in Surf. Sci. **528** (2003) 1, Surf. Sci. **593** (2005) 1, Surf. Sci. **602** (2008) No.20, J. Phys.: Condens. Matter **22** (2010) No. 8 *et al.*
2. M. Hanabusa: Materials Sci. Rep. **2** (1987) 51.
3. S. Tanaka, Y. Kayanuma and K. Ueda: Phys. Rev. A **57** (1998) 3437.
4. D. Menzel and R. Gomer: J. Chem. Phys. **41** (1964) 3311.
5. M. L. Knotek and P. J. Feibelman: Phys. Rev. Lett. **40** (1978) 964.
6. M. L. Knotek: Rep. Prog. Phys. **47** (1984) 1499.
7. R. D. Ramsir and J. T. Yates, Jr.: Surf. Sci. Rep. **12** (1991) 243.
8. K. Mase, M. Nagasono and S. Tanaka: J. Electr. Spects. Rel. Phenom., **101-103** (1999) 13.
9. D. E. Ramaker, T. E. Madey, R. L. Kurtz and H. Sambe: Phys. Rev. B **38** (1988) 2099.
10. H. Kyuragi: J. Vac. Sci. Tech. B **5** (1987) 2644.
11. H. Akazawa and Y. Utsumi: J. Appl. Phys. **78** (1995) 2740.
12. Y. Takakuwa, M. Nogawa, H. Ishida, M. Niwano, H. Kato and N. Miyamoto: Jpn. J. Appl. Phys. **36** (1997) 7699.
13. R. Jaeger, J. Stöhr and T. Kendelewicz: Phys. Rev. B **28** (1983) 1145.
14. Y. Sugita, Y. Nara, N. Nakayama and T. Ito: Photon Factory Activity Reports No. 8 (1990) 331.
15. Y. Takakuwa, Y. Enta and N. Miyamoto: Optoelectronics **11** (1996) 3.
16. M. Niwano, Y. Takakuwa, H. Katakura and N. Miyamoto: J. Vac. Sci. Tech. A **9** (1991) 212.
17. T. Urisu, J. Takahashi, Y. Utsumi and H. Akazawa: Appl. Organometallic Chem. **5** (1991) 229.
18. As a special issue of the synchrotron radiation excited processes, Optoelectronics **11** (1996) No. 1 "Devices and Technologies."
19. B. Zanoni, M. N. Piancastelli, M. Marsi and G. Margaritondo: "Silicon metallization by synchrotron-radiation-induced $W(CO)_6$ surface reaction", Solid State Commun. **76** (1990) 1239.
20. M. Ikejiri, T. Ogata, H. Ogawa, M. Nishio and A. Yoshida: J. Vac. Sci.Tech. A **12** (1994) 278.
21. T. Urisu and H. Kyuragi: J. Vac. Sci. Tech. B **5** (1987) 1436.

22. Y. Nonogaki, H. Hatate, R. Oga, S. Yamamoto, Y. Fujiwara, Y. Takeda, H. Noda and T. Urisu: Mater. Sci. Eng. B **74** (2000) 7.
23. V. Ney and N. Schwentner: J. Phys.: Condens. Matter **18** (2006) S1603.
24. For example, K. Deguchi, K. Miyoshi and T. Matsuda: J. Electr. Spectr. Rel. Phenom. **80** (1996) 321.
25. For example, H. Kinoshita, T. Watanabe and M. Niibe: J. Synch. Rad. **5** (1998) 320.
26. P. Naulleau, K. A. Goldberg, E. Anderson, J. P. Cain, P. Denham, K. Jackson, A. S. Moriens, S. Rekawa and F. Salmassi: J. Vac. Sci. Technol. B **22** (2004) 2962.
27. M. Toyoda, K. Yamasoe, T. Hatano, M. Yanagihara, A. Tokimasa, T. Harada, T. Watanabe and H. Kinoshita: Appl. Phys. Express **5** (2012) 112501.
28. A. Isoyan, A. Wüest, J. Wallace, F. Jiang and F. Cerrina: Opt. Express **16** (2008) 9106.
29. E. W. Becker, W. Ehrfeld, P. Hagmann, A. Mancer and D. Münchmeyer : Microelectronic Engineering **4** (1986) 35.
30. H. Guckel, P. S. Mangat, H. Emmerich, S. Massoud-Ansari, J. Klein, T. Earles, J. D. Zook, T. Ohnstein, E. D. Johnson, D. P. Siddons and T. R. Christenson: Solid-State Sensors and Actuator Workshop (Hilton Head, 1996) p.60.
31. S. Watanabe, M. Esashi and Y. Yamashita: J. Intelligent Material Systems and Structures **8** (1997) 173.
32. T. Katoh and Y. Zhang : Microsystem Technol. **4** (1998) 135.
33. M. Horade and S. Sugiyama: Microsystem Technol. **19** (2013) 351.
34. C. Müller and J. Mohr : Interdisciplinary Science Reviews **18** (1993) 273.
35. Y. Utsumi, T. Kishimoto, T. Hattori and H. Hara:. Microsystem Technol. **13** (2007) 417.
36. V. Saile, U. Wallrabe, O. Tabata and J. G. Korvink eds.: "LIGA and Its Applications" (Wiley-VCH Verlag GmbH & Co. KGaA, 2009).
37. C. G. K. Malek: Proc. SPIE **4234** (2001) 338.
38. M. Kufner and S. Kufner: Microsystem Technol. **2** (1996) 114.

CHAPTER 9 Crystal Structure Analysis by X-Ray Diffraction

The elastic X-ray scattering can provide detailed information on the atomic arrangement of materials and has been widely applied in structure analysis.[1–3] In this chapter, the theoretical basis of X-ray diffraction by periodic materials, *e.g.*, crystals, is presented. For periodic materials, X-ray diffraction gives not only structural parameters such as lattice distances, but also the electronic distributions, which may identify the positions of atoms even in extremely large unit cells. Details of experimental methods, such as powder diffraction and single crystal diffraction are also explained. Researchers have routinely used characteristic X-ray sources for many years and you may ask why synchrotron radiation is particularly useful? The advantages which arise from the access to synchrotron radiation sources originate from the multiple benefits of brilliance, tuneability and polarization among others. As illustrations we describe structural studies using synchrotron radiation diffractometry on novel crystals, such as copper high T_c superconductors, organic materials, ferroelectric materials, charge and orbital ordering systems and proteins. Studies of organic monolayers on water surfaces will be discussed together with studies of crystals under high pressure. For non-periodic materials, the information on distances and the configuration numbers for neighboring atoms can be obtained by X-ray diffractometry. The distribution of atoms in amorphous solids and liquids is introduced. X-ray scattering at small angles provides information on the density distributions within materials. The small angle X-ray scattering methods by which the sizes and shapes of small particles, such as macromolecules of biological materials can be determined is introduced elsewhere.[1–3] Since X-ray diffraction measurements provide the most basic information on materials, *i.e.*, the crystal structure, these measurements are always performed at first when new materials are found or are synthesized. For the latest information on applications in X-ray diffraction, see also Ref. 30 in Chap. **1**.

9.1 ELASTIC SCATTERING BY CRYSTAL LATTICES

In Sect. **4.2.3**, elastic, or Thomson scattering was considered from a classical point of view. When X-rays are incident on any crystal they will be scattered elastically by the regular atomic arrays. This arises from scattering by electrons within the crystal atoms. The diffraction phenomenon will take place whenever the constructive interference condition is satisfied in reflection, that is, whenever the scattered X-rays are in phase. This situation can be calculated using Bragg's law, $n\lambda = 2d\sin\hat{\theta}$, where λ and d are respectively the wavelength of the incident X-ray and the lattice spacing, and $\hat{\theta}$ is the grazing angle of incidence (where $2\hat{\theta}$ is the angle between incident and scattered waves). The integer n represents the order of reflection. When an incident wave of wave vector $\boldsymbol{k}_1$ is scattered and the scattered wave's wave vector is $\boldsymbol{k}_2$, the scattering vector is defined to be $\boldsymbol{q} = \boldsymbol{k}_1 - \boldsymbol{k}_2$ and the momentum transfer vector is $\hbar\boldsymbol{q}$ (see Fig. 9.1, usually in X-ray diffractometry $\boldsymbol{q} = \boldsymbol{k}_2 - \boldsymbol{k}_1$). Since we are considering only elastic scattering the magnitudes of $\boldsymbol{k}_1$ and $\boldsymbol{k}_2$ must be identical and $|\boldsymbol{k}_1| = |\boldsymbol{k}_2| = k$.

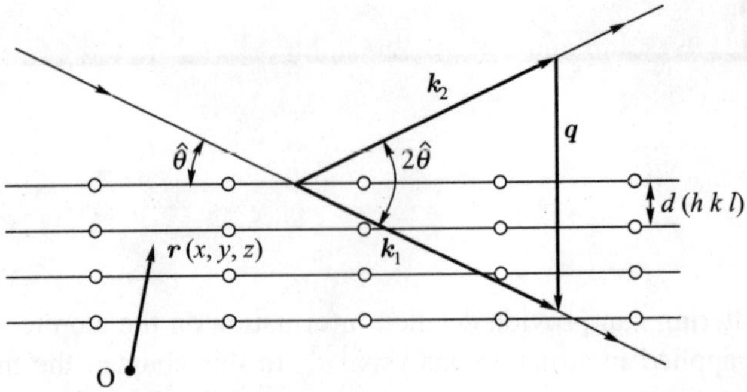

Figure 9.1 X-ray scattering in a crystal and the definition of scattering vectors $\boldsymbol{q}$. $\boldsymbol{k}_1$ and $\boldsymbol{k}_2$ are wave vectors of the incident and scattered X-rays, respectively.

The magnitude of the scattering vector $\boldsymbol{q}$ is given by

$$|\boldsymbol{q}| = q = 2k\sin\hat{\theta} = \frac{4\pi}{\lambda}\sin\hat{\theta}. \tag{9.1}$$

The three-dimensional atomic structure of any crystal can therefore be derived from an analysis of the observed diffraction pattern of X-rays.[1–3]

9.1.1 The diffraction intensity

The intensity of elastic scattering originates from the A^2 term of the radiation with matter interaction described in Sect. **4.3.3**. Since many electrons are distributed within a crystal, the summation term in Eq. (4.111) expressing the transition matrix element can be rewritten as

$$\sum_j e^{i\boldsymbol{q}\cdot\boldsymbol{r}_j} = \int \rho(\boldsymbol{r})\, e^{i\boldsymbol{q}\cdot\boldsymbol{r}}\, dr = \rho_{\boldsymbol{q}}, \tag{9.2}$$

where $\rho(\boldsymbol{r})$ stands for the electron density (the number density) distribution in the crystal. In real crystals, unit cells are aligned periodically in three-dimensional space, so that the crystals have translation symmetry. The electron density $\rho(\boldsymbol{r})$ is invariant under the translation operation $T = p\boldsymbol{a} + q\boldsymbol{b} + r\boldsymbol{c}$ where p, q and r are integers and $\boldsymbol{a}$, $\boldsymbol{b}$ and $\boldsymbol{c}$ are the basic vectors of the unit cell. When X-ray diffraction occurs the scattering vector $\boldsymbol{q}$ can therefore be expressed by $\boldsymbol{q} = -\boldsymbol{G}$, which is the reciprocal vector defined to be $\boldsymbol{G} = h\boldsymbol{a}^* + k\boldsymbol{b}^* + l\boldsymbol{c}^*$, where h, k and l are integers.[4,5] The basic vectors $\boldsymbol{a}^*$, $\boldsymbol{b}^*$ and $\boldsymbol{c}^*$ are then defined by the relationships,

$$\boldsymbol{a}^* = 2\pi\frac{\boldsymbol{b}\times\boldsymbol{c}}{\boldsymbol{a}\cdot(\boldsymbol{b}\times\boldsymbol{c})},\quad \boldsymbol{b}^* = 2\pi\frac{\boldsymbol{c}\times\boldsymbol{a}}{\boldsymbol{b}\cdot(\boldsymbol{c}\times\boldsymbol{a})},\quad \boldsymbol{c}^* = 2\pi\frac{\boldsymbol{a}\times\boldsymbol{b}}{\boldsymbol{c}\cdot(\boldsymbol{a}\times\boldsymbol{b})}. \tag{9.3}$$

The scattering intensity is proportional to $|\rho_{\boldsymbol{G}}|^2$, where $\rho_{\boldsymbol{G}}$ is given by

$$\rho_{\boldsymbol{G}} = \sum_{p,q,r} \int_{\text{unit cell}} \rho(\boldsymbol{r})\, e^{-i\boldsymbol{G}\cdot(p\boldsymbol{a}+q\boldsymbol{b}+r\boldsymbol{c}+\boldsymbol{r})} d\boldsymbol{r} = L(hkl)F(\boldsymbol{G}).$$

The coordination vector $\boldsymbol{r}$ is defined within the unit cell and the integration is made only inside the unit cell. A function L, called the Laue function, is defined by

$$L(hkl) = \sum_{p,q,r} e^{-i\boldsymbol{G}\cdot(p\boldsymbol{a}+q\boldsymbol{b}+r\boldsymbol{c})}. \tag{9.4}$$

In cases of cubic, tetragonal or orthorhombic systems where the basic vectors are orthogonal the Laue function can be expressed by

$$|L(hkl)|^2 = \frac{\sin^2 N_a h\pi}{\sin^2 h\pi} \cdot \frac{\sin^2 N_b k\pi}{\sin^2 k\pi} \cdot \frac{\sin^2 N_c l\pi}{\sin^2 l\pi}, \tag{9.5}$$

where N_a, N_b and N_c are the numbers of unit cells in the a, b and c directions in the volume that scatters the X-rays. The formula Eq. (9.5) is also valid for non-orthogonal lattices. This formula expresses the diffraction intensity in the three-dimensional case while a similar formula for a grating, that in the one-dimensional. $F(\boldsymbol{G})$ is called the 'structure factor' and is represented by

$$F(\boldsymbol{G}) = \int_{\text{unit cell}} \rho(\boldsymbol{r})\, e^{-i\boldsymbol{G}\cdot\boldsymbol{r}} d\boldsymbol{r}. \tag{9.6}$$

When several different atoms are contained in the unit cell, they are numbered by j and the structure factor then becomes

$$F(\boldsymbol{G}) = \sum_j \int_{\text{unit cell}} \rho_j(\boldsymbol{r}-\boldsymbol{R}_j)\, e^{-i\boldsymbol{G}\cdot\boldsymbol{r}} d\boldsymbol{r} = \sum_j \int_{\text{unit cell}} \rho_j(\boldsymbol{r}-\boldsymbol{R}_j)\, e^{-i\boldsymbol{G}\cdot(\boldsymbol{r}-\boldsymbol{R}_j+\boldsymbol{R}_j)} d\boldsymbol{r}$$

$$= \sum_j f_j(\boldsymbol{G})\, e^{-i\boldsymbol{G}\cdot\boldsymbol{R}_j}, \tag{9.7}$$

where

$$f_j(\boldsymbol{G}) = \int_{\text{isolated atom}} \rho_j(\boldsymbol{r}) e^{-i\boldsymbol{G}\cdot\boldsymbol{r}} d\boldsymbol{r}. \tag{9.8}$$

The term $f_j(\boldsymbol{G})$ is an atomic scattering factor and is the form factor, for an isolated j-th atom, being the Fourier transform of the electron distribution of the j-th atom. The electron density distribution in an atom is usually assumed to be spherically symmetric, so that $f_j(\boldsymbol{G}) = f_j(G)$ except in special cases (see Sect **9.4.2**), where $G = 2\pi/d(hkl)$ and $d(hkl)$ is the lattice spacing of planes with index (hkl). The term $f_j(G)$ is identical to that in Sect. **4.2.3**. If the position of the j-th atom is given by $\boldsymbol{R}_j = x_j\,\boldsymbol{a} + y_j\,\boldsymbol{b} + z_j\,\boldsymbol{c}$, the structure factor Eq. (9.7) becomes

$$F(hkl) = \sum_j f_j(G)\, e^{-2\pi i(hx_j+ky_j+lz_j)}. \tag{9.9}$$

The squared absolute value of the structure factor, $|F(hkl)|^2$, is proportional to the intensity of reflection, $I(hkl)$. When crystals are composed of a single element, the structure factor is expressed by

$$F(hkl) = f(hkl)\sum_j \exp\{-2\pi i\,(hx_j + ky_j + lz_j)\} = f(hkl)S. \tag{9.10}$$

S is called the 'geometrical structure factor' and is dependent wholly on the crystal structure. There are cases in which $S = 0$, for example, in bcc crystals $S = 1 + e^{i\pi(h+k+l)}$ and also when $h + k + l$ are odd. The circumstance when $S = 0$ is called the 'general reflection condition' (or 'extinction law'). When crystals contain more than two kinds of atom sites, the structure factors are given as the summation of partial structure factors calculated for atoms situated on the equivalent sites and the reflection conditions then become more complicated.

When the X-ray energy lies close to the absorption edge of any constituent elements, anomalous dispersion will occur as described in Sect. **4.2.3**. This is called the resonant scattering effect and in this case the atomic scattering factor cannot be expressed simply by $f_0(\omega, q)$, but one must also take into account the anomalous dispersion terms, $f'(\omega)$ and $f''(\omega)$, which are the real and imaginary parts of the anomalous scattering factor, respectively.

The atomic scattering factor is given by

$$f(\omega, q) = f_0\,(\omega, q) + f'\,(\omega) + if''\,(\omega), \tag{9.11}$$

where q and ω are the wave number of the scattering vector and the angular frequency of the incident X-ray, respectively. In crystals, $\boldsymbol{q} = -\boldsymbol{G}$, and in general, the imaginary part $f''(\omega)$ is usually small and almost constant below the absorption edge, but increases rapidly above the edge. The real part $f'(\omega)$ shows a large energy dependence below and above the edge (see Fig. 4.7 showing $f_0(\omega, 0) + f'(\omega)$). Then, using the energy dependence of $f'(\omega)$ and $f''(\omega)$ around absorption edge, there is an opportunity to undertake element-specific diffraction experiments from which we can gain information about the locality of the atoms.

9.1.2 The Debye-Waller factor

The intensity distribution patterns of X-ray diffraction observed at low temperatures are usually sharp, but become diffuse with increase in temperature. This is due to the thermal vibrations of constitutent atoms or ions in the sample. The time-averaged distribution of thermally vibrating atoms or ions follows a Gaussian distribution, $\exp(-x^2/2\sigma^2)$, where the constant σ and the variable x are the standard deviation (the amplitude) of the thermal vibration and the displacement, respectively. The cloud of electrons associated with each atom or ion also vibrates, so that the electron distribution function has to be expressed by

$$\rho\,(r) = \rho_0\,(\boldsymbol{r} - \boldsymbol{R}) \exp\left\{-\left(\boldsymbol{R} - \boldsymbol{R}_\mathrm{e}\right)^2 / 2\sigma^2\right\} \Big/ \left(\sqrt{2\pi}\sigma\right), \tag{9.12}$$

where $\boldsymbol{R}$ and $\boldsymbol{R}_\mathrm{e}$ are positions of the atom or the ion at any particular time and at equilibrium, respectively. At finite temperature the atomic scattering factor is calculated using a Fourier transform of the expression Eq. (9.12). If we assume spherically symmetric thermal vibration the temperature dependent form of the atomic scattering factor becomes

$$f(\boldsymbol{G}) = f_0(\boldsymbol{G}) \exp(-\sigma^2 G^2/2). \tag{9.13}$$

The exponential term is the isotropic Debye-Waller factor which is given by $\exp(-W)$, where $W = \sigma^2 G^2/2 = BQ^2$ and Q is defined by $Q = \sin\hat{\theta}_\mathrm{B}/\lambda = G/(4\pi)$. At each Bragg peak, the isotropic atomic displacement parameter, sometimes called the temperature factor B, becomes $8\pi^2\sigma^2$. At room temperature the standard deviation σ is usually of the order of 10^{-2}Å. For crystal structure analyses, B values of constitutent atoms or ions are used rather than σ. B values are relatively large for light elements, being about 1.0 Å^2 but are small for heavy elements and typically about 0.1 Å^2. The values increase almost linearly with increasing temperature, represented by $B \propto T/\Theta_{\mathrm{D}}^2$ where Θ_{D} is the Debye temperature. In the cooling process, the B value decreases linearly down to about 60K, but becomes almost constant below 10K. Since the reflection intensity of each Bragg-reflection is proportional to $|\rho_G|^2$, or $|F(G)|^2$, it is also proportional to the Debye-Waller factor, e^{-2W}. When the temperature is increased, the reduction in the peak height for high-Q Bragg reflections becomes significant compared with those from low-Q reflections.

9.1.3 The electron density

The electron density in any unit cell can be calculated from the observed structure factor $F(\boldsymbol{G})$, using Fourier synthesis or the inverse transform of Eq. (9.6). One can also determine the position of the atoms in the unit cell. The electron density distribution is given by

$$\rho(\boldsymbol{r}) = \frac{1}{v_{\mathrm{c}}} \sum_{\boldsymbol{G}} F(\boldsymbol{G})\, e^{i\boldsymbol{G}\cdot\boldsymbol{r}}, \tag{9.14}$$

where $\boldsymbol{r} = x\boldsymbol{a} + y\boldsymbol{b} + z\boldsymbol{c}$ and v_{c} are the positions in a unit cell and the volume respectively. The structure factors $F(\boldsymbol{G})$ in the formula Eq. (9.14) are generally complex and only the absolute values of $|F(\boldsymbol{G})|$ are observed experimentally. The phase term $\Phi(hkl)$ must be defined for each structure factor in order to undertake the Fourier synthesis. Taking the phase term into account, the electron density distribution can finally be given by

$$\rho(\boldsymbol{r}) = \frac{1}{v_{\mathrm{c}}} \sum_{h,k,l} |F(hkl)| \exp\{i(hx + ky + lz) + i\Phi(hkl)\}. \tag{9.15}$$

One of the problems in attempting a Fourier synthesis arises from the limit in the total number of Fourier terms. This arises because it is usually not possible to observe a sufficiently large number of reflections (hkl). Since observation of diffraction over a large $2\theta_{\mathrm{B}}$ range is important, the use of short wavelength, or high energy X-rays gives a significant advantage by increasing the number of observable structure factors. A further difficulty lies in the choice of a suitable phase angle $\Phi(hkl)$ or phase factor $\exp\{i\Phi(hkl)\}$. The determination of the phase factor is equivalent to the determination of the atomic positions. In centrosymmetric crystals, the phase angles are 0 or π, and the structure factors are real. In these cases, the inversion center is chosen to be the origin of the coordinates for the constituent atom positions. However in non-centrosymmeteric crystals, the structure factors are complex and the phase problem becomes difficult to resolve. There are several methods available to establish $\rho(\boldsymbol{r})$ from Eq. (9.15). One of the favored procedures is to employ a direct method in which hypothetical phase factors are initially used, to be followed by systematic trials in order to establish those phase factors which best fit the experimental results for $|F|$. An alternative procedure is to use the maximum entropy method, MEM,[6)] where 'entropy' is not a technical term from thermodynamics, but is taken from information theory. The entropy is defined in terms of present and previous trial values of the electron density distribution function in a unit cell. It is maximized under a constraint condition given by the differences between the observed and calculated structure factors. After a sufficient number of iteration cycles, reasonable and unbiased electron density maps are determined. However, the MEM technique is not easily applied to very large crystals, because high speed computers are needed to calculate the electron density maps based on large numbers of structure factors. When one replaces $|F|$ in Eq. (9.15) with $|F|^2$ and omits the phase factors, one can obtain the Patterson function. Once calculated, the Patterson function is mapped onto a chosen plane and then the distances between atoms can easily be determined in that plane. Since the scattering factors of heavy atoms are very large the observed structure factors are mostly due to the heavy atoms and therefore the positions of heavy atoms can be estimated relatively easily using this method. Following such estimations, the structure factors and also the atomic positions can be iteratively determined using the difference Fourier synthesis technique (called the heavy atom method). A method for exploiting the anomalous dispersion of specified atoms will be described later.

9.2 DIFFRACTION MEASUREMENT TECHNIQUES

9.2.1 Powder diffraction structure analysis

Since X-ray powder diffraction intensities originate from single scattering by small crystals and not from multiple scattering, structure determination based on powder diffraction is relatively easy. Powder diffraction measurements are frequently used in the determination of the crystal structure of unknown samples, comprising single and multiple phase grains. The method is also applicable to all samples which undergo phase transitions and hence domain structure changes at particular temperatures. A powder diffraction diffractometer for such an experiment, in this case a Bragg-Brentano diffractometer used in conjunction with a monochromatic X-ray line source is shown in Fig. 9.2(a). In the case of focusing geometry, X-rays passing through the entrance slit ES become divergent and irradiate the sample. The sample is rotated by $\hat{\theta}$ while the X-ray detector rotates by $2\hat{\theta}$. The normal operating mode of this apparatus is called the '$\hat{\theta} - 2\hat{\theta}$ mode'. Light diffracted by the powder is detected on the Rowland circle passing through ES and an exit slit RS (see Sect. **5.3.2.c**) at all diffraction angles. Scintillation counters are often used as detectors. The powder samples cover the sample holder up to an area of

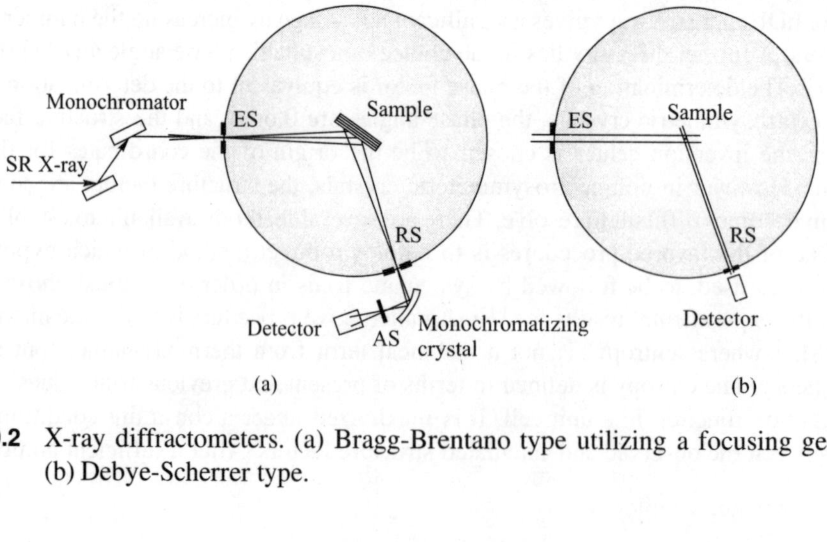

Figure 9.2 X-ray diffractometers. (a) Bragg-Brentano type utilizing a focusing geometry. (b) Debye-Scherrer type.

20 mm × 10 mm. The aperture of both slits is typically set for 0.5° acceptance and may be increased up to 2° without serious deterioration in resolution. Both the surface roughness and the deviation of the sample surface from the rotational center may create a loss in intensity and a systematic deviation in the Bragg peak positions. The reason for this is that the X-ray diffraction intensity is observed on the reflection side rather than on the transmission side. If the powder sample is so thick that X-rays cannot be transmitted then the Bragg-Brentano diffractometer has the advantage that no absorption correction is needed for the diffraction intensity analysis. This advantage is valid only if the sample surface is perpendicular to the scattering vector $\boldsymbol{q}$, that is, the angles of the incident and diffracted X-rays are equal with respect to the sample surface. When the monochromatic incident X-ray penetrates into a sample of absorption coefficient μ, the depth from the top surface at which its intensity decays to $1/e$ does not depend on the angle of incidence, because the equi-amplitude plane is parallel to the top surface (see Sect. **4.1.2.a**). This situation is similar for reflected X-rays. Therefore the Bragg intensity is not affected by the absorption effect even if the diffraction angle is changed. This diffractometer can also be used to characterize thin films using a wide but parallel X-ray beam collimated by a solar slit. Sometimes, diffraction measurements on thin films are made in a mode in which the incident angle $\hat{\theta}$ is fixed at a suitable shallow angle and only the detector rotates.

There are two types of Debye-Scherrer apparatus, a diffractometer type (Fig. 9.2(b)) and a camera type (Fig. 9.3). Usually monochromatic parallel X-rays are used. In the diffractometer method scintillation counters are employed, while in the camera method diffraction patterns have been recorded using photographic plates and X-ray film. Nowadays imaging plates are often used with the X-rays being focused on the imaging plate. X-rays are transmitted through powder samples contained in a glass capillary supported on the glass needle. When the incident X-ray traverses a cylindrical sample with a radius of r, the intensity of the X-rays will decay approximately to $\exp(-2\mu r)$ at every diffraction angle. Therefore an absorption correction is not required in principle. However this is valid only when $\mu r < 0.5$. This is realistic for organic samples but is difficult to realize for metallic samples, and in such cases the absorption correction is needed.

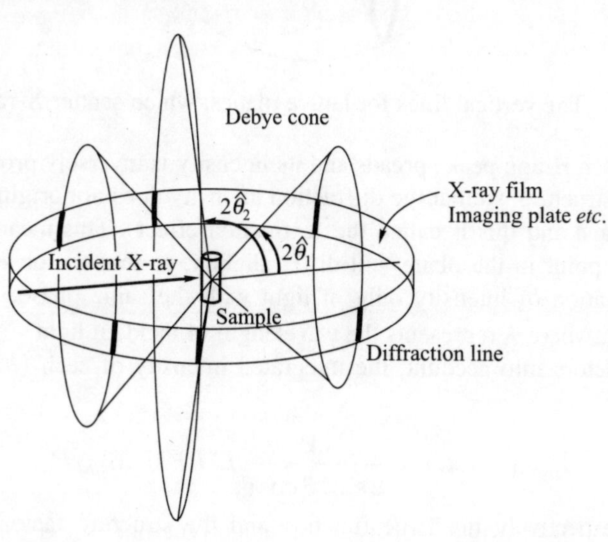

Figure 9.3 A Debye-Scherrer camera drawn with Debye cones and diffraction lines for powder samples.

One parameter used to derive the true diffraction intensities of lattice planes from measured ones is the Lorentz factor (a geometrical correction factor and not $\gamma = E/mc^2$ in Sect. **1.2.2**).[1−3] The following considerations apply both to the diffractometer and to the Debye-Scherrer apparatus. Figure 9.3 shows the geometry of a Debye-Scherrer camera. The incident X-rays coming from the left are collimated into a narrow beam, *i.e.*, effectively point collimation. The diffracted X-rays then spread into a cone, *i.e.*, a Debye cone, with a cone half angle of $2\hat{\theta}$ where the Bragg condition is satisfied. The number of the Debye cones corresponds to the number of different kinds of reciprocal lattice points. The Debye-Scherrer camera records Bragg intensities in the Debye cones using a film which is set horizontally in laboratory. The circumference of the base circle of the cone is maximum at $2\hat{\theta} = 90°$ and minimum at $2\hat{\theta} = 0°$ and $180°$. The diffraction intensity is proportional to $1/\sin 2\hat{\theta}$.

For all powder diffraction measurements, the Bragg conditions are satisfied only for those crystal grains which are appropriately oriented. The normals (vertical lines) of lattice planes in those grains contributing to the diffraction at an angle of $2\hat{\theta}$ are distributed on a cone of $90° + \hat{\theta}$ shown in Fig. 9.4. The number of vertical lines corresponds to the number of contributing grains. The number is proportional to $\cos\hat{\theta}$.

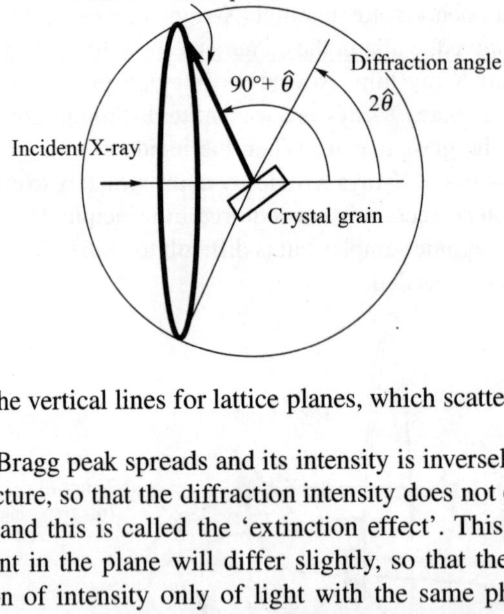

Figure 9.4 The vertical lines for lattice planes, which scatter X-rays by $2\hat{\theta}$.

The peak width of each Bragg peak spreads and its intensity is inversely proportional to $\cos\hat{\theta}$. Real crystals have a mosaic structure, so that the diffraction intensity does not originate from the full extent of the crystal lattice plane and this is called the 'extinction effect'. This means that the phase of the diffracted light at each point in the plane will differ slightly, so that the observed diffracted light is the result of the summation of intensity only of light with the same phase. The coherent region is proportional to $\lambda/\sin\hat{\theta}$, where λ represents the wavelength of incident light.

Taking these four factors into account, the integrated intensity of each (hkl) Bragg peak can be expressed by

$$I_{\text{powder}}(hkl) \propto \frac{1}{2\sin^2\hat{\theta}\cos\hat{\theta}} L^* L F^* F M_{hkl} P, \tag{9.16}$$

where L and F are respectively the Laue function and the structure factor, and P and M_{hkl} are respectively the polarization factor given in Sect. **4.2.3.c** and the multiplicity factor which is the number of equivalent Bragg reflections appearing at the same angle $2\hat{\theta}$. At finite temperature the structure factors include the Debye-Waller factors Eq. (9.13). The trigonometric function in Eq. (9.16) is the Lorentz factor for the Bragg-Brentano diffractometer, and also for the Debye-Scherrer apparatus in the case when no absorption correction is required. Different types of diffractometer have different Lorentz factors.[7] The Compton scattering effect, or the effect of inelastic X-ray scattering, must also be taken into account for accurate structure analyses.

Each type of lattice has its own characteristic diffraction pattern where the relative positions (angles) are unique for each type. When the X-ray powder diffraction patterns of unknown materials are assigned by comparison with known patterns of a Bravais lattice such as body-centered cubic, cI-, all-face-centered cubic, cF- or hexagonal, hP-types *etc.*, then the lattice parameters a, b and c, and the angles between crystal axes, α, β and γ can be determined from the patterns.

When the chemical composition, chemical formula and the density are determined, the number of atoms (or ions) in a unit cell can be estimated. Once the Bravais lattice is established, then the appropriate space group may be selected. This depends on the reflection conditions, chemical formula and Bragg intensities and will give the approximate positions and occupation factors of constituent atoms (ions) in a unit cell. A trial-and-error calculation is often used to fit the calculated results to the experimental data. Lattice parameters a, b and c, angles α, β and γ as well as the positions of atoms x, y and z in any crystals can be obtained by taking the background, line shape and asymmetry parameters for Bragg peaks into account.

The Rietveld method is the most popular algorithm for the calculation of the above parameters using a least squares fitting method, software programs for the calculation of which are freely available.[8,9] A similar least squares algorithm is also applied to the structure determination for the single crystal diffraction data. The reliability parameter is defined to be the sum of the absolute values of the difference between the observed and calculated Bragg intensities divided by a sum of the observed Bragg intensities. Once the reliability parameter is 5% or less, then the simulation is considered to be successful. The crystal structure and electron density distribution in a unit cell can be calculated self-consistently using the Rietveld method and an MEM calculation.[10]

Powder diffraction using white beam X-rays can also be used for crystal structure analyses. In this case, diffraction patterns are measured using an SSD situated at a fixed angle $2\hat{\theta}$. All diffraction data are recorded as a function of X-ray energy, which is then converted to wavelength. The method (energy dispersion method) is particularly useful when the $2\hat{\theta}$ angle used to measure the diffraction intensities is restricted, such as in the case of the using high pressure cell mentioned later.

9.2.2 Single crystal diffraction structure analysis

For an exact determination of the crystal structure and mapping of electron density, accurate integrated Bragg intensities are required. In such cases, diffraction lines with small angular spread and without overlap must be used. The angular spread depends on the angular and wavelength spread of the incident X-rays, and on the crystallinity of the sample. In the laboratory it is essential to use single-grain single crystals with low dislocation and imperfection density, *i.e.*, with small mosaicity as samples. In the case of synchrotron radiation however the circumstances are quite different, because synchrotron radiation is highly collimated, which will be explained in Sect. **9.2.3**.

In single crystal diffraction measurements using monochromatic X-rays, single crystals are rotated to observe the integrated Bragg intensities in three-dimensional space. Figure 9.5 gives an illustration of the rotating crystal method. A tiny single crystal is set on a goniometer so that its crystal axis for a low index, *e.g.*, [001] is parallel to the rotation axis. The point collimation incident X-rays irradiate

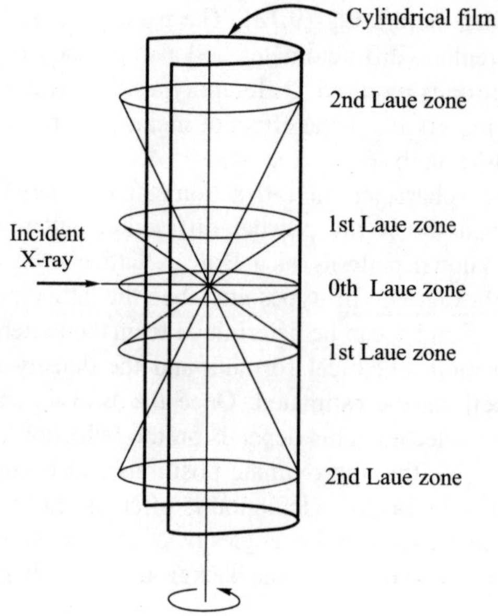

Figure 9.5 A schematic of the rotating crystal method.

the sample perpendicularly to its rotation axis. The X-ray diffraction intensities are recorded using cylindrical film or an imaging plate, where the longitudinal axis is set parallel to the rotation axis. Currently CCD detectors are also used. The zeroth Laue zone is a reciprocal lattice plane including the incident X-ray, *i.e.*, (000) reciprocal point. Lattice planes responsible for the diffraction spots in the zeroth zone lie parallel to the rotation axis. Diffraction will also occur from other planes which are not parallel to the rotation axis. Such diffraction will appear on the reciprocal lattice planes above and below the zeroth zone and are called the first Laue zone, the second Laue zone *etc.*, all parallel to the zeroth zone. The diffraction data collection in the zeroth and also non-zeroth Laue zones is important for a precise determination of the lattice parameters. Since different (hkl) diffraction peaks tend to overlap, simple rotation would be insufficient to distinguish between them and to provide individual integrated intensities. In such cases where lattice planes have similar spacings but different indices, the rotation angles where the diffraction peaks of the planes appear will be different. To distinguish between such overlapping diffraction peaks the sample is not rotated continuously across the whole angular range, but is moved intermittently. First, the sample is rotated through a narrow angular range and the diffraction points recorded on the film. Subsequently the film cartridge is shifted parallel to the rotation axis and then the sample is rotated about the adjacent range and the diffraction peaks is recorded on a new area of the film. By successively repeating such measurements, one can observe the diffraction peaks of the same $2\hat{\theta}$ but with different indices across different angular ranges. For each range the rotation is made in both directions and the procedure called the 'oscillating crystal method' or the 'oscillation method'. When the measurement is restricted to a particular Laue zone using a slit, the integrated Bragg peak measurement becomes more accurate. Apparatus equipped with such a slit is called a Weissenberg camera or Weissenberg diffractometer. The rotation can be made around two or three axes. There are examples of single crystal diffraction equipment with from 4 to 6 independent axes of rotation including those of the sample, detector and the monochromatizing crystal (see Fig. 9.2

(a)). For the absorption correction of the measured diffraction intensities, small samples, or cylindrically or spherically shaped samples are used. In these cases, the Lorentz and multiplicity factors differ from those of powder diffraction, Eq. (9.16), and they are $1/\sin 2\hat{\theta}$ and $M_{hkl} = 1$, respectively.

For single crystal diffraction studies, the Laue method with white X-rays is used. Transmitted diffracted X-ray beams are detected. The white X-rays are incident parallel to one of reciprocal lattice vectors and the diffracted X-ray patterns recorded on planar films or imaging plates placed behind the single crystals. The pattern consists of a number of spots by diffracted X-rays satisfying the Bragg conditions of $n\lambda = 2d\sin\hat{\theta}$. The lattice parameters themselves are not determined accurately by this method, but the (hkl) indices, the symmetry of the crystal, ratios of *a*/*b*, *a*/*c* and *b*/*c* as well as angles α, β and γ can be determined from the positions of diffracted X-rays on the detector plane. If the actual energies of the diffracted X-rays are determined using a solid state detector, then the lattice parameters can be properly established. In that case, the application of the Laue method to time-resolved experiments on proteins using pulsed X-rays and CCD detectors becomes feasible.

9.2.3 The advantages of using synchrotron radiation

The techniques for X-ray structure analysis using single crystals have evolved such that a structure analysis of high quality is possible using characteristic X-rays in the laboratory whenever there are good crystals of sufficiently large size, that is, from 100 to about 300 μm cube with mosaicity of about $0.1°$ or less. Therefore, for which cases is synchrotron radiation essential?

The most common case (case (1)) is that in which crystals with larger than 10 μm side cube cannot be prepared. In this case, powder diffraction should be undertaken, but laboratory X-ray sources (X-ray tubes) are not bright enough for precise powder measurements. The next case is that for samples which contain heavy elements such as Pb and W, making exact measurements difficult due to the strong absorption of the X-rays by those elements (case (2)). A third example is when the absorption edges of elements in the sample lie close to the energies of characteristic X-rays of Fe, Co, Cu, Mo, Ag *etc*., making precise measurements very difficult because of strong absorption or due to a large background of X-ray fluorescence (case (3)).

Many advanced structure studies have become readily feasible using synchrotron radiation, including those concerned with clarification of the origins of the functional properties of materials and are presented in cases from (4) to (7). Case (4) is that of X-ray diffraction experiments under extreme conditions, such as very high pressure or very low temperature. Case (5) refers to one in which the atomic displacements causing or caused by structure phase changes are so small, that the intensity of Bragg reflection or diffuse scattering (scattering originating from the irregularity of lattice periods) is extremely weak compared with a normal Bragg reflection. Case (6) is one in which the change of lattice constant is small. All cases given from (4) to (6) are those in which small signals are very difficult to extract from the large backgrounds originating from surrounding instruments and where it is often impossible to distinguish between the various types of phase change mechanisms or the interaction between atoms or the behavior of bonding electrons. Case (7) is one in which element-specific information is sought for every compound and for each chemical state.

The advantages of synchrotron radiation in the conduct of all such experiments are always those of (a) high intensity, (b) high brilliance, (c) the tuneability of wavelength and photon energy, (d) polarization characteristics, (e) very high energy X-rays and finally (f) its pulsed time structure characteristics. All these different and demanding requirements can be achieved simultaneously at synchrotron radiation sources such as the ESRF (6 GeV), APS (7 GeV) and SPring-8 (8 GeV).

The benefits of high intensity are clear. High brilliance means that the source size of X-rays is very small (because of the small cross section of the electron beam) and that the X-rays are almost exactly parallel (because the angular spread of the electron beam is also very small). Together these yield an extremely high density of photons at the sample. The angular spread of X-rays from undulators is usually only a few μrad, so that even after the X-rays have propagated over a distance of 100 m, the spot will have expanded only to about 1 mm square. Because the source size and the angular spread are so small one can use small mirrors for light collection resulting in a minimization of mirror aberrations, and therefore a further optimization of the X-ray collection efficiency at the sample. Even on a normal dipole beamline extracting synchrotron radiation from a bending section, a 0.1 mm high and 3 mm wide X-ray spot can be obtained by tangential focusing using a mirror in vertical direction and by sagittal focusing using a monochromatizing crystal in the horizontal direction. In this case a beam of 7×10^{11}photons/s with relative band width of $\Delta E/E = 10^{-4}$ available in an area of 0.1×3 mm^2 can be delivered at the sample position. Such high incident photon intensities generate sufficient intensity from reflections from 10 μm cube crystals to permit structure analyses on these and even smaller crystals. These are experiments which are quite impossible using laboratory X-rays. In the focusing case, although the intensity becomes much higher, the resolution of the diffraction angle and the monochromaticity (energy resolution) are not so high. Various focusing new systems have recently been devised and a spot size of 1 μm diameter (present minimum is 0.1 μm) has been achieved at an undulator beamline for diffraction experiments (see Sect. **14.2.1**).

The extraordinary brilliance always yields the merits of high resolution in diffraction angle and high monochromaticity. Focusing is sometimes employed at a level sufficient to maintain good parallelism. However, in any attempt to improve the resolution of laboratory X-ray sources one must always reduce the slit size and use monochromatizing crystals, thus seriously reducing the beam intensity. For example, the intensity will diminish to 1/10 of that of the original characteristic X-ray, even if one halves the

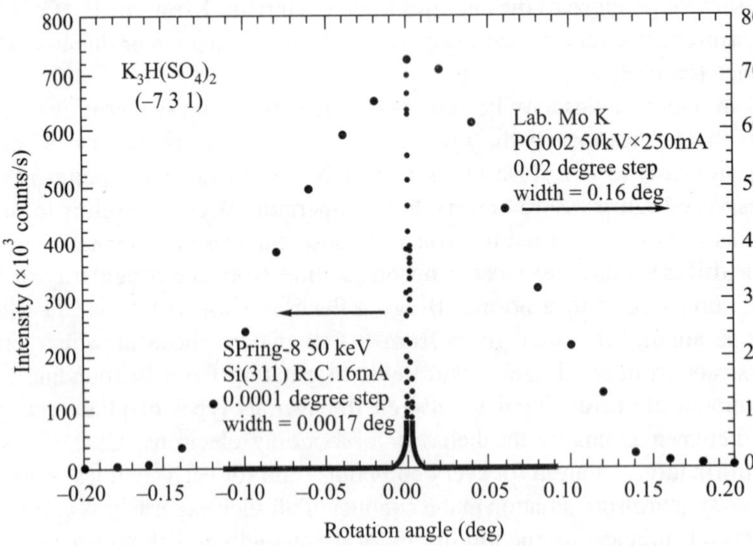

Figure 9.6 Rocking curves for KHS ($K_3H(SO_4)_2$) taken using synchrotron radiation and also laboratory X-rays.

resolution. In contrast, in the case of synchrotron radiation it is actually quite difficult to reduce the resolution and experiments must always be of high resolution. For instance, a rocking curve with full width at half maximum of 0.002° is easily achieved for the dielectric material KHS ($K_3H(SO_4)_2$). In Fig. 9.6, rocking curves taken using laboratory X-rays and synchrotron radiation are compared. The half width measured using synchrotron radiation is 1/100 of that measured using laboratory X-rays. The peak intensity and the width measured using synchrotron radiation are respectively $10^2 \sim 10^4$ times larger and 10^2 times smaller than those measured using laboratory X-rays. Accordingly, the integrated intensity of the former is from 1 to 10^2 times larger than that of the latter. The relative band width is also considerably smaller because $\Delta E/E = 10^{-4}$ even on ordinary beamlines. Therefore, the high brilliance manifests itself as a powerful tool for observation of small changes in lattice constants as described above. Even in the case of powder diffraction experiments, the Bragg reflections can be observed with a width of 0.008°.[11–13] As a result, it has been found possible using synchrotron radiation to study the ferroelectric phase change of hexagonal $BaTiO_3$, which is a phase change associated with small changes of the lengths of a- and b-axes of only 0.005 Å, a value which had never been measurable previously.[12,13] In single crystal diffraction experiments using laboratory X-rays, the density distribution of electrons can be obtained, together with that of nucleus using neutron diffraction on the same sample, so that discussions concerning the existence of electronic polarization has become possible.[14,15] However, when single crystals cannot be obtained, the electron density distribution can always be obtained with sufficient precision using synchrotron radiation powder diffraction methods, described later.

In cases where the photon energies of characteristic X-rays overlap the absorption edges of the sample, one can exploit the advantage of tuneability of synchrotron radiation to make diffraction measurements without this absorption effect problem. The tuneability of the source also gives advantages because the energy of an absorption edge is element-specific and therefore an anomaly in the atomic scattering factor appears close to the edge (resonant elastic scattering/diffraction). Therefore one can extract valuable information about the electronic structure and other information on the atom by tuning the photon energy selectively around the edge. This advantage is particularly important in the determination of phase in diffraction experiments. In addition, since synchrotron radiation from any bending section is intense and has a continuum spectrum, it is useful for the Laue method when applied to single crystals and for energy dispersion experiments on powder samples.

The experiments which exploit the choice of polarization of the radiation are introduced later where the linear polarization is used to identify the orientation of electron orbitals. In the region where quarter wave plates can be used, circularly-polarized X-rays can be obtained from linearly-polarized synchrotron radiation (see Sect. **5.5**), but circularly-polarized synchrotron radiation is also indispensable in other regions. X-ray diffraction experiments which exploit circularly-polarized X-rays, enables the magnetic structures or chirality of many materials to be investigated.[16,17]

The use of very high energy X-rays is advantageous because they pass easily through samples containing heavy elements. When the energies of the incident X-rays are increased, the transmittance of materials becomes larger in general, so that diffraction experiments become simpler. In that case, the magnitude of the wave vector $\boldsymbol{k}$ is large and it becomes possible to measure up to a large magnitude in $\boldsymbol{q}$. However, with an increase in energy, the general Lorentz factor (factor in Eq. (9.16) multiplied by λ^3) will decrease, resulting in a decrease in scattering intensity, while the background due to Compton scattering increases. Furthermore, Bragg reflections tend to overlap in materials with large lattice constants, so that measurements with sufficiently high resolution are required. Therefore it is not necessarily true that the higher the energy is, the better.

The large transmittance is advantageous not only for measurements of samples containing heavy elements, but also for the experiments at high pressures and low temperatures.[18] In these cases high energy X-rays can pass easily through a diamond anvil cell designed for extremely high pressure experiments (see Fig. 9.19). For measurements at low temperatures, the sample shroud is cooled to liquid helium temperatures. The outer shroud also is cooled to liquid nitrogen temperatures, thus preventing heat entering from outside, while the outermost shroud is that sealing the vacuum (the vacuum shroud). They are cylindrical or half spherical, usually made from Be to permit access to incident and scattered X-rays. For high energy X-rays, one can use a cylinder made from Al, which is readily available and easily machined, instead of Be. In the case of 30 keV X-rays the intensity decreases only down to 63% after passing through 2 mm thick Al.

The Al shroud is sometimes made with a hole of from 2 to 3 cm diameter covered by a Kapton film through which one can see the sample. If the vacuum shroud is made using transparent 1 mm thick, acrylic resins, one can view the samples from all directions. In one case, a large scale cryostat has incorporated a diamond anvil cell with sample shroud made from Al, with a vacuum shroud from acryl.[13] When the sample temperature changes during experiments at low temperature under high pressure, for instance at 10K at 10^5 atm (about 10 GPa), the cold head contracts causing sample movements in the cryostat. By the use of this cryostat, it is possible to adjust the sample position while viewing the sample directly through a telescope. It is possible to adjust the sample by observing its X-ray shadow directly using a CCD when the transmittance is not large or when using X-rays close to an absorption edge by exploiting the tuneability.

The use of synchrotron radiation is unique because it makes feasible the time-resolved analysis of every possible type of experiment simply by taking advantage of the pulsed structure of the radiation from the storage ring, giving time modulation from the ns to the μs time domain. The principle of such analysis is given in Sect. **5.7**. Many experiments of this type have been undertaken. The time-resolved structure analysis of biological materials will be presented later.

9.3 THE STRUCTURAL ANALYSIS OF HIGH T_c SUPERCONDUCTORS, ORGANIC CONDUCTORS AND FERROELECTRIC MATERIALS

9.3.1 High T_c superconductors

The high T_c superconductors considered here are copper oxides of perovskite type similar to those in Sect. **7.6**. There are four constituent elements one of which is always a heavy element. Figure 9.7 shows photographs of diffraction patterns of LSCO ($La_{2-x}Sr_xCuO_4$) taken by the oscillating crystal method. The lower one was taken using 35 keV X-rays and the upper one with 70 keV.[18] Even for crystals containing heavy elements, such as La and Sr, the diffraction patterns corresponding to reciprocal lattices taken using 70 keV X-rays, shown in the photograph, are observed as clearly as those taken by electron diffraction. (The left photograph taken using 70 keV X-rays is an expanded version of the right one). Figure 9.8 shows the crystal structure of LSCO.

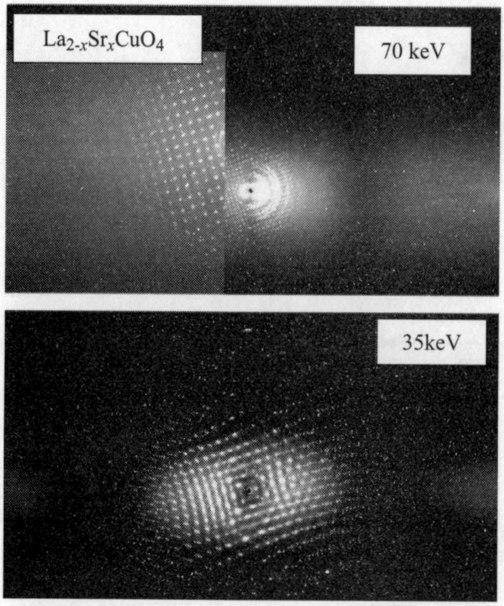

Figure 9.7 A photograph of the diffraction pattern of LSCO ($La_{2-x}Sr_xCuO_4$) taken using the oscillating crystal method with high energy X-rays.[18)]

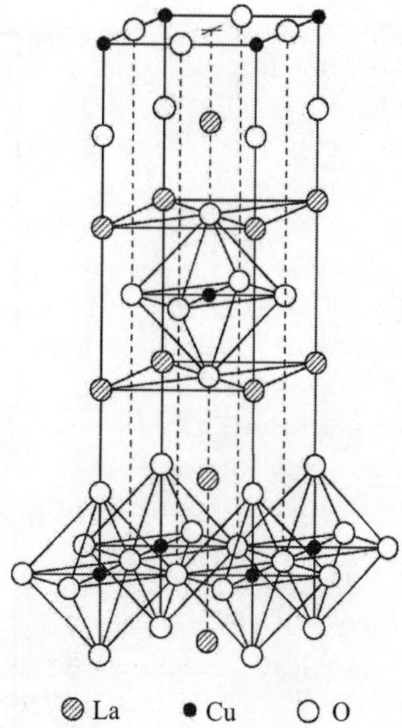

Figure 9.8 The crystal structure of LSCO ($La_{2-x}Sr_xCuO_4$).

In high T_c superconductors, it is well-known that the superconductivity is strongly suppressed when the concentration of holes is 1/8. This phenomenon is explained by the appearance of a stripe structure, which is caused by the separation of the magnetic ordering region and the charge ordering region. Evidence also arises from neutron diffraction experiments in which the magnetic reflection and the superlattice reflection due to modulation of the lattice have been observed. Figure 9.9 shows additional evidence derived from X-ray diffraction. Here the diffraction pattern of LBSCO ($La_{1.875}Ba_{0.125-x}Sr_xCuO_4$) gives the profile of a superlattice reflection exhibiting the stripe structure.[19)] Therefore the stripe structure is confirmed both by neutron and by X-ray diffraction experiments. The upper profile is that of the fundamental reflection of (200) and the lower one, that of the superlattice (*n.b.* 4 minutes were required to take each experimental point). The count rate at the fundamental reflection was 200 Mcps, while that of the superlattice was about 3 cps at most and therefore it has become possible to identify the existence of the very weakest reflections whose intensity may only be as low as 10^{-8} of that of the fundamental, by making use of synchrotron radiation. In the laboratory, the minimal detectable intensity of any reflection is about 10^{-5} of that of the fundamental thus giving a measure of the immense advantage in using synchrotron radiation for the study of very small changes of atomic position and charge.

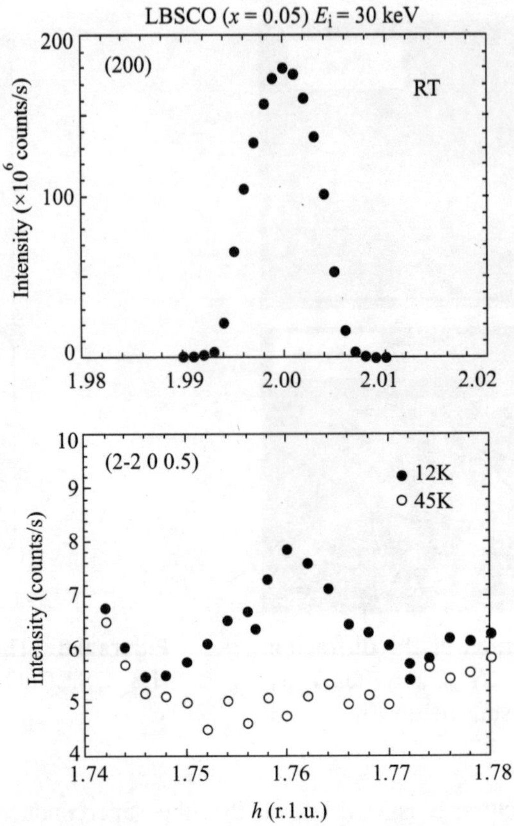

Figure 9.9 A profile of the superlattice reflection of LBSCO ($La_{1.875}Ba_{0.125-x}Sr_xCuO_4$) caused by stripe structure.[19]

9.3.2 Organic conductors

a. λ-BETS: Figure 9.10 shows the diffraction pattern of λ-(BEDT-TSF)$_2$FeCl$_4$ (λ-BETS) taken using the oscillating crystal method at 20K using a vacuum camera with 30 keV X-rays.[20] From this photograph, diffraction intensities were derived for each of the Bragg reflection indices (hkl) (where the total number of data points is 21,800) and a normal structure analysis was performed. The positions of 57 independent atoms were determined with a reliability factor (residue) of about $R = 5\%$. In organic conductors, the roles of donor and acceptor molecules are very important, so that it is true to say that an understanding of such molecules will define the development of organic conductors. BEDT-TSF is a good donor molecule, which has two pairs of 5- and 6-membered rings, shown in the inset of Fig. 9.11. This figure shows the electron density distribution of the molecule, in which the regions of high density in the 5- and 6-membered rings correspond to the positions of the Se and S atoms and in other places, to C atoms. This molecule and FeCl$_4$ comprise a salt of λ-BETS, which is metallic at room temperature. The material exhibits various kinds of anomalous behavior at low temperature. For example, it has a large dielectric constant in the metallic phase while the dielectric constant decreases drastically

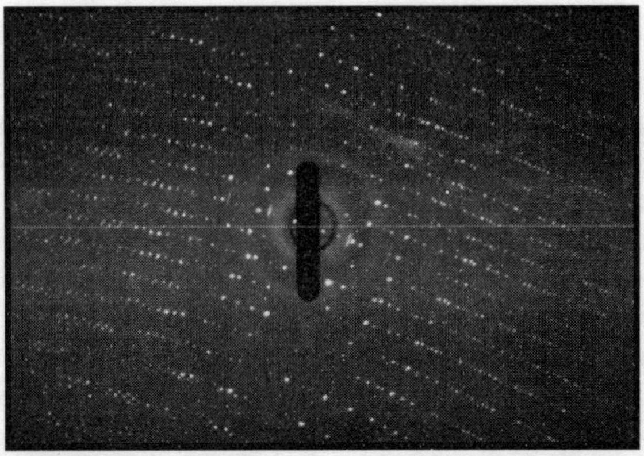

Figure 9.10 A photograph of the diffraction pattern of λ-BETS (λ-(BEDT-TSF)$_2$FeCl$_4$) taken by the oscillating crystal method at 20K.

with the metal-insulator transition. Structural changes during such a transition had never before been observed experimentally, until the small change in lattice constant and anomaly in the structure were recently observed using synchrotron radiation. Figure 9.11 illustrates the dependence of the lattice constant on temperature in λ-(BEDT-TSF)$_2$FeCl$_4$. There is a discontinuity in the lattice constant at 8K, which represents the temperature of the metal-insulator transition simultaneously accompanying an antiferromagnetic phase change and an unknown dielectric phase change. The excellent resolution permits the small change in lattice constant of 0.0008 Å to be observed at 8K. It is concluded that the transition is of the first order.[21] The lattice constant can be determined with an accuracy of better than 0.0001 Å (fm) using highly monochromatic light dispersed using the back scattering method (see

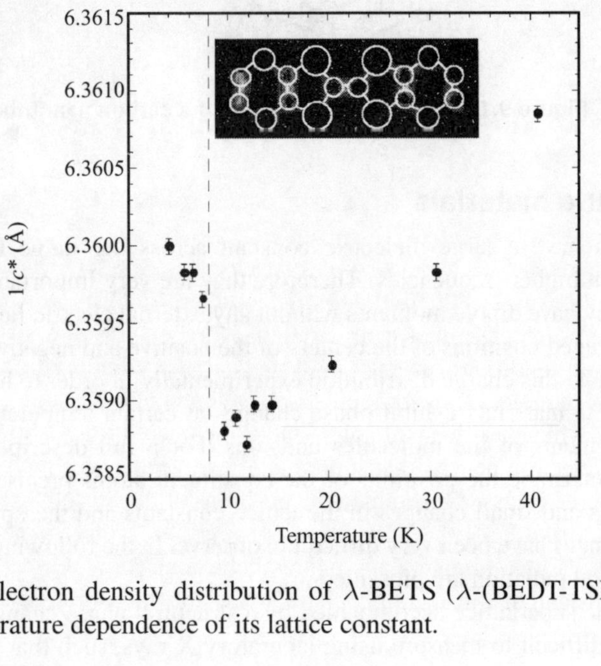

Figure 9.11 The electron density distribution of λ-BETS (λ-(BEDT-TSF)$_2$FeCl$_4$) and the temperature dependence of its lattice constant.

Sect. **5.3.2.b**).[22] It is anticipated that research on novel structural anomalies in organic conductors will be much extended through the measurement of such very small changes in lattice constants.[23]

b. Carbon nanotube: The fullerenes are spherical molecules comprised solely of C atoms (see Sect. **7.7.2.d**). There are also other molecules formed solely of C atoms to be found in which the C atoms lie in hexagonal networks and which form tubes with radii of from 1 to a few nm, and with lengths of up to 1 μm.[24] There are further types of tube structures, in which the walls may be single, double or multiple. Figure 9.12 shows the schematic structure for a single-walled carbon nanotube (SWNT). Currently there are many trials in progress to develop functional materials by accommodating various kinds of materials within the tubes. One example consists of an attempt to make an FET (field effect transistor) by accommodating donor or acceptor molecules and thus making n- and p-semiconductors in the tubes. Recently, TCNQ, a typical acceptor molecule, was successfully contained in SWNT[25] and this was confirmed by the observation that the intensity of the (10) reflection of X-ray ($q \approx 0.45$ Å^{-1}) became much weaker than that from SWNT alone.

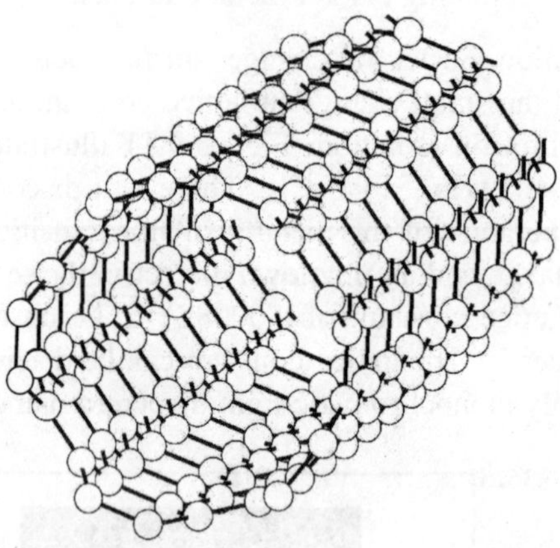

Figure 9.12 Schematic structure of a carbon nanotube.

9.3.3 Ferroelectric materials

Ferroelectric materials have a large dielectric constant across the range from DC to normal AC frequencies and to even higher frequencies. Therefore they are very important for use as capacitors in electronic devices. They have dipole moments without any external electric field. This is a consequence of the fact that the averaged positions of the centers of the positive and negative ions do not coincide. It is important to investigate this charge distribution experimentally in order to help understand the origin of ferroelectricity. These materials exhibit phase changes at certain temperatures and this property is related to the displacements of the molecules and ions. For a full description of ferroelectricity, it will be important to determine the positions of the constituent atoms precisely. However, even small displacements of atoms and small changes in the lattice constants and the appearance of superlattices arising from phase changes have been very difficult to observe. In the following some new experimental results using synchrotron radiation are presented.

A reflection from the superlattice accompanied by a structural phase change in the antiferroelectric $Rb_3D(SeO_4)_2$ is very difficult to measure using laboratory X-rays, such that important features of the

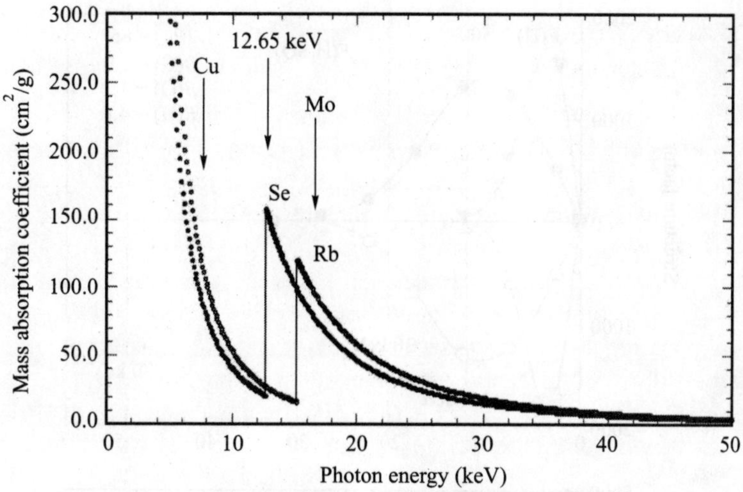

Figure 9.13 An example of the window of absorption (K absorption spectra of Rb and Se).

phase change have never before been observable. The main reason is that the characteristic X-rays of the Cu Kα and the Mo Kα lines are located within strong absorption regions just above the core edges of Se and Rb, respectively (see Fig. 9.13) and this causes large backgrounds to appear in all experimental diffraction data, smearing out the true signals. In this case, measurements can be obtained which are almost free from any background, if the energy of the X-rays are chosen to be at 12.6 keV where absorption is small. By exploiting the tuneability of synchrotron radiation, therefore, the superlattice reflection (0 3.5 0.5) of $Rb_3D(SeO_4)_2$ which cannot be observed in the laboratory has been observed clearly at the Photon Factory, PF.[26] The photon energy region where absorptions are minimal, as in the above case, is called the 'window of absorption'.

A further illustration of the advantage of access to such high energy X-rays is given in the mapping of electron distribution functions in ferroelectric materials. Detailed mapping is important in clarification of the origin of ferroelectricity. Figure 9.14 shows the dependence of the structure factor on the magnitude of the reciprocal lattice vector in squaric acid ($C_4O_4H_2$) having hydrogen bonds. The upper figure presents a simulation result at room temperature, where the temperature factor (B) is large and the lower figure is at low temperature, where the temperature factor can be assumed to be 0.[27] The abscissa represents the magnitude (wave number) of the reciprocal lattice vector, $G = 2\pi/d(hkl)$ in units of Å^{-1}. For all planes of (hkl) around $\boldsymbol{G}$, the structure factors can be treated in two groups, with positive and negative signs. The positive and negative factors around a certain range of $\boldsymbol{G}$ are summed separately and results of the summation are plotted against G in the figure. The ordinate represents the summed factors. The structure factor corresponds to the Fourier coefficient in the formula expressing electron distribution. If all structure factors are measured, the errors in the Fourier synthesis become infinitesmally small in principle.

Providing that the maximum measurable scattering angle $2\theta_{\max} = 86°$, the measurement using 30 keV X-rays will correspond to the measurement up to the wave number $G = 20$ Å^{-1} and that using 70 keV X-rays, up to $G = 48$ Å^{-1}. The number of non-measured reflections is large in room

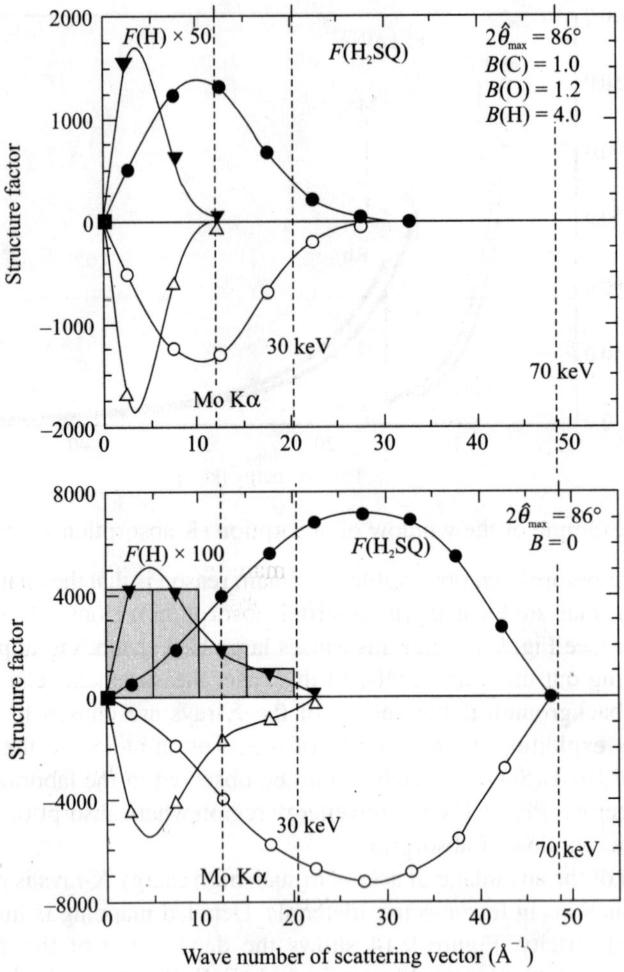

Figure 9.14 The dependence of the magnitude of the structure factor on the magnitude of reciprocal lattice vector ($G = 2\pi/d(hkl)$) (wave number of scattering vector) measured on squaric acid, with inversion symmetry, using high energy X-rays. The upper plot is at room temperature with large temperature factor; the lower, at low temperature with temperature factor $= 0$. Ordinate of closed circle at a certain wave number (G) represents sum of magnitudes of structure factors for all (hkl) existing around G giving positive values and that of open circle, negative values. Triangles represent the contribution of H atoms.

temperature measurements as indicated on the figure, unless X-rays of energy higher than 30 keV are used. In particular, the number of non-measured reflections becomes much larger in low temperature measurements, so that X-rays of energy above 70 keV are required to measure all reflections at low temperature. Note that, $F(\mathrm{H})$ in the figure represents solely the contribution of H atoms. It is suggested that in the case where the atomic scattering factor decreases rapidly with increase of wave number as in the case of H atoms, one can measure a sufficient number of reflections using the characteristic X-rays of Mo in the laboratory. This is the reason why clear mapping of the distribution of the electrons of H atoms is possible even using the classical method of difference Fourier synthesis.

Figure 9.15 shows the striking change in the covalent bond of Ti–O in the ferroelectric material $PbTiO_3$, undergoing a phase change from cubic to tetragonal observed directly by powder diffraction.[28] It provides an example of the mapping of the electron density distribution, which has so far proved very difficult to achieve using powder diffraction. Figures 9.15(a) and (b) show the electron distribution in the planes formed by the Pb and O atoms and also the Ti and O atoms in the cubic structure and in Figs. 9.15 (c) and 9.15 (d), in the tetragonal structure. This data was obtained using the MEM and Rietveld methods. They provide fundamental information which aids in understanding of the origin of the ferroelectric phase change in ABO_3 type perovskite materials in terms of the difference in the amount of covalency and ionicity in bonds between atoms. As shown in Fig. 9.15 (b), the distribution of the covalent electrons of Ti–O in the cubic phase at high temperature is isotropic around the four O atoms, while the covalent bonds become anisotropic and some covalent bonds are cut in tetragonal phase as shown in Fig. 9.15 (d). By contrast, Pb ions seem to be spherical in the cubic phase, while the Pb and O atoms form bonds of zig-zag chains in the tetragonal phase. This experiment helps to check and clarify theoretical calculations made from first principles. It also helps to identify the origin of the phase change in perovskite type ferroelectric materials which has been an unsolved problem for some time.

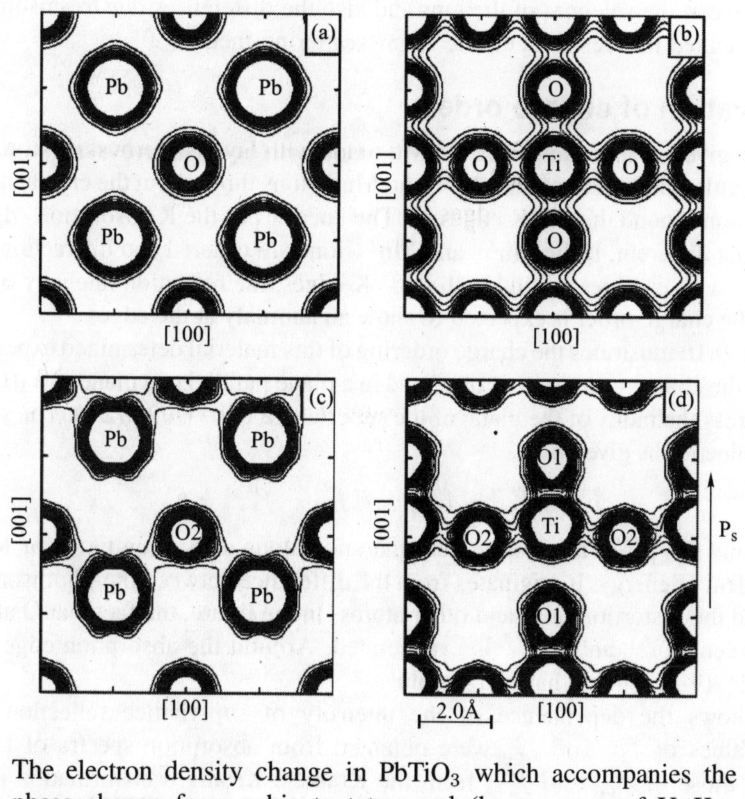

Figure 9.15 The electron density change in $PbTiO_3$ which accompanies the ferroelectric phase change from cubic to tetragonal (by courtesy of Y. Kuroiwa, © The American Physical Society 2001).[28] See details in text.

Some organic materials which exhibit novel ferroelectricity have been studied intensively to establish their structural origins.[29,30]

9.4 CHARGE AND ORBITAL ORDERING SYSTEMS—PEROVSKITE Mn OXIDE

Recently, perovskite Mn oxide, which is one of the strongly correlated electron systems, is considered to be a charge and orbital ordering system. The reason is that various novel properties induced by a magnetic field have been found, such as a 'colossal magneto-resistance', where electric resistance decreases drastically with the application of a magnetic field and a structural phase transition induced by a magnetic field. It has now been recognized that such materials have considerable potential as new functional electronic materials. As a result of the study of such strongly correlated electron systems, it is now accepted that the electronic degrees of freedom (charge, spin and orbital) and their combination undoubtedly influence the characteristics of these materials. The observation of spin order has been carried out using neutron diffraction but that of charge or orbital order has not been so easy. In this section, the observation of charge order and also of the orbital order is described through the exploitation of synchrotron radiation X-rays in terms of their high brilliance, energy tuneability and excellent polarization characteristics. The method described here takes advantage of the anomalous dispersion of the atomic scattering factors around the core absorption edges. In other words, this method exploits the differences between the valences of the ions and also the differences due to anisotropy even for the same valence. It is called the 'resonant elastic X-ray scattering method'.

9.4.1 Observation of charge order

$La_{0.5}Sr_{1.5}MnO_4$ is given as an example. It is a Mn oxide with layered perovskite structure. The charge order, that is the regular alignment of the Mn^{3+} and Mn^{4+} ions throughout the crystal, is observed using anomalous dispersion around the Mn K edges.[31] The energies of the K absorption edges of Mn^{3+} and Mn^{4+} are somewhat different. If the Mn^{3+} and Mn^{4+} ions are ordered and diffraction experiments are made using X-rays with energies around both ions' K edges, the reflection intensity of the superlattice originating from the charge order is expected to show an anomaly at the edges.

The inset in Fig. 9.16 illustrates the charge ordering of this material determined experimentally. When the superlattice in the charge order state is observed in a plane parallel to a plane $(hh\,0)$ in the disordered state, one can express the index of the plane of the superlattice to be $(h/2\ h/2\ 0)$. The structure factor of the superlattice reflection is given by

$$F = (f'_{3+} - f'_{4+}) + i(f''_{3+} - f''_{4+}) + C, \tag{9.17}$$

where f'_{3+}, f''_{3+} and f'_{4+}, f''_{4+} are respectively the anomalous dispersion terms of Mn^{3+} and Mn^{4+}, and C is independent of energy. It originates from the difference between the Thomson scattering terms $(f_{03+} - f_{04+})$ and the distortion of O and other atoms. In the figure, the fact that O atoms do not exist at mid-point between Mn^{3+} and Mn^{4+} is exaggerated. Around the absorption edge the first and the second terms of Eq. (9.17) should have large values.

Figure 9.16 shows the dependence of the intensity of superlattice reflection (3/2 3/2 0) on photon energy. Values of f''_{3+} and f''_{4+} were obtained from absorption spectra of $LaSrMn^{3+}O_4$ and $Sr_2Mn^{4+}O_4$, and those of f'_{3+} and f'_{4+}, from the Kramers-Krönig transformation of the absorption spectra. The solid curve in Fig. 9.16 shows the scattering intensity calculated by substitution of these anomalous dispersion terms for Eq. (9.17) and explains the experimental results well. This agreement is regarded as good evidence for the regular alignment of Mn^{3+} and Mn^{4+} (charge order) at low temperatures.

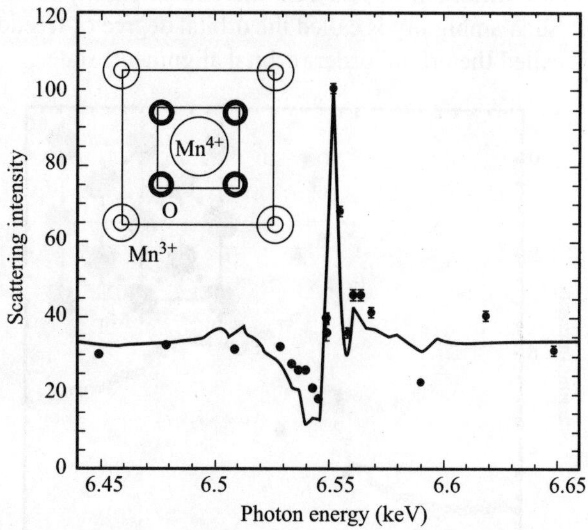

Figure 9.16 The intensity spectrum of the reflection from superlattice (3/2 3/2 0) of the charge order state measured close to the Mn K absorption edge in $La_{0.5}Sr_{1.5}MnO_3$ ($T = 29.6$K). The inset illustrates the charge order model.[31]

However it is not certain that each ion having 1 electron unit charge (valance) of 3 or 4 is located at each Mn site. An alternative possibility would be for ions having a charge difference less than 1 electron unit to be located at Mn sites. Recently, it has been found that such a type of charge order is related to colossal magneto-resistance and for many Mn oxides which exhibit such colossal magneto-resistance. It has been suggested that charge ordering occurs in the state of large electric resistance under no magnetic field, while the charge ordering becomes random in the state of small resistance under magnetic field.

9.4.2 Observation of orbital order

The orbital degree of freedom, that is the orientation of the lobes of the orbitals has been regarded as a hidden order parameter, previously unrecognized in solid state physics because observation of the order was difficult. The importance of this observation was appreciated only recently from studies on transition metal oxides and it has been noted in terms of the alignment of electric quadrupole moment in f electron compounds. An example of observation of the orbital order in $LaMnO_3$ is introduced here since it is regarded as a typical material showing the orbital order.[32]

The Mn^{3+} ion in $LaMnO_3$ is located at the center of an octahedron formed by six O atoms. From the crystal field effect originating from the O atoms, the 5-fold degenerate atomic d orbitals split into the 3-fold degenerate t_{2g} orbitals (d_{xy}, d_{yz}, d_{zx}) and the 2-fold degenerate e_g orbitals ($d_{x^2-y^2}, d_{3z^2-r^2}$). The long lobes of the t_{2g} orbital are situated to minimize interaction with the O atoms resulting in a reduction of the potential energy of the electrons in these orbitals. The lobes of the e_g orbital are directed to the O atoms, so that the potential energy increases. In the Mn^{3+} ion, four 3d electrons are accommodated in the above mentioned d orbitals. Owing to the strong Hund coupling, the most stable state is one in which electrons with parallel spin are accommodated from the lowest state. As a consequence, one electron is accommodated in a 2-fold degenerate e_g orbital whose shapes (or configuration of the lobes) are different. In this case there is ambiguity in terms of which orbital the electron actually occupies. Such ambiguity is called the orbital degree of freedom. The spatially aligned state of orbital direction is called the orbital order (orbital alignment) state.

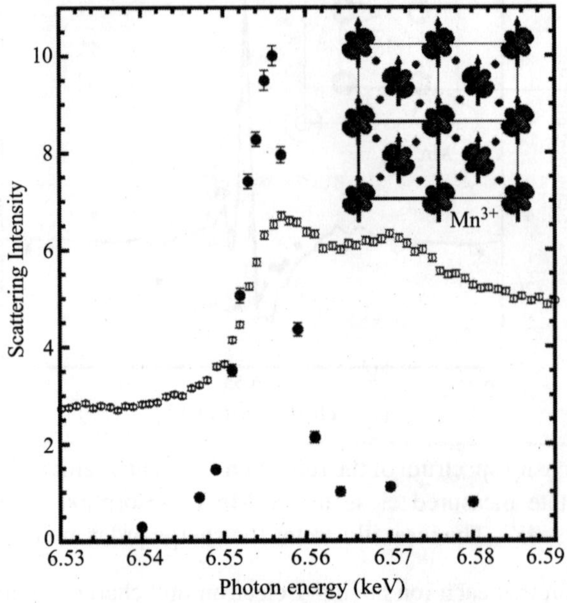

Figure 9.17 The intensity spectrum of ATS (300) reflection in the orbital order state of $LaMnO_3$ measured close to the Mn K absorption edge (closed circles) and also the X-ray fluorescence yield spectrum (open circles) (T = 10.0K). The inset illustrates orbital order model.[32)]

The inset in Fig. 9.17 illustrates the orbital order state observed in $LaMnO_3$. The X-ray structure analysis method should help to identify the electron density distribution. However, it is not easy to observe the orbital order because the number of electrons contributing to it is, at most, 1 per Mn atom. An alternative method to observe the orbital order is one which utilizes ATS (that is, the anisotropy of the tensor of susceptibility) scattering. ATS scattering is a phenomenon in which a reflection is observed even at the forbidden position only in the case when photon energy is tuned to the absorption edge of constituent atoms. This type of scattering is characterized by the following fact. Atoms usually regarded as equivalent atoms in the derivation of the extinction law, become no longer equivalent owing to any difference in symmetry of the chemical bond. In this case the atomic scattering factor should be expressed by a tensor. This is because the deviation of the electron distribution from spherical symmetry generates a large anisotropy in the anomalous dispersion term around the absorption edges. The ATS scattering measurement enables us to observe the deviation of electron distribution from spherical due to orbital order. In Fig. 9.17, closed circles show the measured intensity spectrum of the ATS reflection, that is the (300) reflection, around the Mn K absorption edge. Open circles show the X-ray fluorescence yield spectrum which has a spectral shape similar to the absorption spectrum. As expected, a large enhancement of intensity is observed around the absorption edge. The peak energy of the enhancement is about 3 eV higher than the Mn^{3+} K absorption edge. From this fact, it is inferred that the ATS scattering is due to resonant scattering in the dipole transition.

The atomic scattering factors of Mn corresponding to two kinds of orbitals ($3x^2 - r^2$ and $3y^2 - r^2$) illustrated in the inset of Fig. 9.17 are given by

$$f_{3x^2-r^2} = \begin{pmatrix} f_{\|} & 0 & 0 \\ 0 & f_{\perp} & 0 \\ 0 & 0 & f_{\perp} \end{pmatrix}, \quad f_{3y^2-r^2} = \begin{pmatrix} f_{\perp} & 0 & 0 \\ 0 & f_{\|} & 0 \\ 0 & 0 & f_{\perp} \end{pmatrix}. \tag{9.18}$$

Here $f_{\|}$ represents the atomic scattering factor in the direction parallel to the long lobe of the orbital and $f_{\perp}$, that in the vertical direction. By calculation of the structure factor using these tensors, the reflection intensity of the orbital order state is derived to be

$$I\left(\hat{\theta}, \psi, \phi\right) \propto \left(f_{\|} - f_{\perp}\right)^2 \cos^2\hat{\theta} \sin^2\psi \sin^2\phi, \tag{9.19}$$

where $\hat{\theta}$ is the Bragg angle, ψ is the rotation angle around the scattering vector (called the azimuth angle) and ϕ is the rotation angle of the polarization analyzer crystal around the scattered X-rays. When $\phi = 0°$, only σ-polarized X-rays are observed, while when $\phi = 90°$, π-polarized X-rays are observed. In the inset of Fig. 9.18, these angles are illustrated and also shown are the incident and scattered wave vectors and the electric vectors. The scattering vector lies in the plane of the inset and its direction is upward. When the sample is rotated around the axis of the scattering vector, it is predicted from Eq. (9.19) that the reflected X-ray from the orbital order state will have π-polarization and show 2-fold

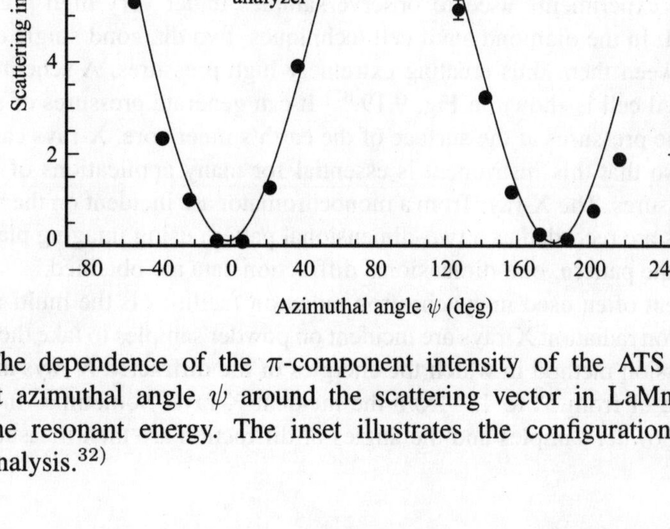

Figure 9.18 The dependence of the π-component intensity of the ATS (300) reflection at azimuthal angle ψ around the scattering vector in $LaMnO_3$ measured at the resonant energy. The inset illustrates the configuration of polarization analysis.[32)]

symmetry in its azimuthal angular dependence. Normally reflected X-rays from spherically symmetric atoms have π-polarization and do not show an intensity change in their azimuthal angle dependence. Figure 9.18 shows the azimuthal angular dependence for the π-component intensity for the (300) reflection at resonant energy. In this case, the rotation angle of the analyzer is fixed at 90°. As expected, the intensity change for 2-fold symmetry is observed. The solid curve in the figure shows the calculation fitted to the experimental result by changing only a factor in Eq. (9.19), which agrees well with the experimental result.

Discussion about the orbital order state becomes possible by comparison with the proposed model for this state and the experimental results concerned with the polarization state of ATS scattering and its azimuthal angular dependence. Such orbital order can give rise to large anisotropy in electron transfer and can dominate the electric and magnetic properties of the material. Therefore, the resonant X-ray scattering method offers us valuable information on the origins of such novel properties. This method has the further advantage that the dependence of the order parameter on differing conditions of temperature and pressure can be determined easily by measuring the resonant scattering intensity under such conditions. The microscopic mechanism for this scattering is not yet fully understood, but remains under discussion.

In this section, the observation of the 'antiferro orbital order' has been introduced, in which two different orbitals align alternatively on the lattice sites. Recently, the analyses of the antiferro orbital order state in some materials have been made more precisely by comparison with results from more than two resonant scattering experiments on the material.[33] The experiments on the 'ferro orbital state' in which atoms with the same orbital alignment on every lattice site in a material have been performed by a new method called 'interference resonant X-ray scattering'.[34,35] These two techniques represent further advances in the study of orbital order state using X-ray diffractometry. For recent studies in this field, see Ref. 36.

9.5 STRUCTURE ANALYSES UNDER HIGH PRESSURE

9.5.1 The generation of high pressure

In X-ray diffraction experiments used to observe samples under very high pressures, two kinds of instruments are used. In the diamond anvil cell techniques, two diamond single crystals compress the sample inserted between them thus creating extremely high pressures. A schematic of an instrument using a diamond anvil cell is shown in Fig. 9.19.[37] It can generate pressures of up to 300 GPa which are comparable to the pressures at the surface of the earth's inner core. X-rays can easily pass through the diamond anvil, so that this instrument is essential for many applications of X-ray diffractometry under very high pressures. The X-rays from a monochromator are incident on the sample in the cell and the diffracted X-rays are recorded as a two-dimensional pattern using imaging plates. After integration of the intensities of the pattern, one-dimensional diffraction data are obtained.

Another instrument often used in synchrotron radiation facilities is the multi-anvil press.[38] In this case white synchrotron radiation X-rays are incident on powder samples to take their diffraction patterns by the energy dispersion method in which the energies of the diffracted X-rays are measured using an SSD, set at an angle of from 5° to 15° from the incident X-rays. Sometimes monochromatic X-rays are incident on the powder samples and the angles of diffraction are then measured by rotation of the

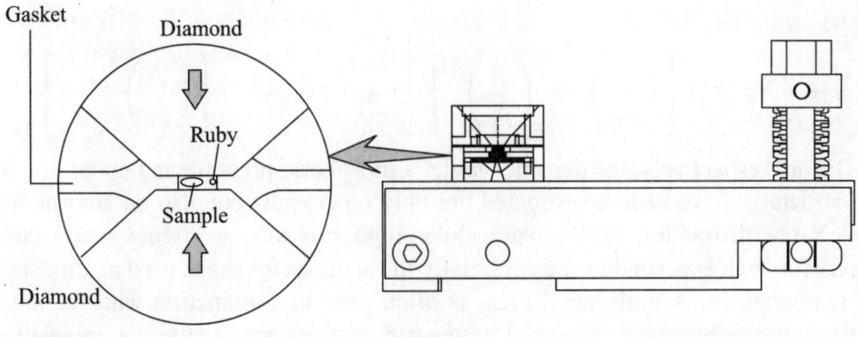

Figure 9.19 High pressure instrumentation incorporating a diamond anvil cell.

detector (the angle dispersion method). In this technique, the pressure medium used is usually a mixture of boron and epoxy or magnesia, which has high transparency to X-rays. Nowadays, *in situ* observations have become possible under extremely high pressures and temperatures up to about 30 GPa and 2000°C, respectively.[39] If a sintered diamond is used as the anvil in this instrument, *in situ* observations are feasible up to pressures exceeding 90 GPa.[40,41]

9.5.2 The pressure scale

The pressure scale is the crucial issue in diffractometry at high pressures. The usual pressure scale is based on the volume contraction of pressure standard materials such as NaCl, Au, Pt *etc.*, and application of the equation of state.[42] These methods have become increasingly important and popular since X-ray diffractometry under high pressure and high temperature has become readily accessible using synchrotron radiation. In the case of an instrument using a diamond anvil cell, the ruby fluorescence method is used. In this method a laser is used to irradiate ruby powder situated near the sample and wavelength shifts of the fluorescence from the R1 and R2 lines are observed. The relation between pressure p and wavelength shift $\Delta\lambda$ of the R1 line (wavelength: λ_0) is given by

$$p\,(\text{GPa}) = \left(\frac{A}{B}\right)\left[\{1 + (\Delta\lambda/\lambda_0)\}^B - 1\right], \tag{9.20}$$

where $A = 1904$ GPa and $B = 5$.[43] Between the pressure scales mentioned above, there exist disagreements which exceed 10% above 100 GPa and those about 2 GPa in the pressure range from 20 to 30 GPa when the temperature rises above 1000°C. Therefore reliable scales must be established as a matter of necessity.

9.5.3 X-ray diffraction experiments under high pressure

a. Determination of the equation of state and the observation of structural phase changes: Powder diffraction experiments under high pressure are of increasing importance. Stable phases of various materials at high pressure and at high temperature have been observed, and phase diagrams determined. In addition, compression curves for solids have been obtained at high pressures and temperatures and the equations of state established. In the determination of the equations of state from the compression curves of solids, the Birch-Murnaghan equation of state is often used, which incorporates the theory of elasticity at finite strain. This equation includes the bulk modulus (K_0) and its pressure derivative (K_0')

and is given by

$$p=\frac{3}{2}K_0\left(T_0\right)\left\{\left(\frac{\hat{\rho}}{\hat{\rho}_0}\right)^{\frac{7}{3}}-\left(\frac{\hat{\rho}}{\hat{\rho}_0}\right)^{\frac{5}{3}}\right\}\left[1+\frac{3}{4}\left(K_0'-4\right)\left\{\left(\frac{\hat{\rho}}{\hat{\rho}_0}\right)^{\frac{2}{3}}-1\right\}\right], \tag{9.21}$$

where $\hat{\rho}_0$ and $\hat{\rho}$ are respectively, the densities under atmospheric pressure and under pressure p.

Recently, structures have been investigated not only on crystals but also on amorphous and molten materials by X-ray diffraction experiments under high pressure and temperature conditions. The diffractometers for such experiments are essentially the same as for those used at atmospheric pressure and at high temperatures. A multi-anvil press is often used in conjunction with the angle dispersion technique using monochromatic X-rays. Using such instruments, scattering intensities have been measured on molten phosphor (P), iodine (I), germanium (Ge), *etc.*, at about 5 GPa and at high temperature. Pair distribution functions, $g(r)$, given in Sect. **9.8.3** for non-periodic systems have then been determined. Clearly the analysis of structures as a function of pressure has now become possible even on molten materials, therefore yielding a comprehensive understanding about their structural phase changes with increase in pressure and temperature.[44)]

X-ray diffraction experiments using synchrotron radiation are vital for the observation of the kinetics of structural phase changes. Synchrotron radiation is sufficiently brilliant to permit time-resolved diffraction patterns to be observed, so that intensity changes in the time domain make it feasible to determine reaction rates quantitatively. Figure 9.20 shows measurements on the speed of the phase change of Mg_2SiO_4 (forsterite) from an olivine structure to a spinel structure, as an example.[45)] In this figure, solid curves represent X-ray diffraction patterns measured with accumulating signals during 10 s for every 45 s at 13.4 GPa and at 980°C. It reveals that the peaks of the initial material forsterite (α phase) become weaker with the passage of time, while the peaks of modified spinel structure (β phase, wadsleyite) become stronger. In an analysis of this experimental data, the nucleation rate of the modified spinel phase and its growth rate associated with the phase change of Mg_2SiO_4, an important mineral

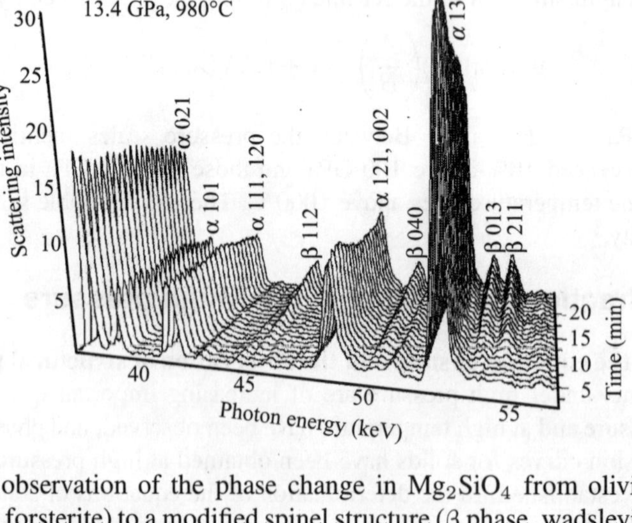

Figure 9.20 *In situ* observation of the phase change in Mg_2SiO_4 from olivine structure (α phase, forsterite) to a modified spinel structure (β phase, wadsleyite) by X-ray time-resolved diffractometry.

in the earth's mantle can be determined. Additionally, from the temperature dependence and excess pressure dependence of the growth rate, new information on the activation energy and activation volume of the phase change can be obtained. This study on the speed of phase changes in silicate minerals is useful not only in materials science, but also in earth science.

From such work, it has become possible to consider the relationships between the phase changes of the minerals inside the earth, and convection within the mantle or in deep earthquakes. The phase change of Mg_2SiO_4 is considered to be a possible cause of the discontinuity in the seismic wave velocity which exists at a depth of 410 km in the earth's mantle. The mantle, which is solid, flows (convects) with a velocity of from a few cm to a few 10 cm per year. This convection current interacts with the phase change in a complicated way. Inside the sinking plate at low temperature, the phase change does not proceed efficiently, so that a low pressure phase exists metastably. The rapid phase change from this metastable phase under excess pressure has been postulated to be a trigger mechanism for deep earthquakes. For recent progress involving high pressure studies in the field of earth sciences using X-ray diffraction, see Refs. 46 and 47.

b. The electron density distribution of β-O_2 and CeP under high pressure: Figures 9.21 and 9.22 show the electron density distribution of solid oxygen (β-O_2)[48] and CeP[49] under high pressure at room temperature as examples of measurements on single crystals. The sample of β-O_2 shown in Fig. 9.21 is of a single crystal grown in a diamond anvil cell at 6.1 GPa. Its structure analysis was undertaken using X-rays of 30 keV and 50 keV. The lattice parameters are $a = b = 2.83$ Å, and

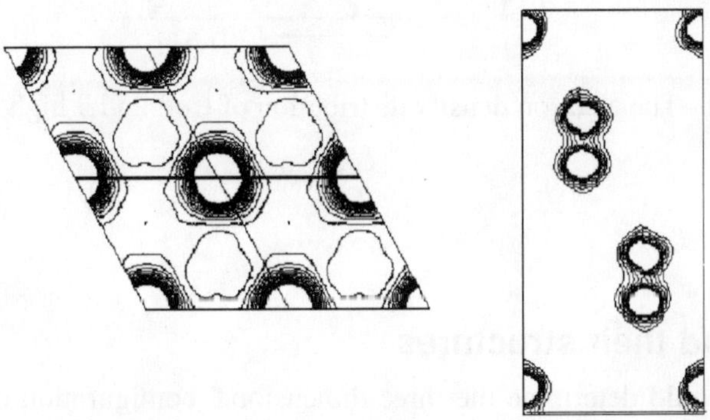

Figure 9.21 Electron density distribution of β-O_2 under high pressure.[48]

$\alpha = \beta = 90°$ and $\gamma = 120°$. The most intense reflection is the (003) reflection, where the count rate was 1.6 Mcps and the time required to gather enough data for the structure analysis was about 8 hours. If one uses high energy X-rays, Bragg reflections will gather in a small region of $2\hat{\theta}$, so that it becomes possible to collect many Bragg reflections even in such high pressure experiments. However, in this particular experiment, the temperature factor is large, so that the intensities at large diffraction angles are considerably reduced. Therefore, only from 60 to 70 reflections are measured. The reliability factors (residues) were $R = 3.4\%$ and 1.8% for X-rays of 30 keV and 50 keV, respectively. The use of 50 keV X-rays has the great advantage that the absorption effect in diffraction measurements becomes 0. The left hand side in the figure represents the electron density distribution projected onto the *ab*-plane and the right side figure gives the cross section including the *c*-axis, in which one can see that molecular axis is parallel to c-axis and that the O_2 molecules form the crystal. The O_2 molecules have magnetic moments, so that the crystal of β-O_2 usually shows paramagnetism. Solid O_2 has various phases and it is suggested that the neighboring O_2 planes align in an antiferromagnetic manner in δ phase. The X-ray diffraction and magnetic scattering measurements using synchrotron radiation are expected to help to interpret and understand such structures. Figure 9.22 shows the electron density distribution for a single crystal containing the heavy element Ce at 1.2 GPa generated in a diamond anvil cell. One can see the distribution clearly. The present maximum pressure for this experiment is restricted, but if it can be increased it may become possible to observe directly the change in the valency of the 4f electrons causing a change of electron distribution.

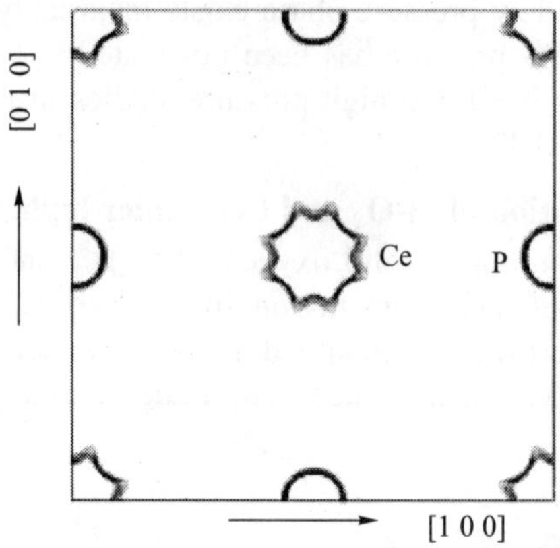

Figure 9.22 The electron density distribution of CeP under high pressure.[49)]

9.6 PROTEINS

9.6.1 Proteins and their structures

X-ray crystallography could determine the three-dimensional configuration of atoms within proteins at atomic resolution (*i.e.* from 1 to 5 Å or better) for our understanding relationships between their structure and function. Synchrotron radiation has already made it possible to determine the structures of biological macromolecules with molecular weight exceeding 10^6, precisely and quickly. Since the 1980s, the utilization of synchrotron radiation for structure biology has become, possibly, its most important single application (leading to Nobel prizes in Chemistry for the analysis of the structure of ATP synthase in 1997, for structural and mechanistic studies of ion channels in 2003, for studies of the molecular basis of eukaryotic transcription in 2006, for the structure and function of the ribosome in 2009, and for studies of G-protein-coupled receptors in 2012). The development of detectors, computers and other hardware, and related software have all drastically accelerated the pace in the determination of protein and other complex molecular structures such as viruses. In the year 1980, the number of registered protein structures determined by X-ray crystallography to Protein Data Bank was 20, while, by 2002, it had grown to 3381 and had reached 94336 by October 1, 2013. For recent advances in protein crystallography, see Refs. 50 and 51.

A protein molecule consists of amino acids from a repertoire of 20 linked by peptide bonds as a polymer chain without branching. The number of amino acids and their three-dimensional combination or assembly is specific for each protein. Its molecular weight ranges from a few thousand to over a few million. There are four levels representing the structure of a given protein. The primary structure is the linear sequence of the component amino acids. The secondary structure identifies special structures created by hydrogen bonds which shape the polypeptide chain, for example, α helix or β sheet. The tertiary structure represents a three-dimensional structure compactly folded, in which α helices, β sheets and other secondary structures are combined resulting from the interaction of amino acid residues and which are sometimes separated relatively far apart in the sequence. The quaternary structure represents the special arrangement of polypeptides in a protein consisting of more than two polypeptides. X-ray crystallography enables us to observe the tertiary and the quaternary structure of proteins down to atomic resolution. The structure determined at a resolution better than about 2 Å would visualize interactions between the atoms in the protein, so that one could deduce how the protein structure relates to its particular function. These studies not only enhance our understanding of life, but also provide invaluable information for medicine and pharmacy, helping the analysis of the origins of diseases and the development of treatments and curatives for them.

9.6.2 Overview of protein crystallography

a. Sample preparation: For the structure determination of the proteins by X-ray crystallography, preparation of high quality single crystals is essential. To prepare such good crystals, one must first extract a large amount of a target protein from some organisms and purify the protein up to high purity level. More than 10 mg of the protein is usually required to achieve its crystallization conditions. Although it was not always easy to obtain the target protein in such quantity, it has now become somewhat easier because of the development in molecular biology techniques which enable larger amounts of the target protein to be obtained from selected bacteria, yeasts or insect cells. However, difficulties remain in the preparation of some kinds of molecules such as super macromolecules composed of many subunits, membrane proteins, or proteins with post-translational modifications. There is no established method for the routine crystallization of proteins in spite of the fact that the crystallization is the fundamental starting point for all protein crystallography.[52] The method is often a trial-and-error procedure and needs many successive attempts and iterations to obtain the best crystallization conditions, even though occasionally the conditions can be found using commercial screening kits. Crystallization is usually the bottleneck in protein crystallography research, and crystallization robots have now been devised to find the crystallization conditions for many types of proteins rapidly and automatically.

Usually the characterization of the crystals and preliminary measurements are performed in advance by the use of laboratory X-ray sources to minimize the use of synchrotron radiation beam time. Initially there is a need to investigate up to what angle the crystal diffracts X-rays, then to estimate the mosaicity. From the preliminary work one can estimate the degree of spatial resolution and the quality of the electron density map. After the provisional measurements one repeats the crystallization in order to prepare better crystals by altering the crystallization conditions such as temperature, nature of additives, precipitants, salt concentration, pH, *etc.* These crystallization and characterization processes must be repeated until one can obtain crystals of appropriate quality. Some protein crystals are not stable at room temperature and must be kept frozen in liquid nitrogen until they are finally brought to synchrotron radiation facilities for diffraction measurement.

b. Experiment: Protein crystals suffer radiation damage during X-ray irradiation, resulting in diminishing the diffraction intensity during data collection. Radiation damage becomes particularly serious when using the most brilliant X-ray sources. Samples are usually frozen at liquid nitrogen temperature to minimize damage, but because the protein crystals contain water in their solvent region, which occupies 30 ~ 70% of the crystal volume, crystalline ice is formed during cooling process. This often destroys the crystals or perturbs their original diffraction patterns. Counter measures may be applied against this problem by replacing the water in the buffer solution with anti-freeze solution (cryo-protectant) such as high concentration glycerol, sucrose, *etc*., and then by cooling the crystals quickly by a flash-cooling method, which does not give sufficient time for water to form crystalline ice. The crystal in cryo-protectant is picked up by the nylon loop (cryo-loop) and put into liquid nitrogen or gaseous nitrogen at 90 ~ 100K immediately. In one particular method the crystal with mother liquor is packed in a capillary and then flash-frozen. This method has been applied to X-ray crystallography on viruses because some virus crystals deteriorate when exposed in air. After the freezing process, the quality of the diffraction data often deteriorates such that the intensities at high angle decrease or else the mosaic spreads expand, due to the deterioration of the crystallinity caused by the physical shock of freezing or from the bad influence of cryo-protectant. A search for effective cryo-protectants and for optimal conditions for the freezing procedure is essential.

For intensity measurements at the diffraction points, imaging plates or CCDs have been used as detectors. The advantages of imaging plates are to have a large detection area permitting the registration of many diffraction spots with high sensitivity and spatial resolution, and also large dynamic range. However, their drawback is a very long reading-time, which precludes real time measurements. Real time measurements have the advantage to identify the status of the experiment quickly, and an additional advantage in obtaining reliable data because the measurement time is always short and therefore many errors caused by changes in experimental conditions can be minimized. It also helps effective use of the precious synchrotron radiation beam time. Therefore, CCDs are extensively used at synchrotron radiation facilities. The usual method used for protein diffraction measurements is the oscillation method, in which crystals are rotated with oscillations of from 0.1° to a few degrees during the measurement of diffraction patterns.

c. Phase determinations: The principal methods used to determine the phase angles of structure factors are those of MIR (multiple isomorphous replacement), MR (molecular replacement), MAD (multi-wavelength anomalous dispersion), and SAD (single-wavelength anomalous dispersion). Direct method has been used for small proteins or proteins including metal atoms, although the cases are very limited. The MIR method needs many crystals other than the native crystal, that is, derivative crystals containing heavy metal atoms. The derivative crystals must be isomorphous with the native crystal. However they often show non-isomorphism caused by replacements by heavy metal atoms. Even when they are isomorphous at room temperature, non-isomorphism may occur when they become frozen. For phase determination, the MIR method has been the method of choice so far, but recently the methods utilizing anomalous dispersion effects have become popular because of the availability of synchrotron radiation by which one can choose appropriate wavelength. A popular choice is the MAD method.[53] It is useful for proteins containing atoms which show appreciable anomalous dispersion (Fe and Zn for instance). Because all data can be collected from a single crystal by choosing the appropriate wavelengths, the isomorphism is always guaranteed. Moreover, it is also useful for isomorphous crystals of proteins containing special non-metal atoms, which can be replaced by homologous atoms. The highly popular method is the selenomethione (Se-Met) method, where the target protein is expressed in a medium containing Se-Met but free from methionine so as to prepare the target protein whose S atoms of methione are replaced by Se atoms of Se-Met. The crystals of the Se-Met protein show larger anomalous dispersion than the original one. In this case, the differences between $|F_{hkl}(+)|^2$ and $|F_{hkl}(-)|^2$ in Bijvoet pairs (reflections of (hkl) and $(-h\ -k\ -l)$, and their equivalent reflections in space group) can be observed, which are useful to determine the phase angles. If one can introduce other heavy atoms showing anomalous dispersion into protein crystals, one can determine the phase angles in a manner equivalent to a combination of MIR and MAD methods. A disadvantage of the MAD method arises from radiation damage to the crystal caused by the repeated diffraction data collection using multiple wavelength X-rays on one crystal. This disadvantage can be limited to some extent by cooling the crystal.

Recently, SAD method has grown in favor. It utilizes the anomalous dispersion of atoms and is applicable even to S atoms in proteins and with a single wavelength of X-ray.[54,55] An S atom has 16 electrons (as reference, an O atom has 8 electrons). The K absorption edge of the S atom is at 5.02 Å, so that the imaginary part of atomic scattering factor (f'') at 1.54 Å (Cu Kα line) is 0.56 electrons. It is not a large value. However, if one can distinguish the differences in structure factors of Bijvoet pairs by the anomalous dispersion of the S atoms, their positions can be determined. This is a useful clue in structure determination. Using synchrotron radiation, diffraction data with a sufficiently large number of high angle diffraction spots can be collected with good S/N ratio, and in addition, the data can be collected in short time. Therefore, the experimental errors could be suppressed considerably. These facts enable us to distinguish the small differences in Bijvoet pairs. The time required for the structure determination is also reduced by utilizing the SAD method. New developments in detectors and in data-processing programs also considerably enhance the quality of all experimental measurements.

d. Analysis: Once the phases are determined, one can calculate and obtain the electron density distribution within the crystal. The structure model of the protein is built in such a way that the amino acid residues are appropriately positioned into the cages of electron density on the computer screen using a three-dimensional graphic program. This positioning is performed manually and is based on experience, because the automatic model-building programs can sometimes make wrong results. The model created is further refined with the help of molecular dynamics through the use of energy minimization methods and by checking its appropriateness from the point of view of stereochemistry.

9.6.3 Research in protein crystallography

In general, growing large crystals of proteins is always not easy. Protein molecules themselves are large, so that the unit cell volumes are much larger than those of small molecules. The diffracted X-ray intensity goes in proportion to the crystal volume, while it is inversely proportional to the square of the unit cell volume, so that the crystallographic resolution diminishes for crystals of large macromolecules. They also have large lattice constants. This makes it difficult to separate the adjacent diffraction spots when using laboratory X-ray generators of rotating Cu anode which have large divergent angles.

In contrast, synchrotron radiation is highly parallel and its emitting point source size is small, so that it remains possible to obtain good diffraction data even for crystals of super macromolecules. Typical challenges are to derive structures for viruses and ribosomes, which are complexes of RNA and protein. The former are well-known pathogens and the latter is a macromolecule which synthesizes proteins in cells. As examples, the structure of a reovirus with molecular weight of 5.2×10^7 (cubic, lattice constant: 1255 Å) was determined at 3.6 Å resolution,[56] while that of 70S ribosome with molecular weight of 2.5×10^6, at 7.8 Å resolution (S: Svedberg, unit of sedimentation velocity).[57] The structure determination of the 70S ribosome was significant, because it represented success in overcoming many difficulties. The ribosome lacks any intramolecular symmetry, unlike viruses, so that its crystallization is difficult even compared to viruses and also the non-crystallographic averaging technique cannot be applied to the phase determination.

The proteasome is a super macromolecule, which contains only protein molecules and does not contain nuclear acids. The X-ray structure analysis of mammalian 20S proteasome has been successfully performed at 2.75 Å resolution.[58] The 20S proteasome is a large protease complex with a molecular weight of 7.5×10^5 composed of 2 pairs of 14 kinds of (total 28) subunits. The 20S proteasome is an active core of 26S proteasome, which selectively breaks down the poly-ubiquitinated substrates into short peptides. The proteasome cleaves the peptide bond of non-terminal amino acids (*i.e.* within the molecule). The 6 of 28 subunits of the 20S proteasome possess the activity to degrade proteins and their two active sites cleave the protein chain to produce peptides with a defined length.

When the higher vertebrates are infected by a pathogen such as a virus, they ingeniously utilize the peptides produced by the proteasome for the immune system. That is, the peptides derived from viruses are delivered to a certain protein, which activates the killer T cells. The activated killer T cell attacks and kills the infected cells before the virus goes out from the cell. In this manner, the higher animals protect themselves against infection by viruses. The structural analysis of the mammalian 20S proteasome has greatly contributed to the understanding of this mechanism. As shown in Fig. 9.23, the shape of the 20S proteasome is cylindrical. It has a chamber, in which the substrate is degraded. Usually the entrance to the chamber, through which the substrates enter from outside, is closed. The active subunits are separated by appropriate distances to generate peptides with the required length. Among 28 subunits, the 14 subunits including the active 6 subunits form the chamber. Except for viruses, the proteasome is one of the largest molecules for which the structure has been determined at a resolution better than 3.0 Å.

Proteins consist of C, N, O, S and H and sometimes contain metal atoms. The H atom has only 1 electron, so that its diffraction intensity is very small in comparison with other atoms. Therefore,

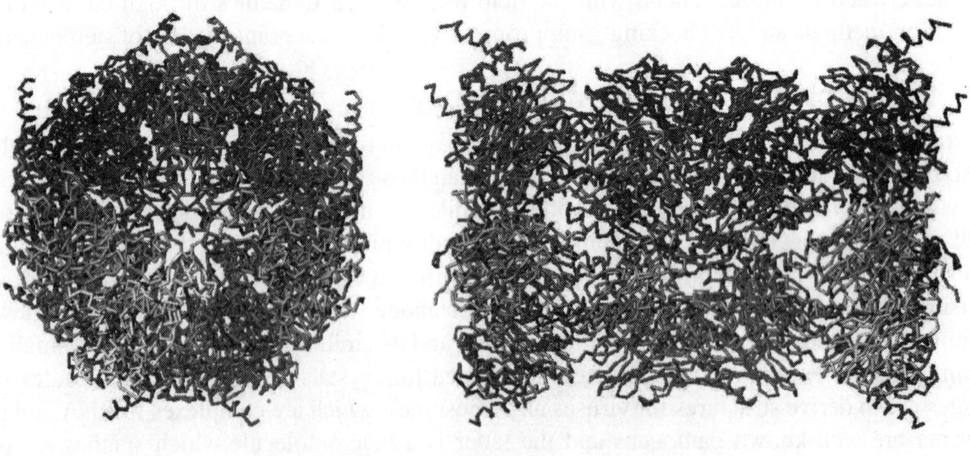

Figure 9.23 The structure of 20S proteasome from bovine liver. Its shape is cylindrical. Right: side view, left: front view.

determination of its position by X-ray diffraction is very difficult. However, by the use of synchrotron radiation it is sometimes possible even to identify the positions of H atoms. For instance, the structure of subtilisin, one of the serine proteases, has been determined at 0.78 Å resolution.[59] Data were collected using synchrotron radiation X-rays of 0.77 Å. The positions of some H atoms have been determined, and even the hydrogen-bond network, which could not be found in the previous structure analysis at 2.0 Å resolution, has been clearly identified. Furthermore, C, N and O atoms can be distinguished from differences in the electron densities obtained at this high resolution and even the sides of histidine in the catalytic triad, could be uniquely determined.

If the positions of H atoms can be determined, then the hydrogen-bond network in heme oxygenase (HO) that degrades heme, the iron-porphyrin complex, could be identified. The Fe atom is recycled for the synthesis of new heme and Fe-S proteins, and CO is known to be a physiological messenger molecule that transmits a neural signal. The biliverdin and its reduced form, bilirubin, are antioxidants, which react with the reactive oxygen species (ROS) accumulated in the human body changing the toxic ROS to become inactive. Therefore, HO not only metabolizes the heme, but also defends the cells. This enzyme binds one molecule of heme *via* several interactions including a coordinate bond between histidine and the heme iron, electrostatic interaction and hydrogen bond between amino acid residues and the side chains of heme. HO utilizes heme both as the substrate and the active center. The heme degradation starts when O_2 is bound to the heme iron and an electron (e^-) is supplied. Subsequently, H^+ is bound to the O_2 to form the ferric hydroperoxide (Fe^{3+}–OOH) intermediate, then a further supply of H^+, O_2 and e^- eventually opens the porphyrin ring, and Fe is released. (Without provision of H^+, O_2 and e^-, the heme degradation will not proceed.)

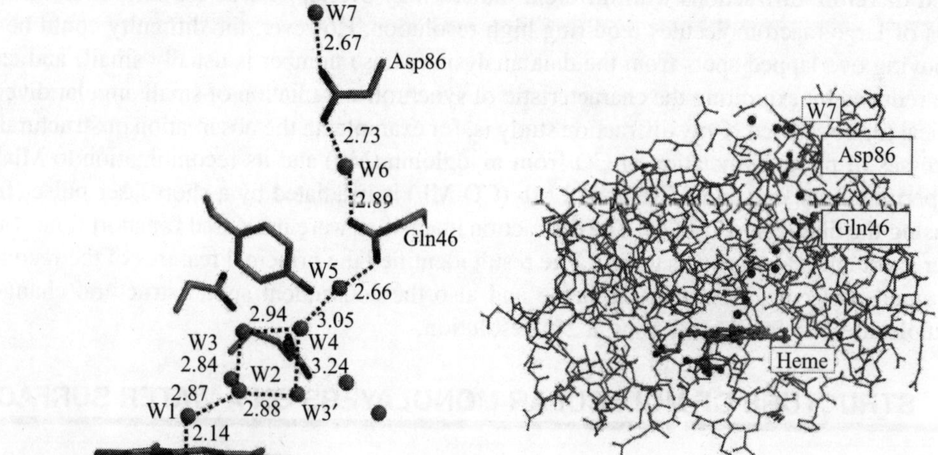

Figure 9.24 The hydrogen bond network in heme oxygenase from *Corynebacterium diphtheriae*. (a) Around the active center. Orange-colored balls and other balls represent Fe and water molecules respectively. Numerals represent distances in Å. (b) The structure of the molecule. (© The American Society for Biochemistry and Molecular Biology).[60]

Figure 9.24 shows the crystal structure of HO at 1.4 Å resolution.[60] In the figure, W stands for a water molecule. W7 is a water molecule located at the surface of the enzyme, that is, located in the solvent region. A pathway linked from W7 to heme iron is found, which consists of water molecules and charged amino acid residues connecting as W7 → aspartic acid (Asp) 86 → W6 → gultamine (Gln) 46 → W5 → W4 → W3(W3′) → W2 → W1 → Fe. These water molecules and amino acid residues are suitably located to form a hydrogen-bond network. This hydrogen-bond network could be regarded as a conduit for a pathway of H^+ delivery from solvent to the active site. However, whether the hydrogen bonds really exist or not is yet to be established due to the insufficient resolution for visualization of H atoms of the water molecules and the amino acid residues in this structure analysis. Once H atoms are visualized, then the relation between the donor and the acceptor of H^+ in hydrogen bonds will be clarified. This will determine the H^+ pathway in the enzyme, thereby further enhancing an understanding of the heme degradation mechanism by HO.

9.6.4 Time-resolved structure analysis

Using monochromatic X-rays, diffraction measurements must be repeated many times to collect data to cover all the diffraction data required for structural determination. On the other hand, using the pulsed characteristics of synchrotron radiation, structural changes within proteins could sometimes be observed during the reaction processes as a sequence of 'snapshots'. When a crystal is illuminated with white synchrotron radiation, from one to a few shots are sufficient to collect one set of Laue diffraction data for structure analysis. This method cannot easily be adapted to the laboratory. The problems with the normal Laue method are that higher order diffractions from a set of planes are included in the same Laue spot and therefore diffractions with different indices may overlap. These are serious in the structural analysis of large macromolecules requiring high resolution. However, the difficulty could be resolved by removing overlapped spots from the data analysis (whose number is usually small) and can in any case be reduced by exploiting the characteristic of synchrotron radiation of small angular divergence.

Typical time-resolved X-ray diffraction study is, for example, in the observation of structural changes in processes of photodissociation of CO from myoglobin (Mb) and its recombination to Mb.[61–63] In this experiment, the crystal of CO-bound Mb (CO-Mb) is irradiated by a short laser pulse (from fs to ns) causing the dissociation of CO. The diffraction intensities were measured for short times (about 100 ps) over short intervals (from ns to ms). The result identified the principal features of the recombination of the small molecule, CO, with the heme and also the consequent subtle structural changes in the macromolecule, Mb for about 100 ps at 2 Å resolution.

9.7 STRUCTURE OF MOLECULAR MONOLAYERS ON A WATER SURFACE

X-ray diffractometry can be used to explore the structure of matter down to only a few nm in thickness including surface layers, such as amphiphilic molecules and soaps which have both a partial affinity to water (the hydrophilic group) and a partial affinity to oil (the hydrophobic group).

When a solution of amphiphilic molecules in an organic solvent is spread on a water surface, the hydrophilic group tends to orient towards water and the hydrophobic group aligns towards the air, forming a surface monolayer at the air-water interface. Lipids which make up biological membranes are also amphiphilic molecules. It is well known that the basic structure of the biological membrane is a molecular layer of lipid (a lipid bilayer).

A molecular monolayer on water surface and a monolayer transferred from the water surface on to a solid substrate are called a Langmuir film and a LB (Langmuir-Blodgett) film, respectively. They have been studied intensively in order to elucidate the detailed functioning of biological membranes and of phenomena which occur in such two-dimensional structures, and in their three-dimensional configurations, with the primary objective being to develop them as functional materials. However, even the most basic information such as the molecular structure in the film is not yet understood. We expect X-ray surface reflection methods to become frequently used in future for the determination of structures on surface monolayers on water.[64)]

An example is the investigation of the structure of the phospholipid (L-α-dimyristoyl-phosphatidic acid, DMPA) monolayer which is a model biological membrane at an air-water interface.[65)] The water phase contains the double valence metal ions, Ca^{2+} and Mg^{2+} ions, which have differing effects on the aggregation or fusion of the phospholipid bilayer dispersions. Figure 9.25 shows a schematic of a grazing incidence X-ray diffraction experiment for the structure analysis of the monolayer on a water surface. X-rays of 0.1488 nm were incident on the monolayer on a water surface at a grazing angle 0.10° a little smaller than a total reflection grazing angle of 0.13°, where its diffracted rays were measured using a one-dimensional detector. The monolayers were prepared on a surface of 0.1 mM (M: mol/l) aqueous $CaCl_2$ or $MgCl_2$ using a Langmuir trough. Molecular packing of the monolayer was regulated by a barrier bar and the diffraction experiments were performed at the Photon Factory.

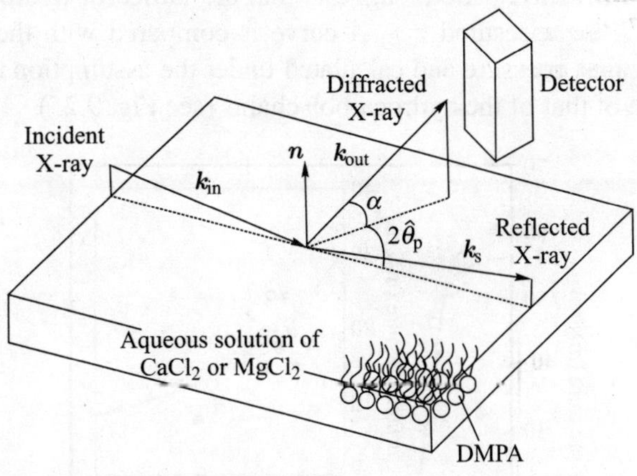

Figure 9.25 Grazing incidence X-ray diffraction measurement of monolayer on water surface.

Figure 9.26 shows diffraction intensity curves from DMPA films on water during compression of the film. The barrier bar was moved horizontally to compress the monolayer. Here the surface pressure is defined to be the reduction in the surface tension from that of pure water. It increases under compression and reflects the degree of compression (packing concentration of molecules). The surface pressure (π) was measured for each area (A). In the presence of Mg^{2+} ions, no diffraction peak was found at $\pi = 10$ mN/m, while in the presence of Ca^{2+} ions, a broad diffraction peak was found even at $\pi = 7.5$ mN/m. In both cases, sharp diffraction peaks were found above $\pi = 15$ mN/m. The full width at half maximum of the peaks and the lattice parameters given by diffraction experiments decreased with increasing surface pressure. This lattice parameter (0.420 $\sim$ 0.425 nm) corresponds to that of a two-dimensional lattice of DMPA molecules. The lattice parameter is smaller in the presence of Ca^{2+}

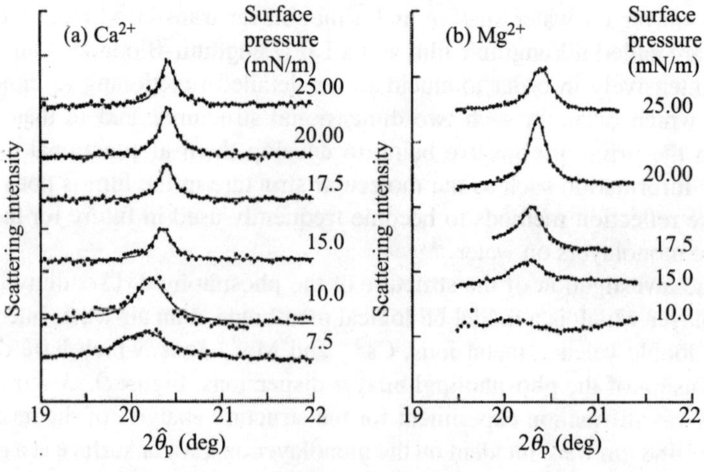

Figure 9.26 X-ray diffraction curve for DMPA monolayer[65].

ions than in the presence of Mg^{2+} ions. In addition, it has been found from an analysis of Bragg rods that the hydrocarbon chains are not inclined, but are oriented vertical to the surface.

The symmetry of the diffraction pattern suggests that the lattice of hydrocarbon chains of DMPA is hexagonal. In Fig. 9.27, the measured $\pi - A$ curve is compared with the area obtained from the diffraction data plotted against pressure and calculated under the assumption that the area occupied by DMPA molecules is twice of that of the hydrocarbon chains (see Fig. 9.25).

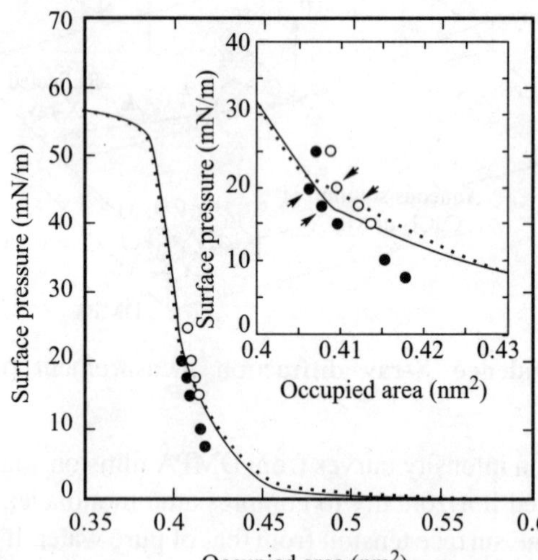

Figure 9.27 The relation between the surface pressure and molecular area in a DMPA monolayer.[65] Solid and dotted curves represent the measured areas *versus* pressure in the presence of Ca^{2+} and Mg^{2+} ions, respectively, at 20.0 ± 0.1°C. • and ○ represent calculated areas from X-ray diffraction data for Ca^{2+} and Mg^{2+} ions, respectively.

The important conclusions derived from Figs. 9.26 and 9.27 are as follow. First, the $\pi - A$ curves observed in the presence of both ions are almost the same, although a small expansion of the film has been observed with Mg^{2+} ions. The occupied area calculated from the diffraction data is apparently larger in the presence of Mg^{2+} ions than that of Ca^{2+} ions. In the former case, a large affinity for water is anticipated. Second, it is noted that in the presence of Ca^{2+} ions, diffraction peaks are found even at low surface pressures, suggesting that the solid phase grows even at low surface pressures and that the hydrocarbon chains are regularly packed at lower concentrations of Ca^{2+} ions. On the other hand, the solid phase does not grow at low surface pressures in the presence of Mg^{2+} ions. Finally, it shows that above $\pi = 20$ mN/m, the occupied area given in the $\pi - A$ curve decreases, while that calculated from the diffraction data remains almost constant. This suggests that the former represents the area averaged over the entire film including the destroyed parts.

Clearly, the X-ray surface reflection methods enable us to distinguish fine structural differences in monolayers in the presence of different divalent cations, such as Ca^{2+} and Mg^{2+} ions in this particular case. This analysis could readily be extended to structural research for any type of biological membrane.

9.8 NON-CRYSTALLINE SYSTEMS

9.8.1 X-ray diffraction by non-periodic systems

In all crystals, atoms are arranged in a regular ordered pattern in three dimensions and this periodic arrangement will generate X-ray diffraction patterns exhibiting discrete sharp peaks. However these diffraction patterns cannot be observed in non-crystalline systems, *e.g.*, liquids and glasses, because of their lack of long-range order. Nevertheless, an averaged ordering in near neighbor regions is well-recognized in non-crystalline systems. This contrasts with the gaseous state where the atoms distribute randomly and their motion is also random. From typical atomic distributions X-ray diffraction patterns of non-crystalline systems are usually concentric, where the radial intensity distribution shows a relatively sharp first peak and subsequent small oscillations toward higher angles. It should be noted that the measured diffraction pattern in reciprocal space can provide an atomic distribution in real space *via* use of a Fourier transform.

The amplitude of the scattered X-rays may be expressed in the static approximation as $\sum_j f_j \exp(i\boldsymbol{q} \cdot \boldsymbol{r}_j)$, similar to F for the crystalline system. The total scattering intensity of I_{T} is then given as follows.

$$I_{\mathrm{T}} = \sum_j \sum_k f_j^* f_k \exp\{-i\boldsymbol{q} \cdot (\boldsymbol{r}_j - \boldsymbol{r}_k)\} = \sum_{j,k} f_j^* f_k \exp(-iqr_{jk} \cos\alpha). \tag{9.22}$$

Here $r_{jk} = |\boldsymbol{r}_j - \boldsymbol{r}_k|$, α is the angle formed by $\boldsymbol{r}_j - \boldsymbol{r}_k$ and $\boldsymbol{q}$. f_k is the atomic scattering factor for atom k located at position of $\boldsymbol{r}_k$. Since the atomic distribution in any non-crystalline system can be considered to be homogeneous, I_{T} can be obtained by integration with respect to α. The total scattering intensity can subsequently be written in the following form,

$$I_{\mathrm{T}} = \sum_{j,k} f_j^* f_k \frac{\sin qr_{jk}}{qr_{jk}}. \tag{9.23}$$

For a one-component system including the $\hat{N}$ atoms, the total scattering intensity can be given, using the relation $f_j = f_k = f$, by

$$I_{\mathrm{T}} = \hat{N} f^2 \left(1 + \sum_j{}' \frac{\sin qr_{jk}}{qr_{jk}} \right). \tag{9.24}$$

Here, the summation of Σ' is made by excluding the $j = k$ case.

The scattering intensity per atom is defined by the relation $I(q) \equiv I_{\mathrm{T}}/\hat{N}$ and Σ' is replaced by integral. Thus one obtains the following equation

$$I(q) = f^2 \left\{ 1 + \int_0^{\infty} 4\pi r^2 \hat{\rho}(r) \frac{\sin qr}{qr} dr \right\}, \tag{9.25}$$

where $\hat{\rho}(r)$ is the radial number density function. $I(q)$ can be rewritten as follows.

$$I(q) = f^2 \left[1 + \int_0^{\infty} 4\pi r^2 \left\{ \hat{\rho}(r) - \hat{\rho}_0 \right\} \frac{\sin qr}{qr} dr \right], \tag{9.26}$$

where $\hat{\rho}_0$ is the average number density of atoms. In this formula, an additional term is neglected, which includes the derivative of delta function ($\delta'(q)$) and is only distinct at $q = 0$. $I(q)$ can be expressed using Fourier transformation by

$$I(q) = f^2 \left[1 + \int \left\{ \hat{\rho}(r) - \hat{\rho}_0 \right\} e^{i\mathbf{q}\cdot\mathbf{r}} d\mathbf{r} \right]. \tag{9.27}$$

The radial density function of $\hat{\rho}(r)$ can be obtained by using the inverse Fourier transformation, as follows.

$$\hat{\rho}(r) = \hat{\rho}_0 + \frac{1}{2\pi^2 r} \int \frac{I(q) - f^2}{f^2} q \sin qr \, dq, \tag{9.28}$$

where $I(q)$ corresponds to the absolute intensity and the measured intensity will be given in electron units.

The diffraction intensity for non-crystalline systems is usually measured under the condition where the anomalous dispersion effect is excluded by selecting only those X-rays whose energy lies away from absorption edge of the constituent elements. The atomic scattering factor f is then attributed to the atomic number Z in forward scattering, so that the intensities of raw data can be expressed in electron units by normalizing them to the intensity at $q = 0$ which is attributed to Z^2. The Fourier integral in Eq. (9.28) is taken from zero to infinity $q = \infty$. However, the available experimental data are restricted to a finite upper q value. The larger the q value is, the more precise will be the result of the Fourier transform.

Let us extend this idea to non-crystalline systems of n-components. The analysis and its interpretation are more complicated in a system containing more than two kinds of components. For the A-B binary case, the average structural information $\hat{\rho}(r)$ obtained from measured intensity data is usually the convolution of three pairs of radial density functions, $\hat{\rho}_{AA}(r)$, $\hat{\rho}_{AB}(r)$ and $\hat{\rho}_{BB}(r)$. For the n-component system, $\hat{\rho}(r)$ can be given by

$$\hat{\rho}(r) = \sum_{j=1}^{n} \sum_{k=1}^{n} c_j f_j f_k \hat{\rho}_{jk}(r) / \langle f \rangle^2, \tag{9.29}$$

where c_j is the atomic fraction of j-th kind atom and $\hat{\rho}_{jk}(r)$ is called the partial pair density function, corresponding to the number of the k-th kind atoms found at a radial distance of r from a j-th kind atom at the origin. Let us start the calculation on the multi-component system from Eq. (9.23). The scattering intensity per atom, $I(q)$, for a non-crystalline system containing more than two kinds of atoms is given as follows.

$$I(q) = \langle f^2 \rangle + \langle f \rangle^2 \int 4\pi r^2 \{\hat{\rho}(r) - \hat{\rho}_0\} \frac{\sin qr}{qr} dr, \tag{9.30}$$

where $\langle f \rangle = \sum_{j=1}^{n} c_j f_j$ and $\langle f^2 \rangle = \sum_{j=1}^{n} c_j f_j^2$. The so-called environmental pair density function around j-th kind atom is defined by $\hat{\rho}_j(r) = \sum_k f_k \hat{\rho}_{jk}(r)$.

For a description of atomic distributions which lack long-range atomic periodicity, such as in liquids and glasses, the so-called 'radial distribution function' (RDF) expressed by $4\pi r^2 \hat{\rho}(r)$ is widely employed. The RDF may be defined by considering a spherical shell of radius r around an origin atom and can be considered to be the average probability of finding another atom as a function of distance alone. The RDF can be estimated from the scattering intensity per atom, $I(q)$, by using the following Fourier transform

$$4\pi r^2 \hat{\rho}(r) = 4\pi r^2 \hat{\rho}_0 + \frac{2r}{\pi} \int \frac{I(q) - \langle f^2 \rangle}{\langle f \rangle^2} q \sin qr dq. \tag{9.31}$$

The distance and coordination number in the nearest neighbor region can be estimated from the position and area of the first peak of the RDF for a one-component system. On the other hand, the surroundings of specific atoms in multi-component systems are generally different from those of other kinds of atoms. Therefore, the average atomic distribution of $4\pi r^2 \hat{\rho}(r)$ given by Eq. (9.31) is unable to describe completely the positions of neighbors with identified kinds of atoms. To overcome this difficulty, a knowledge of the structure of individual pairs, $\hat{\rho}_{ij}(r)$, is required. There are $n(n + 1)/2$ possible pairs in a system containing n-components. Three partial RDFs in a binary system, six, in a ternary system and ten, in a quaternary system are required to describe their atomic scale structure completely. To obtain each partial RDF, the anomalous dispersion effect of the specific atom is useful. When the incident X-ray energy is chosen to be on the lower energy side of the absorption edge for the j-th kind atom, the environmental structure around j-th kind atom can be established. This is readily seen from Eqs. (9.28) and (9.29). Specifically, one can derive the partial number density function of $\hat{\rho}_j(r)$. $4\pi r^2 \hat{\rho}_j(r)$ is usually called the 'environmental radial distribution function' around j-th kind of atom. When the anomalous X-ray scattering method is applied at energies close to different absorption edges, we obtain more than two $\hat{\rho}_{jk}(r)$ for one sample.

In contrast to this experimental method, the atomic correlation, distances and coordination numbers for non-crystalline systems in the near neighbor region are frequently estimated by calculating the scattering intensity from a chosen model structure.[66] This is based on the fact that averaged ordering often exists in near neighbor regions and that local basic units can be seen in non-crystalline systems. The distances and coordination numbers for near neighbors can also be obtained using a least-squares analysis, which will reproduce the experimental intensity from a calculated one based on Eq. (9.29). In this approach, a certain number of k-type atoms are assumed to be located at an average distance around a j-type atom at an origin and with a distribution which is approximately Gaussian-like. The local basic units are not strongly correlated and the positional atomic correlation disappears with increasing distance.[67] Therefore, the distribution of higher neighbor atoms is expressed by a continuous distribution with an average number density. Such conditions are well accepted for non-crystalline systems consisting of metal and oxygen or semimetal and oxygen, as exemplified by silicate glasses. It is noteworthy that improved reliability could be obtained by fitting not only of the ordinary intensity data but also of the environmental intensity data around a specific element determined by the anomalous X-ray scattering method. This is particularly advantageous when the diffraction experiments are carried out using synchrotron radiation.[68] High energy synchrotron radiation is also ideal for the structure analysis of metallic glasses containing heavy elements.[69] Structure analyses on some aqueous solution and molten material are given here as an illustration.

9.8.2 Zinc chloride aqueous solutions

Consider the environmental radial distribution function of $4\pi r^2 \hat{\rho}_j(r)$ around the j-th kind atom obtained from anomalous X-ray scattering measurements using synchrotron radiation. Figure 9.28 shows results for the hydration structure around a Zn ion in $ZnCl_2$ in aqueous solution.[70] The broken lines in this figure correspond to the ordinary RDFs, which mainly reflect the contribution from water molecules. For example, the peak at about 0.3 nm in the ordinary RDF is ascribed to O–O pairs of water molecules. It is then very difficult to extract information about hydration around a Zn ion from the normal RDF alone. On the other hand, the oxygen coordination numbers for a Zn ion, corresponding to water molecules around $ZnCl_2$ in aqueous solution, are readily realized in the environmental RDF_{Zn}, because the contribution from the solvent water molecules is completely eliminated in the anomalous X-ray scattering measurement at the absorption edge of Zn.

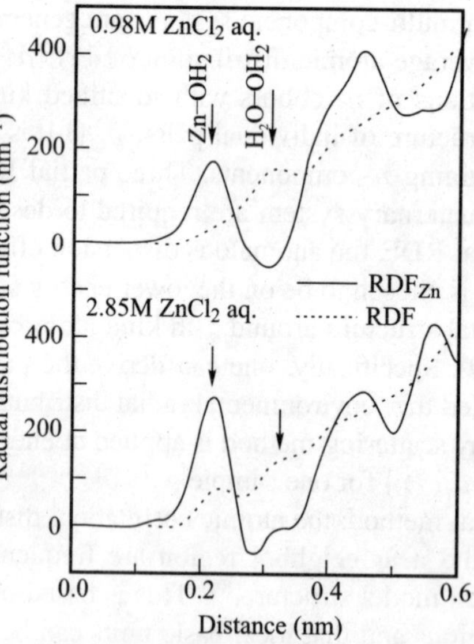

Figure 9.28 The environmental RDF_{Zn} around a Zn ion (solid line) obtained from anomalous X-ray scattering measurements at the absorption edge of Zn for $ZnCl_2$ aqueous solution and its ordinary RDF (dotted line).

When applying a least-squares analysis to these results, the hydration numbers around a Zn ion are found to be about 6 at 0.21 nm. The relatively strong interaction between hydrated water molecules and a zinc ion in aqueous solution is considered to be realistic. Zn ions and hydrated water molecules will then move together when a voltage is applied to the electrodes. Therefore, knowledge about the hydration structure around any specific element is essential for theoretical calculation of the mobility of the element in the aqueous solution.

9.8.3 Correlations in molten noble metal halides

One additional example of the determination of three partial pair density functions, $\hat{\rho}_{jk}(r)$, is given below, using the results from CuBr.[71] Some noble metal halides show good ionic mobility at room temperature, which is almost identical to the electrolyte case. This particular ionic conducting property may be attributed to the distribution of the noble metal cation, but information which includes partial functions has not yet been fully obtained.

Figure 9.29 shows the intensity profiles of molten CuBr at 810K in a quartz cell in the transmission mode at incident energies of 8.680, 8.955, 13.170 and 13.445 keV, corresponding to 300 eV and 25 eV below the Cu K (8.980 keV) and Br K (13.470 keV) absorption edges, respectively. Figure 9.29 also includes the ordinary intensity profile obtained at single energy of 17.0 keV, away from the absorption edges of both Cu and Br. The anomalous dispersion terms used for Cu and Br are listed in Table 9.1.

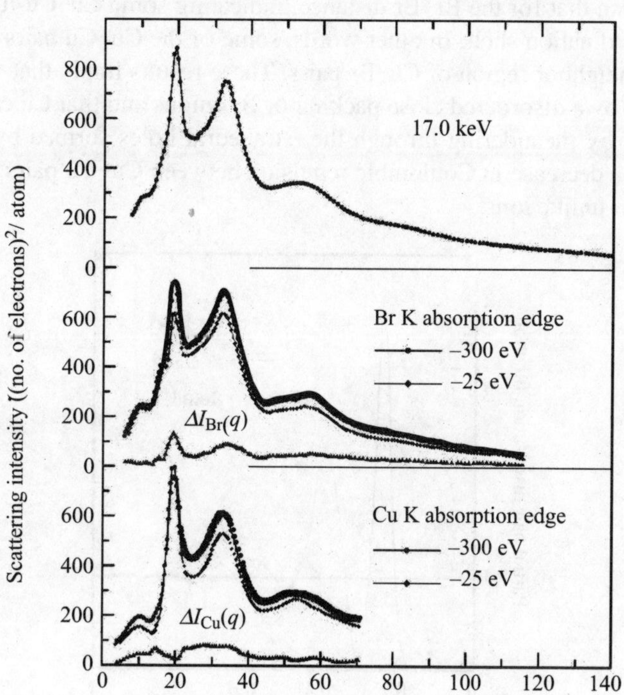

Figure 9.29 Intensity profiles of scattered X-rays for molten CuBr as a function of wave-vector measured at the low energy sides of K edges of Cu and Br.[71]

Table 9.1 Values of the X-ray anomalous dispersion terms used in the study of molten CuBr

Energy	$f'_{Cu}(E)$	$f''_{Cu}(E)$	$f'_{Br}(E)$	$f''_{Br}(E)$
8.680 keV (Cu K – 300 eV)	–3.062	0.514	–0.900	1.121
8.955 keV (Cu K – 25 eV)	–5.578	0.485	–0.961	1.061
13.170 keV (Br K – 300 eV)	0.002	2.064	–3.366	0.527
13.445 keV (Br K – 25 eV)	0.040	1.993	–5.815	0.507
17.0 keV	0.255	1.329	–0.481	2.572

Three partial pair density functions, $\hat{\rho}_{CuCu}(r)$, $\hat{\rho}_{CuBr}(r)$, and $\hat{\rho}_{BrBr}(r)$ can be estimated from the experimental results given in Fig. 9.29 using the anomalous dispersion terms given in Table 9.1. Figure 9.30 shows the so-called partial pair distribution functions, $g(r)$, which are the partial pair density functions, $\hat{\rho}_{jk}(r)$ normalized to the average value of $c_k\hat{\rho}_0$. The crystalline phase of CuBr is zinc blend type structure and the distances are 0.246 nm for Cu–Br pairs and 0.402 nm both for Br–Br and Cu–Cu pairs. On the other hand in the molten phase the distances for Br–Br pairs and Cu–Br pairs are similar to those in the crystalline phase. As shown in the results in Fig. 9.30, the closest Cu–Cu distance is significantly smaller than that for the Br–Br distance, indicating some Cu–Cu-like ion penetration into the first unlike-ion coordination shell. In other words, some of the Cu–Cu pairs in the molten state are located in the nearest neighbor region of Cu–Br pairs. These results imply that the structure of molten CuBr can be described by a disordered close packing of Br anions and that Cu cations adopt a strongly disordered distribution by meandering through the tetrahedral holes formed by Br ions. This can be interpreted in terms of a decrease in Coulombic repulsion between Cu–Cu pairs arising from a reduced charge transfer between unlike ions.

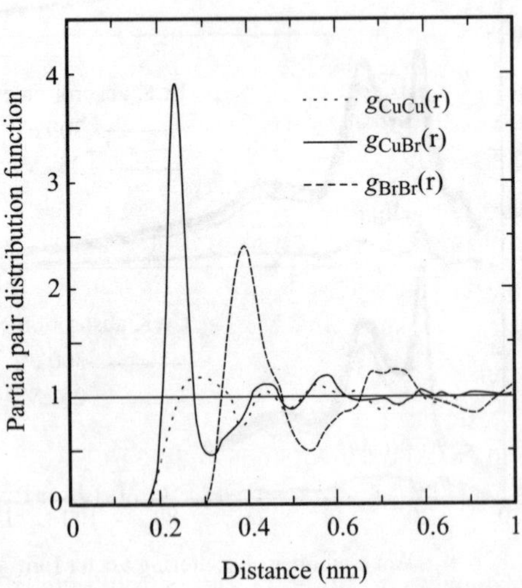

Figure 9.30 The partial pair distribution functions for molten CuBr.[71)]

REFERENCES

1. B. D. Cullity: "Elements of X-Ray Diffraction", 2nd ed., (Addison-Wesley Publishing Company, 1977).
2. J. Als-Nielsen and D. McMorrow: "Elements of Modern X-Ray Physics" (John Wiley & Sons, 2011).
3. T. A. Ezquerra, M. C. Garcia-Gutierrez, A. Nogales and M. Gomez eds.: "Applications of Synchrotron Light to Scattering and Diffraction in Materials and Life Sciences" (Springer-Verlag, 2009).
4. F. Seitz: "The Modern Theory of Solids" (McGraw-Hill, 1940).
5. C. Kittel: "Introduction to Solid State Physics" 8th ed. (John Wiley & Sons, 2005).
6. M. Sakata and M. Sato: Acta Crystallogr. A **46** (1990) 263.
7. A. J. C. Wilson ed.: "International Tables for Crystallography" Vol. C (Kluwer Academic Publisher, 1992) p. 518.
8. Laboratoire de Mineralogie Cristallographie de Paris, http://www1.iucr.org/sincris-top/logiciel/result.html.
9. Collaborative Computational Project No. 14; CCP14, http://www.ccp14.ac.uk/index.html.
10. M. Takata, E. Nishibori, M. Shinmura, H. Tanaka, K. Taniguchi, M. Kosaka and M. Sakata: Mater. Sci. Engn. A **312** (2001) 66.
11. Y. Noda, K. Ohshima, H. Toraya, K. Tanaka, H. Terauchi, H. Maeta and H. Konishi: J. Synch. Rad. **5** (1998) 485.
12. Y. Noda, K. Akiyama, T. Shobu, Y. Kuroiwa, H. Nakao, Y. Morii and H. Yamaguchi: Ferroelectrics **217** (1998) 1.
13. Y. Noda, K. Akiyama, T. Shobu, Y. Kuroiwa and H. Yamaguchi: Jpn. J. Appl. Phys. S **38-1** (1999) 73.
14. Y. Noda, R. Kiyanagi, H. Kimura, M. Watanabe, A. Kojima, T. Hayashide, T. Mochida and T. Sugawara: Ferroelectrics **269** (2002) 327.
15. R. Kiyanagi, A. Kojima, T. Hayashide, H. Kimura, M. Watanabe,Y. Noda, T. Mochida, T. Sugawara and S. Kumazawa: J. Phys. Soc. Jpn. **72** (2003) 2816.
16. H. Sagayama, N. Abe, K. Taniguchi, T. Arima, Y. Yamasaki, D. Okuyama, Y. Tokura, S. Sakai, T. Morita, T. Komesu, H. Ohsumi and M. Takata: J. Phys. Soc. Jpn. **79** (2010) 043711.
17. Y. Tanaka, S. P. Collins, S. W. Lovesey, M. Matsumami, T. Moriwaki and S. Shin: J. Phys.: Condens. Matter **22** (2010) 122201.
18. Y. Noda, T. Shobu and N. Ikeda: J. Crstallographic Soc. Jpn. **2** (2000) 12, in Japanese.
19. H. Kimura, H. Goka, M. Fujita, Y. Noda, K. Yamada and N. Ikeda: Phys. Rev. B **67** (2003) 140503.
20. S. Komiyama, M. Watanabe, Y. Noda, E. Negishi and N. Toyota: J. Phys. Soc. Jpn. **73** (2004) 2385.

21. M. Watanabe, S. Komiyama, R. Kiyanagi, Y. Noda, E. Negishi and N. Toyota: J. Phys. Soc. Jpn. **72** (2003) 452.
22. I. Kagomiya, S. Matsumoto, K. Kohn, Y. Fukuda, T. Shobu, H. Kimura, Y. Noda and N. Ikeda: Ferroelectrics **286** (2003) 889.
23. F. Sawano, I. Terasaki, H. Mori, T. Mori, M. Watanabe, N. Ikeda, Y. Nogami and Y. Noda: Nature **437** (2005) 522.
24. S. Iijima: Nature **354** (1991) 56.
25. T. Takenobu, T. Takano, M. Shiraishi, Y. Murakami, M. Ata, H. Kataura, Y. Achiba and Y. Iwasa: Nature Materials **2** (2003) 683.
26. Y. Noda, I. Tamura, Y. Shiuchi, T. Shobu and R. Matsuo: Photon Factory Activity Report **14** (1996) 67.
27. Y. Noda and H. Kimura: J. Korean Phys. Soc. **35** (1999) S1137.
28. Y. Kuroiwa, S. Aoyagi, A. Sawada, J. Harada, E. Nishibori, M. Takata and M. Sakata: Phys. Rev. Lett. **87** (2001) 217601.
29. R. Kumai, S. Horiuchi, H. Sagayama, T. Arima, M. Watanabe, Y. Noda and T. Tokura: J. Am. Chem. Soc. **129** (2007) 12920.
30. K. Kobayashi, S. Horiuchi, R. Kumai, F. Kagawa, Y. Murakami and Y. Tokura: Phys. Rev. Lett. **108** (2012) 237601.
31. Y. Murakami, H. Kawada, H. Kawata, M. Tanaka, T. Arima, Y. Moritomo and Y. Tokura: Phys. Rev. Lett. **80** (1998) 1932.
32. Y. Murakami, J. P. Hill, D. Gibbs, M. Blume, I. Koyama, M. Tanaka, H. Kawata, T. Arima, Y. Tokura, K. Hirota and Y. Endoh: Phys. Rev. Lett. **81** (1998) 582.
33. H. Nakao, Y. Wakabayashi, T. Kiyama, Y. Murakami, M. v. Zimmermann, J. P. Hill, Doon Gibbs, S. Ishida, Y. Taguchi and Y. Tokura: Phys. Rev. B **66** (2002) 184419.
34. T. Kiyama, Y. Wakabayashi, H. Nakao, H. Ohsumi, Y. Murakami, M. Izumi, M. Kawasaki and Y. Tokura: J. Phys. Soc. Jpn. **72** (2003) 785.
35. H. Ohsumi, M. Kubota, Y. Murakami, T. Kiyama, H. Nakao, Y. Wakabayashi, Y. Konishi, M. Izumi, M. Kawasaki and Y. Tokura: J. Phys. Soc. Jpn. **72** (2003) 1006.
36. T. Matsumura, H. Nakano and Y. Murakami: J. Phys. Soc. Jpn. **82** (2013) 021007.
37. H. K. Mao and P. M. Bell: Science **200** (1978) 1145.
38. W. Utsumi, K. Funakoshi, S. Urakawa, M. Yamakata, K. Tsuji, H. Konishi and O. Shimomura: Rev. High Press. Sci. Technol. **7** (1998) 1484.
39. E. Ohtani, M. Toma, T. Kubo, T. Kondo and T. Kikegawa : Geophys. Res. Lett. **30** (2003) 1029.
40. E. Ito: J. Mineralogical and Petrological Sci. **191** (2006) 118.
41. E. Ito, D. Yamazaki, T. Yoshino, H. Fukui, S. Zhai, A. Shatzkiy, T. Katsura, Y. Tange and K. Funakoshi: Earth and Planetary Sci. Lett. **293** (2010) 84.
42. T. Sakai, E. Ohtani, N. Hirao and Y. Ohishi: J. Appl. Phys. **109** (2011) 8.
43. H. K. Mao, J. Xu and P. M. Bell: J. Geophys. Res. **91** (1986) 4673.
44. Y. Katayama and K. Tsuji: J. Phys.: Condens. Matter **15** (2003) 6085.
45. T. Kubo, E. Ohtani and K. Funakoshi: Am. Mineral. **89** (2004) 285.

46. D. C. Rubie, T. S. Duffy and E. Ohtani eds.: "New Developments in High-Pressure Mineral Physics and Applications to the Earth's Interior" (Elsevier, 2004).

47. E. Ohtani, D. Andrault, P. D. Asimow, L. Stixrude and Y. Wang eds.; "Advances in High Pressure Mineral Physics: from Deep Mantle to the Core" (Elsevier, 2009).

48. E. Uemura, Y. Akahama, H. Kawamura, T. Le Bihan, T. Shobu, Y. Noda and O. Shimomura: J. Phys.: Condens. Matter **14** (2002) 10423 .

49. T. Shobu, Y. Noda, K. Iwasa, A. Hannan, M. Kohgi, N. Ishimatsu and O. Shimomura: J. Phys. Soc. Jpn. **70** (2001) 1162.

50. A. Wlodawer, W. Minor, Z. Dauter and M. Jaskolsk: FEBS Journal **275** (2008) 1.

51. S. Doublie ed.; "Macromolecular Crystallography Protocols" Vols.1 and 2 (Springer-Verlag, 2007).

52. A. Duruix and R. Giegĕ: "Crystallization of Nucleic Acids and Proteins: A Practical Approach" (IPL Press/Oxford University Press, 1992).

53. W. A. Hendrickson, J. L. Smith, R. P. Phizackerley and E. A. Merritt: Proteins **4** (1988) 77.

54. B. C. Wang: Methods Enzymol. **115** (1985) 90.

55. E. J. Gordon, G. A. Leonard, S. McSweeney and P. F. Zagalsky: Acta Crystallogr. D **57** (2001) 1230.

56. K. M. Reinisch, M. L. Nibert and S. C. Harrison: Nature **404** (2000) 960.

57. J. H. Cate, M. M. Yusupov, G. Z. Yusupova, T. N. Earnest and H. F. Noller: Science **285** (1999) 2095.

58. M. Unno, T. Mizushima, Y. Morimoto, Y. Tomisugi, K. Tanaka, N. Yasuoka and T. Tsukihara: Structure **10** (2002) 609.

59. P. Kuhn, M. Knapp, S. M. Soltis, G. Ganshaw, M. Thoene and R. Bott: Biochem. **37** (1998) 13446.

60. S. Hirotsu, G. C. Chu, M. Unno, D.-S. Lee, T. Yoshida, S.-Y. Park, Y. Shiro and M. Ikeda-Saito: "The crystal structures of the ferric and ferrous forms of the heme complex of HmuO, a heme oxygenase of corynebacterium diphtheriae", J. Biol. Chem. **279** (2004) 11937.

61. V. Šrajer, T. Teng, T. Ursby, C. Pradervand, Z. Ren, S. Adachi, W. Schildkamp, D. Bourgeois, M. Wulff and K. Moffat: Science **274** (1996) 1726.

62. V. Šrajer, Z. Ren, T-Y. Teng, M. Schmidt, T. Ursby, D. Bourgeois, C. Pradervand, W. Schidkamp, M. Wulff and K. Moffat: Biochem. **40** (2001) 13802.

63. F. Schotte, M. Lim, T. A. Jackson, A. V. Smirnove, J. Soman, J. S. Olson, J. N. Phillips Jr., M. Wulff and P. A. Anfinrud: Science **300** (2003) 1944.

64. D. Jacquemain, S. Grayer Wolf, F. Leveiller, M. Deutsch, K. Kjyaer, J. Als-Nielsen, M. Lahav and L. Leiserowitz: Angew. Chem. Int. Ed. Engl. **31** (1992) 130.

65. H. Takahashi, Y. Murase, K. Kurihara, I. Hatta, E. Arakawa, K. Takeshita and T. Matsushita: Jpn. J. Appl. Phys. **35** (1996) L1092.

66. Y. Waseda, E. Matsubara and K. Shinoda: "X-Ray Diffraction Crystallography –Introduction, Examples and Solved Problems–" (Springer-Verlag, 2011).

67. A. H. Narten and H. A. Levy: Science **160** (1969) 447.

68. Y. Waseda: "Anomalous X-Ray Scattering for Materials Characterization", STMP-No.179 (Springer-Verlag, 2002).

69. E. Matsubara, T. Ichitsubo, J. Saida, S. Kohara and H. Ohsumi: J. Alloys and Compounds **434-435** (2007) 119.

70. E. Matsubara and Y. Waseda: J. Phys.: Condens. Matter **1** (1989) 8575.

71. M. Saito, C. Park, K. Omote, K. Sugiyama and Y. Waseda: J. Phys. Soc. Jpn. **66** (1997) 633.

CHAPTER 10

Local Structure Analysis using EXAFS

In addition to X-ray diffractometry discussed earlier, there is another important method for the structural analysis of materials called EXAFS and which is based on X-ray absorption spectroscopy, having the advantage to be able to identify the local structure of materials, even of those with no periodicity. We now present an overview and an explanation of X-ray absorption spectroscopy including EXAFS. This is followed by a theoretical outline of the EXAFS method illustrated by actual applications to the local structure analysis of crystals including metalloproteins. The merits and demerits of structure analysis using EXAFS and X-ray diffraction for non-crystalline materials will also be discussed.

10.1 X-RAY ABSORPTION SPECTROSCOPY

X-ray absorption spectroscopy is sometimes abbreviated to either XAS, or XAFS. There is of course no difference in principle between XAS and VUV or soft X-ray absorption spectroscopy from the point of view of a spectroscopic measurement (see Sect. **7.3**). However in the hard X-ray absorption region the absorption due to valence electrons is small, while that due to core electrons is predominant. The core absorption spectra exhibit prominent structures with sharp peaks near to the core absorption edges, called X-ray absorption near edge structures (or XANES). On the high energy side of XANES these structures decrease in intensity with the weak oscillatory structure called EXAFS (see Figs. 10.3(a) and 10.4(a)). XANES can provide information on the electronic structure of empty states in a similar manner to VUV and soft X-ray spectra. It also shows a characteristic spectral feature reflecting the valency of the atom and of the ligand field surrounding it. Therefore, it can be used to identify the chemical states of compounds and the nature of the chemical species involved. The XANES structures are analyzed using ligand field theory or multiple scattering theory in which the level of scattering is greater than that in the simpler theory used to explain EXAFS structures.

EXAFS structures were well-known and their origins had been discussed for many years,[1] until 1970 when it was first demonstrated that EXAFS could be applied to structural analyses.[2] The spectroscopic features in EXAFS are a consequence of the interference between the electron wave from the excited core electron of an atom and electron waves scattered by atoms surrounding the initially excited atom. That is, EXAFS embodies the relationship between the wavelength of the electron wave and the interatomic distances. The abscissa chosen for the absorption spectra is usually photon energy. The wave number of the excited electron can be obtained from the photon energy by regarding the electron as a free electron. When the spectrum is plotted against the wave number and is Fourier-transformed, the distance between the excited atom and surrounding atoms can be deduced. Therefore it is widely used for obtaining the local structure around any selected atom. The structures obtained using EXAFS yield the structures immediately surrounding the selected atom and periodicity is not required in contrast to usual X-ray diffractometry.[3] Currently EXAFS is applied to numerous problems ranging from studies of local disorder from periodicity in crystals, the average structure of non-crystalline materials, the local configuration around any specific atom in organic molecules and bio molecules, the local structures at adsorbed surfaces with atoms and molecules and many others. Clearly, EXAFS is a vital analytical tool in the study of materials,[4] including even bio molecules,[5], soil[6] and others.[7] In what follows, only EXAFS will be described.

Absorption spectra are mostly measured using double crystal monochromators scanning with wavelength. Monochromatic light from the monochromator is introduced to a sample chamber. The incident light intensity, I_0, and that passing through the sample, I, are measured at each wavelength and absorption spectra are obtained. An alternative method called the wavelength or energy dispersive method uses the white light from a bending section and spectrometer with a specially-curved monochromatizing crystal equipped with a one-dimensional detector mentioned in Sect. **5.3.2.c**. With such a spectrometer, time-resolved EXAFS studies become possible.[8] The sample thickness is larger in the hard X-ray region than it is in the VUV and soft X-ray regions. The absorption coefficients differ from material to material and are strongly dependent on photon energy. Typically the coefficients are $1 \sim 10^3$ cm^{-1} in the hard X-ray region which means that the required sample thicknesses are from a few μm to a few cm. In many cases, samples are studied in powder form and in such cases, the sample powders are mixed with powders consisting of lighter elements such as BN (which does not have an absorption edge in the energy region of the measurement) and which can be pressed into tablet form. When samples suitable for absorption measurements cannot be prepared the total electron yield spectra or fluorescence yield spectra are taken instead of absorption spectra as mentioned in Sect. **7.3.1**. For insulators, the sample powders are often mixed with graphite powders to prevent charge-up during the total electron yield measurements.

10.2 EXAFS

Figure 10.1 illustrates the scattering of an outgoing wave from a core-excited electron of atom A in presence of a nearest neighboring atom B. Consider the superposition of the three waves mentioned below. The first (1) is the electron wave expanding from atom A. The second (2) is wave (1) which reaches the neighboring atom B and is then scattered by it. The third (3) is the wave (2) scattered by atom B returning to atom A and becoming re-scattered by it. The superposition of these three waves provides the new final state. The transition matrix between the initial and the final states of the isolated atom A is modulated by the interference of these three waves. This in turn manifests itself in the oscillatory structure (EXAFS) which is defined by the distance between two atoms and their scattering amplitudes. The EXAFS formula is derived by calculating the transition matrix between the initial and the new final states. The simplest case is that for K absorption where the angular momentum $l = 0$ (an s state), which is usually used for structure analysis. The wave number k of an excited electron which can be regarded as a free electron is calculated using $\hbar\omega - E_0 = \hbar^2 k^2/2m$. Here E_0 is the core absorption edge energy. The absorption coefficient for core transitions including EXAFS oscillations is expressed by $\mu(\omega) = \mu_0(\omega)\{1 + G(k)\}$, where $\mu_0(\omega)$ is the absorption coefficient of the isolated atom A and

$$G(k) = \frac{1}{k} \sum_j \frac{N_j}{R_j^2} |f_j(\pi, k)| \sin\{2kR_j + \Phi_j(k)\} e^{-2\sigma_j{}^2 k^2} e^{-2\gamma R_j}. \tag{10.1}$$

Here j represents the neighboring atoms B. R_j and N_j are respectively the distance and number of the neighboring atoms. $|f_j(\pi, k)|$ is the back scattering amplitude of the neighboring atoms, $\Phi_j(k)$ is the phase change, $\exp(-2\sigma_j{}^2 k^2)$ is the Debye-Waller factor (see Sect. **9.1.2**) and $\exp(-2\gamma R_j)$ is the damping factor of electron waves.

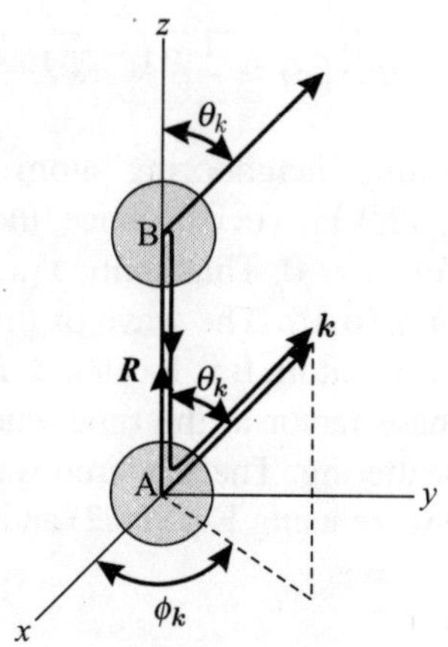

Figure 10.1 The propagation features of an excited electron wave and its back-scattered waves.

The derivation of Eq. (10.1) except for exponential functions is explained below. Assume for atom A, the s electron (wave function $\Psi_s(r)$) is excited to a continuum state. Using the selection rule, the excited state is of $l = 1$, *i.e.*, the p state.

The waves in the excited state are those contained in the above three waves. They can be approximated by plane waves at large distances from atoms A and B. The wave function of a free electron is expressed by a plane wave $e^{i\boldsymbol{k}\cdot\boldsymbol{r}}$, which is expanded with spherical Bessel function j_l to be

$$e^{i\boldsymbol{k}\cdot\boldsymbol{r}} = 4\pi \sum_{l=0}^{\infty} i^l j_l(kr) \sum_{m=-l}^{l} Y_{lm}(\theta_{\boldsymbol{k}}, \phi_{\boldsymbol{k}}) Y_{lm}{}^*(\theta, \phi). \tag{10.2}$$

Here, Y_{lm} represents the spherical harmonic function, $(\theta_{\boldsymbol{k}}, \phi_{\boldsymbol{k}})$ and (θ, ϕ), the directions of $\boldsymbol{k}$ and $\boldsymbol{r}$, respectively. The wave function of the first wave (1) which represents the final continuum state of $l = 1$ is extracted from Eq. (10.2) to give

$$\Psi_0 = 4\pi i j_1(kr) \sum_{m=-1}^{1} Y_{1m}(\theta_{\boldsymbol{k}}, \phi_{\boldsymbol{k}}) Y_{1m}{}^*(\theta, \phi). \tag{10.3}$$

The second wave (2) is the electron wave expanding from atom A and scattered by atom B. The scattered wave is usually described by the Green function.[9,10] However for simplicity, we express the scattered wave function in a form similar to that in Eq. (10.3). The z direction is defined to be the direction from atom A to atom B (see Fig. 10.1). The electron wave scattered by atom B in the $\theta_{\boldsymbol{k}}$ direction with scattering amplitude $f_B(\theta_{\boldsymbol{k}}, k)$ is given by

$$\Psi_1 = 4\pi i \frac{e^{ikR(1-\cos\theta_{\boldsymbol{k}})}}{R} f_B(\theta_{\boldsymbol{k}}, k)\, j_1(kr)\, Y_{10}{}^*(\theta, \phi)\, Y_{10}(0, 0)\,. \tag{10.4}$$

The wave function Eq. (10.4) can be explained in the following way. Using the first kind of spherical Hankel function $h_l^{(1)}$, the wave function of the excited electron propagating from the origin to outside is given by $ih_l{}^{(1)}(\rho)\, Y_{lm}(\theta, \phi)$, where $\rho = kr$ (see App. **F**). For $\rho \gg 1$, the asymptotic formula of $ih_l^{(1)}$ is

$$ih_l^{(1)}(\rho) \approx \frac{1}{\rho} e^{i\left(\rho - \frac{l\pi}{2}\right)} \tag{10.5}$$

Therefore, the wave of $l = 1$ reaching neighboring atom separated by distance R is given by $ih_l{}^{(1)}(kR)\, Y_{10}(0, 0) = -i\,(e^{ikR}/kR)\, Y_{10}(0, 0)$. Since the angular coordinate for atom B is $(\theta_{\boldsymbol{R}}, \phi_{\boldsymbol{R}}) = (0, \phi_{\boldsymbol{R}})$, $Y_{1m} \neq 0$ only for $m = 0$. Thus, only Y_{10} appears in Eq. (10.4). $\phi_{\boldsymbol{R}}$ is arbitrary, so that one can adopt $Y_{10}(0, \phi_{\boldsymbol{R}}) = Y_{10}(0, 0)$. The wave of the electron scattered by atom B can be expressed as a plane wave $e^{i\boldsymbol{k}\cdot(\boldsymbol{r}-\boldsymbol{R})}$. Since atom B is located at $\boldsymbol{R}$ from the origin, $e^{-i\boldsymbol{k}\cdot\boldsymbol{R}}$ of the above plane wave can be regarded as the phase factor at the time when the electron wave propagates from atom B in the $\boldsymbol{k}$ direction following scattering. The scattered wave is isotropic with respect to $\phi_{\boldsymbol{k}}$. Let us expand $e^{-i\boldsymbol{k}\cdot\boldsymbol{r}}$ for the above plane wave using Eq. (10.2) and multiply the wave function at atom B by it. One can obtain their product as

$$4\pi i \frac{e^{i(kR - \boldsymbol{k}\cdot\boldsymbol{R})}}{kR} j_1(kr)\, Y_{10}(\theta_{\boldsymbol{k}}, 0) Y_{10}{}^*(\theta, \phi)\, Y_{10}(0, 0)\,. \tag{10.6}$$

The scattering amplitude of atom B is given by $f_B(\theta_{\boldsymbol{k}}, k)$. It is assumed that $Y_{10}(\theta_{\boldsymbol{k}}, 0)$, which represents the $\boldsymbol{k}$ dependence of the scattering amplitude is included in $f_B(\theta_{\boldsymbol{k}}, k)$ (see Eqs. (10.16) and (F.8)), though it is not exact. Finally, by multiplying Eq. (10.6) by the appropriate factor, one derives the function for the wave (2) to be as Eq. (10.4).

The wave function for wave (3) is given by

$$\Psi_2 = 4\pi i \frac{e^{2ikR}}{R^2} f_B(\pi, k)\, f_A(\pi - \theta_{\boldsymbol{k}}, k)\, j_1(kr)\, Y_{10}{}^*(\theta, \phi)\, Y_{10}(0, 0)\,. \tag{10.7}$$

This is explained as follows. The wave (1) from atom A is back-scattered ($\theta_{\boldsymbol{k}} = \pi$) by atom B, becomes a spherical wave to atom A ($r \approx 0$), so that $f_B(\theta_k, k)$ becomes $f_B(\pi, k)$ and the wave function reaching atom A becomes $e^{ik|\boldsymbol{r}-\boldsymbol{R}|}/|\boldsymbol{r} - \boldsymbol{R}| \approx e^{ikR}/R$. This wave is scattered by atom A and propagates in the $\theta_{\boldsymbol{k}}$ direction as a plane wave which is the wave (3). The wave function is derived by multiplying the plane wave expanded using Eq. (10.2) as

$$4\pi i \frac{e^{ikR}}{R} f_B(\pi, k)\, j_1(kr)\, Y_{10}^*(\theta, \phi)\, Y_{10}(0, 0)\, \frac{e^{ikR}}{R}. \tag{10.8}$$

Here $Y_{10}(\theta_{\boldsymbol{k}}, 0)$ is omitted for the above-mentioned reason. After the scattering by atom A, Eq. (10.8) is multiplied by scattering amplitude of atom A and the wave function of the wave (3) can be expressed as Eq. (10.7).

Here a_1 and a_2 are defined to be

$$a_1 = \frac{e^{ikR(1-\cos\theta_{\boldsymbol{k}})}}{R} f_B(\theta_{\boldsymbol{k}}, k), \tag{10.9}$$

$$a_2 = \frac{e^{2ikR}}{R^2} f_B(\pi, k) f_A(\pi - \theta_{\boldsymbol{k}}, k). \tag{10.10}$$

The final state wave function Ψ_f is the superposition of Ψ_0, Ψ_1 and Ψ_2. Using a_1 and a_2, the final state wave function is given by

$$\Psi_f = 4\pi i j_1(kr)\sqrt{\frac{3}{4\pi}}\left\{\sqrt{\frac{4\pi}{3}}\sum_{m=-1}^{1} Y_{1m}(\theta_{\boldsymbol{k}}, \phi_{\boldsymbol{k}}) Y_{1m}^*(\theta, \phi) + (a_1 + a_2) Y_{10}^*(\theta, \phi)\right\}, \tag{10.11}$$

where $Y_{10}(0,0) = \sqrt{3/4\pi}$.

The square of the absolute value of the transition matrix element $|\langle f|\boldsymbol{e}\cdot\boldsymbol{r}|i\rangle|^2$ is required to explain EXAFS, where $\boldsymbol{e}$ is the unit vector of the electric field of the incident light whose direction is (θ_e, ϕ_e), and inner product $\boldsymbol{e}\cdot\boldsymbol{r}$ is expressed by

$$\boldsymbol{e}\cdot\boldsymbol{r} = \frac{4\pi}{3} r \sum_{m=-1}^{1} Y_{1m}^*(\theta_e,\ \phi_e) Y_{1m}(\theta,\ \phi). \tag{10.12}$$

Using Eq. (10.12), the transition matrix element is calculated to be

$$\begin{aligned}\langle f|\boldsymbol{e}\cdot\boldsymbol{r}|i\rangle &= -4\pi i\sqrt{\frac{4\pi}{3}}\sum_{m,m'=-1}^{1} Y_{1m}^*(\theta_e, \phi_e)\int j_1(kr) \\ &\quad\times\left\{\sqrt{\frac{4\pi}{3}} Y_{1m'}(\theta_{\boldsymbol{k}}, \phi_{\boldsymbol{k}}) Y_{1m'}^*(\theta, \phi) + (a_1 + a_2) Y_{10}^*(\theta, \phi)\right\}^* \\ &\quad\times r\Psi_s(r) Y_{1m}(\theta, \phi) r^2 \sin\theta dr d\theta d\phi \\ &= -i\sqrt{\frac{4\pi}{3}}\left\{\sqrt{\frac{4\pi}{3}}\sum_{m=-1}^{1} Y_{1m}^*(\theta_e, \phi_e) Y_{1m}(\theta_{\boldsymbol{k}}, \phi_{\boldsymbol{k}}) + (a_1^* + a_2^*) Y_{10}(\theta_e, \phi_e)\right\} M,\end{aligned} \tag{10.13}$$

$$M = 4\pi\int j_1(kr)\Psi_s(r) r^3 dr. \tag{10.14}$$

The average of $|\langle f|\boldsymbol{e}\cdot\boldsymbol{r}|i\rangle|^2$ is used in the analysis of measurements on powders crystals and amorphous solids, *etc.*, and is widely carried out with unpolarized light. The order of the averaging procedure with respect to the directions of $\boldsymbol{e}$ or $\boldsymbol{k}$ does not matter. Let us integrate with respect to θ_e and ϕ_e and then divide the result by 4π. In this process, the terms of R^{-3} and R^{-4}, that is, those of $a_1 a_2^*$, $a_1^* a_2$ and $|a_2|^2$ are neglected. The next procedure is to obtain the average with respect to direction of $\boldsymbol{k}$ which is given by

$$\langle|\langle f\,|\boldsymbol{r}|\,i\rangle|^2\rangle_{\text{av}} = \frac{4\pi M^2}{3}\frac{1}{4\pi}\int\left\{1+|a_1|^2+2Re\sqrt{\frac{4\pi}{3}}Y_{10}^*\left(\theta_{\boldsymbol{k}},\phi_{\boldsymbol{k}}\right)\left(a_1+a_2\right)\right\}\sin\theta_{\boldsymbol{k}}d\theta_{\boldsymbol{k}}d\phi_{\boldsymbol{k}}. \tag{10.15}$$

Here the main part of the average with respect to direction of $\boldsymbol{e}$ is given in { } of the integrand. Hereafter we express scattering amplitude f explicitly. For instance for atom B

$$f_B(\theta_{\boldsymbol{k}},k) = \sum_{l=0}^{\infty} f_{Bl}\left(k\right)P_l\left(\cos\theta_{\boldsymbol{k}}\right) = \sum_{l=0}^{\infty}\frac{2l+1}{2ik}\left(e^{2i\delta_{Bl}}-1\right)P_l\left(\cos\theta_{\boldsymbol{k}}\right), \tag{10.16}$$

where P_l is the Legendre polynomial and

$$|f_{Bl}(k)|^2 = \frac{(2l+1)^2}{k^2}\sin^2\delta_{Bl}\,,\quad Re\left\{if_{Bl}\left(k\right)\right\} = -\frac{2l+1}{k}\sin^2\delta_{Bl}. \tag{10.17}$$

Let us calculate Eq. (10.15). Note that the factor $4\pi M^2/3$ in Eq. (10.15) is omitted in the following presentation. The integral of the second term in { } in Eq. (10.15) is given by

$$\frac{1}{4\pi}\int|a_1|^2\sin\theta_{\boldsymbol{k}}d\theta_{\boldsymbol{k}}d\phi_{\boldsymbol{k}} = \frac{1}{2R^2}\sum_{l,l'=0}^{\infty}f_{Bl}^*\left(k\right)f_{Bl'}\left(k\right)\int_{-1}^{1}P_l\left(x\right)P_{l'}\left(x\right)dx$$

$$= \frac{1}{2R^2}\sum_{l=0}^{\infty}\frac{2}{2l+1}\left|f_{Bl}\left(k\right)\right|^2 = \frac{1}{k^2R^2}\sum_{l=0}^{\infty}\left(2l+1\right)\sin^2\delta_{Bl}. \tag{10.18}$$

Here the orthogonality of Legendre polynomial is applied. The third term is calculated using $Y_{10}^*\left(\theta_{\boldsymbol{k}},\phi_{\boldsymbol{k}}\right) = \sqrt{3/4\pi}\cos\theta_{\boldsymbol{k}}$ as

$$\frac{1}{2\pi}Re\int\cos\theta_{\boldsymbol{k}}\left(a_1+a_2\right)\sin\theta_{\boldsymbol{k}}d\theta_{\boldsymbol{k}}d\phi_{\boldsymbol{k}}. \tag{10.19}$$

The first term in the integral Eq. (10.19) is calculated by setting $x = \cos\theta_{\boldsymbol{k}}$ as

$$Re\int a_1\cos\theta_{\boldsymbol{k}}\sin\theta_{\boldsymbol{k}}d\theta_{\boldsymbol{k}} = Re\frac{e^{ikR}}{R}\sum_{l=0}^{\infty}f_{Bl}\left(k\right)\int_{-1}^{1}e^{-ikRx}P_l\left(x\right)xdx$$

$$\approx Re\frac{ie^{ikR}}{kR^2}\sum_{l=0}^{\infty}f_{Bl}\left(k\right)\left\{P_l\left(1\right)e^{-ikR}+P_l\left(-1\right)e^{ikR}\right\}$$

$$= Re\frac{i}{kR^2}\left\{\sum_{l=0}^{\infty}f_{Bl}\left(k\right)+f_B\left(\pi,k\right)e^{2ikR}\right\}. \tag{10.20}$$

Here, all terms smaller than R^{-3} are neglected in the procedure of integration by parts. The second term in the integral Eq. (10.19) is calculated to be

$$Re\int a_2\cos\theta_k\sin\theta_k d\theta_k = Re\frac{e^{2ikR}}{R^2}f_B\left(\pi,k\right)\sum_{l=0}^{\infty}f_{Al}\left(k\right)\int_{-1}^{1}P_l\left(-x\right)xdx$$

$$= Re\frac{e^{2ikR}}{kR^2}f_B(\pi,k)\,f_{A1}\frac{-2}{3} = Re\frac{ie^{2ikR}}{kR^2}f_B(\pi,k)\left(e^{2i\delta_{A1}}-1\right). \tag{10.21}$$

From $P_1(x) = x$ and the orthogonality of the Legendre polynomial, only those terms with $l = 1$ are included in the integration of Eq. (10.21). The integral Eq. (10.19) is sum of Eqs. (10.20) and (10.21). Using Eq. (10.17), the real part of Eq. (10.19) is obtained as

$$\frac{1}{kR^2}\left\{-\sum_{l=0}^{\infty}\frac{2l+1}{k}\sin^2\delta_{Bl} + |f_B(\pi,k)|\sin\left(2kR+2\delta_{A1}+2\phi_B+\pi\right)\right\}, \tag{10.22}$$

here $f_B(\pi,k) = |f_B(\pi,k)|\,e^{2i\phi_B}$ is used to calculate the second term of Eq. (10.22). The summation of Eqs. (10.18) and (10.22) yields the cancellation of the first terms in Eqs. (10.18) and (10.22). As a result, Eq. (10.15) is given by

$$\langle|\langle f\,|\boldsymbol{r}|\,i\rangle|^2\rangle_{\text{av}} = \frac{4\pi M^2}{3}\left\{1+\frac{1}{kR^2}|f_B(\pi,k)|\sin\left(2kR+2\delta_{A1}+2\phi_B+\pi\right)\right\}. \tag{10.23}$$

Finally, the integral of the first term of { } in Eq. (10.15) gives the absorption coefficient μ_0 without scattering by surrounding atoms and the sum of the integrals of the second and the third terms gives the EXAFS oscillation. Replacing the subscript B by j, summing with respect to scattering atoms j and setting $\varPhi_j(k) = 2\delta_{A1} + 2\phi_j + \pi$ in Eq. (10.23), one can obtain

$$\mu = \frac{4\pi M^2}{3}\left[1+\sum_j\frac{N_j}{kR_j^2}|f_j(\pi,k)|\sin\left\{2kR_j+\varPhi_j(k)\right\}\right] = \mu_0\left\{1+G(k)\right\}, \tag{10.24}$$

which is equal to Eq. (10.1) except for factors expressed by exponential functions.

This theoretical explanation of the EXAFS structures in the absorption spectra shows that an important advantage of EXAFS is that $G(k)$ can be extracted from the experimental results and that a local structure analysis around any specific atom can also be derived. That is, R_j can be obtained using the Fourier transform of $G(k)$, where k is wave number of the electron wave regarded as the free electron wave. The Fourier transform is given by

$$F(R) = \frac{1}{2\pi}\int G(k)\,e^{-2ikR}dk. \tag{10.25}$$

Here, the values of E_0, $|f_j(\pi,k)|$ and $\varPhi_j(k)$ are obtained using those of standard material. In many cases, simulations are made using the most plausible values in $F(R)$ in order to reproduce $G(k)$. From the Fourier-transformed $F(R)$, the distance to the nearest neighbors and their number and often those of the second nearest neighbors can be obtained. Since the EXAFS oscillation becomes weak with increase in k, $k^nG(k)$ (n = 1, 2, 3 …) is sometimes Fourier-transformed to make the result clearer. Analyzed results from experimental EXAFS data are often drawn in figures using $|F(R)|$ called the 'radial structure function'. In the above explanation multiple scattering effects have not been taken into account. In a recent EXAFS analysis program called FEFF however, multiple scattering up to the fourth shell (the fourth-nearest neighbor) is taken into account.[11)]

10.3 LOCAL STRUCTURE ANALYSIS OF CRYSTALS

10.3.1 Site occupation of metallic atoms in ferrites

It is well-known that EXAFS is a useful tool for the local structure analysis of non-crystalline materials such as amorphous solids and liquids, about which conventional X-ray diffractometry (XRD) can provide only limited information. Importantly, EXAFS can often add invaluable data also about the structure of crystalline materials. The following is an example of EXAFS applied to a crystal in the determination of the occupation sites for two kinds of metal ions in a compound.

In some materials the same element can occupy two or more different sites. Spinels are a class of minerals of general formulation $A^{2+}B^{3+}_{2}O^{2-}_{4}$, the crystal structure of which is shown in Fig. 10.2. In normal spinels one kind of element can be found only in tetragonally coordinated A sites and the other kind in octahedrally coordinated B sites,[12] but sometimes the two elements replace each other and occupy both sites. In this case, since XRD is based on the periodicity of the structure, it cannot determine the structure uniquely, but EXAFS can. An example of how the site occupation can be investigated by using EXAFS is given using zinc ferrite ($ZnFe_2O_4$), one of the spinel compounds.

In the structure of the bulk Zn ferrite, Zn and Fe ions are considered to exclusively occupy A and B sites respectively. The bulk Zn ferrite synthesized by thermal diffusion at high temperature (ceramic method) is antiferromagnetic with a Néel temperature of about 10K. This anomalous magnetism is not explained by the structure above and has aroused much interest, leading to intensive studies as to the origin of the magnetism.[13,14] It has been suggested that a small number of Fe ions occupy A sites instead of Zn ions and magnetically couple with Fe ions at B sites. Because the A–B distance is shorter than that of B–B, the magnetic interaction between Fe(A) and Fe(B) is much stronger than that of Fe(B)–Fe(B) and accordingly governs the magnetic properties.[15] Recently, it was found that the $ZnFe_2O_4$ nano particle, prepared by the co-precipitation method, shows large magnetization compared with the bulk $ZnFe_2O_4$ prepared by the ceramic method. The finding suggests that A sites are occupied by more Fe ions in the nano particles than in the bulk, but no direct experimental evidence was then

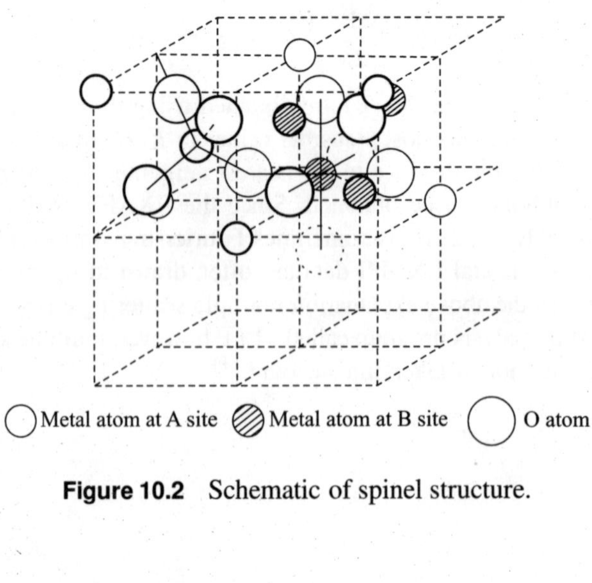

Figure 10.2 Schematic of spinel structure.

reported.[16] Under those circumstances an EXAFS study was attempted to clarify the difference in local structures around Fe and Zn ions in both $ZnFe_2O_4$ prepared by ceramic and by the co-precipitation methods.[17]

The $ZnFe_2O_4$ samples were prepared using the following procedure. (1) Ceramic method: Powders of ZnO and Fe_2O_3 were mixed at a mole ratio of 1 : 1. And the mixture was calcined at 1373K for 3 hours in an oxidizing atmosphere and cooled slowly. Then the product was ground and calcined again at 1373K for 3 hours. (2) The co-precipitation method: Powders of $ZnCl_2$ and $FeCl_3$ were mixed at a mole ratio of 1 : 1 and then a 5N aqueous solution of NaOH was added with constant stirring to reach pH 12 ~ 13. Subsequently the solution was kept at 368K for 30 minutes with stirring.[17] The colloidal zinc ferrite (nano particles) thus prepared were filtered, washed and dried.

Lattice constants obtained by XRD were 0.844 nm and 0.849 nm for the samples (1) and (2), respectively, which are within the experimental error. Hence these two samples are identical from the viewpoint of XRD. Therefore in order to specify the local structures around the Fe and Zn ions by EXAFS, Fe K and Zn K absorption measurements were undertaken. For the measurements a laboratory X-ray instrument was employed. Figure 10.3(a) shows the Fe K absorption spectra of both samples, which reflect the local structure around the Fe ions. By comparison, one can see that the EXAFS oscillation periods are equal, but that the amplitude is somewhat smaller in sample (2) than in sample (1). The two curves of radial structure functions in Fig. 10.3(b) are obtained from the Fourier transform of the EXAFS oscillations[11,18] extracted from the data of Fig. 10.3(a). The abscissa and ordinate of the peaks can yield respectively the interatomic distances and coordination numbers around the Fe ions. It is evident from the radial structure function that peak intensities of sample (2) are much lower than those of sample (1). The distance between the central atom and the second nearest atom is marginally greater in sample (2) than in sample (1). Zn K-edge EXAFS oscillations as well as radial structure functions of the two samples are shown in Fig. 10.4. It is evident even from a qualitative inspection of radial structure functions that local structures around Zn ions are very different in the two samples.

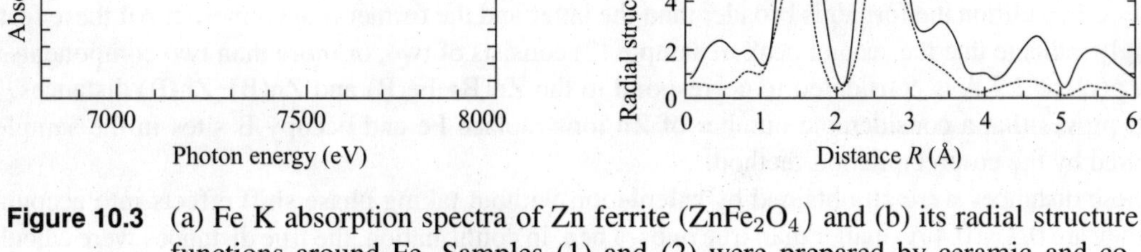

Figure 10.3 (a) Fe K absorption spectra of Zn ferrite ($ZnFe_2O_4$) and (b) its radial structure functions around Fe. Samples (1) and (2) were prepared by ceramic and co-precipitation methods, respectively.

In spinels there are two possibilities for site occupation. They are (a) two different sites are occupied by different kind of ions, and (b) both the ions can occupy both sites. If (a) is the case, then the radial structure functions around the two ions must be different, that is, one exhibits features characteristic of

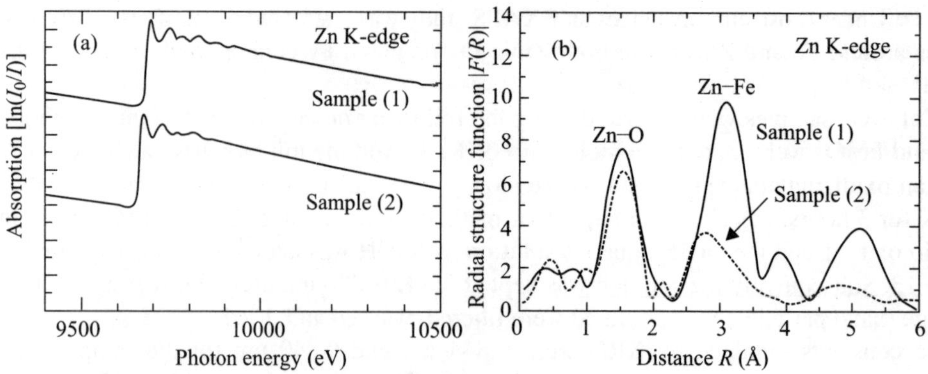

Figure 10.4 (a) Zn K absorption spectra of Zn ferrite ($ZnFe_2O_4$) and (b) its radial structure functions around Zn. Samples (1) and (2) were prepared by ceramic and co-precipitation methods, respectively.

tetragonal coordination and the other those of octahedral coordination. On the other hand, if (b) is the case then the radial structure functions must be mixtures of the two. It is concluded that samples (1) and (2) correspond respectively to the cases (a) and (b) from the quantitative analyses described below.

First, observe sample (1) and examine the local structures around the Fe and Zn ions in more detail. Figure 10.3(b) reflects the local structure around the Fe ions. Two intense peaks are seen at about 1.5 Å and 2.6 Å. Evidently the former corresponds to Fe(B)–O and the latter to Fe(B)–Fe(B). Minor contributions from Fe(B)–Zn(B) might have been included in the latter. The third peak at 3.4 Å can be attributed to Fe(B)–Zn(A). Peaks are found at distances greater than 4.6 Å, but these peaks cannot be used for the determination of the structure because there are several bonds having these bond lengths. Similarly, the peak at 1.5 Å can be assigned to Zn(A)–O and that at 3.1 Å to Zn(A)–Fe(B) in Fig. 10.4(b). Hence all the EXAFS results demonstrate that sample (1) is a normal spinel in which most of Zn and Fe ions occupy A and B sites respectively.

The radial structure functions for sample (2) show considerable differences from those for sample (1). That is, the intensities of all the peaks are reduced in both Figs. 10.3(b) and 10.4(b). The reduction in peak intensities in the radial structure function, which is directly related to the coordination number is often observed in nano particles and is attributed to the increase in the number of surface atoms. As seen in Fig. 10.4(b), the second peak of the sample (2) is found at 2.7 Å, while in that of sample (1) it is at 3.1 Å. In addition the former is broader than the latter and the former is asymmetric. All these features strongly indicate that the second peak of sample (2) consists of two, or more than two components. The main peak at 2.7 Å is considered to correspond to the Zn(B)–Fe(B) and Zn(B)–Zn(B) distances. This result proves that a considerable number of Zn ions replace Fe and occupy B sites in the sample (2) prepared by the co-precipitation method.

These distances were all obtained by calculation without taking phase shift effects into account, so that they are 0.3 $\sim$ 0.4 Å smaller than true ones. Then, in confirmation, the true distances were calculated taking all effects into account. An illustration of the result is that the above-mentioned distance of 3.1 Å assigned to Zn(A)–Fe(B) was determined to be 3.48 Å, which agrees well with 3.51Å, the known Zn(A)–Fe(B) distance in the bulk $ZnFe_2O_4$.[16] This supports the above-mentioned assignment.

Thus quite subtle differences between the local structures can easily be observed using EXAFS *n.b.*, it is an intrinsic advantage of EXAFS to be able to select any absorbing elements and therefore to extract information of local structure only around that particular element.

10.3.2 Interatomic Distances in Mixed Crystals of Magnetic Semiconductors

In solid solutions of two kinds of materials having the same crystal structure, the lattice constants will change in proportion with the concentration. For instance in Si_xGe_{1-x} alloy, when the Si content x is changed from 0 to 1, the lattice constants measured by X-ray diffraction changes linearly in proportion to x from the lattice constant of single crystal of Ge to that of Si. However, there are many cases in which the lattice constants measured by X-ray diffraction change linearly in proportion to the concentration, while the local structure does not change significantly. Here, we introduce, as an example, a study of the configuration of Te atoms around Zn atom and Mn atom in a magnetic semiconductor $Zn_{1-x}Mn_xTe$.[19] In pure ZnTe, its crystal structure is zinc-blende type, while in pure MnTe it is NiAs type. However in the case of $Zn_{1-x}Mn_xTe$ mixed crystals in which concentration of Mn, x is changed from 0 to 0.86, their crystal structure is zinc-blende type. The EXAFS measurements were made on Mn K and Zn K absorptions. In the following, the absorption spectra and their Fourier transform are omitted and only the dependence of the obtained lattice constants on the concentration is explained.

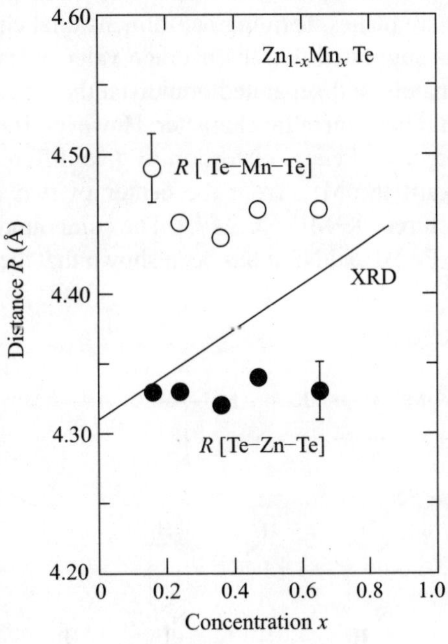

Figure 10.5 Lattice constant (—) of $Zn_{1-x}Mn_xTe$ measured by X-ray diffraction (XRD) and distances of Te–Zn–Te (●) and Te–Mn–Te (○) measured by EXAFS (by courtesy of M. Taniguchi, © IOP Publishing 1996).[19]

In Fig. 10.5, plotted against Mn concentration are R [Te–Zn–Te] and R [Te–Mn–Te] obtained by EXAFS, which are the distance between two nearest neighbors Te (Te–Te) located at opposite sides of

Zn and Mn, respectively. In addition, the Te–Te distance determined from X-ray diffraction is given by a straight line. As shown in Fig. 10.5, Te–Te distance determined from X-ray diffraction changes with change in proportion to x, while R [Te–Zn–Te] and R [Te–Mn–Te] do not change with change in x. Each value is essentially the same as those obtained from distances of Zn–Te or Mn–Te in zinc-blende type. This means that the average lattice constants change in proportion to x, while the nearest distance between Zn and Te, and that between Mn and Te are kept even in the mixed crystals. That is, Zn and Te atoms, and Mn and Te atoms are bound strongly by covalent bond, and those atoms form locally $ZnTe_4$ and $MnTe_4$ clusters. This conclusion is supported by analysis of valence band photoelectron spectroscopy.[19] EXAFS is indispensable to obtain such information on local structure. In many mixed crystals, the above-mentioned phenomena are often found, in which lattice constants measured by X-ray diffraction change in average, while their local structures do not change substantially.[20–22]

10.3.3 Metal-halogen distances in one-dimensional halogen-bridged mixed valence complexes

Halogen-bridged mixed valence compounds (metal: Pt, Pd, Ni) that are an analogues of Wolfram's red salt are known to be one-dimensional semiconductors. Figure 10.6 shows the one-dimensional structure of these complexes, consisting of alternating octahedral and square-planar complexes linked by halogen bridges. Structures of the organic ligands are also shown in Fig. 10.6. The metal atoms (M) are located in the center of a square comprised of four nitrogens of the ligand. Halogen atoms (X) are located at both sides of the square planes, forming one-dimensional chain, –X–M–X–M–. For an ideal one-dimensional structure, it is suggested that the average valence state of M is +3, and that the two units should be arranged alternately and separated equidistantly *i.e.*, for all the M–X distances to be equal so that the complex should have metallic character. However, this ideal ordered one-dimensional structure is unstable and undergoes a Peierls distortion to give different valence states, M^{II} and M^{IV}. The halogen atoms move toward the M^{IV} from the center of two M atoms, giving an alternately arranged one-dimensional structure, $\cdot\cdot X{-}M^{IV}{-}X\cdot\cdot M^{II}\cdot\cdot$. The compound is then semiconductive. Among this type of compounds having Pt, Pd and Ni, it has been shown that the Ni compounds have the highest conductivities.

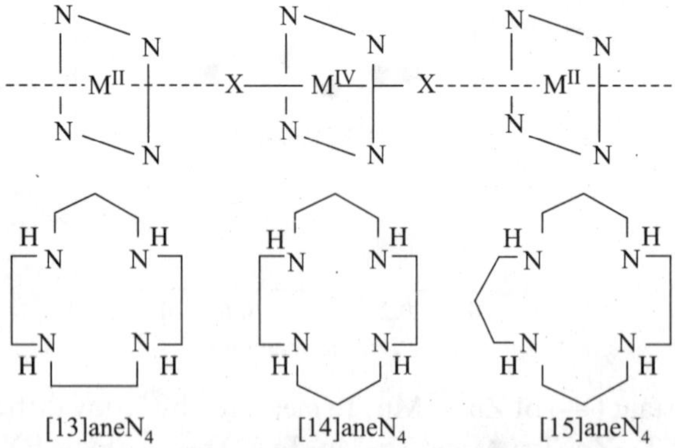

Figure 10.6 Schematic structures of the halogen-bridged mixed valence complexes and ligand structures. $[13]aneN_4$ = 1,4,7,10-tetraazacyclotridecane, $[14]aneN_4$ = 1,4,8,11-tetraazacyclotetradecane, $[15]aneN_4$ = 1,4,8,12-tetraazacyclopentadecane.

In these structures it is unclear how far the halogen atoms displaced from the center between the M^{II} and M^{IV} atoms. It is difficult to analyze the structure of Ni compounds using single crystal X-ray analysis, because the available single crystals are too small. Therefore, EXAFS analysis was used. The target compounds were a series of bromo-bridged Ni^{II}–Ni^{IV} mixed valence complexes, $[Ni([n]aneN_4)][NiBr_2([n]aneN_4)](ClO_4)_4$, $n = 13, 14, 15$ (see Fig. 10.6).[23] Ni K absorption and Br K absorption spectra of the compounds were measured and their EXAFS spectra analyzed using a Fourier method. The structures of reference compounds, *i.e.*, discrete Ni^{II}, Ni^{III} and Ni^{IV} complexes with similar coordination environments were also studied using the same method. From the results for the one-dimensional mixed valence compound of $n = 14$, the Ni^{IV}–Br and Ni^{II}–Br distances were 2.61 $\sim$ 2.66 Å and 2.73 $\sim$ 2.77 Å, respectively, and the ratio of the distances between Ni^{IV}–Br and Ni^{II}–Br was 0.94 $\sim$ 0.97. On the other hand, Ni^{IV}–Br and Ni^{III}–Br distances in discrete compounds were 2.64 and 2.63 Å, respectively. These results indicate that, in a one-dimensional mixed valence complex, significant charge transfer from Ni^{II} to Ni^{IV} takes place and that the Ni^{IV}–Br distance becomes longer. Accurate structural parameters for the one-dimensional mixed valence complexes of $n = 13$ and 15 were not obtained. The ratios of the distances between M^{IV}–Br and M^{II}–Br in one-dimensional Pd and Pt mixed valence complexes have been known to be about 0.86, which is clearly smaller than that of the Ni analogues.[24] Thus this effectively shows why the electrical conductivities of the Ni complexes are higher than those of Pd and Pt analogues.

10.3.4 Local distortion in perovskite-type proton conductors

In the perovskite type oxide of $SrZrO_3$, when a few mol% of Zr^{4+} is substituted for Yb^{3+}, protons are introduced in it for charge compensation. The protons are supplied from water vapor in their processing atmosphere. Since the above-mentioned material in which protons are introduced shows large proton conduction at high temperature, it could be applied to a high temperature fuel cell and has thus attracted the attention of scientists and engineers.[25,26] The proton conductivity is characterized by the relation between introduction and diffusion of protons, both of which originate from a local interaction between the host crystal and Yb^{3+} ions. In other words, the existence of dopant ions greatly alters the property of the host crystal of $SrZrO_3$. A Zr ion is normally located at the center of a regular octahedron formed by oxygen ions. A Yb ion with the different ionic radius should make the oxygen octahedron locally distort. The proton diffusion process has been known to consist of a sequence of free diffusion and trapping events by Yb ions.[27] It is important to understand the behavior of Yb^{3+} as a dopant ion in the host crystal in order to appreciate the proton diffusion process. The local distortion of the oxygen octahedron can be inferred from the analysis of the structure around the dopant Yb ion using the EXAFS method.[28]

Figure 10.7(a) shows radial structure function of $SrZrO_3$ around Zr, Figs. 10.7(b) and 10.7(c), those of Yb-doped proton conductors of $SrZr_{0.95}Yb_{0.05}O_3$ and $SrZr_{0.9}Yb_{0.1}O_3$ around Yb, respectively, and Fig. 10.7(d), that of Yb_2O_3 around Yb. In the figures, peaks at 1.0 $\sim$ 2.0 Å are due to O ions at the first-nearest neighbor sites. In Figs. 10.7(a) and 10.7(d), the structures show clear contributions from the second-nearest ions at 3.0 $\sim$ 4.0 Å, while in Figs. 10.7(b) and 10.7(c), the structures are not observed. The decrease of the signal from the second-nearest ions corresponds to the decrease in correlation between Yb and the second-nearest ions. Such a phenomenon is well-known in amorphous or non-crystalline materials as mentioned later.[30] It suggests that the crystalline order of the second-nearest neighbors around Yb is lost in $SrZr_{1-x}Yb_xO_3$.

In both Figs. 10.7(b) and 10.7(c), it is found that a peak corresponding to the distribution of the first-nearest neighbors of O ions has a strong side peak (see the arrow in each figure). It suggests that

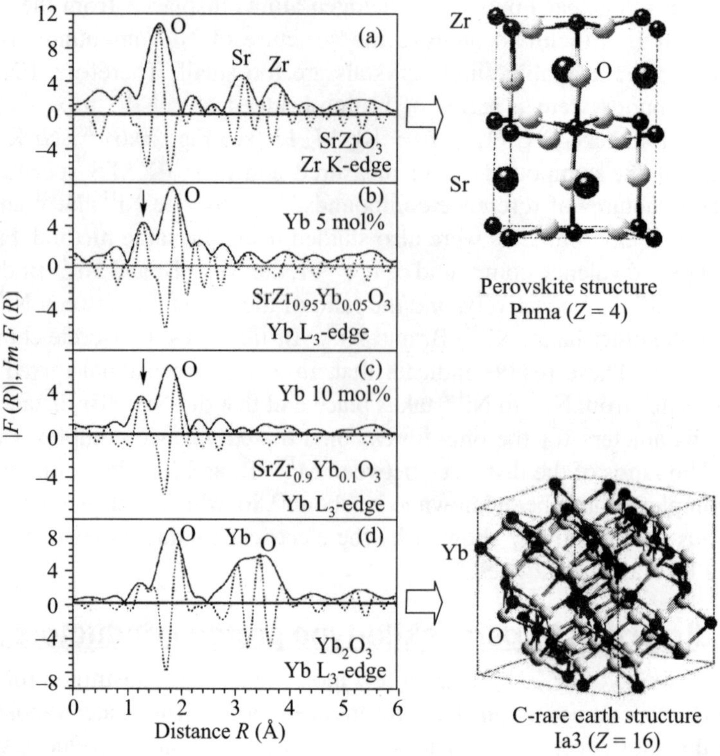

Figure 10.7 Radial structure functions of (a) $SrZrO_3$, (b) $SrZr_{0.95}Yb_{0.05}O_3$, (c) $SrZr_{0.9}Yb_{0.1}O_3$ and (d) Yb_2O_3 around Yb, and crystal structures of $SrZrO_3$ and Yb_2O_3 (© IOP Publishing 2001).[28)]

there exist more than two kinds of Yb–O ion pairs having different interatomic distances. Then a least square calculation was performed with assumption that there are two kinds of Yb–O distances in the oxygen octahedron. Yb–O distances are estimated to be 1.96 and 2.18 Å. The Zr–O distance in $SrZrO_3$ measured by X-ray diffraction was 2.09 Å,[30)] which corresponds to a value intermediate between the above-mentioned two values for Yb–O distances and seems to be an average of Yb–O distances in the oxygen octahedron. It suggests two structure models around Yb as shown in Figs. 10.8(a) and 10.8(b).

In model 1, the oxygen octahedron is distorted by substituting a Zr ion for the Yb ion, having a larger ionic radius resulting in two kinds of Yb–O distances. Owing to the large distortion of the oxygen octahedron, the $SrZrO_3$ crystal surrounding the octahedron is distorted and the crystalline order of the second-nearest neighbors is lost. In the model 2 the Yb ion is located at an off-center position in the oxygen octahedron as the case for the Nb ion in perovskite $KNbO_3$,[31)] resulting in two kinds of Yb–O distances. In this case, the Yb ion is located randomly at off-center positions, yielding two Yb–O distances, so that the radial structure function was observed as if the order is lost at the second-nearest neighbors. Thus information from EXAFS gives the structure around Yb ion but it is not clear whether the Yb ion itself is located randomly or the surrounding structure is random. Therefore, it should be clarified by other kinds of experiments. In both models, there exist large local distortions of the crystal, which generate deep potentials preventing free proton diffusion.

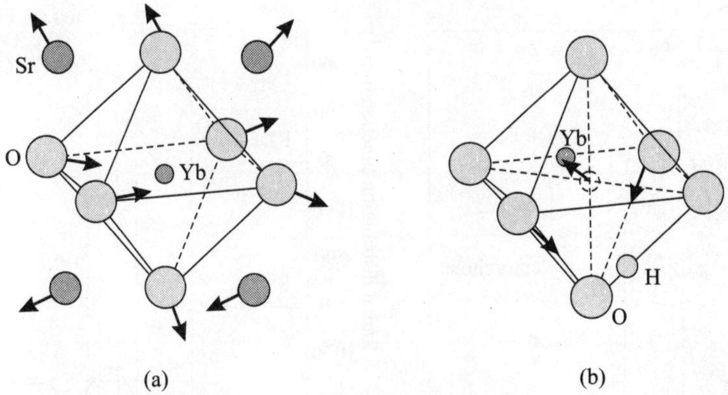

Figure 10.8 Oxygen octahedron models of $SrZr_{1-x}Yb_xO_3$ around Yb. (a) model 1, (b) model 2.

10.4 A COMPARISON BETWEEN EXAFS AND X-RAY DIFFRACTION METHODS IN THE ANALYSIS OF NON-CRYSTALLINE MATERIALS

X-ray diffraction is the method of choice for the analysis of the periodic structure of crystals. On the other hand, EXAFS is a powerful and versatile tool for the analysis of local structure of even non-crystalline or amorphous materials around specific atoms because EXAFS possesses element selectivity. However X-ray diffraction is needed to find the mean distances between atoms in non-crystalline materials and also can have element selectivity if X-ray anomalous dispersion is applied. What therefore in the structure analysis of non-crystalline materials are merits and demerits of EXAFS and X-ray diffraction? The two methods are contrasted below.

Figures 10.9(a) and 10.9(b) show the results of the structure analysis of Zn ferrite glass by EXAFS and X-ray diffraction, respectively.[32)] Figure 10.9(a) shows the radial structure functions obtained by Fourier transformation of Fe and Zn K absorption spectra. The radial structure functions of the Zn ferrite crystal obtained by EXAFS in Fig. 10.9(a) are the same as those of sample (1) in Figs.10.3(b) and 10.4(b). As seen in Fig. 10.9(a), once the sample becomes amorphous, the peaks corresponding to the distances of Zn–O and Fe–O remain, but those of Zn–Fe and Fe–Fe/Zn disappear. This implies that the first-nearest neighbor atom of Zn and Fe is O in the amorphous, but that the second-nearest neighbor atom cannot be assigned. Fig. 10.9(b) shows the environmental radial distribution function around Fe and Zn obtained from X-ray diffraction using anomalous dispersion at the low energy side of the Fe and Zn K edges. In Fig. 10.9(b) the first-nearest neighbor atom of both Zn and Fe is O, which is the same result obtained using EXAFS, although, in addition, a peak which indicates the existence of the second-nearest neighbor atoms appears clearly. In the figure the pairs of atoms for which interatomic distances are included in the width of the second peak are shown by vertical bars. Note that the position of the peak in the radial structure function of the EXAFS and that of corresponding peak in the radial distribution function obtained by the X-ray diffraction do not exactly coincide, probably because the former was a preliminary result calculated by Fourier transformation only.

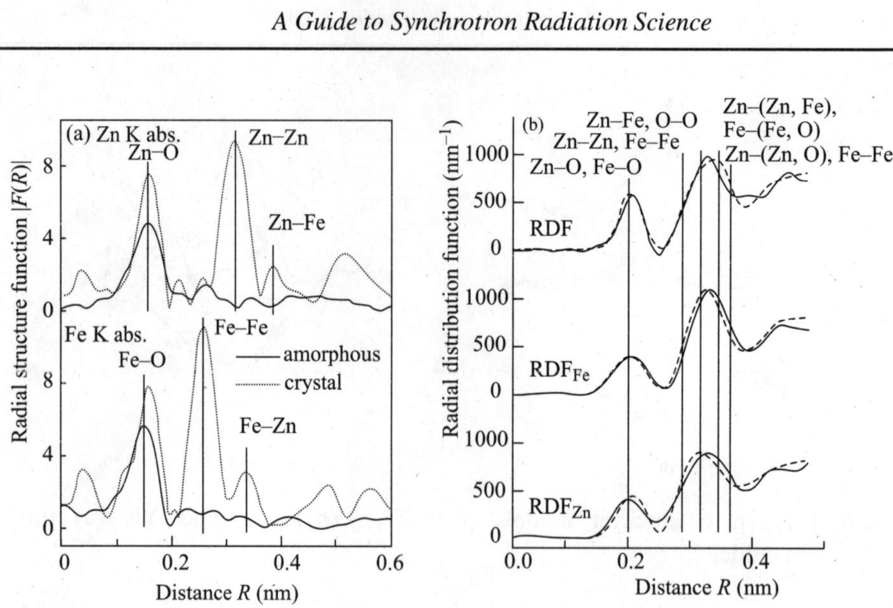

Figure 10.9 (a) Radial structure functions obtained by EXAFS and (b) environmental radial distribution function obtained by X-ray diffraction utilizing anomalous dispersion, around Fe and Zn of Zn ferrite glass.

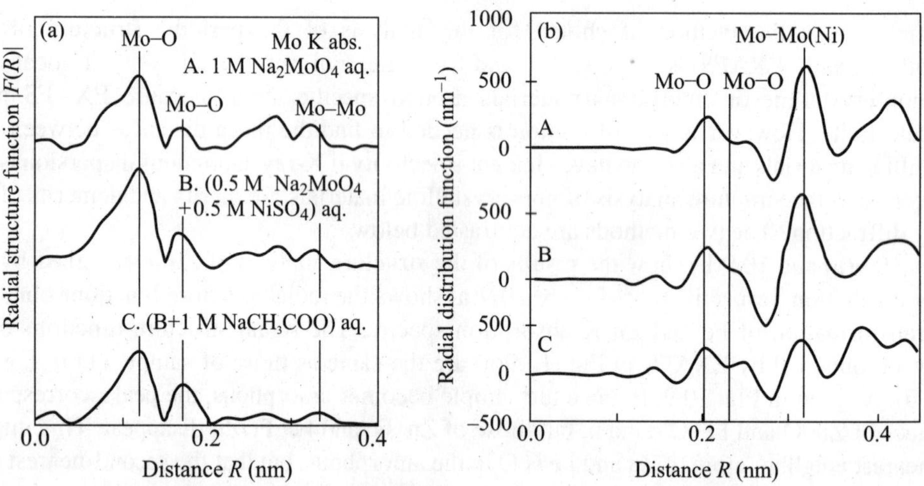

Figure 10.10 (a) Radial structure functions obtained by EXAFS and (b) environmental radial distribution function obtained by X-ray diffraction utilizing anomalous dispersion, around Mo ion. Samples are aqueous solutions of A : 1 M Na_2MoO_4, B: 0.5 M Na_2MoO_4 + 0.5 M $NiSO_4$ and C: B + 1 M $NaCH_3COO$.

Figures 10.10(a) and 10.10(b) show radial structure function and environment radial distribution function around Mo ion in aqueous solutions including sodium molybdate.[33] As seen in the figures, the EXAFS gives information about the position of the first-nearest neighbor more clearly than does X-ray diffraction, while the latter gives information about atoms at greater distances more clearly than the former. This is explained as follows. In the case of the EXAFS, the interference of electron waves is measured and it is easy to analyze local structure of a selected atom and its first-nearest neighbors, even though the concentration of the atoms is very small. However it is difficult to get structural information concerning the second-nearest neighbors and more distant atoms in non-crystalline materials. But in the case of the X-ray diffraction, which measures the interference of X-rays, it is rather easy to get structural information such as distances and the correlations between atoms at larger distances, even though the detectability is low compared with that of the EXAFS.[34] Therefore the advantages of the two methods are rather different but they are complementary.

10.5 METALLOPROTEINS

Metalloproteins containing Fe, Cu, Mn, Ni, Mo *etc.*, support various physiological functions, such as catalytic oxidation-reduction chemistry, oxygen transfer, electron transfer, sensing *etc.* From configurations of atoms in a unit cell determined by X-ray crystallography on single crystals, structures of proteins have been resolved at atomic levels as described in Sect. **9.6**. Further information around any metal site could be obtained by XANES and EXAFS. Here a study on myoglobin is presented as an example of a study of Fe proteins by X-ray crystallography, XANES and EXAFS.[35] Myoglobin is a protein having a heme similar to hemoglobin. In the heme, an Fe atom is located at the center of a planar porphyrin ring with four pyrrole N atoms coordinated to the Fe atom. In addition, the Fe atom is bound to axial ligands with two bonds vertical to the plane of the porphyrin. One of the axial ligands is a histidine residue, the N atom of which directly binds to the Fe atom, and the other a H_2O molecule the O atom of which binds to the Fe atom with one of the H atoms forming a hydrogen bond with a histidine residue nearby.

Figure 10.11 shows a stereo view of the myoglobin structure determined by X-ray crystallography (H_2O is omitted from the figure for greater clarity). The figure displays two myoglobin molecules. One is the wild-type pig myoglobin, which is the same as the native pig myoglobin molecule but expressed in and purified from *E. coli* with the native pig myoglobin gene. The other is one genetically modified at the gene level so that the positions of histidine at the amino acid sequence position 64 and valine at 68 are reversed, termed as H64V/V68H. In the figure, the thin lines represent the structure of the wild-type and the thick lines H64V/V68H. The X-ray crystal structures show changes in the coordination structure and the position of Fe associated with the amino acid replacements. In the double mutant, H64V/V68H, histidine now at position 68 directly binds to the Fe atom and pushes out the H_2O that is the original axial ligand bound to the Fe atom in the wild-type protein. It then inhibits oxygen storage, which is the primary function for the myoglobin.

XANES and EXAFS studies using Fe K absorption also provide details of the coordination structure. The measurements were made by the fluorescence yield method at 10K using a double crystal monochromator mounted by Si (220) crystal at Stanford Synchrotron Radiation Laboratory. XANES has shown that the valence of Fe is 3 (trivalent), as expected from the results of other spectroscopic methods. The EXAFS analysis, whose resolution is about 0.11 Å, indicates that the distance of Fe-surrounding N is 2.01 Å and that the difference in the average distances of Fe-histidine N between the two axial

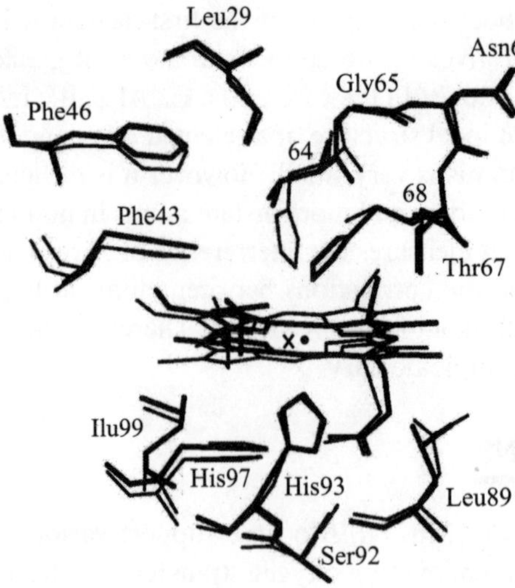

Figure 10.11 Stereo view of pig myoglobin from above the plane of porphyrin. H_2O molecules are omitted. Thin lines: wild-type, thick lines: double mutant H64V/V68H. × and • represent Fe atoms in wild-type and mutant, respectively (© The American Society for Biochemistry and Molecular Biology 1995).[35]

histidine ligands is less than 0.11 Å. These features are not apparent in the 2.07 Å resolution X-ray crystal structure of the mutant.

XANES and EXAFS are now extensively used to establish the detailed coordination structures of many metalloproteins of physiological importance so greatly assisting us in the understanding of relationships between structure and biological function.[36,37]

References

1. L. V. Azaroff: Rev. Mod. Phys. **35** (1963) 1012.
2. D. E. Sayers, F. W. Lytle and E. A. Stern: Advances in X-Ray Analysis **13** (1970) 248.
3. For instance, "EXAFS Spectroscopy. Techniques and Applications", B. K. Teo and D. C. Joy eds. (Plenam, 1981).
4. D. C. Koningsberger and R. Prins eds: "X-Ray Absorption: Principles, Applications, Techniques of EXAFS, SEXAFS, and XANES", Chemical Analysis **92** (John Wiley & Sons, 1988).
5. J. M. Arber, B. R. Dobson, R. R. Eady, P. Stevens, S. S. Hasnain, C. D. Garner and B. E. Smith: Nature **325** (1987) 372.
6. S. D. Kelly, D. Hesterberg and B. Ravel: "Analysis of Soils and Minerals Using X-Ray Absorption Spectroscopy" in Methods of Soil Analysis, Part 5 –Mineralogical Methods, A. L. Ulery and L. R. Drees eds. (Soil Science Society of America, 2008.) Chap. 14.

7. Int. Conf. on X-Ray Absorption Fine Structure has been held every 3 years. The proceedings are published in J. Synch. Rad. **8** (2001), Phys. Scr. T **115** (2005), AIP. Conf. Proc. **882** (2007), J. Phys.: Conf. Ser. **190** (2010), J. Phys.: Conf. Ser. **430** (2013) *et al.*
8. Moniek Tromp: "Developments of Time-Resolved XAFS Spectroscopy Techniques – Applications in Homogeneous Catalysis" (Ponsen & Looijen, 2007), http://eprints.soton.ac.uk/id/eprint/47084.
9. P. A. Lee: Phys. Rev. B **13** (1976) 5261.
10. T. Ishii: Prog. Theor. Phys. **72** (1984) 412.
11. As for use of FEFF, see http://leonardo.phys.washington.edu/feff/.
12. W. Shumann: "Gemstones of the World" 3rd ed. (Sterling, 2006) p. 116.
13. V. Konig, E. F. Bertaut, V. Gros, M. Mitrikov and G. Chol: Solid State Commun. **8** (1970) 759.
14. F. K. Lotgering: J. Phys. Chem. Solid **27** (1966) 139.
15. S. Ligenza: Phys. Status Solidi B **75** (1976) 315.
16. T. Kamiyama, K. Haneda, T. Sato, S. Ikeda and H. Asano: Solid State Commun. **81** (1992) 563.
17. B. Jeyadevan, K. Tohji and K. Nakatsuka: J. Appl. Phys. **76** (1994) 6325.
18. A. L. Ankudinov, B. Ravel, J. J. Rehr and S. D. Conradson: Phys. Rev. B **58** (1998) 7565.
19. N. Happo, H. Sato, T Mihara, K. Mimura, S. Hosokawa, Y. Ueda and M. Taniguchi: "Zn, Mn and Te K-edge EXAFS studies of the diluted magnetic semiconductor $Zn_{1-x}Mn_xTe$", J. Phys.: Condens. Matter **8** (1996) 4315.
20. R. N. Mayanovic, W.-F. Pong and B. A. Bunker: Phys. Rev. B **42** (1990) 11174.
21. J. C. Mikkelsen and J. B. Boyce: Phys. Rev. B **28** (1983) 7130.
22. T. Murata, T. Matsukawa and S. Naoe: Solid State Commun. **66** (1988) 787.
23. K. Toriumi, T. Kanao, Y. Umetsu, A. Ohyoshi, M. Yamashita and T. Ito: J. Coord. Chem. **19** (1988) 209.
24. M. Yamashita, K. Toriumi and T. Ito: Acta Cryst. C **41** (1985) 876.
25. H. Iwahara, T. Esaka, H. Uchida and N. Maeda: Solid State Ionics **3/4** (1981) 359.
26. S. Shin, H. H. Huang, M. Ishigame and H. Iwahara: Solid State Ionics **40/41** (1990) 910.
27. R. Hempelmann, Ch. Karmonik, Th. Matzke, M. Cappadonia, U. Stimming, T. Springer and M. A. Adams: Solid State Ionics **77** (1995) 152.
28. O. Kamishima, K. Ohta, Y. Chiba and T. Hattori: "Local structure around Yb in $SrZr_{1-x}Yb_xO_3$", J. Phys.: Condens. Matter **13** (2001) 2455.
29. C. E. Bouldin, E. A. Stern, B. von Roedern and J. Azoulay: Phys. Rev. B **30** (1984) 4462.
30. A. Ahtee, M. Ahtee, A. M. Glazer and A. W. Hewat: Acta Cryst. B **32** (1976) 3243.
31. R. Comes, M. Lambert and A. Guinier: Acta Cryst. A **26** (1970) 244.
32. Y. Waseda, E. Matsubara, K. Okuda, K. Omote, K. Tohji, S. N. Okuno and K. Inomata: J. Phys.: Condens. Matter **4** (1992) 6355.
33. K. Shinoda, E. Matsubara, M. Saito, Y. Waseda, T. Hirato and Y. Awakura: Z. Naturforsch. **52a** (1997) No.12, 855.

34. E. Matsubara and Y. Waseda: "Resonant Anomalous X-Ray Scattering, Theory and Applications", G. Materlik, C. J. Sparks and K. Fischer eds. (Elsevier Science, 1994) p. 345.

35. Y. Dou, S. J. Admiraal, M. Ikeda-Saito, S. Krzywdal, A. J. Wilkinson, T. Li, J. S. Olson, R. C. Prince, I. J. Pickering and G. N. George: "Alteration of axial coordination by protein engineering in myoglobin. Bisimidazole ligation in the His^{64} → Val/Val^{68} → His double mutant", J. Biol. Chem. **270** (1995) 15993.

36. T. Glaser, B. Hedman, K. O. Hodgson and E. I. Solomon: Acc. Chem. Res. **33** (2000) 859.

37. R. A. Scott: "Physical Methods in Bioinorganic Chemistry –Spectroscopy and Magnetism", L. Que ed. (University Science Books, 2000) Chap. 9.

CHAPTER 11 Elemental and Structural Analysis using X-Ray Fluorescence

Following irradiation with X-rays there are interactions in which X-rays are emitted with energies differing from those of the incident X-rays. These interactions also differ from those described in the previous two chapters. One mechanism is in the emission of X-ray fluorescence described here and a second is the inelastic scattering of X-rays to be considered in the next chapter. Once X-rays have been absorbed by any material, X-ray fluorescence may be produced when some of the absorbed energy is released. It is created when an electron in an outer shell moves to a core hole created by the absorption event. It generates characteristic X-rays whose energies are element-specific. Two fields of X-ray fluorescence research are described.

The first is X-ray fluorescence analysis which has long been important as an analytical tool. Using the technique it is possible to identify many elements in trace amounts and to assess their quantity. The second is the field of X-ray fluorescence holography in which it has become possible to observe interference patterns derived from the fluorescence emitted by specific atoms and which is subsequently scattered by the surrounding atoms, thus making it possible to establish the spatial atomic arrangement.

11.1 X-RAY FLUORESCENCE ANALYSIS

11.1.1 Overview

The characteristic X-rays emitted from any sample make it possible to identify all the elements contained in the sample and also to determine the sample elemental concentrations down to extremely low levels. Beams of X-rays or charged particles such as electrons and ions are usually used as excitation sources for the generation of these characteristic X-rays. Among types of commercial equipment which can be used there is EPMA (or 'electron probe microanalysis') in which an electron beam is used to generate the characteristic X-rays. In this case, all samples must be contained in a vacuum. In X-ray fluorescence analysis when X-rays are used as the excitation source, samples can be studied in air at atmospheric pressure. The analysis is also non-destructive, because radiation damage to the samples is minimal. The equipment for such analysis comprises an X-ray source, X-ray optical system, sample stage and an X-ray energy analyzer. In the laboratory one can always use an X-ray tube, but it is then difficult to select the energy of the excitation X-rays. It is obviously a major advantage to be able to undertake X-ray fluorescence analysis using synchrotron radiation because it is a simple matter to select the appropriate energy X-rays to excite any chosen element. Furthermore, the high brilliance and parallelism of synchrotron radiation makes 'super-microanalysis' possible using the technique of total reflection X-ray fluorescence analysis, in which the X-rays are incident at grazing angles and the fluorescence is detected at large angles. In the following method X-rays are incident normal to the sample and the fluorescence detected at grazing angles. The use of an X-ray microbeam combined with detection of the X-ray fluorescence obtained while scanning across the sample will yield an element-specific image of the sample giving element mapping or chemical mapping and this technique will be described in Sect. **14.2.2**. The use of high energy synchrotron radiation sources makes the X-ray fluorescence analysis of heavy elements perfectly feasible, a technique which is extremely difficult to undertake in the laboratory.

Two kinds of apparatus are used for X-ray fluorescence analysis, the wavelength dispersion type and the energy dispersion type. The wavelength dispersion type uses a monochromatizing crystal (see Sect. **5.3.2.c**) and has the advantage of high energy-resolution. In this way, it is feasible to investigate the chemical state of the sample by analyzing the spectral shape and also the energy shift of all spectral structure in detail. The energy dispersion method employs detectors which can measure X-ray energies directly (see Sect. **5.6.1**) and therefore has the advantage of high detection efficiency and compactness since it is relatively small. The energy resolution is fundamentally restricted by the detector to around 130 eV and it is always necessary to cool the detector using liquid nitrogen to minimize the noise. Recently, silicon drift detectors (SDD) with very low noise have been developed.[1)] These detectors are used at temperatures around –20°C, temperatures which are easily achieved using a Peltier cooling device and they can then detect high energy X-rays at high count rates without count loss. This is clearly a great asset at synchrotron radiation facilities. At present a band width of almost 140 eV is achievable, so that SDD are ideal for use in X-ray fluorescence analysis.

The intensity of the X-ray fluorescence is proportional to that of the incident X-rays and also to the concentration of the element generating the fluorescence. Following excitation by X-rays all absorbed energy is not totally converted to X-ray fluorescence, but part of the energy will go to generate Auger electrons in the relaxation process. As stated in Sect. **4.3.4**, the yield of X-ray fluorescence is high in atoms of heavy elements and the yield is low in the light elements. The minimum detection limit (MDL) for X-ray fluorescence is determined by the relation between the intensity of the signal and the background. It is clearly important in the assignment of dilute elements, to be able to maximize the signal intensity and to minimize the background. It is important in all quantitative measurements of elements to take into account absorption effects of incident X-rays and of the X-ray fluorescence from the sample. The contribution of secondary excitation by X-ray fluorescence emitted from other elements must also be considered. The use of synchrotron radiation to maximize the intensity of X-ray fluorescence is certainly an advantage but the background problem will not necessarily be resolved. The total reflection method described in the next section is very effective in reduction of the background.

There are various procedures for X-ray fluorescence analysis used in different fields, which are still developing.[2)] Three of them will be described to illustrate the significant advantages obtained using synchrotron radiation.

11.1.2 Total reflection X-ray fluorescence

Total reflection will occur when X-rays are incident on the surface of any material at small grazing angles (see Sect. **4.1.2**). The critical angle for total reflection is dependent on the optical constant of the material at the energy of the incident X-ray. For instance in the case of X-rays of energy 17.4 keV, the critical angle is 0.1° (the grazing angle) for Si. The penetration depth of X-rays under this total reflection condition is several nm, so that X-ray fluorescence analysis at the surface is possible by detecting fluorescence from the few nm deep region. In this analysis only the fluorescence emitted from those atoms close to the surface is measured effectively and the background from bulk is minimal. This technique is called 'total reflection X-ray fluorescence' (TXRF).[3] Since the critical angle for total reflection depends on the energy of the incident X-ray it is clearly effective for suppression of the background to use monochromatic X-rays which give rise to total reflection.[4] For this purpose, the angle of incidence should be controlled within an angular accuracy better than 0.01°. Clearly the exceptional brilliance and high degree of collimation of synchrotron radiation is ideal for TXRF analysis.

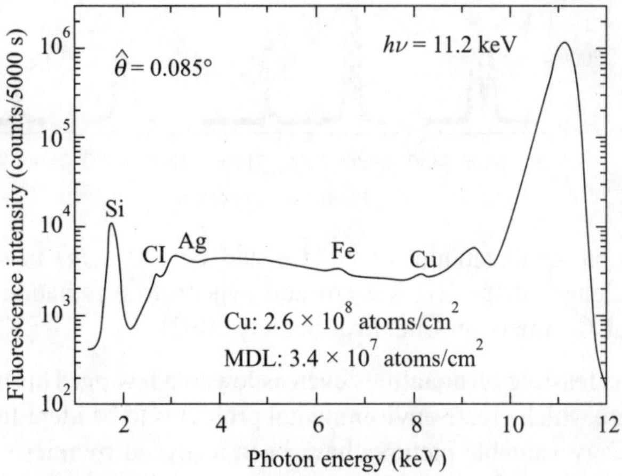

Figure 11.1 An example of TXRF on a Si wafer using synchrotron radiation (by courtesy of K. Baur, © American Chemical Society 2002).[5]

TXRF has been frequently applied in the analysis of impurities on Si wafers. At SSRL TXRF has been performed using SPEAR as light source and an energy dispersive detector. For the impurity atom Ni on a Si wafer, an MDL of 10^7 atoms/cm^2 has been achieved. This value is about 50 times more sensitive than can be obtained in the laboratory.[5] Figure 11.1 shows the fluorescence spectrum of a clean surface of a Si wafer. Peaks can be observed which originate from small amounts of the contaminants, Cu and Fe, both of which are difficult to measure in laboratory. The area density of the contaminant was 2.6×10^8 atoms/cm^2 for Cu and 6.4×10^8 atoms/cm^2 for Fe.

When the X-ray fluorescence is analyzed using wavelength dispersive detection systems, a further reduction of background and separation of overlapping fluorescence peaks becomes possible, resulting in an analysis with even greater sensitivity and improved element selectivity.[6,7] Figure 11.2 shows an X-ray fluorescence spectrum measured using a TXRF apparatus equipped with a wavelength dispersive system of the Johansson type attached to a synchrotron radiation beamline. The sample was a Si wafer covered by a dropped 0.1 μ*l* solution. As seen in the figure, the fluorescence peaks are well separated. The MDL for Ni was 0.3 fg (3.1 ppt, 0.1 μ*l*).[7] Such methods for the microanalysis of light elements have also been developed at SSRL and BESSY II and an absolute value of the MDL smaller than 100 fg has been achieved for Na, for example.[8]

In addition to the surface analyses of Si wafers the TXRF method can be used for the analysis of micro or even nano samples. In such cases samples must be prepared on a clean flat substrate, for example a metallic sample can be prepared from a droplet of its acid solution placed on the substrate

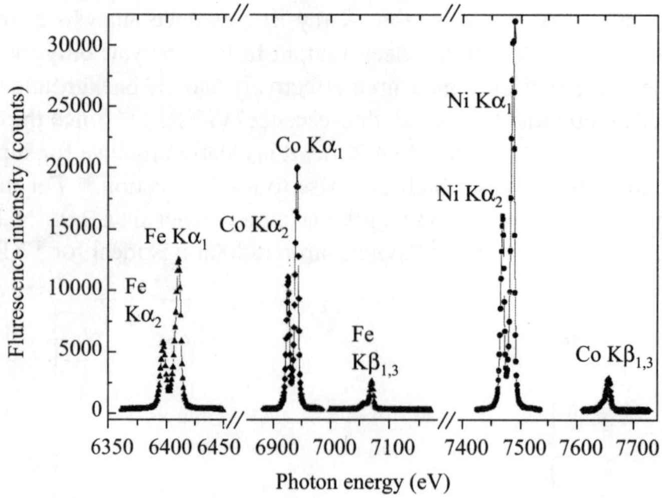

Figure 11.2 TXRF on small amount of Ni, Co and Fe (20 ng/g) in 0.1 μ*l* droplet using a wavelength dispersive system and synchrotron radiation (by courtesy of K. Sakurai, © American Chemical Society 2002).[7]

and then dried. Analysis is feasible on quantities even as low as a few pg. This has enabled small floating particles in the atmosphere which create environmental problems to be identified and assayed. Contents of paints used in historically valuable pictures have been analyzed by micro-sampling. By measuring the angular dependence of fluorescence intensity, an analysis of the surface regions of bulk materials and of thin films in a direction normal to the surface becomes feasible. The applications of the TXRF method have become very extensive.[9–11]

11.1.3 Small volume analysis experiments undertaken by X-ray fluorescence emitted at grazing angles

TXRF employs irradiation by X-rays at grazing incidence to the sample surface. An alternative method is to detect X-ray fluorescence from the sample surface at grazing angles,[12,13] when using irradiation by the X-ray microbeam at normal incidence to the sample. This permits the analysis of small areas of bulk materials and thin films. The angular dependence of X-ray fluorescence intensity is measured by rotation of the detector. Figure 11.3 shows the angular dependence of the Ni Kα line emitted from Ni films on a Si wafer (film thickness: 20 and 50 nm) as an example.[14] Characteristic structures which depend on the film thickness are clearly visible. The oscillatory structure observed in the angle dependence curve of the 50 nm thick sample is due to an interference effect. It is thus possible to estimate the actual thickness and the density of the film from such data.

Information below the surface layer is obtained by changing the grazing angle. Combining this feature with two-dimensional scanning of the microbeam, makes possible, in principle, a full three-dimensional analysis of the region around the surface.[15] Many reports are now available on X-ray fluorescence analysis of small regions.[16,17]

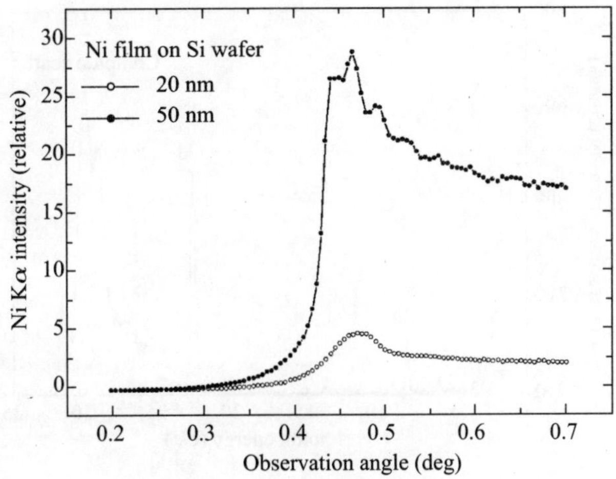

Figure 11.3 Dependence of intensity of Ni Kα line emitted from Ni films on Si wafer on the grazing exiting angle.

11.1.4 The X-ray fluorescence analysis of heavy elements

In the laboratory, the analysis of heavy elements is usually made using the characteristic X-ray line from Mo at 17.5 keV and then detecting the L lines which appear below 15 keV. The L lines of the heavy elements and the K lines of the light elements often overlap in samples containing both elements. In such cases, the fluorescence analysis is difficult. However, the analysis can be made using K lines of the heavy elements located in the higher energy region.[18–21] The energy of K lines of the heavy elements are separated from those of the lighter elements, so that the problem of overlap of fluorescence lines will not occur.

The method of using high energy X-rays is advantageous for the assignment and quantitative study of heavy elements, such as Cd, Hg, Pb *etc.* Figure 11.4 shows an example in which the Cd (K line: 23 keV) was detected in the kidney tissue of a patient of *Itai-itai* disease (a painful disease of bone caused by a poisonous Cd compound) using X-rays of 32 keV.[22]

Following excitation using 116 keV X-rays, more than 30K lines of heavy elements have been found from the X-ray fluorescence spectrum of the standard sample of glass (SRM612, NIST).[20] The MDL of 0.1 ppm is estimated from the analysis on W of 1.7 ppm on a standard sample of rock (JG-1).[20] High energy X-ray fluorescence analysis has even been used in criminal forensic investigations, for instance in the quantitative assay of trace heavy elements such as Sb, Bi, Sn and Mo contained in arsenious acid reagent in attempts to identify their place of origin and production.[21]

X-ray fluorescence analysis using synchrotron radiation is extremely sensitive, so that the samples can be small. It is also a non-destructive method, and the sample is preserved and can be studied repeatedly. In order to assign the origin of archaeological artifacts unearthed at archaeological sites, it is important to analyze their heavy element content, which can give information on the location of the site of origin for the raw materials used and sometimes even of the fabrication methods used in their production. Figure 11.5 shows the X-ray fluorescence spectrum from part of an old *Kutani* pottery.[21] By a fluorescence intensity comparison of heavy elements, such as Ba/Ce-Nd/Ce in Fig. 11.5, a distinction between the production sites of unearthed pottery such as *Kutani*, *Arita*, *Himetani* and other places becomes possible.

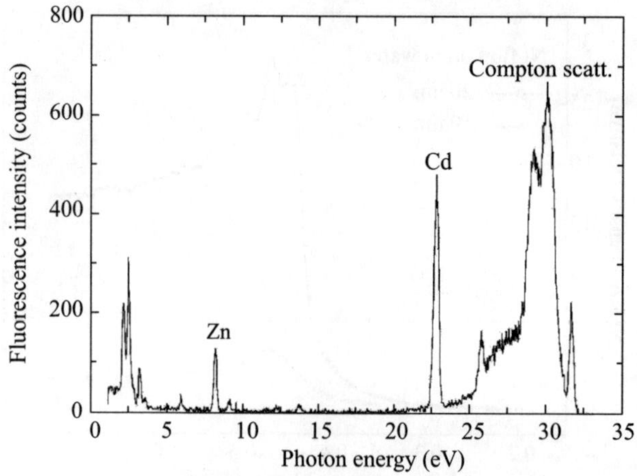

Figure 11.4 X-ray fluorescence spectrum of kidney tissue of a patient of *Itai-itai* disease excited by 32 keV X-rays (by courtesy of S. Hayakawa and SPring-8).[22)]

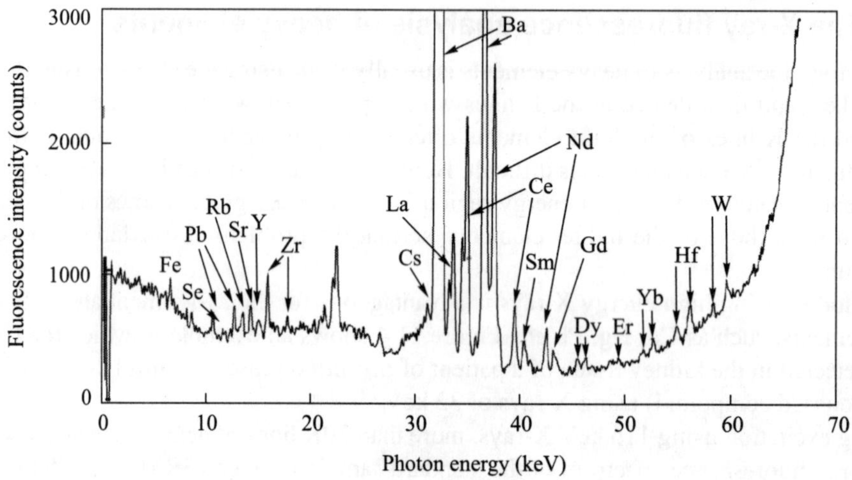

Figure 11.5 X-ray fluorescence spectrum of a piece of pottery unearthed from No. 1 kiln at *Kutani* excited by 115 keV X-rays (by courtesy of I. Nakai, © John Wiley & Sons 2004).[21)]

11.2 X-RAY FLUORESCENCE HOLOGRAPHY

11.2.1 Principle

Before X-ray fluorescence exits a material, a small part of the fluorescence signal is scattered. The non-scattered and the scattered X-ray fluorescence components can be expressed in terms of a reference wave and an object wave emitted from the atom as a point source scattered by objects, as shown in

Fig. 11.6 (a). As a result of their interference, a very weak hologram pattern is formed, which contains information about the detailed arrangement of atoms around the atom emitting the fluorescence.[23] A three-dimensional Fourier-transform-like analysis of the pattern gives an image of the atomic structure around the specific atom.[24] This kind of structural analysis is called X-ray fluorescence holography. To obtain the hologram pattern, the sample should be single-crystal-like around the specific atom emitting the fluorescence, with the lattice direction being well-defined although, it does not mean that a perfect single crystal is necessary. This method has the advantage that it is applicable to samples which show a middle-range order for their atomic arrangement. Although X-ray fluorescence holography was first demonstrated only in 1996,[25] its application has grown steadily and has now spread to the local structure analyses of dopants,[26] films,[27–29] quasicrystals[30] *etc*. The most recent development is given in Ref. 31.

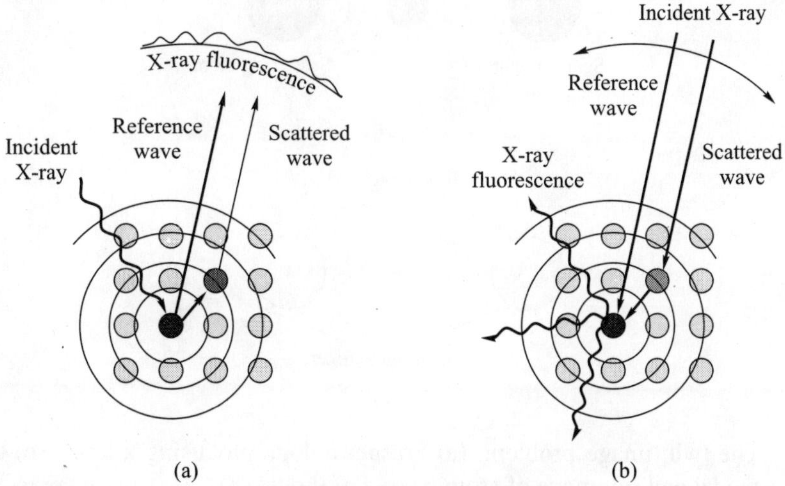

Figure 11.6 The principle of X-ray fluorescence holography. (a) Normal type and (b) inverse type.

In holography using lasers what we see is the real image which is a part of the reconstruction image. In addition, an image called conjugated image appears in different space. This is not a problem in optical holography, because the real and the conjugated images are reconstructed in separate spaces as shown clearly in Fig. 11.7(a). However, X-ray fluorescence holography is a method which records information from all atoms around the specific atom. Therefore, in the case of three atoms (one emitter and two scatterers) shown in Fig. 11.7(b), the conjugated and the real images overlap as shown in Fig. 11.7(c). Therefore the image of the atomic arrangement is often distorted, especially in the case of crystals with complicated structures and this is called the 'twin image' problem. It is a major obstacle when the atomic location has to be determined precisely. One of the most effective ways to resolve twin image problem is to record holograms using light of several wavelengths for processing or removal of the conjugated image contributions.[32,33] It is important to obtain as many holograms as possible with an energy span of about 2 keV, where the central energy of the range is chosen for each experiment according to the X-ray fluorescence efficiency.

The method shown in Fig. 11.6(a) exploits the interference of X-ray fluorescence and is called 'normal X-ray fluorescence holography'. In this case, characteristic X-ray lines such as Kα and Kβ are

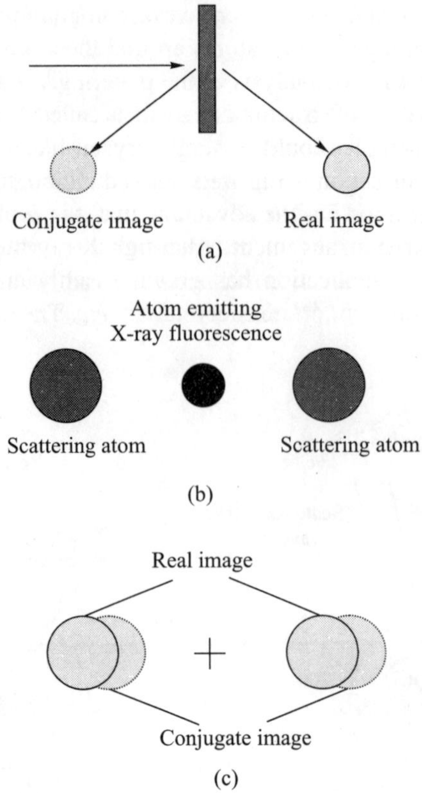

Figure 11.7 The twin image problem. (a) Fresnel holography using a laser, (b) three-atom model and (c) image of atomic array of model (b).

observed, so that the number of the X-ray fluorescence lines is limited and removal of the conjugated image may be incomplete. There is also 'inverse X-ray fluorescence holography' given by analogy with the optical reciprocity theorem.[34] As illustrated in Fig. 11.6(b), an X-ray incident on any material and the X-rays scattered from atoms in the material excite atoms emitting X-ray fluorescence whose intensity reflects the interference effect between the incident and scattered X-rays. Therefore in the inverse type, the intensity change in the total yield of the X-ray fluorescence will create the hologram pattern when the direction of the incident X-ray is scanned. Using synchrotron radiation it is possible in principle to choose any number of X-ray wavelengths so that it should be feasible to reduce the intensity of the conjugated image to a negligible level. Of course, when incident X-rays are diffracted by a crystal lattice, a so-called X-ray standing wave line appears in the hologram.[35] It corresponds to a Kossel line, in the normal type, originating from X-ray fluorescence diffracted by the crystal lattice.

The intensity of the hologram pattern is about 0.1% of the background. Therefore, X-ray fluorescence should be detected for each measuring point (angle) up to at least by 10^6 counts to reduce the statistical error to less than 0.1%. Obviously many tens of thousands of points are needed to achieve a perfect hologram pattern. Accordingly, it is necessary both to use high brilliance X-rays and also to improve the detection efficiency in order to complete the measurement in a reasonable time (from several hours to days). When using conventional X-ray semiconductor detectors, a few months may be necessary for the measurement even when using high brilliance synchrotron radiation, because the upper limit of their count rates is only about 5000 cps.

Alternatively an avalanche photodiode (APD) can be used which is able to detect up to 10^2 Mcps.[36) In this case, a monochromatizing crystal is essential to disperse the X-ray fluorescence because of the absence of energy dispersion in the APD. In the system shown in Fig. 11.8, the fluorescence is dispersed and then focused on the APD using a cylindrical shaped crystal (LiF or graphite). Using such a system it is possible to get a hologram pattern within 2 to 3 hours, so it is then possible to obtain several holograms within a few days. However, the system has the disadvantage that the acceptance angle for fluorescence by the monochromatizing crystal is relatively small in comparison with that of semiconductor detectors. If the crystal shape is changed from cylindrical to toroidal one can obtain greatly enhanced intensity.

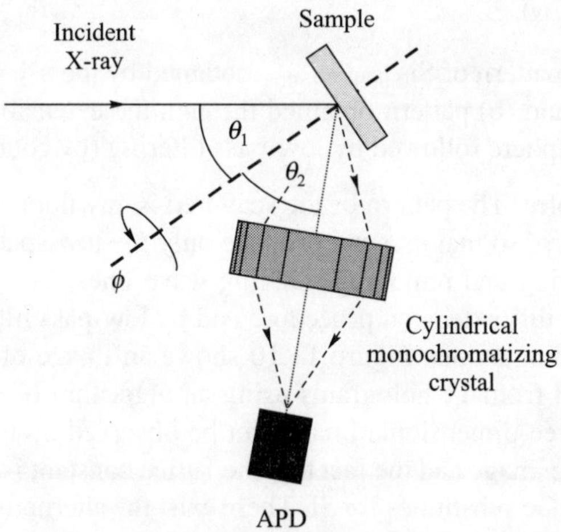

Figure 11.8 Schematic of high-speed detection system for X-ray fluorescence holography. APD: avalanche photodiode.

11.2.2 Local structure around the dopant

Measurement of the local structure around dopants in single crystals is one of the important applications of X-ray fluorescence holography.[28)

In an experiment using a $Si_{0.999}Ge_{0.001}$ single crystal performed in SPring-8,[37) the X-ray fluorescence hologram for Ge was obtained with the condition of $0 \leq \phi \leq 360°$(0.3° steps), $0 \leq \theta_1 \leq 76°$(1° steps) and $\theta_2 = 45°$ using X-rays of from 14.5 to 17.0 keV (0.25 keV steps). The measuring time for each hologram was about 2.5 hours, and a total of 11 holograms were recorded. An example of the hologram pattern recorded using 15.75 keV X-rays is given in Fig. 11.9. In the raw data given in Fig. 11.9(a), the length of the pattern in the k_z direction is short compared with those in k_x and k_y directions. Therefore the resolution of the image of the atomic array along the z-axis is poor and any precise estimate of the local lattice distortion is hard to achieve using the raw data directly. In order to reconstruct the image precisely the data must be extended to a perfect sphere. This can be achieved by transferring some raw data to the region of no data, guided by the positions of X-ray standing wave lines, which in the case of normal type, is the Kossel line, and the crystal symmetry. However, this extension is limited to materials of good symmetry with long-range periodicity, and

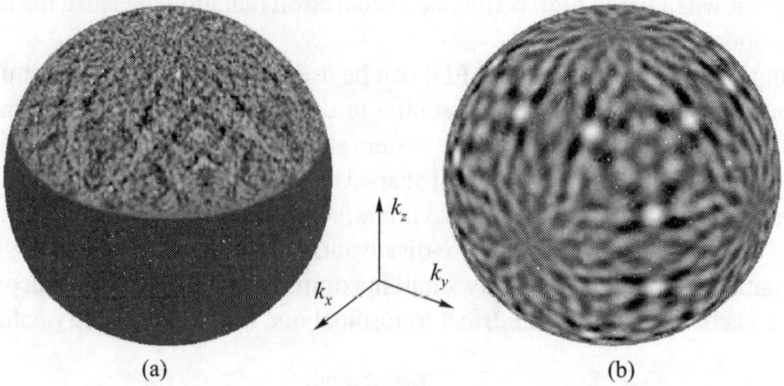

Figure 11.9 Hologram pattern of $Si_{0.999}Ge_{0.001}$ obtained by the use of 15.75 keV X-rays. (a) Raw data and (b) pattern obtained through the extension procedure of raw data to perfect sphere followed by low-pass filtering (by courtesy of K. Hayashi).[37)]

cannot be applied to all samples. The pattern of the scattered X-ray fluorescence is broad, while that of the standing wave line is sharp, so that one can produce only the low spatial frequency contribution in the pattern by low-pass filtering and removing standing wave lines. The pattern given in Fig. 11.9(b) is a hologram obtained using this extension procedure and by low-pass filtering. It shows contributions to the scattering by surrounding atoms. Figure 11.10 shows an image of the atomic structure around Ge, which was reconstructed from 11 holograms using an algorithm of multi-energy holography. As seen in the figure, a clear three-dimensional image can be observed up to the fourth nearest neighbor atoms. From the pattern of the image and the fact that the lattice constant is very close to that of Si single crystal, it was concluded that Ge substitutes for Si. There exist the alternative techniques of transmission electron microscopy (TEM), scanning probe microscopy (SPM) and atomic force microscopy (AFM)

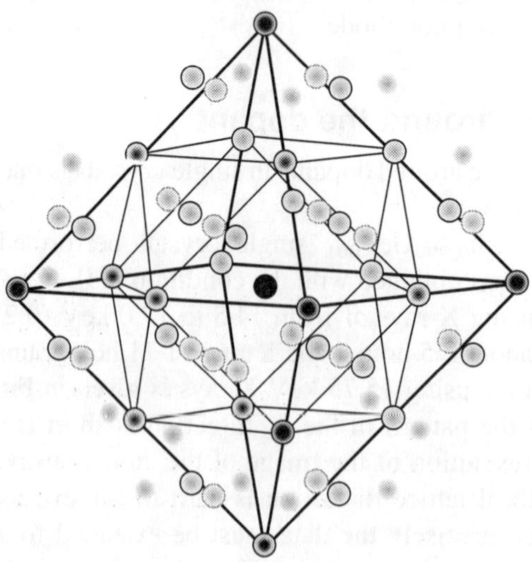

Figure 11.10 Image of atomic structure around Ge in $Si_{0.999}Ge_{0.001}$ (by courtesy of K. Hayashi).[37)]

as general tools by which atoms can be observed directly. Transmission images of thin samples can be obtained by the use of TEM, while profiles of surfaces are observed by the use of SPM and AFM. Three-dimensional images of atomic structures within a radius of about 10 Å around the specific atom can be observed with good resolution using X-ray fluorescence holography. Therefore, it is the method of choice for making local structural analyses around impurities in semiconductor single crystals.

REFERENCES

1. P. Lechner, C. Fiorini, R. Hartmann, J. Kemmer, N. Krause, P. Leutenegger, A. Longoni, H. Soltau, F. Stotter, R. Stotter, L. Struder and U. Weber: Nucl. Instr. Meth. A **458** (2001) 281.
2. B. Beckhoff, B. Kanngiesser, N. Langhoff, R. Wedell and H. Wolff eds.: "Handbook of Practical X-Ray Fluorescence Analysis" (Springer-Verlag, 2006).
3. Y. Yoneda and T. Horiuchi: Rev. Sci. Instr. **42** (1971) 1069.
4. A. Iida and Y. Gohshi: Jpn. J. Appl. Phys. **23** (1984) 1543.
5. K. Baur, S. Brennan and P. Pianetta: Anal. Chem. **74** (2002) 609A.
6. N. Awaji, S. Ozaki, J. Nishino, S. Noguchi, T. Yamamoto, T. Syoji, M. Yamagami, A. Kobayashi, Y. Hirai, M. Shibata, K. Yamaguchi, K. Y. Liu, S. Kawado, M. Takahashi, S. Yasuami, I. Konomi, S. Kimura, Y. Hirai, M. Hasegawa, S. Komiya, T. Hirose and T. Okajima: Jpn. J. Appl. Phys. **39** (2000) L1252.
7. K. Sakurai, H. Eba, K. Inoue and N. Yagi: Anal. Chem. **74** (2002) 4532.
8. C. Streli, P. Wobrauschek, P. Kregsamer, G. Pepponi, P. Pianetta, S. Pahlke and L. Fabry: Spectrochim. Acta B **56** (2001) 2085.
9. R. Klockenkämper: "Total Reflection X-Ray Fluorescence Analysis" (John Wiley & Sons, 1997).
10. P. Kregsamer, C. Streli and P. Wobrauschek: "Handbook of X-Ray Spectrometry", 2nd ed., R. Van Grieken and A. A. Markowicz eds. (Marcel Dekker, 2002) Chap. 9.
11. Int. Conf. on Total Reflection X-Ray Fluorescence and Related Methods has been held every 2 years. The proceedings are given in Spectrochimica Acta **58** B (2003) No.12, Spectrochimica Acta **61** B (2006) No.10, 11, Spectrochimica Acta **63** B (2008) No.12, Spectrochimica Acta **65** B (2010) No. 6 *et al.*
12. T. Noma and A. Iida: Rev. Sci. Instr. **65** (1994) 837.
13. K. Tsuji: "X-Ray Spectrometry: Recent Technological Advances", K. Tsuji, J. Injuk and R. Van Grieken eds. (John Wiley and Sons, 2004) Chap. 5.2.
14. K. Tsuji, S. Hayakawa, T. Chou, I. Yamane, N. Kometani and Y. Yonezawa: Nanotechnology Support Project, SPring-8 Activity Report, Vol. 1 (2003) A, p.151, in Japanese.
15. K. Tsuji and F. Delalieux: J. Anal. At. Spectrom. **17** (2002) 1405.
16. K. Janssens, F. Adams and A. Rindby: "Microscopic X-Ray Fluorescence Analysis" (John Wiley & Sons, 2000).

17. A. Rindby and K. Janssens: "Handbook of X-Ray Spectrometry", 2nd ed., R. Van Grieken and A. A. Markowicz eds. (Marcel Dekker, 2002) Chap. 11.
18. V. B. Baryshev, A. E. Gilbert, O. A. Kozmenko, G. N. Kulipanov and K. V. Zolotarev: Nucl. Instr. Meth. A **261** (1987) 272.
19. M. Harada and K. Sakurai: Jpn. J. Appl. Phys. **37** (1998) 2740.
20. I. Nakai, Y. Terada, M. Itou and Y. Sakurai: J. Synch. Rad. **8** (2001) 1078.
21. I. Nakai: "High energy X-ray fluorescence" in "X-Ray Spectrometry: Recent Technological Advances", K. Tsuji, J. Injuk and R. Van Grieken eds. (John Wiley and Sons, 2004) Chap. 5.5.
22. K. Takagawa and S. Hayakawa : SPring-8 Research Frontier 2000/2001, p. 71.
23. A. Szöke: AIP Conf. Proc. **147** (1986) 361.
24. J. J. Barton: Phys. Rev. Lett. **61** (1988) 1356.
25. M. Tegze and G. Feigel: Nature **380** (1996) 49.
26. K. Hayashi, M. Matsui, Y. Awakura, T. Kaneyoshi, H. Tanida and M. Ishii: Phys. Rev. B **63** (2001) R41201.
27. Y. Takahashi, K. Hayashi, E. Matsubara, T. Shima, K. Takanashi, T. Mori and M. Tanaka: Scripta Mater. **48** (2003) 975.
28. K. Hayashi, Y. Takahashi, E. Matsubara, K. Nakajima and N. Usami: J. Mater. Research **18** (2003) 1471.
29. S. Marchesini, M. Tegze, G. Faigel and M. Belakhovsky: Nucl. Instr. and Meth. A **457** (2001) 601.
30. S. Marchesini, F. Schmithüsen, M. Tegze, G. Faigel, Y. Calvayrac, M. Belakhovsky, J. Chevrier and A. Simionovici: Phys. Rev. Lett. **85** (2000) 4723.
31. K. Hayashi, N. Happo, S. Hosokawa, W. Hu and T. Matsushita: J. Phys.: Condens. Matter **24** (2012) 093201.
32. P. M. Len, T. Gog, C. S. Fadley and G. Materlik: Phys. Rev. B **55** (1997) R3323.
33. J. J. Barton: Phys. Rev. Lett. **67** (1991) 3106.
34. T. Gog, P. M. Len, G. Materlik, D. Bahr, C. S. Fadley and C. Sanchez-Hanke: Phys. Rev. Lett. **76** (1996) 3132.
35. M. W. Batterman: Phys. Rev. Lett. **22** (1969) 7038.
36. S. Kishimoto, N. Ishizawa and T. P. Vaalsta: Rev. Sci. Instr. **69** (1998) 384.
37. The original result was reported in K. Hayashi: J. Jpn. Soc. Synch. Rad. Res. **15** (2002) 267, in Japanese, which is cited in K. Hayashi: Adv. Imaging Electr. Phys. **140** (2006) 119.

CHAPTER 12

Inelastic X-Ray Scattering

Compton first described inelastic X-ray scattering (IXS) now known as Compton scattering, in the 1920s. This discovery has played a major role in history of science because it provided a firm basis for the particle concept of electromagnetic waves. Compton scattering has since been the subject of much research. In X-ray diffraction experiments inelastic scattering creates problems due to undesirable background and blurs signals. However inelastic scattering actually provides vital information about the electronic structure of matter because it arises from exchanges of energy and momentum between X-ray photons and electrons. The low intensity of inelastic scattering has so far precluded its wider application but now that synchrotron radiation is widely available, the field of inelastic X-ray scattering has transformed to an area of vigorous growth and development.[1,2] What we can learn from Compton scattering and from X-ray Raman scattering will now be discussed.

12.1 THE PRINCIPLES

In all inelastic scattering processes, the incident monochromatic X-ray beam with energy $\hbar\omega_1$, momentum $\hbar\boldsymbol{k}_1$ and polarization vector $\boldsymbol{e}_1$ is scattered by the sample and the intensity of the scattered beam with energy $\hbar\omega_2$, momentum $\hbar\boldsymbol{k}_2$ and polarization vector $\boldsymbol{e}_2$ is measured at the scattering angle 2θ. The scattering spectrum is obtained as a function of transfer momentum $hq = h\boldsymbol{k}_1 - h\boldsymbol{k}_2$ and energy loss $\Delta E = \hbar\omega_1 - \hbar\omega_2 = \hbar\omega$.[3] In inelastic X-ray scattering spectroscopy, it is the convention to employ atomic units for momentum (a. u.). 1 a. u. $= me^2/\hbar = 1.993 \times 10^{-24}$ kg · m/s. Since $p = \hbar k$ and $k = 2\pi/\lambda$, 1 a. u. corresponds to a wavelength of 3.32 Å or to 18.9 nm^{-1} in wave number.

Compton scattering is a type of scattering in which an incoming X-ray photon transfers large energy and momentum to a weakly bound electron in the sample. After scattering the electron recoils and is ejected. It usually gains such high energy that the potential energy from nuclei and from other electrons can be neglected. Therefore the excited or recoil electron can be treated as a free electron. Hence scattering spectra or profiles are governed primarily by the motion of the electron before scattering or by that in the ground state.

Figure 12.1 illustrates the principle of Compton scattering. Here $\boldsymbol{p}$ is the momentum of the electron in the initial state and $\hbar\boldsymbol{q}$ is the transfer momentum. Assume that the electron momentum distribution is described by a sphere with a radius of $\hbar k_{\mathrm{F}}$, or the Fermi sphere. First, consider the case where an electron is at the center of the Fermi sphere and at rest ($\boldsymbol{p} = 0$). The wavelength of X-rays scattered from the electron at rest ($\lambda_2(p = 0) \equiv \lambda_{20}$) is given by the well-known Compton formula $\lambda_{20} - \lambda_1 =$

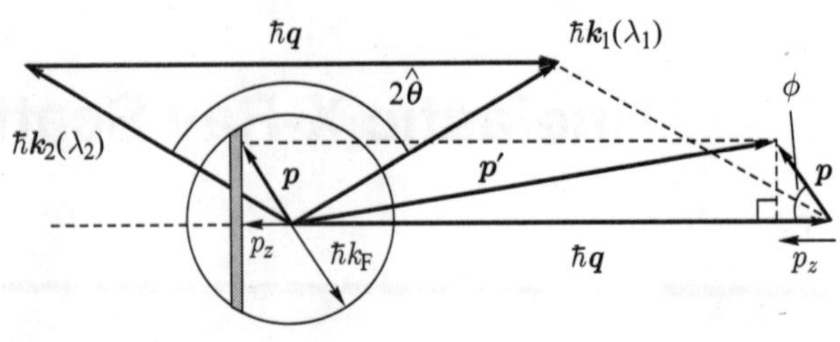

Figure 12.1 Schematic diagram of Compton scattering.

$\lambda_c(1 - \cos 2\hat{\theta})$, where λ_1 is the wavelength of the incident X-rays and λ_c is the Compton wavelength, 0.0243 Å. Next, consider the case where an electron with a momentum $\boldsymbol{p}$ ($p > 0, p_{\max} = \lambda k_F$) collides with an X-ray photon. Take the z-axis as the direction of transfer momentum as shown in Fig. 12.1 and assume that $\boldsymbol{p}$ makes angle ϕ with the z-axis. Then, the momentum distribution of scattered X-ray, $\hbar k_2$, has a width of $\hbar dq = 2(\lambda_c/\lambda_{20})p\sin\hat{\theta}\cos\phi$ centered at $2\pi\hbar/\lambda_{20}$ where $p\cos\phi = p_z$. Since the cross sectional area of the Fermi sphere cut at p_z gives the momentum distribution of the electrons, the profile of the Compton scattering forms a quadratic curve ('parabola') opening downward.[4,5] The width of the profile represents the momentum distribution of the electrons prior to scattering and can be regarded as the width due to the Doppler effect. This model holds fairly well for valence electrons of simple metals such as alkali metals. However, even for such metals, the profile should not be expected to be a true parabola since the effects of binding and correlation in electrons cannot be neglected. A deviation of the profile from parabolic becomes prominent for transition metals, semiconductors and insulators, in that order because of their increasing binding energies. Deviation from a parabola reflects the detailed structure of Fermi surfaces, that is, how distorted the Fermi surface becomes from that of a perfect sphere.

If the transfer momentum $\hbar q$ and energy loss $\hbar\omega$ are not large compared with momentum and energy of the electrons in the ground state, the scattering process must be treated differently from Compton scattering. In this case, an electron is excited by the transferred energy $\hbar\omega$. Even though $\hbar\omega$ is larger than the ionization energy, the excited electrons remain under the influence of the potential energy of the nuclei and cannot be regarded as free electrons. This final state has a close resemblance to that of photoabsorption and it can be anticipated that similar information may be obtained and this is correct as shown later. Such scattering is often called X-ray Raman scattering and its spectrum is subject to the characteristics of both the ground state and the excited state just as in photoabsorption. Accordingly, if the electrons involved are in a well-defined state before scattering such as a 1s state, then information on excited electronic states can easily be extracted from the X-ray Raman scattering, just as in core absorption spectroscopy (XAFS).

Whether experimental results of the scattering should be analyzed as Compton scattering or as X-ray Raman scattering depends on relationship between the potential energy of the electron involved in the scattering and the kinetic energy of the electron transferred from X-ray photon. That is, on the difference between the average radius of the electron orbital in the ground state and the wavelength of the excited electron after scattering, which determines the characteristics of the inelastic X-ray scattering. In the limit of small q, inelastic scattering by any electron, including core and valence electrons, should be regarded as X-ray Raman scattering. In the limit of large q, even the scattering by core electrons of heavy elements should be analyzed as Compton scattering. In the intermediate region between these two extremes, if the momentum transferred is small, the scattering is X-ray Raman-like while if the momentum transferred is large, the scattering is Compton-like.

One of the main characteristics of inelastic X-ray scattering is based on the fact that the interaction of hard X-ray photons with matter is generally weak. This weak interaction leads to a weak scattering signal, which is an obvious disadvantage. However, it also brings advantages. For example, data analysis is straightforward because multiple scattering is negligible and surface effects can be neglected, thus easily enabling the extraction of the bulk properties. In addition, measurements can be carried out under atmospheric conditions and also measurements on liquids and on samples under high pressure become possible because window materials are available. All these advantages make it possible to circumvent those experimental difficulties encountered when the probe used is electrons or soft X-rays.

Many experiments in the past have been devoted to Compton scattering for studying the electron momentum distribution of samples in the ground state. Recently attention has focused on the fact that inelastic scattering can provide information about wide $(\omega, \boldsymbol{q})$ ranges and therefore various studies under conditions where the Compton scattering picture is not adequate have started.[3,6] For example, the valence band structure of solids can be analyzed using the dynamic structure factor $S(\omega, \boldsymbol{q})$ or dielectric functions described later.[7] As will be described, information relating to the very wide range of absorption spectra from the far infrared ($\sim$ 1 meV) to the soft X-ray region, covering several hundred eV, can be deduced from analyses of X-ray Raman scattering. Of additional importance is the fact that $\boldsymbol{q}$ can also become a variable, which is impossible with conventional absorption spectroscopy where the momentum transferred is almost zero. A knowledge of the $\boldsymbol{q}$ and ω dependence of $S(\omega, \boldsymbol{q})$ is indispensable for an understanding of radiation chemistry or of radiation damage effects.[8] In order to investigate how the energies of X-rays or of charged particles incident on any sample are dispersed, the energy loss function which can be derived from $S(\omega, \boldsymbol{q})$, must be known for a wide $(\omega, \boldsymbol{q})$ range. A variety of novel applications are currently being developed which include studies on the phonons in inelastic X-ray scattering, on the magnons in resonant inelastic X-ray scattering and on the phonons in inelastic resonant nuclear scattering including excitation of nuclei using ultra-high resolution spectrometers, described elsewhere.[9–11]

Some examples of instrumentation are given here. For Compton scattering experiments (at large q) the SPring-8 multipole wiggler line makes use of 115 keV X-rays which are monochromatized by a doubly-bent crystal, then made incident on the sample and the scattered X-rays then introduced into a second monochromator.[12] The design of the whole apparatus is such as to allow acceptance of X-rays at large scattering angles. This, together with the high incident X-ray energy, makes it possible to measure at high q. The band width is 80 eV, corresponding to a momentum resolution of 0.02 a. u. For X-ray Raman scattering (at small q) experiments can be carried out on the NSLS multipole wiggler line, where the 8 keV X-rays used are monochromatized by a four-crystal monochromator and then focused using a toroidal mirror onto the sample.[3] The scattered X-rays are analyzed using a spherical crystal at a fixed energy while the scattering spectra can be obtained by scanning the energy of the incident X-rays. The techniques employed for inelastic X-ray scattering are currently in a state of change and development. For example, a high throughput monochromator, has been equipped with up to 8 crystals to collect scattered X-rays over wide solid angle and is now available at APS.[13] At both the ESRF and SPring-8, very high resolution monochromators have been constructed where monochromatizing crystals are used at Bragg angles of almost 90°. Such monochromators are used for both excitation and for the analysis of scattered X-rays, making it possible to obtain inelastic scattering spectra with a band width of less than a few meV (see Sect. **5.3.2**).

12.2 THE CROSS SECTION OF INELASTIC X-RAY SCATTERING

As in Sect. **4.3.3.a**, the first order perturbation by the square of vector potential, $\boldsymbol{A}^2$, is responsible for the inelastic scattering. Its cross section is given from the first term of Eq. (4.112) by[3)]

$$\begin{aligned}\frac{\partial^2\sigma}{\hbar\partial\Omega\partial\omega} &= (\boldsymbol{e}_1\cdot\boldsymbol{e}_2)^2{r_0}^2\left(\frac{\omega_2}{\omega_1}\right)\sum_f|\langle f|\sum_j^N\exp(i\boldsymbol{q}\cdot\boldsymbol{r}_j)|i\rangle|^2\delta(\hbar\omega-E_{fi})\\ &= (\boldsymbol{e}_1\cdot\boldsymbol{e}_2)^2{r_0}^2\left(\frac{\omega_2}{\omega_1}\right)S(\omega,\,\boldsymbol{q}),\end{aligned} \tag{12.1}$$

where r_0 is the classical electron radius, N, the number of electrons, and E_{fi}, the energy difference between the initial state $|i\rangle$ and the final state $|f\rangle$. $S(\omega,\boldsymbol{q})$ is the dynamic structure factor which governs the transition for inelastic scattering.

For isotropic samples such as gases and liquids, the cross section can be related to the generalized oscillator strength ($d\hat{f}(\omega,q)\big/\hbar d\omega$) given below.

$$\frac{d\hat{f}(\omega,q)}{\hbar d\omega} = \left(\frac{2\hbar\omega}{q^2}\right)\langle\sum_f|\langle f|\sum_j^N\exp(i\boldsymbol{q}\cdot\boldsymbol{r}_j)\,|i\,\rangle\,|^2\rangle_\Omega\,\delta(\hbar\omega-E_{fi}), \tag{12.2}$$

where $\langle--\rangle_\Omega$ means orientation average.[14)]

The inverse of the imaginary part of the complex dielectric function $Im\{-1/\varepsilon(\omega,q)\}$, called the energy loss function, is related to the generalized oscillator strength by the following equation.

$$Im\left(\frac{-1}{\varepsilon(\omega,q)}\right) = \left(\frac{4\pi^2\rho}{2\hbar\omega}\right)\times\left(\frac{d\hat{f}(\omega,q)}{\hbar d\omega}\right), \tag{12.3}$$

where $\varepsilon(\omega,q)=\varepsilon_1(\omega,q)+i\varepsilon_2(\omega,q)$ and ρ is the electron number density (see App. **L**). From Eqs. (12.1) to (12.3), it is found that the shape of the loss function is equivalent to that of the inelastic scattering spectrum. By applying the Kramers-Krönig transformation (see Sect. **4.1.3**) to $Im(-1/\varepsilon)$, one can obtain ε_1 and ε_2.

We consider the cross section given by Eq. (12.1) in some detail. When q is large, *i.e.*, when the wavelength of electron wave after scattering is much smaller than that of the spread of valence electrons ($q\gg 1/r$), the correlation of the positions of electrons can be neglected and scattering occurs independently by individual electrons. Therefore all of $\exp(i\boldsymbol{q}\cdot\boldsymbol{r}_j)$ can be replaced by $\exp(i\boldsymbol{q}\cdot\boldsymbol{r})$. As a result, the sum of the square of absolute value of matrix elements over the final states is determined only by the delta function in Eq. (12.1). Following inelastic scattering, the electron momentum changes from $\boldsymbol{p}=\hbar\boldsymbol{k}$ to $\boldsymbol{p}+\hbar\boldsymbol{q}$, and hence, the energy of the final state becomes $E_f=(\boldsymbol{p}+\hbar\boldsymbol{q})^2/2m$. From these considerations, the dynamic structure factor is given by

$$S(\omega,\boldsymbol{q}) = \int\rho_{\boldsymbol{p}}\,\delta\left\{\hbar\omega-\frac{(\boldsymbol{p}+\hbar\boldsymbol{q})^2}{2m}+E_i(\boldsymbol{p})\right\}d\boldsymbol{p},$$

where $\rho_{\boldsymbol{p}}$ and $E_i(\boldsymbol{p})$ represent the probability that the electron has momentum $\boldsymbol{p}$ at the initial state and the energy of the initial state, respectively. By taking $\boldsymbol{q}$ along the z-axis, the above integral can be rewritten as

$$S(\omega, \boldsymbol{q}) = \int \rho_{\boldsymbol{p}} \, \delta\left(\hbar\omega - \frac{\hbar^2 q^2}{2m} - \frac{\hbar q p_z}{m} - \frac{p^2}{2m} + E_i(\boldsymbol{p})\right) d\boldsymbol{p} = \frac{m}{\hbar q} \int [\rho_{\boldsymbol{p}}] \, dp_x dp_y, \quad (12.4)$$

where $[\rho_{\boldsymbol{p}}]$ represents the probability of $\rho_{\boldsymbol{p}}$ in which p_z satisfies the condition that the argument in the delta function is 0. If the initial state is described by the free electron approximation, the fourth and fifth terms in the argument of the delta function cancel out each other. In this case, p_z becomes simply $p_z = m\omega/q - \hbar q/2$. Using the wave function in momentum representation $\chi(\boldsymbol{p})$, Eq. (12.4) including $[\rho_{\boldsymbol{p}}]$ is rewritten as[15]

$$S(\omega, q \gg 1/r) \propto J(p_z)/q, \quad (12.5)$$

$$J(p_z) = \iint |\chi(\boldsymbol{p})|^2 dp_x dp_y. \quad (12.6)$$

Therefore, the dynamic structure factor can be expressed by the Compton profiles when $q \gg 1/r$.

In the case of $qr \ll 1$, the electric dipole approximation is valid and the generalized oscillator strength is given by

$$\frac{d\hat{f}(\omega, q \ll 1/r)}{\hbar d\omega} = 2\hbar\omega \langle \sum_f |\langle f | \sum_j^N \boldsymbol{r}_j |i\rangle|^2 \rangle_\Omega \delta(\hbar\omega - E_{fi}). \quad (12.7)$$

This equation shows that the generalized oscillator strength is equivalent to the (optical) oscillator strength. This means that one can derive the dielectric function, reflectivity and optical constants such as the refractive index, by using the energy loss function obtained from the inelastic scattering experiment (see Sect. **4.1.1**). By replacing the matrix elements in Eq. (12.1) by those in Eq. (12.7), the cross section for X-ray Raman scattering can also be obtained.

12.3 SPECTRA AND PROFILES OF INELASTIC X-RAY SCATTERING

12.3.1 The dependence of the spectral shape of benzene on the transfer momentum

Figure 12.2 presents the inelastic scattering spectrum, or energy loss spectrum of liquid benzene as an example.[3] The energy loss function is described by two variables, the energy and the momentum. In the figure, two abscissas represent $\hbar\omega$ and log (q^2) and the ordinate represents the intensity. Three-dimensional plots such as given in Fig. 12.2, are called Bethe surface.[14] The broad bands with maxima from a few to several tens eV originate from valence electron excitation, which then gradually shift to the high energy and change shape into a symmetric broader pattern with increase in q. As discussed, the inelastic scattering spectra at large q are described by the Compton profile $J(p_z)$.

The prominent profile change with decrease in q corresponds to a transition in the scattering process from Compton to X-ray Raman scattering.[3] The profile finally converges to X-ray Raman scattering spectrum as q decreases. The spectra at small q shown in Fig. 12.2 are equivalent to the absorption spectrum of benzene in the vacuum ultraviolet region. The sharp absorption band at 7 eV is the well-known $\pi - \pi^*$ transition of benzene. For core electrons whose spatial spreads are small, the condition $qr \ll 1$ holds well even at fairly large q. Hence one can obtain information equivalent to core absorption

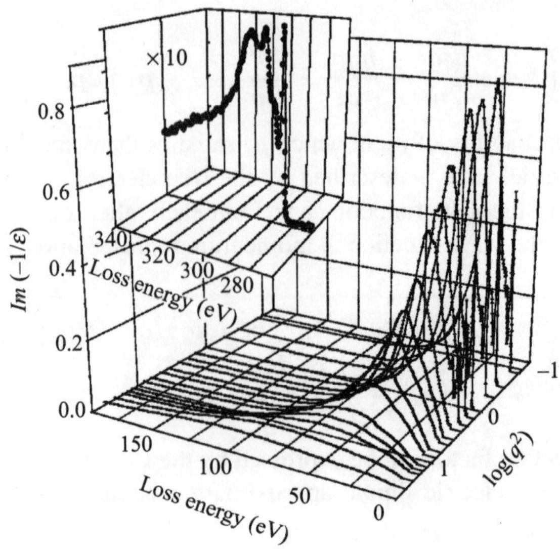

Figure 12.2 Energy loss function of liquid benzene. Inset: X-ray Raman scattering spectrum of C 1s electron of benzene.

spectra rather easily. The inset in Fig. 12.2 shows the X-ray Raman spectrum of benzene, which exhibits the fine structure equivalent to that in the C 1s absorption spectrum. If the measurements are made up to the large loss energy region, one can even observe EXAFS oscillations! Indeed, EXAFS analysis has been made on diamond using X-ray Raman scattering.[3)]

12.3.2 The Compton scattering of Li

As an illustration Fig. 12.3 shows Compton profiles for a typical simple metal, Li.[16)] The observed $J(p_z)$ function along three directions of a single crystal is compared with calculated results using the local density approximation. The $J(p_z)$ of Li consists of contributions from valence electrons which can be regarded as free-electron-like and from 1s core electrons. The $J(p_z)$ originating from valence electrons forms a parabola whose distribution is within about 0.6 a. u. The value of p_z at the foot of the parabola corresponds to the radius of the Fermi sphere. A close look at the profiles reveals that the radius of the Fermi sphere varies a little, depending on direction and that the anisotropy is found to be about 4.6%. On the other hand, since the 1s electrons are well localized, their distribution in the momentum space is much broader than that of delocalized valence electrons, extending to above 3 a. u. The calculated profiles almost exactly reproduce the observed ones except for $p_z \approx 0$ a. u., where the contribution from valence electrons is dominant. This discrepancy is due to electron correlation effects, which are not taken into account in this calculation.

One of the difficulties encountered in the analysis of Compton scattering is that the obtained $J(p_z)$ is the integrated value with respect to p_x and p_y, and hence a large part of three-dimensional information contained in anisotropic $|\chi(\mathbf{p})|^2$ is lost. Anisotropy can be retrieved to some extent, however, by measuring $J(p_z)$ along several directions as demonstrated in Fig. 12.3.[17)] To resolve this problem, attempts have been made to observe directly the anisotropy of $|\chi(\mathbf{p})|^2$ by coincidence measurements of recoil electrons and scattered X-rays.[18,19)]

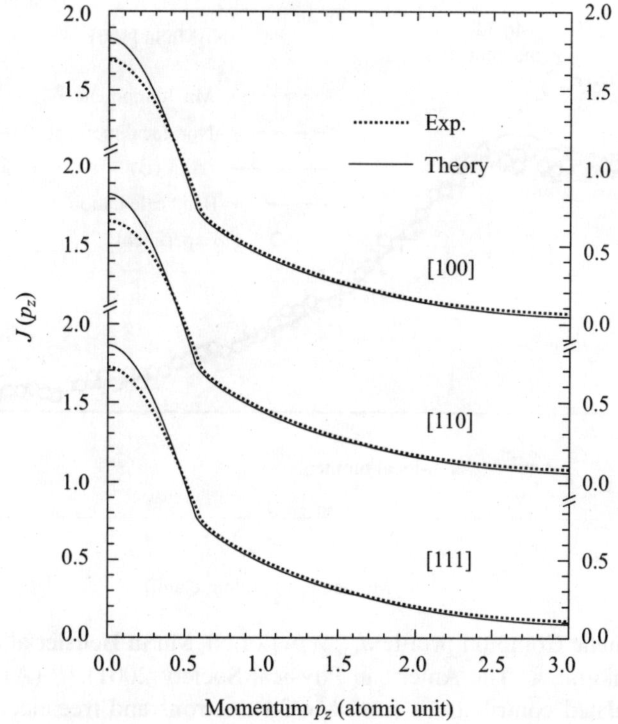

Figure 12.3 Compton profile $J(p_z)$ of Li (by courtesy of Y. Sakurai, © The American Physical Society 1995).[16)]

12.3.3 Magnetic Compton scattering for a Heusler alloy

Compton scattering can give not only information on the momentum distribution of all valence electrons, but also on the spin-dependent electron momentum distribution (or, in principle on the spin-dependent wave function of the electrons). In such 'magnetic Compton scattering' experiments, a circularly-polarized X-ray beam is incident on a sample maintained under a magnetic field. The sample magnetization is subsequently alternately reversed and the differences in the Compton scattering intensity are measured. Figure 12.4 shows the magnetic Compton scattering profile for a Heusler alloy Ni_2MnSn together with theoretical calculations.[20)] The chain represents the result of a simple approximation, in which contributions from localized 3d electrons (dots) of a Mn atom and those from free electrons (broken line) in the alloy are added. Whereas the solid line is the profile calculated using band theory, both reproduce the observed profile well. Hence it is concluded that most of the contribution to the magnetic moment of this alloy comes from the Mn 3d electrons which spread wide in the momentum space. Free electrons near the origin in the momentum space have only a small magnetic effect, whose direction is reverse to that of the Mn 3d electrons. That is, the contribution to ferromagnetism from the localized Mn 3d electrons is the most important, while the delocalized electrons have some antiferromagnetic contribution. There is no contribution to magnetic properties from Ni 3d electrons.

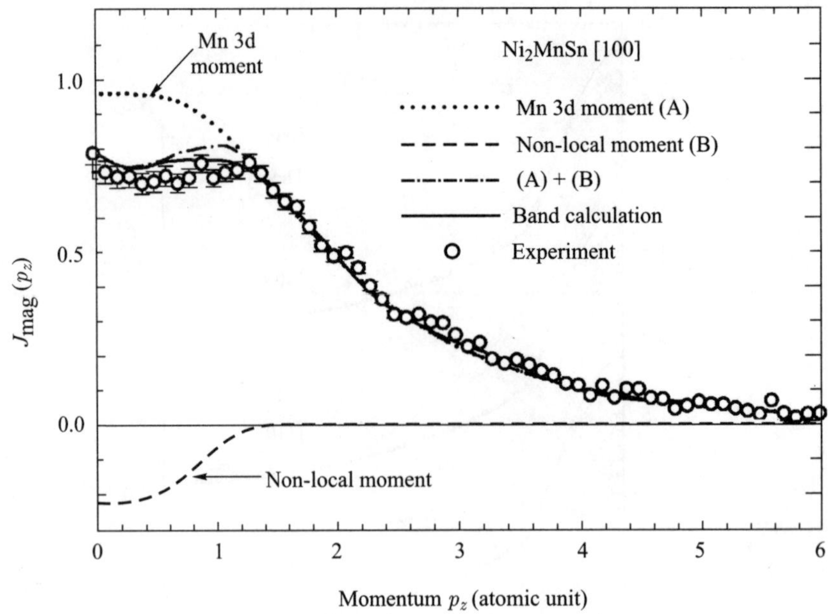

Figure 12.4 Magnetic Compton profile $J_{\text{mag}}(p_z)$ of Ni_2MnSn Heusler alloy (by courtesy of Y. Sakurai, © The American Physical Society 2001).[20] (A) and (B) represent calculated contributions from Mn 3d electrons and free electrons, respectively. The result from band calculations is also shown.

12.3.4 The X-ray Raman scattering of water

Inelastic X-ray scattering can be regarded as X-ray Raman scattering if qr is small. Hence, if q is small enough, even for delocalized valence electrons with large r, the scattering spectra can be considered equivalent to absorption spectra from which the dielectric functions can then be calculated. Figure 12.5 shows $\varepsilon_1(\omega, 0)$, $\varepsilon_2(\omega, 0)$ and $Im\{-1/\varepsilon(\omega, 0)\}$ for water derived from X-ray Raman scattering.[6] In this figure, shown by solid squares, are several ε_1 values calculated from the refractive indices in the visible and ultraviolet regions. Wherever comparisons can be made, the ε_1 values obtained from each method agree well, thus reinforcing the accuracy of the results derived from X-ray Raman scattering. The inset in Fig. 12.5 contrasts the optical oscillator strength $d\hat{f}(\omega, 0)/d\omega$ obtained in this way with those of water, water vapor and solid water (amorphous ice). The phase dependence of the spectral profile is striking. Recently, X-ray Raman scattering measurements on water and ice have been extended to the core energy region and have been analyzed using the EXAFS formula.[21] Water is the most fundamental substance for all life, and any of the information derived from the spectra in Fig. 12.5 may help to play a crucial role in understanding radiation damage effects in DNA and other materials *in vivo*.[8]

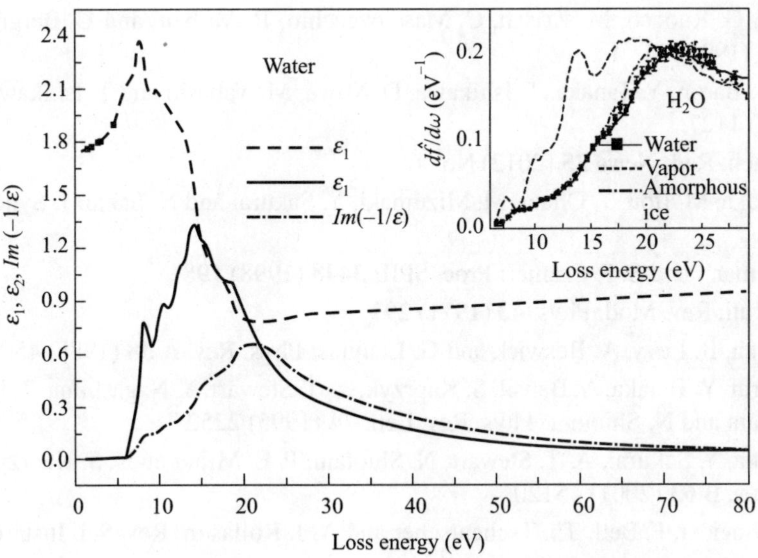

Figure 12.5 The real (ε_1) and imaginary (ε_2) part of the dielectric function, and loss $Im(-1/\varepsilon)$ function for water obtained from X-ray Raman scattering. Closed squares are ε_1 derived from refraction measurements. Inset: usual oscillator strengths $d\hat{f}/d\omega$ of water, water vapor and amorphous ice.

REFERENCES

1. W. Schülke: "Electron Dynamics by Inelastic X-Ray Scattering" (Oxford University Press, 2007).
2. P. Glatzel, F. M. F. de Groot and U. Bergmann: Synch. Rad. News **22** (2009) No. 2, 12.
3. H. Hayashi, Y. Udagawa, J.-M. Gillet, W. A. Caliebe and C.-C. Kao: "Chemical Applications of Synchrotron Radiation" Part 2, T.-K. Sham ed. (World Scientific, 2002) Chap. 18.
4. T. Suzuki: Butsuri **23** (1968) 12, in Japanese.
5. R. H. Stuewer and M. J. Cooper: "Compton Scattering", B. Williams ed. (McGraw-Hill International, 1977) Chap. 1.
6. H. Hayashi, N. Watanabe, Y. Udagawa and C.-C. Kao: Proc. Nat. Acad. Sci. USA **97** (2000) 6264.
7. W. Schülke: "Handbook on Synchrotron Radiation" Vol. 3, G. Brown and D. E. Moncton eds. (Elsevier, 1991) Chap. 15.
8. H. Nikjoo and S. Uehara: "Charged Particle and Photon Interactions with Matter. Chemical, Physicochemical and Biological Consequences with Applications", A. Mozumder and Y. Hatano eds. (Marcel Dekker, 2004) Chap. 17.

9. F. Sette, G. Ruocco, M. Krisch, C. Masciovecchio, R. Verbeni and U. Bergmann: Phys. Rev. Lett. **77** (1996) 83.
10. A. Q. R. Baron, Y. Tanaka, T. Ishikawa, D. Miwa, M. Yabashi and T. Ishikawa: J. Synch. Rad. **8** (2001) 1127.
11. See Synch. Rad. News **25** (2012) No. 4.
12. N. Hiraoka, M. Itou, T. Ohata, M. Mizumaki, Y. Sakurai and N. Sakai: J. Synch. Rad. **8** (2001) 26.
13. U. Bergmann and S. P. Cramer: Proc. SPIE **3448** (1998) 198.
14. M. Inokuti: Rev. Mod. Phys. **43** (1971) 297.
15. A. Issolah, B. Levy, A. Beswick and G. Loupias: Phys. Rev. A **38** (1988) 4509.
16. Y. Sakurai, Y. Tanaka, A. Bansil, S. Kaprzyk, A. T. Stewart, Y. Nagashima, T. Hyodo, S. Nanao, H. Kawata and N. Shiotani: Phys. Rev. Lett. **74** (1995) 2252.
17. Y. Tanaka, Y. Sakurai, A. T. Stewart, N. Shiotani, P. E. Mijnarends, S. Kaprzyk and A. Bansil: Phys. Rev. B **63** (2001) 45120.
18. J. R. Schneider, F. Bell, Th. Tschentscher and A. J. Rollason: Rev. Sci. Instr. **63** (1992) 1119.
19. M. Itou, S. Kishimoto, H. Kawata, M. Ozaki, H. Sakurai and F. Itoh: J. Phys. Soc. Jpn. **68** (1999) 515.
20. A. Deb, N. Hiraoka, M. Itou, Y. Sakurai, M. Onodera and N. Sakai: Phys. Rev. B **63** (2001) 205115.
21. D. T. Bowron, M. H. Krisch, A. C. Barnes, J. L. Finney, A. Kaprolat and M. Lorenzen: Phys. Rev. B **62** (2000) R9223.

CHAPTER

13

Infrared Spectroscopy

Synchrotron radiation has primarily been exploited as a source of light for the higher energy regions of the VUV and soft and hard X-rays, where sources with a continuum spectral output cannot easily be obtained. Nevertheless synchrotron radiation is also valuable in the much lower energy regions including the infrared, far infrared and sub-millimeter wave regions. In fact synchrotron radiation sources have greater brilliance than blackbody radiation sources which yield a continuum spectrum. However synchrotron radiation has in addition a unique pulsed time structure, whereas blackbody radiation sources are difficult to modulate at high frequencies. Infrared studies are currently active at many synchrotron radiation facilities around the world specifically to take advantage of the spectral brilliance and time modulation.[1,2)] These regions cover the energy range of from 1 meV to 1 eV corresponding to the wavelength region of from 1 mm to about 1 μm. Research has been extensively undertaken on free electrons, shallow electronic levels, lattice vibrations, molecular vibrations and so on. In this chapter, the characteristic qualities of synchrotron radiation in the infrared region are explained and the researches using it are introduced.

13.1 THE CHARACTERISTICS OF SYNCHROTRON RADIATION IN THE INFRARED

A comparison between the intensity and the brilliance of synchrotron radiation and of blackbody radiation, such as from a Globar lamp which is widely used in conventional infrared spectroscopy is given in Fig. 13.1. This reveals (a) the flux and (b) the brilliance of 1400K blackbody radiation and also of synchrotron radiation from the infrared beamlines at UVSOR and SPring-8.[3,4)] The flux of synchrotron radiation is much higher than that of a blackbody source in the extreme infrared, while being comparable in mid infrared. In cases where radiation is sought for a large sample, blackbody radiation may offer higher intensity than synchrotron radiation, however in the far infrared region, synchrotron radiation always offers greater broad-band brilliance than a blackbody source. This is because the power of blackbody radiation diminishes by λ^{-2} as the wavelength increases towards the extreme infrared, while that of the synchrotron radiation decreases according to $\lambda^{-1/3}$. In those cases where small samples or small parts of materials are the object of the measurements, the source brilliance becomes the dominant factor since only high brilliance beams can be focused to a small spot. As shown in Fig. 13.1(b), the brilliance of synchrotron radiation is from two to four orders of magnitude higher than that from a blackbody source. Synchrotron radiation has the additional advantage that it can

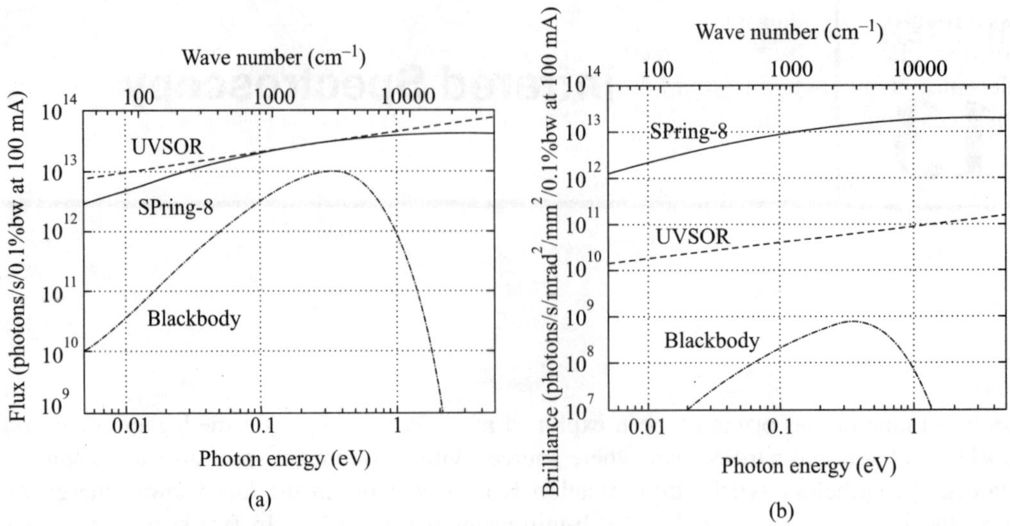

Figure 13.1 (a) The calculated flux and (b) brilliance for blackbody radiation (1400K, ϕ (radius) $= 1$ mm, $\Delta\Omega = 0.015$ sr) and synchrotron radiation from SPring-8 ($\Delta\psi = 36.5$ mrad, $\Delta\hat{\theta} = 10.5$ mrad, $\varepsilon_x = 6.2$ nm$\cdot$rad, $\varepsilon_y = 0.35$ nm$\cdot$rad) and UVSOR ($\Delta\psi = 80$ mrad, $\Delta\hat{\theta} = 60$ mrad, $\varepsilon_x = 31.2$ nm$\cdot$rad, $\varepsilon_y = 15.6$ nm$\cdot$rad) (by courtesy of Toshiharu Takahashi).

be transformed either into a bright parallel beam or focused on a small area. Synchrotron radiation offers the further advantage that it provides linearly-polarized light in the orbital plane ($\hat{\theta} = 0$) and circularly-polarized light outside the plane ($\hat{\theta} \neq 0$). Synchrotron radiation is produced as a train of light pulses with time widths of from 10^{-11} to 10^{-9} s (see Sect. **1.2.1**). It is difficult to obtain pulsed light with short time widths combined with a continuous spectrum, in the region between the near and far infrared in laboratory.

Alternative light sources with high intensity and high brilliance for the infrared region are those which generate coherent synchrotron radiation, including free electron lasers (see Sect. **3.5** and App. **E**). 'Edge radiation' which is emitted due to the steep gradient of the magnetic field near to the entrance and the exit of bending magnets has also been used as a source of infrared light.[5)]

13.2 AN INFRARED BEAMLINE

13.2.1 The front end

Figure 13.2 shows the outline of an infrared beamline consisting of (a) the front end and pre-mirrors, and (b) the measuring system. Any pre-mirrors will be exposed to all the emitted radiation and therefore heat load problems are inevitable. However when selecting infrared radiation the heat load problem can be minimized by using a pre-mirror designed with a slit aperture cut along the median (horizontal) plane. Since synchrotron radiation spreads across a wide range of vertical angles in the infrared region, the infrared radiation can be gathered effectively by this mirror, while the most energetic X-rays simply pass through the slit, thus minimizing the heating problems at the mirror. The mirrors are usually water-cooled and fabricated from light metals such as Be, because their X-ray absorption is small. Since the reflectance of any metal surface is close to 1 in the infrared region, any intensity loss by the mirrors can be neglected and many metal-coated mirrors can be used behind the pre-mirror.

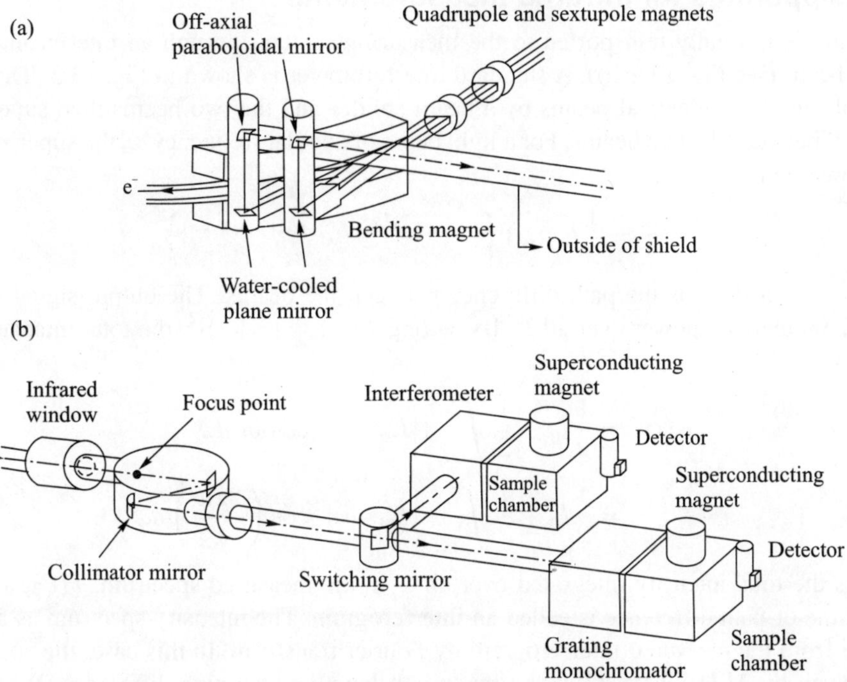

Figure 13.2 An outline of the infrared beamline consisting of (a) front end and pre-mirrors, and (b) measuring system. Radiation from the bending magnets and the straight sections are both available (by courtesy of Y. Shibata).

Synchrotron radiation is of course not emitted from a single point, but is emitted continuously by the electrons in orbit, along all directions tangential to their circular orbit. To focus such radiation, an aspherical mirror which can condense the radiation to a point in the orbital plane has been used in SPring-8 for example.[6)] A problem in infrared beamline design originates from the need to capture such a large vertical acceptance angle. It is always difficult to install an infrared beam pipe with wide vertical aperture close to the vacuum pipe of the storage ring, since it can create instabilities in the stored electron beam (see Sect. **3.4.3**). If the beam pipe is too narrow then the extraction efficiency for the infrared will be low, however when a beam pipe wide enough to minimize losses due to beam diffraction effects is feasible, then the synchrotron radiation source can be used effectively out into the far infrared and sub-millimeter regions.

Ultra-high vacuum is essential in the chamber of the pre-mirror, which is directly linked to the storage ring (see Fig. 13.2(a)). However for the infrared measuring system itself ultra-high vacuum is not essential since window materials are available to transmit infrared light and the vacuum system of the infrared measuring system can easily be separated from that of the storage ring using a window beyond the primary mirror chamber made from diamond or alkali fluorides, for example (see Fig. 13.2(b)).

These windows are wedge-shaped to minimize interference effects. In infrared spectroscopy, a unit of cm^{-1} (wave number) is normally used as the unit of energy, where it expresses the number of waves per 1 cm and 1 cm^{-1} = 0.124 meV.

13.2.2 Apparatus for infrared measurements

The radiation is normally transported to the measuring system through an interferometer as a time-correlated beam (see Fig. 13.2(b)). A standard interferometer is shown in Fig. 13.3. The incident light beam is split into two identical beams by a beam splitter and the two beams then superimposed with delay time t' between the two beams. For a light beam of angular frequency ω, the superimposed electric field is expressed by

$$E = \frac{1}{2} E_0(\omega) \left\{ e^{-i(\omega t - kr)} + e^{-i(\omega t + \omega t' - kr)} \right\}, \tag{13.1}$$

where $t' = 2x/c$ and $2x$ is the path difference between the beams. The output signal of the detector indicates the integrated power over all ω. By setting $I(\omega) = |E(\omega)|^2$, the total intensity $I(t')$ can be expressed as

$$\begin{aligned} I(t') &= \frac{1}{2} I_0 + \frac{1}{2} \int_{-\infty}^{+\infty} |E_0(\omega)|^2 \cos \omega t' d\omega \\ &= \frac{1}{2} I_0 + \frac{1}{2} \int_{-\infty}^{+\infty} |E_0(\omega)|^2 \cos \left(\frac{2x\omega}{c} \right) d\omega, \end{aligned} \tag{13.2}$$

where I_0 is the total intensity integrated over all ω in the measured spectrum. $I(t')$, as a function of the delay time or path difference is called an interferogram. The intensity spectrum as a function of ω is obtained from the measured interferogram by Fourier transform. In this case, the spectral resolving power is given by $2L/\lambda$ where $2L$ is the maximum length of the path difference. When synchrotron radiation is used as the light source for Fourier spectroscopy, the divergence of the light is small owing to its high brilliance and a long path difference is feasible yielding high resolution spectroscopy.

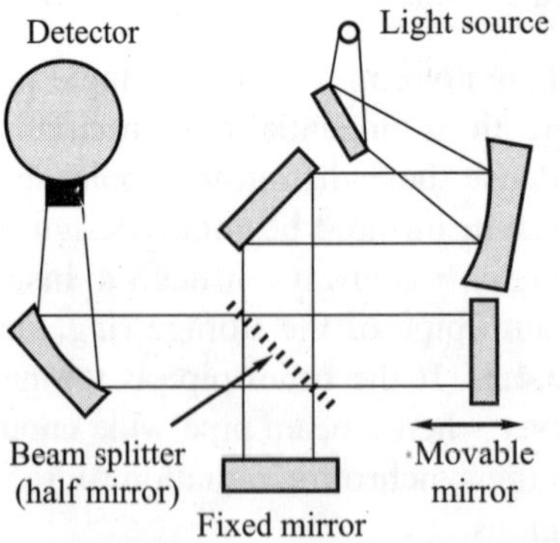

Figure 13.3 Schematic diagram of the interferometer. For transmission measurements, the sample is placed in front of the detector.

The high brilliance also offers advantage in that the light can be focused within a small area even with a large F-value system (see App. **A**). Thus light can be transported along long pipes and delivered to the sample with minimal intensity loss. This is especially useful for measurements using high-field magnets where the light path is long but must still be focused on a point within a small area at the bottom of a narrow and deep pipe. For example, a spot size of 1 mm^2 can be attained around wavelengths of 10 μm even with a large F-value system.[4] A beamline with large acceptance angle at a bending section enables precise measurements to be made in the region around wave number of 10 cm^{-1} in comparison with a conventional instrument using a high pressure mercury lamp.

Various detectors are used, appropriate to the energy region of interest. Hot-electron bolometers cooled down to liquid helium temperature are used in the region lower than 10 meV, Si or Ge bolometers cooled down to liquid helium temperature, in the 0.5 $\sim$ 80 meV region, HgCdTe photodiodes, in the 50 meV $\sim$ 1 eV region and CdS photodiodes, in the 0.5 $\sim$ 2 eV region. The response times of infrared detectors are usually slow and are typically μs (photodiodes) to ms (bolometers). No detectors with as fast a response time as those used in visible region are available in the infrared. However, by exploiting the pump-and-probe technique using pulsed synchrotron radiation, time-resolved studies can be achieved with better than ns time resolution as described in Sect. **5.7**.

13.3 INFRARED SPECTROSCOPY OF SOLIDS AND GAS-ADSORBED SOLID SURFACES

With infrared spectroscopy, one can study the absorption or the reflection originating from free electrons in metals. It is also possible to investigate lattice vibrations and molecular vibrations. In the following section the topics considered include magnetooptical studies of a rare earth compound and a high temperature superconductor, time-resolved spectroscopy on a superconducting metal, the pressure effect on lattice vibrations in an alkali halide and the adsorption of atoms and molecules on semiconductor surfaces. Recent researches on strongly correlated electron systems are given in Ref. 7. For details of an infrared microscope, see Sect. **14.2.3**.

13.3.1 Electronic structure studies

a. Magnetooptical spectroscopy of rare-earth compounds: In Sect. **7.5.3.b**, a photoelectron study on CeSb, a strongly correlated electron system with NaCl type structure was described. The spectral change caused by the phase change of this material has also been observed using infrared spectroscopy. CeSb has a total of 16 phases under different combinations of temperature and magnetic field. These phases are paramagnetic (P) where the material is composed only of paramagnetic layers, antiferro-paramagnetic phases (AFP) where antiferromagnetic and paramagnetic layers appear, and so on. In this work the reflection spectra were measured under a magnetic field aligned along the [001] direction by using linearly-polarized light with electric vectors parallel and perpendicular to the magnetic field in the 0.1 $\sim$ 0.8 eV region.[8] The ranges of temperature and magnetic field were respectively 6.5 $\sim$ 30K and 0 $\sim$ 6 T. Figure 13.4 shows the electric (optical) conductivity $\sigma(\omega)$ spectra obtained from reflection spectra using the Kramers-Krönig transformation. In the figure, the conductivity decreases with increase in photon energy from 0.1 eV to 0.2 eV. This feature is due to absorption by free electrons. In the P phase and antiferromagnetic (AF) phase, the polarization dependence of $\sigma(\omega)$ was not observed, which suggests that the magnetic and electronic structures are isotropic against the applied field in both phases. However in the AFP phase or AFF phase in which antiferromagnetic and ferromagnetic layers appear, the polarization dependence of $\sigma(\omega)$ was observed, reflecting the anisotropy of the magnetic structure. From an analysis using the Drude formula, it was found that the effective masses of free electrons parallel and perpendicular to the magnetic field are different. This reflects the anisotropy of the hybridization of Ce4f-Sb5p. Reflection spectra were also measured using circularly-polarized light incident along the direction parallel to the magnetic field. These measurements suggested that Sb 5p level is raised and mixes with the Ce 4f level, and that Sb 5p and Ce 5d hybridize, which would reinforce the conclusion of Sect. **7.5.3.b**.

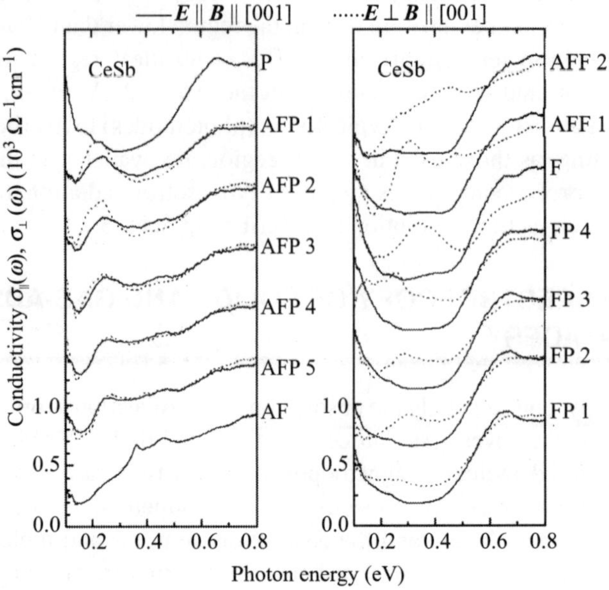

Figure 13.4 Polarization dependence of the optical conductivity of CeSb under a magnetic field. Its magnetic structure is given in the text (by courtesy of S. Kimura).[8)]

b. Reflection spectra of a high T_c superconductor and the effect of a magnetic field: In high T_c superconductors such as $Bi_{1.80}Pb_{0.38}Sr_{2.01}CuO_{6-\delta}$, it is known that superconduction occurs in the CuO_2 plane (*ab*-plane) and that a superconductivity gap opens in the diagonal direction of the *ab*-plane (see Sect. **7.6**). In the direction of the *c*-axis which is perpendicular to this plane, these materials have the property of a semiconductor or an insulator. Figure 13.5 shows far infrared reflection spectra (< 5 meV) of $La_{2-x}Sr_xCuO_4$ $(x \approx 0.1)$ using linearly-polarized light.[9)] Its electric vector was parallel to *c*-axis. Below T_c, Drude type reflection spectra were observed in the photon energy region less than the energy of the superconductivity gap. When this material is regarded as a Josephson junction (which consists of two superconducting layers sandwiching a thin insulating layer) this observation means that a tunneling effect is observed optically. When the magnetic field is applied perpendicularly to the *c*-axis, the spectra do not change significantly but while when parallel, Drude type reflection disappears with an increase in the magnetic field. This suggests that in the perpendicular case, superconductivity remains in a vortex state with superconducting current flows outside the vortex under the magnetic field. However in the parallel case, superconductivity is easily destroyed by the magnetic field and normal conduction appears outside the vortex. It is noteworthy that the superconductivity gap itself was not observed in the

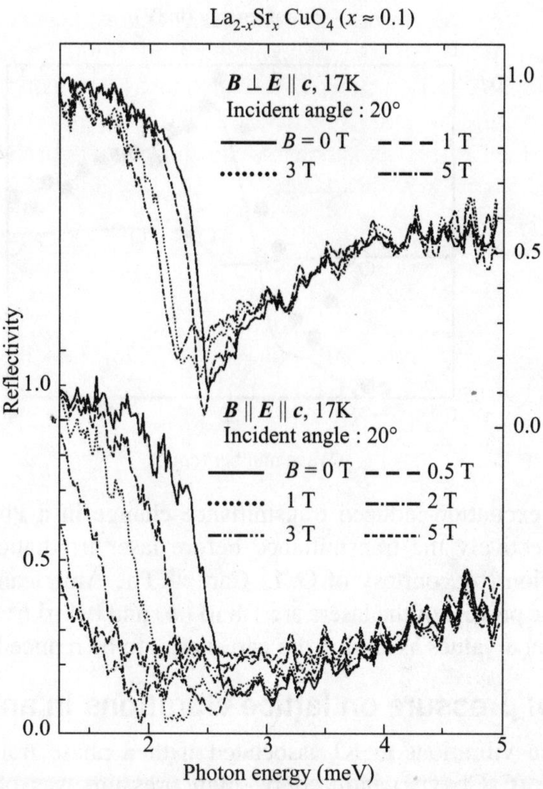

Figure 13.5 Magnetic field dependence of far infrared reflection spectra for $La_{2-x}Sr_xCuO_4$ $(x \approx 0.1)$ $(\boldsymbol{E}||\boldsymbol{c})$ (by courtesy of S. Kimura).

far infrared region,[10] as observed for normal superconducting materials. Perhaps this is due to the fact that the symmetry of the superconducting state is not normal s-like but is d-like (see Sect. **7.6**).

c. Time-resolved spectroscopy on superconducting metal: An example of time-resolved spectroscopy using the pump-and-probe technique is provided when the source of the excitation pulses is a ps pulse laser driven synchronously with the RF oscillator of the storage ring. The excited states generated by the laser pulses are subsequently synchronously probed using the synchrotron radiation pulses. An outline of the method is given in Sect. **5.7**. In this case, the change in the transmittance was observed for a thin Pb film in a superconducting state, created by quasi-particles generated from excitation of the Cooper pairs by the laser.[11] When the Pb film has been cooled down sufficiently to below the critical temperature T_c, the energy of superconductivity gap Δ changes by $\delta\Delta/\Delta \approx 2n_{qp}/n_{pairs}$, where n_{qp} is the density of the quasi-particles and n_{pairs} is that of Cooper pairs. The quasi-particles of the excited states generated by the laser relax to lower energy states, and later return to Cooper pairs by recombination. Since the relaxation time of this process is about 100 ns in superconducting metals, the change in the transmittance must be measured within the relaxation time in the infrared region around the gap energy. Figure 13.6 shows the far infrared spectra of the transmittance change induced by laser irradiation for a Pb thin film. The initial energy for the transmittance change shifts to lower energy with increase in the laser intensity, which means that the gap energy shifts with increase in laser intensity.

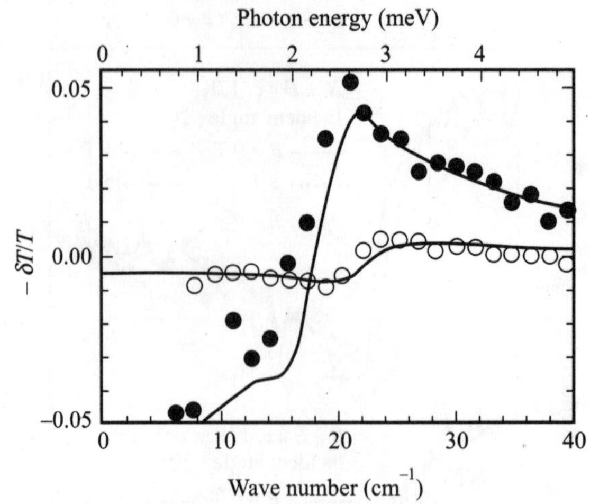

Figure 13.6 The optical excitation-induced transmittance change in a Pb thin film. T and δT are respectively the transmittance before laser irradiation and its change after irradiation (by courtesy of G. L. Carr, © The American Physical Society 2000).[11] The powers of the lasers are 1.8 nJ (●) and 0.4 nJ (○). The solid curves show calculated values assuming the gap energy to be reduced by 3% or 0.6%.

13.3.2 The effect of pressure on lattice vibrations in an alkali halide

The energy shift of lattice vibrations in KI associated with a phase transition from the B1 phase (NaCl type) to B2 phase (CsCl type) induced by high pressure was observed using far infrared radiation.[12] In many cases, such experiments are performed by laser Raman spectroscopy. However in centrosymmetric materials such as alkali halides in both B1 and B2 phases are Raman-inactive and infrared-active,[13] and therefore this experiment was carried using the far infrared. The high pressure apparatus is similar to that used for X-ray diffraction experiments (see Sect. **9.5.1**) employing a diamond anvil cell, but one which is transparent to the far infrared and, of course whose optical path is very narrow. The pressure is monitored using ruby as in the X-ray diffraction experiments and is determined from the wavelength shift of the R1 fluorescence line. Figure 13.7 shows the transmission spectra under high pressure. The position of the arrow gives the energy position of TO (transverse optical) phonon and the sharp absorption at 130 cm^{-1} originates from the quartz filter of the detector. The oscillating structure appearing in the spectra is caused by interference in the diamond anvil cell. In Fig. 13.8, the energy position of the TO phonon is plotted against the pressure. When the pressure is not applied, the energy position of the TO phonon is 102 cm^{-1}. As the pressure is increased, the position moves towards the high energy side, but when the pressure exceeds 1.8 GPa (about 1.8×10^4 atm), the position shifts abruptly to the low energy side. As the pressure is increased further, the position moves again to the high energy side. This was the first observation of lattice vibration changes caused by pressure-induced phase changes for centrosymmetric materials. They became possible only because of the high brilliance of synchrotron radiation in the far infrared.

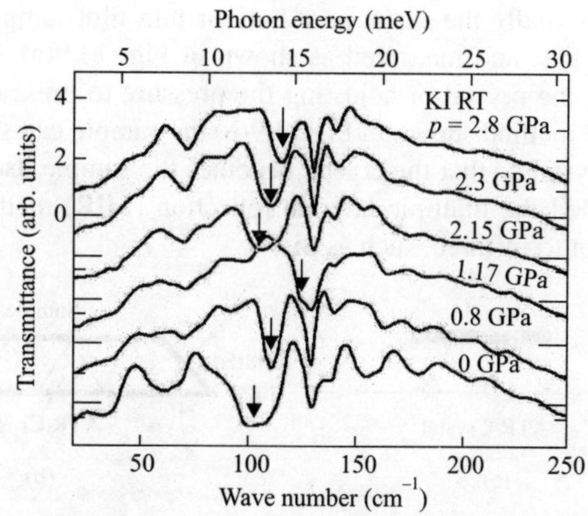

Figure 13.7 Far infrared spectra for a KI film (2 μm thick) at room temperature under high pressure. Arrows indicate the energy positions of a TO phonon. The sharp peak at 130 cm^{-1} originates from absorption by the quartz filter used in the detector.[12)]

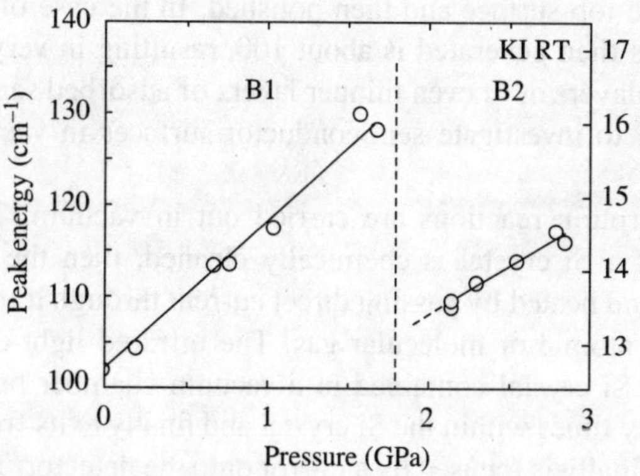

Figure 13.8 The pressure dependence of the lattice vibration energy (TO phonon) of KI.[12)]

13.3.3 The adsorption of atoms and molecules on semiconductor surfaces

The adsorption of atoms and molecules on solid surfaces has been studied in various ways (see Sects. **7.9, 8.1**). Here, the attenuated total reflection (ATR) technique using infrared radiation is applied to the observation of surface chemical reactions. This method can also be exploited to assist in the development of monitors for environmental security, of devices for molecular recognition, and also for the characterization of bio molecules.

a. The attenuated total reflection technique: In this technique, atoms or molecules are adsorbed on semiconductor thin crystals and infrared light is used to illuminate the crystals under the condition of total reflection with multiple reflections at their internal surfaces resulting in the generation of evanescent waves (see Sect. **4.1.2**). The evanescent waves can yield information about the surface or interface regions leading to surface chemical analyses which may detect adsorbed atoms or molecules with high sensitivity.[14–16] Usually the surfaces of bulk or thin film samples are pressed tightly onto a prism-shaped crystal for ATR and measured as shown in Fig. 13.9(a). Difficulties may arise such as surface contamination of the crystal or adjusting the pressure to press the sample onto the crystal surface. With the alternative method shown in Fig.13.9(b) the sample can simply be adsorbed onto the clean surface of the ATR crystal so that the crystal becomes the sample itself and these problems then avoided. The method is called the multiple internal reflection (MIR) method and is used for surface analyses of solids transparent to infrared, such as Si.

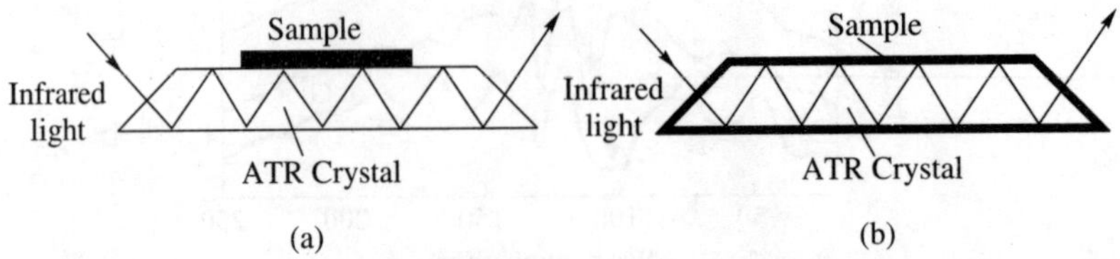

Figure 13.9 Schematics of (a) normal ATR method and (b) MIR method. Thick lines represent samples to be measured.

An analysis of surface reactions on Si surfaces will be described. As an ATR crystal for MIR, a 0.5 mm thick Si wafer is cut into the shape of a rectangular plate whose short edges are tailored so as to have an angle of 45° to the top surface and then polished. In the case of a 50 mm long crystal, the number of internal reflections then generated is about 100, resulting in very high sensitivity detection of adsorbed samples of monolayers or of even thinner layers of adsorbed samples. Using this technique it becomes possible not only to investigate semiconductor surfaces in vacuum but also in air and in liquids *etc.*

The experiments on adsorption reactions are carried out in vacuum. The sample is prepared as follows. First, the surface of a Si crystal is chemically cleaned, then the crystal introduced into an ultra-high vacuum chamber and heated by passing direct current through it for further cleaning. Second, the surface is exposed to an atomic or molecular gas. The infrared light exiting an interferometer is focused onto an edge of the Si crystal contained in a vacuum chamber behind a CaF_2 window. The infrared light is reflected many times within the Si crystal and finally exits from another edge to vacuum through a CaF_2 window and finally is focused by a mirror onto the detector. The infrared light is focused onto 1 mm wide edge of the ATR crystal.

b. The adsorption of atomic hydrogen on Si surfaces and etching reactions: The surface chemical reaction between atomic hydrogen H and Si surfaces is an important one for semiconductor processing. This reaction can be investigated by measuring the infrared absorption spectra of a Si (100) 2 × 1 surface exposed to H atoms in the region of the Si–H stretching vibration mode.[17–19] Figure 13.10 shows some experimental results. Atomic hydrogen H was prepared by thermal dissociation of molecular hydrogen H_2 at a filament. The number in the figure is a measure of the amount of exposure in Langmuir units ($1\ L = 1 \times 10^{-6}$ Torr · s $= 133 \times 10^{-6}$ Pa · s)) of H estimated from the partial pressure of H_2. Many peaks found in the 2000 cm^{-1} ~ 2200 cm^{-1} region originate from various kinds of Si hydride SiH_x ($x = 1 \sim 3$) at the surface. In the figure, one can see the change of the type of hydride and its amount with increase in the amount of exposure to H.

In the sample of 800 L, there exist only two peaks around 2100 cm^{-1}. The structure of the Si (100) 2 × 1 surface is dimeric as shown in the left side of Fig. 13.11. Each dimer has two dangling bonds. In the

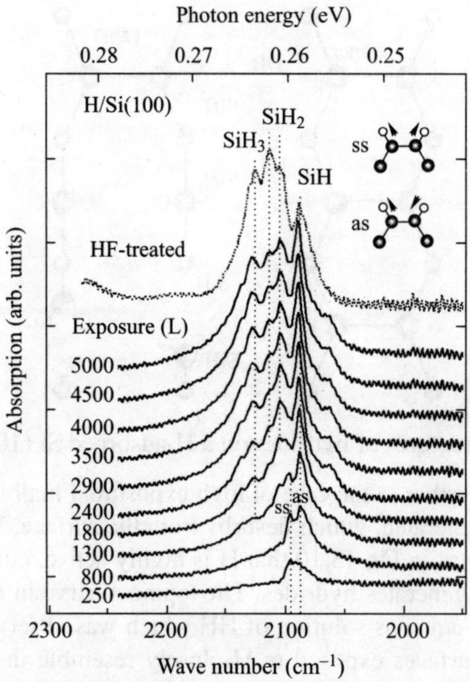

Figure 13.10 Infrared absorption spectra of a Si (100) 2 × 1 surface exposed to atomic H at room temperature.[18]

case of low exposure, H adsorbs to these two dangling bonds and the doubly-occupied dimer (DOD) is formed as shown upper part of right side of Fig.13.11. In DOD, there exist two Si–H bonds. The stretching vibration of the two bonds with the same phase is called the symmetric stretching vibration (ss) and that with opposite phase the asymmetric stretching vibration (as). The two peaks in Fig. 13.10 correspond to these two kinds of vibration modes. The symmetric stretching vibration cannot be excited by s-polarized light, because it has only vibration components vertical to the surface. In fact, one of the two peaks at the higher wave number side cannot be observed in the absorption spectrum measured using s-polarized light.

As shown in Fig. 13.10, a new peak appears around 2110 cm^{-1} rather than the above-mentioned DOD component dependent on the increase in the amount of exposure. As exposure is increased, the intensity of the peak around 2080 cm^{-1} increases and a new peak appears around 2130 cm^{-1}. The peak around 2110 cm^{-1} originates from SiH_2, while that at 2080 cm^{-1} is associated with SiH and the peak around 2130 cm^{-1} originates from SiH_3.[20] With further increase in exposure, almost of all dangling bonds adsorb H and form DOD and the H atoms coming to the surface start to cut Si–Si bonds of the surface Si dimer. As schematically illustrated in Fig. 13.11, when the bond of the dimer is cut, dangling bonds are generated at two Si sites which then adsorb H and form SiH_2. When the bonds of even deeper Si atoms on the edge of DOD (back bonds) are cut, then SiH and SiH_2 are also generated. This explains the appearance of the SiH_2 peak. When exposure is further increased, the reaction with H also progresses further. The target attacked by H is the Si–Si bond in SiH_2, and once this bond is cut, SiH_3 and SiH are both generated as shown in Fig. 13.11. In fact, the intensities of the peaks corresponding to

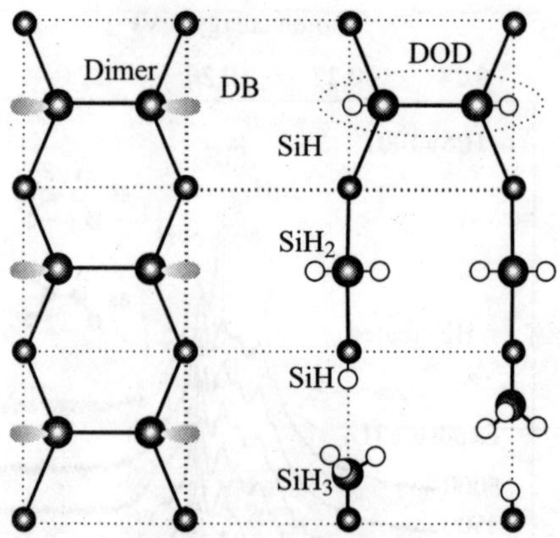

Figure 13.11 Structures of hydrides on a H-adsorbed Si (100) 2 × 1 surface.

these hydrides increase even further in the case of high exposure. Finally, when the back bond of SiH_3 is cut, the SiH_4 molecule is generated, which desorbs from the surface. Therefore it is shown from an analysis of the change in spectra in Fig.13.10 that H is highly active, cuts the Si–Si bonds near the Si surface successively and then generates hydrides. The topmost curve in Fig.13.10 is the spectrum of a Si (100) surface treated by an aqueous solution of HF, which was observed to be etched at the atomic level.[19] The spectra of the surfaces exposed to H closely resemble those of an HF-treated surface, suggesting that the surface exposed to H is etched and becomes rough at the atomic level similarly to the HF-treated surface.

c. The adsorption of water molecules on Si surfaces and the oxidation reaction: In processes used to form an oxidized layer of gate in SiMOS (Si metal-oxide-semiconductor) devices, water can be used as the oxidizer in place of oxygen. It is anticipated that the presence of hydrogen would modify the microscopic process of oxidation from that using O_2 alone. In an attempt to understand this process of the oxidation of a Si surface by H_2O, observations were made on the infrared spectra of Si (100) 2 × 1 surfaces adsorbed by H_2O heated at various temperatures.[18,20]

The infrared absorption spectrum of a Si (100) 2 × 1 surface with adsorbed H_2O at room temperature in the Si–H stretching vibration region is shown at the top of Fig. 13.12. Two peaks can be observed which originate from the stretching vibrations of Si–H in the Si–H bond and from O–H in the Si–OH bond around 2100 cm^{-1} and 3640 cm^{-1}, respectively (the latter is not shown in the figure). At adsorption, H_2O is dissociated into H and OH and each of them respectively reacts with the dangling bonds forming Si–H and Si–OH bonds.

The sample was then heated up to 310°C, kept at 310°C for 1 s and was suddenly cooled down to room temperature, and the absorption spectrum again measured. Similar measurements were repeated with increase in temperature and the results are presented in Fig. 13.12. At the beginning of oxidation (heating temperature < 400°C), four peaks appear around 2100 cm^{-1}. The two peaks at the low wave number side correspond to DOD. The two peaks at high wave number side are considered to be due to an oxidation radical in which an O atom is inserted in DOD (=HSi–O–SiH=). This oxidation radical is formed according to a process where the OH radical at the surface is dissociated. The O atom enters into a dimer bond and the H atom is adsorbed on the dangling bond of the Si atom. When the O atom does not enter the dimer bond, DOD is formed. The radical =HSi–O–SiH= is stable and survives

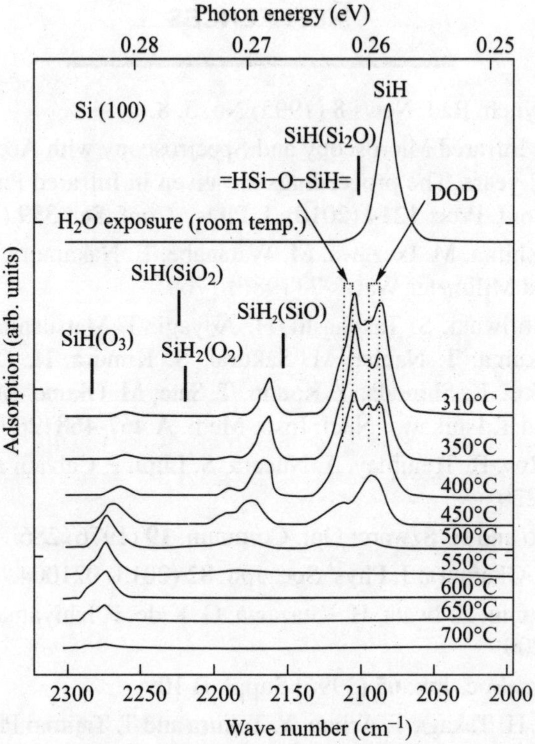

Figure 13.12 SiH stretching vibrational spectra of Si (100) 2 × 1 surface adsorbed by water molecules after heating.[18] The heating temperature is given at each spectrum. All measurements were made at room temperature after heating.

to about 400°C together with DOD. When the temperature is around 300°C, a new peak appears at 2170 cm^{-1} in addition to the above-mentioned peaks. The vertical lines in Fig.13.12 correspond to the frequencies of the Si–H stretching vibrations in the oxidation radical $SiH_x(Si_yO_z)$ $(x + y + z = 4)$ at intermediate oxidation states calculated semiempirically.[21] By comparison of the peak positions and calculated values, it is suggested that the peak at 2170 cm^{-1} originates from an oxidation radical of $SiH_2(SiO)$. This radical is composed of a central Si atom connected with two H atoms, one O atom and one Si atom.

When the heating temperatures exceed 450°C, the intensities of the four peaks around 2100 cm^{-1} decrease considerably, while that of the peak around 2170 cm^{-1}, diminishes only slightly. This is due to thermal desorption of H from the surface. New peaks appear between 2260 cm^{-1} and 2270 cm^{-1}, and according to the semiempirical calculation, these peaks would appear to correspond to $SiH_2(O_2)$, $SiH(SiO_2)$ or $SiH(O_3)$. Finally, when the heating temperature is 650°C, only the peak corresponding to $SiH(O_3)$ survives, suggesting that this radical is thermally stable. For low temperature oxidization, several kinds of radicals remain stably in the oxidized layers.

References

1. G. P. Williams: Synch. Rad. News **8** (1995) No. 5, 8.
2. Int. Workshop on Infrared Microscopy and Spectroscopy with Accelerator-Based Sources has been held every 2 years. The proceedings are given in Infrared Phys. and Technol. **51** (2008) No.5, 361, AIP Conf. Proc. **1214** (2010), J. Phys.: Conf. Ser. **359** (2012) *et al.*
3. T. Nanba, Y. Urashima, M. Ikezawa, M. Watanabe, E. Nakamura, K. Fukui and H. Inokuchi: Int. J. Infrared and Millmeter Waves **7** (1986) 1769.
4. H. Kimura, T. Moriwaki, S. Takahashi, H. Aoyagi, T. Matsushita, Y. Ishizawa, M. Masaki, S. Oishi, H. Ohkuma, T. Nanba, M. Sakurai, S. Kimura, H. Okamura, H. Nakagawa, T. Takahashi, K. Fukui, K. Shinoda, Y. Kondo, T. Sata, M. Okuno, M. Matsunami, R. Koyanagi, Y. Yoshimatsu and T. Ishikawa: Nucl. Instr. Meth. A **467-468** (2001) 441.
5. Y.-L. Mathis, P. Roy, B. Tremblay, A. Nucara, S. Lupi, P. Calvani and A. Gerschel: Phys. Rev. Lett. **80** (1998) 1220.
6. R. Lopez-Delgado and H. Szwarc: Opt. Commun. **19** (1976) 286.
7. S. Kimura and H. Okamura: J. Phys. Soc. Jpn. **82** (2013) 021004.
8. S. Kimura, M. Okuno, H. Iwata, H. Kitazawa, G. Kido, F. Ishiyama and O. Sakai: J. Phys. Soc. Jpn. **71** (2002) 2200.
9. S. Kimura: J. Phys. Soc. Jpn. **65** (1996) Suppl. B 109.
10. S. Uchida, T. Ido, H. Takagi, T. Arima, Y. Tokura and T. Tajima: Phys. Rev. B **43** (1991) 7942.
11. G. L. Carr, R. P. S. M. Lobo, J. LaVeigne, D. H. Reitze and D. B. Tanner: Phys. Rev. Lett. **85** (2000) 3001.
12. T. Nanba and M. Watanabe: J. Phys. Soc. Jpn. **58** (1989) 1535.
13. For instance, M. Born and K. Huang: "Dynamical Theory of Crystal Lattices" (Oxford, 1954) Chap.VII.
14. N. J. Harrick: "Internal Reflection Spectroscopy" (Harrick Scientific Corporation, 1987).
15. Y. J. Chabal: Surf. Sci. Rep. **8** (1988) 211.
16. F. M. Mirabella Jr. and N. J. Harrick: "Internal Reflection Spectroscopy: Review and Supplement" (Harrick Scientific Corporation, 1985).
17. M. Terashi, J. Kuge, M. Shinohara, D. Shoji and M. Niwano: Appl. Surf. Sci. **130-132** (1998) 260.
18. M. Niwano: Surf. Sci. **427-428** (1999) 199.
19. Y. J. Chabal, G. S. Higashi, K. Raghavachari and V. A. Burrows: J. Vac. Sci. Tech. A **7** (1989) 2104.
20. M. Niwano, M. Terashi, M. Shinohara, D. Shoji and N. Miyamoto: Surf. Sci. **401** (1998) 364.
21. G. Lucovsky: J. Vac. Sci. Tech. **16** (1979) 1225.

CHAPTER 14

Imaging

Imaging is normally associated with photographic focusing and registration on film or some other photosensitive detector, of the scattered, reflected or transmitted light from objects in visible region. Röntgen photographs register the image resulting from the passage of X-rays through objects. Similar images can be obtained also in infrared region. Following new developments in imaging with synchrotron radiation, the whole field of imaging has undergone a dramatic expansion (see Ref. 8 in Chap. **1**). In this chapter imaging in the soft and hard X-ray regions and also the infrared region are considered. In any wavelength region, a high quality imaging optical system is necessary to observe small objects. In addition, good contrast is required to identify and distinguish features in complex objects. This overview of imaging includes non-focusing imaging methods and the spatial resolution of detectors, far field and near field microscopes, contrast mechanisms, fluorescence and photoelectron microscopes and imaging using coherent synchrotron radiation. Far field focusing systems are finally explained in some detail. Examples are then given of microscopic imaging applications in research on materials used in electronic devices and also in studies of biological tissues. X-ray diffraction microscopy, also called X-ray topography used for the investigation of defects and dislocations in crystals by exploiting differences in diffraction intensities is also described from first principles.

14.1 IMAGING AND CONTRAST

The ratio of magnitudes of an object and its image in Röntgen photography is usually 1:1, where the X-rays are regarded as being in a parallel beam. The situation is also similar in CT (computed tomography). In both cases the minimum distinguishable distance between two points, that is the spatial resolution, is limited by the grain size of the photographic film or the pixel size of the CCD (see Sect. **5.6.3**) even if the X-rays are perfectly parallel. When very small objects are to be studied then a magnified image will be required. A second method for which a parallel beam of X-rays is essential is called the 'contact imaging' method. Here, the object is placed in direct contact with the photosensitive detecting medium material (photoresist or film) and then irradiated from the sample side. Subsequently the film or resist is developed and the image is then magnified and studied using optical or electron microscopes. The spatial resolution is thus dependent on the grain size. The technique of X-ray topography is one in which the image of the light diffracted by crystals is registered using photographic (or nuclear emulsion) plates and in this case, the spatial resolution is defined by the grain size. Whenever a CCD is used for detection, the spatial resolution is always dependent on the pixel size of the device which is typically about 25 microns. CCDs with sub-micron resolution have now been developed as mentioned in Sect. **5.6.3.c.**

Magnified images may be obtained without any focusing devices by using the 'projection method'. The principle resembles that of a pinhole camera. The magnification factor is given by the ratio of distances between object and pinhole and the distance between the pinhole and the screen. The resolution is defined by the size of the pinhole. Since any light emitted or scattered from the object directly passes through the pinhole, the quantity of light introduced in this system is small in comparison with that of the total light emitted and the technique is therefore relatively inefficient. However, the method can be usefully used to observe emitting objects of high intensity such as plasmas.

Currently, microscopes with focusing optics have been developed in many centers to derive maximum benefit from the exceptional brilliance of synchrotron radiation which is now so widely accessible.[1)] In the visible region, the principal focusing elements are glass convex lenses. In the ultraviolet region (UV) and near VUV down to about 165 nm quartz or fused silica lenses are needed. For the shorter wavelength regions, lenses made from CaF_2 and LiF can be used. However their limitations are mentioned in Sect. **5.1**. In the X-ray region, focusing is achieved using concave lenses, which exploit the fact that the refractive indices are a little less than 1 and these are called refractive lenses.[2–6)] Using them it is possible to obtain a light spot size of a few μm. The most commonly used optics in the soft and hard X-ray regions are mirror systems of either the Schwarzschild, Wolter, or Kirkpatrick-Baez (K-B) types. Nowadays using a K-B mirror, a spatial resolution of less than 10 nm has been achieved in the hard X-ray region using superpolished mirrors.[7)] Zone plates are also used and the minimum resolution which can be achieved using them is better than 20 nm in the soft X-ray region and better than 0.1 μm in the hard X-ray region. In the infrared region the Schwarzschild method is mainly used and the resolution is then of the order of 1 μm depending on the wavelength of the light. Recently, microbeam techniques have been developed which exploit thin glass capillaries[8)] and also narrow line beams have been made using wave guides constructed from multilayers,[9)] and used in near field regime. There are two major kinds of microscope. First, is the 'imaging type' which generates a magnified image and registers the image immediately. Second, is the 'scanning type' which uses a microbeam. In this second type the microbeam illuminates each point on the sample and is then scanned across the entire sample while registering the data, point by point, using a computer. After sample scanning has been completed the image is reconstructed by computation. In the infrared region a scanning near field optical microscope (SNOM) has been developed with narrow capillaries to achieve a resolution less than the wavelength (diffraction limit) in conjunction with synchrotron radiation and FEL.[10)] In many cases, microscopy and spectroscopy can, of course, be combined, for example, in a spectro microscope or a micro spectrometer.

Using synchrotron radiation, images in transmission are often obtained. The difference between absorption coefficients is then creates the contrast. In this case the energy of the incident light is chosen to lie at the absorption edge of a specific element or at the absorption peak specific to the compound. In the observation of magnetic domains, magnetic circular dichroism is utilized as the contrast which is derived from the differences in absorption between by right and left circularly-polarized light. In the water window region, the contrast in biological tissues derives from the C K absorption. However in the hard X-ray region, it is difficult to obtain a clear transmission image of any organism which is composed solely of elements of C, N, O and H, because of their very weak absorption contrast. In some cases, improved contrast can be obtained by adding strongly absorbing elements. One such example is in angiography, used to observe coronary arteries. In this case a harmless contrast medium is added, for example, a salt of iodine (I), which when injected into the human body gives images of blood vessels, heart and arteries with greatly enhanced contrast arising from the I K absorption.

In the hard X-ray region, the refractive index is given by $n = 1 - \delta + i\kappa$ and since the differences in δ are about 1000 times larger than those in κ, imaging making use of phase contrast rather than absorption contrast has been developed.[11] When monochromatic light passes through areas different in δ its wavelength changes. Therefore, after transmission through the material the light phase will not be uniform but will differ from point to point and therefore an intensity modulation due to the interference with the reference light will appear. This can be exploited to give clarification and enhancement in image contrast. X-rays are also refracted slightly and so the image of the refracted light may sometimes be clearer than that of the simple transmission image.[12] A novel technique to improve contrast is called 'diffraction-enhanced imaging (DEI)', in which monochromatic X-rays traversing an object are diffracted by a monochromatizing crystal when its image is registered.[13] Medical diagnostic facilities using synchrotron radiation have been developed to some extent at NSLS, HASYLAB and the PF[14] and are operational in the ESRF, CLS, Australian Synchrotron and at other centers.[15] In X-ray topography, the differences in intensity of diffracted light are utilized whereas in the infrared region it is the absorption due to molecular vibrations that provide the contrast. All vibrations have absorption energies which are specific to a particular chemical bond so that one can obtain images which can, in principle, highlight any specific bond in the molecule.

There are alternative types of microscope which detect either fluorescence, or perhaps photoelectrons from the sample rather than simply the transmitted light. By detecting fluorescence, one can obtain information from sites located deep inside the sample, while by detecting photoelectrons, the information derives from volumes close to the surface. In each case one can identify specific elements or compounds by reference to the characteristics of the spectra. To obtain fluorescence images, one needs focusing optics. A microscope which can detect photoelectrons has a special focusing electron system and is called a PEEM (photoelectron emission microscope).[16] At present, the resolution of a PEEM is about 10 nm and it will be improved in the future. A PEEM may also be used as a detector to magnify transmission image. In this case a uniform Au foil is placed immediately behind the sample to act as a photocathode emitting photoelectrons and the magnified two-dimensional image of the photoelectrons is then obtained effectively yielding a transmission microscopic image of the sample.[17]

Another imaging method, but one which uses coherent radiation, is called holography.[18] In this technique, interference patterns are generated between the light beam scattered by the object and the incident light beam or the reference light beam. It is possible only because of the high level of coherence of synchrotron radiation. From their interference patterns, an image can be produced. In the case of laser holography, imaging is achieved by the use of the same wavelength light, but in the case of synchrotron radiation holography it is obtained *via* the analysis of the holographic patterns. In the case without the use of a reference light beam, the interference pattern of the object is recorded as an 'in-line hologram'. The intensity of the pattern is then inversely Fourier-transformed and the image of the object finally obtained.[19] In this case the detectors limit the spatial resolution and therefore the distance between the object and detector should be sufficient to magnify the interference pattern so as to easily distinguish the structures of interest within the pattern. When the object is a material consisting of small particles or possessing a non-uniform structure, the hologram is given the name 'speckle', from which one can extract the grain structure of crystals, information which cannot be observed using usual visible light.

The interference pattern obtained using reference light is called a 'Fourier hologram'. However, when viewing transmission images of small samples using coherent light, there is always the possibility that interference may affect normal imaging adversely. In such cases, incoherent light should be generated by letting the coherent light pass through some scatterer, such as graphite powder sandwiched by two Kaptan films.

14.2 IMAGING USING FOCUSING OPTICS

14.2.1 Focusing optics

Examples of focusing optics are illustrated in Fig. 14.1. Figures 14(a) and 14(b) present optics using mirrors, and are respectively the Schwarzschild and Wolter types. The K-B type is shown in Fig. 5.7. Figures 14(c) and 14(d) are schematics of zone plate and refractive lens instruments, respectively. In those using mirrors, there is no achromatic aberration. In the Schwarzschild type the light is incident on the mirrors at near normal incidence. The objective is composed of two spherical mirrors one of which is convex and the other concave. They are situated so as to cancel the spherical aberrations of the third order. In visible and infrared microscopes, Schwarzschild optics is often used instead of lenses. To obtain an image with few aberrations, the focusing optics must satisfy Abbe's condition (see App. **A**) and also the condition that the path differences for light emitted from the same point should be smaller than $\lambda/4$ which is met by the Schwarzschild type. There is always the problem that the shorter the wavelength, the lower the reflectivity of the mirrors, but this difficulty can be overcome using multilayer mirrors as mentioned in Sect. **5.2.2**. Objectives coated with multilayers can be used effectively up to about 100 eV. In Wolter type instruments the light is incident at grazing incidence angles. The objective is then a cylindrical mirror combining the surfaces of a hyperboloid of revolution and an ellipsoid of revolution, both located so as to have a common focal point to satisfy Abbe's condition. Note that

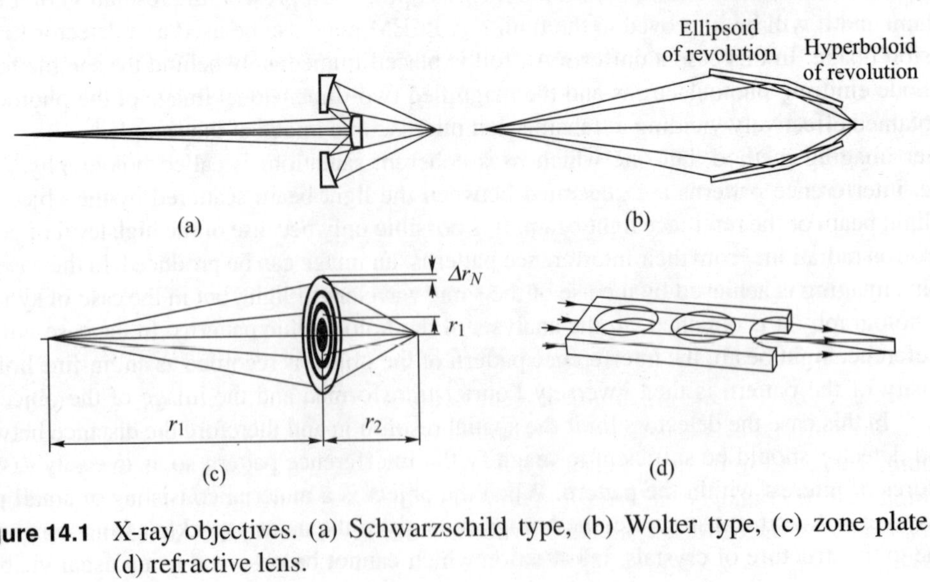

Figure 14.1 X-ray objectives. (a) Schwarzschild type, (b) Wolter type, (c) zone plate and (d) refractive lens.

the Wolter systems used in telescopes are also mirrors which combine a paraboloid of revolution with hyperboloid of revolution to focus the parallel light. The K-B type is composed of two mirrors. Each one successively focuses the light, first in the horizontal plane and then the vertical plane. Abbe's condition is not satisfied, but if the acceptance angle is small, there is no problem. This type has the advantage that the focusing is achieved by the two mirrors independently for the horizontal and vertical planes, resulting in a minimization of the aberrations.

Zone plates are essentially circular transmission diffraction gratings composed of alternate transparent and opaque concentric zones. The radius of n-th zone is made to be

$$r_n = \sqrt{nf\lambda}, \quad n = 1, 2, \ldots\ldots N, \tag{14.1}$$

where λ and f stand for wavelength and the focal length of the diffracted light of the first order, respectively and N, the total number of zones. The outermost zones have the narrowest widths. The focal length of diffracted light of m-th order, f_m is given by

$$f_m = \frac{f}{m} = \frac{r_1^2}{m\lambda}, \quad m = 1, 3, 5, \ldots\ldots. \tag{14.2}$$

Here $f_1 \equiv f$. The light from a point source on the optical axis apart from the zone plate by r is focused on the optical axis oppositely apart from the zone plate by r', which satisfies

$$\frac{1}{r} + \frac{1}{r'} = \frac{1}{f_m}. \tag{14.3}$$

The spatial resolution is expressed using the width Δr_N of the outermost zone as $1.22\Delta r_N/m$. There is also a special zone plate, in which the surface of a single crystal is manufactured as a zone plate. This combines the functions of Bragg reflection and the focusing of a zone plate and is called a 'Bragg-Fresnel zone plate'.

The first refractive lens to be fabricated was simply an array of cylindrical holes made in an Al plate using a drill. Generally, materials with a small absorption coefficient are desirable for the lens. Therefore materials such as acryl, diamond, *etc*., are commonly used.[4,5] In the case of lenses with cylindrical holes, the focal plane is perpendicular to the cylindrical axis. Therefore, in order to focus in both the horizontal and vertical directions, two kinds of hole whose axes are orthogonal, are required. It is possible to obtain spherical holes using liquid glycerin which can focus in both planes. This is sometimes called a 'bubble lens'.[6] The number of holes (or micro lenses) in such an array may be from a few 10 to a few 100 depending on the kind of materials used, the photon energy and the focal length. Assuming that the shape of the micro lens is a sphere of radius of R, the number of lenses is N and refractive index of the material is n, the focal length F will be given by $F = R/\{2N(1-n)\}$. Lenses without any spherical aberration have been developed by choosing the shape to be parabolic.[5]

14.2.2 Soft and hard X-ray microscopes and their applications

In the soft X-ray region, the predominant technical problem is that of working in vacuum. That is, the optical paths must be evacuated, and therefore the sample must also, in general, be contained within a vacuum. There are some exceptions. In the observation of wet organisms in the transmission mode, the sample can be contained in water in a thin wet cell made using two windows from thin films such as Si_3N_4 and with the cell itself situated in the vacuum (see Sect. **7.3.1**). An alternative method makes use of a very narrow gap in air in the optical path using similar thin film windows where the sample is located. If the gap is only of a few microns then the measurement can be performed with minimal losses of intensity even in the soft X-ray region. Of course, in the hard X-ray region, the sample can be maintained in air and the windows, if needed, can be relatively thick.

Microscopes were developed first in the hard X-ray region but it was always difficult to suppress mirror surface figure or shape errors and obtain magnified images or small spot sizes with few aberrations using reflection optics at grazing incidence. For these reasons, the first significant advances in X-ray microscopy using synchrotron radiation were initiated in the soft X-ray region. However, grazing incidence mirrors can now be precisely fabricated for use even in the hard X-ray regions.[7)] Zone plates made from thin films have been prepared using electron beam lithography for the use in soft X-ray region, but this type of thin zone plate cannot be used in hard X-ray region, because the X-rays are readily transmitted. Relatively thick zone plates have now been developed and their usable range has been extended well into the region of short wavelengths, up to energies of 10 keV or more. For example, at 80 keV, a microbeam of a diameter less than 1 μm has been obtained.[20)] There is continuing progress in microscopic imaging using refractive lenses although systems using mirrors and zone plates are still widely used.

At Elettra, a microbeam using Schwarzschild type objectives has been created and microscopic photoelectron spectroscopy carried out.[21)] The scanning photoelectron microscopy performed using this microbeam combined with photoelectron spectroscopy has enabled the electronic structures of surfaces and the bulk near the surface to be obtained. This microscope must, of course, be kept in a chamber at ultra-high vacuum. Multilayers with periodic structures will enhance the reflectivity at chosen wavelengths. Therefore many sets of mirrors are required to produce incident light at different energies. At Elettra, three Schwarzschild mirrors coated with multilayers Mo/Si (74 eV), Mo/Si (95 eV) and Ru/B_4C (110 eV) are interchangeable without exposing them to air. The minimum diameter of the incident pinhole is 5 μm, through which 10^{12}/s photons enter. The demagnifying factor is 1/50. We give the description of a study by an imaging microscope for the detection of soft X-ray fluorescence which was made to investigate the diffusion of Au on Si wafers (see Fig. 14.2).[22)] In this case, excitation was made not using synchrotron radiation, but by using an electron beam. The sample was prepared by deposition

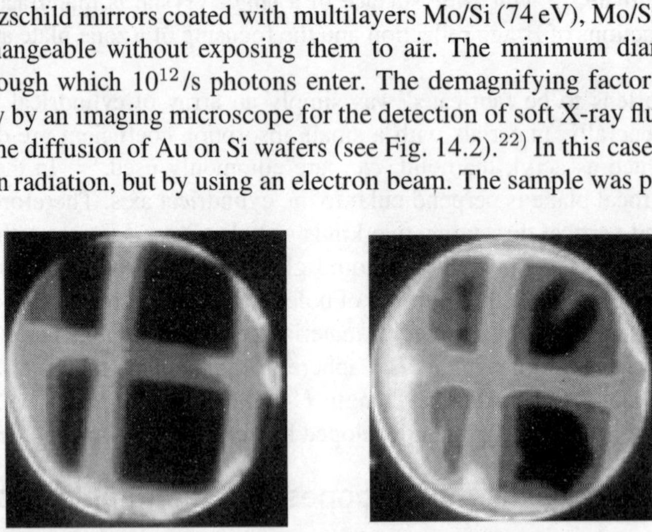

(a) (b)

Figure 14.2 The image of a gold pattern deposited on a Si wafer through a Ni mesh-screen by detecting the Si L emission and using a Schwarzschild objective. (a) For the sample as deposited and (b) heated at 450°C for 3 hours. The high intensity area corresponds to the uncovered Si area and weaker intensity area, the area covered with Au, through which the Si L emission passed.

of Au on Si wafer which was covered with a mesh-screen. Part of the surface of the Si wafer corresponding to the lines of the mesh-screen was not coated with Au, but the other square shaped parts were coated with Au. In that part, the Si L emission passing through Au was measured. When the Si wafer was heated up, the boundary of the square became diffuse and its shape became unclear. From such results it was deduced that Au moved to an uncoated part of the Si wafer with increase in temperature.

An example of the observation of magnetic domains using a zone plate microscope at ALS is shown in Fig. 14.3.[23] In this microscope, two zone plates were used. The first one was a condenser zone plate (diameter, 9 mm) to gather and monochromatize the light, and then to focus it on a pinhole of 10 μm diameter. The sample was situated 100 μm apart from the pinhole. The resolving power of the system is 700 in the 250 $\sim$ 900 eV region. The second is a micro zone plate, to make magnified images, whose diameter and width of the outermost zone are respectively 63 μm and 25 nm. The detector was a back-illuminated CCD with 1024 $\times$ 1024 pixels. The sample was a Fe(4 Å)/Gd(4 Å) magnetic multilayer fabricated on Si_3N_4. The circularly-polarized light from bending section was used. The direction of the magnetic field was perpendicular to the sample, *i.e.*, parallel to the direction of the light. Figures 14.3(a) and 14.3(b) show the microscopic images obtained using the light at the Fe L_3 and L_2 absorption edges. The contrast of the magnetic domains was obtained using the fact that the sign of the magnetic circular dichroism at the L_3 and L_2 edges are opposite and these are clearly indicated in Fig. 14.3(a) and Fig. 14.3(b), where the resolution was 25 nm. This method is very useful in the observation of the magnetic domains.

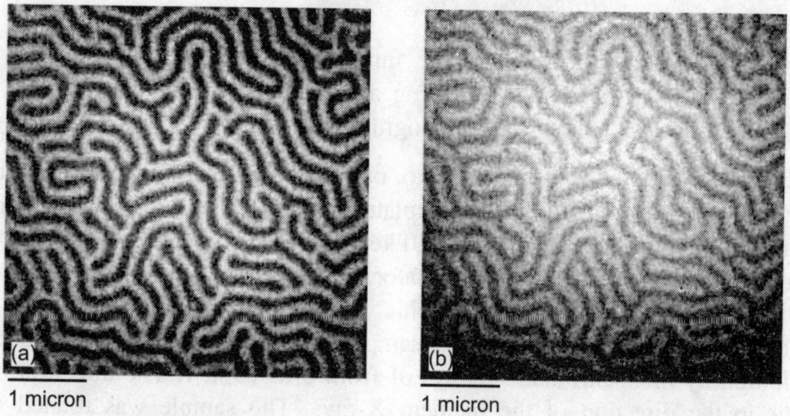

Figure 14.3 The transmission image of a Fe/Gd multilayer using right- and left-circularly polarized light at (a) the L_3 and (b) the L_2 absorption peaks of Fe using a zone plate (by courtesy of D. Attwood, © Elsevier 2001).[23]

Figure 14.4 gives an example of images of cells obtained using zone plates.[24] The sample was a cell called 'COS-7', derived from African green monkey's kidney, cultured, fixed and then measured in a wet condition. Its thickness was less than 10 μm. The image was taken using light in the water window region of 2.4 nm at BESSY II without the use of a contrast medium. In the figure, N, Nu and B indicate the nucleus, nucleolus and border of the cell, respectively. The arrow indicates a mitochondrion. The nucleus is found to be surrounded by small granules, which seem to include proteins, based on their contrast. This figure clearly demonstrates that one can observe fine structures in cells using soft

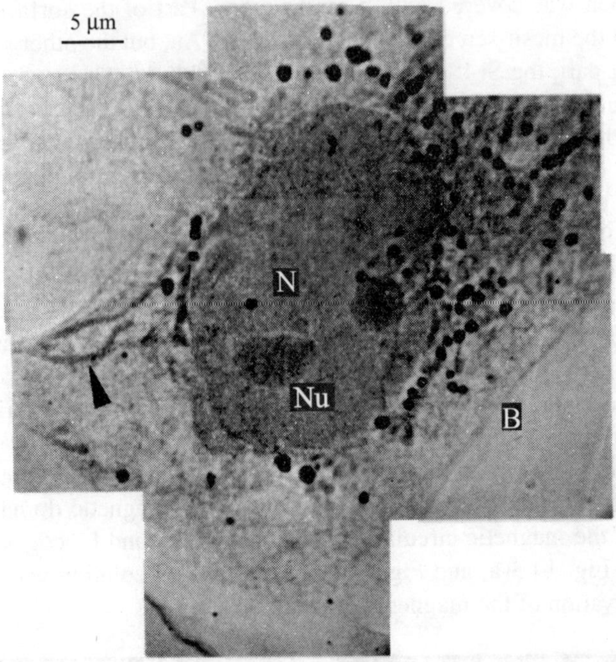

Figure 14.4 The transmission microscopic image of a cell derived from the kidney of an African green monkey using a zone plate (by courtesy of H. Kihara, © International Union of Crystallography 1998).[24]

X-ray microscopes with resolution comparable to, or better than that which can be obtained using a conventional optical microscope. Currently zone plates have been developed for the hard X-ray region and a microbeam of 1 μm has been obtained at 20 keV, delivering an intensity of 10^{13} photons/s/mm^2 in combination with a double crystal monochromator.[20]

An example is also given of the tomographic imaging of small objects using a Wolter type objective.[25] The 10 times magnified image of a sample detecting the X-ray fluorescence was obtained following irradiation by monochromatic X-rays of from 6 to 10 keV. The direction of the detection is perpendicular to the direction of the incident X-rays. The sample was rotated around an axis perpendicular to the directions of both the incident X-rays and the detection of X-ray fluorescence. The images were taken at each step of a rotation of 7.2° and eventually 50 magnified images were registered at different angles. The detector used was a CCD. Figure 14.5 shows a schematic of the sample and the results of the analysis. The sample used was for demonstration purposes and was a quartz capillary accommodating narrow lines of Fe, Ni and Cu as shown in Fig. 14.5(a). Figures 14.5(b), 14.5(c) and 14.5(d) are three-dimensional images reproduced from two-dimensional images taken by incident X-rays whose energies are a little higher than the K absorption edges of Cu, Ni and Fe, respectively. One can see that, when the excitation energy was low, only the Fe line could be observed (Fig. 14.5(d)) but as the energy was increased, the Ni and Cu lines could also be measured (Fig. 14.5(c) → Fig. 14.5(b)).

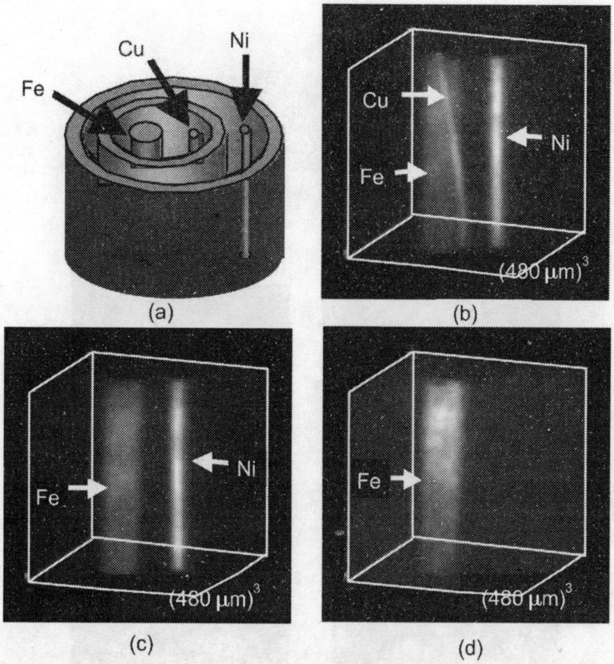

Figure 14.5 An X-ray fluorescence tomographic image using a Wolter type objective. (a) A sample of Fe, Ni and Cu narrow wires accommodated in a quartz capillary. Images from (b) to (d) present the X-ray fluorescence tomographic images excited by X-rays of different energies (see text) (by courtesy of N. Watanabe, © Elsevier 2001).[25)]

Finally, imaging by X-ray fluorescence microscopy using the K-B configuration is described. Using synchrotron radiation, it is possible to irradiate samples with X-ray microbeams with diameters of only a few microns and whose energy is also tunable. The sample is set on an *XYZ* stage and the stage then moved to scan the sample for analysis. The X-ray fluorescence is usually measured using energy dispersive detectors. The spatial resolution of this method is sufficient to permit analyses to be made on small samples, such as single aerosol particles in the atmosphere.[26)] Using this equipment, imaging of the two-dimensional distribution of the elements in the samples (element mapping) can be achieved. Figure 14.6 gives an example of the mapping of trace elements in hair with a microbeam which has a diameter of only about 5 μm and was studied at the PF.[27)] The mapping was made on cross sections selected at the positions 2 mm apart from the root of hair. The hair sample was embedded in thin epoxy (10 μm thick) for reduction of background and improvement of the spatial resolution in cross section. Figure 14.6(a) shows a photograph taken using an optical microscope and Figs. 14.6(b), 14.6(c), 14.6(d) and 14.6(e) show the X-ray fluorescence mapping with respect to the elements S, K, Cu and Zn, respectively. It can be seen that S, Cu and Zn are distributed uniformly over the hair-shaft, while K is located in the cortex. The X-ray elemental mapping method is excellent for use in studies of biological samples, for instance, in the analysis of trace elements in annual rings of trees,[28)] microbes in sea and in the analysis of poisonous metal elements in brain tissue[29)] and so on.

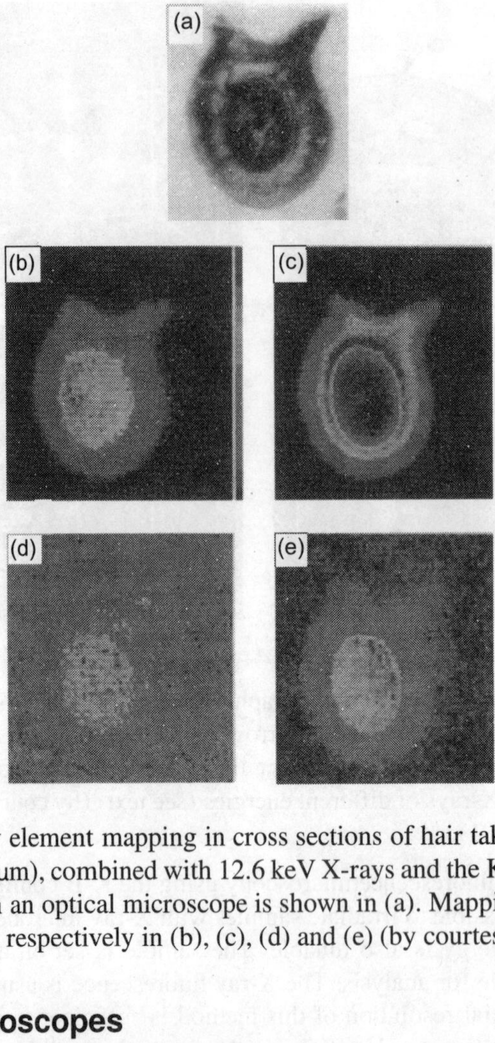

Figure 14.6 The X-ray element mapping in cross sections of hair taken using 51 × 51 steps (1 step: 3 μm), combined with 12.6 keV X-rays and the K-B mirrors. Photograph taken with an optical microscope is shown in (a). Mappings on S, K, Cu and Zn are shown respectively in (b), (c), (d) and (e) (by courtesy of A. Iida, © Elsevier 1993).[27)]

14.2.3 Infrared microscopes

The exceptional brilliance of synchrotron radiation offers many advantages even in infrared micro spectroscopy. In several synchrotron radiation facilities scanning microscopes of such types using microbeams made by Schwarzschild objectives, having large numerical aperture, *i.e.*, large acceptance angle are used.[30)] However an increase in the numerical aperture will not allow the focal size to be reduced without limit because the focal size is ultimately determined by the wavelength of light used, due to light diffraction effects (see App. **A**). Figure 14.7 shows the layout of the infrared microscope at SPring-8 as an example. Its numerical aperture is 0.5. Figure14.8 shows an image of the light source at the sample position in this system using light in the 1 to 10 micron wavelength region.[30)] The incident light is focused to a spot with FWHM of about 10 μm. In the figure is also shown the intensity distribution of a standard infrared light source (a Grobar lamp) focused to a spot with FWHM of about 50 μm for comparison. Clearly, the infrared microbeam using synchrotron radiation is about 100 times more intense at the maximum than that of the standard source and the former is much narrower than the latter (about 1/4). Infrared micro spectroscopy has much potential in measurements on organisms and cells,[31] non-uniform materials in which domains of different phases coexist and also new materials obtained at the early stages of new material synthesis *etc*.

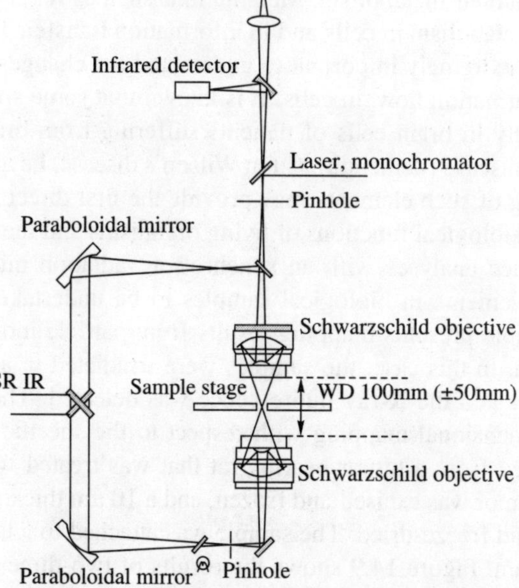

Figure 14.7 An infrared microscope using a Schwarzschild objective. The light is incident on the sample from below in transmission measurements and from above in reflection measurements. WD: working distance. SR: the synchrotron radiation infrared passing through the interferometer (by courtesy of SPring-8).

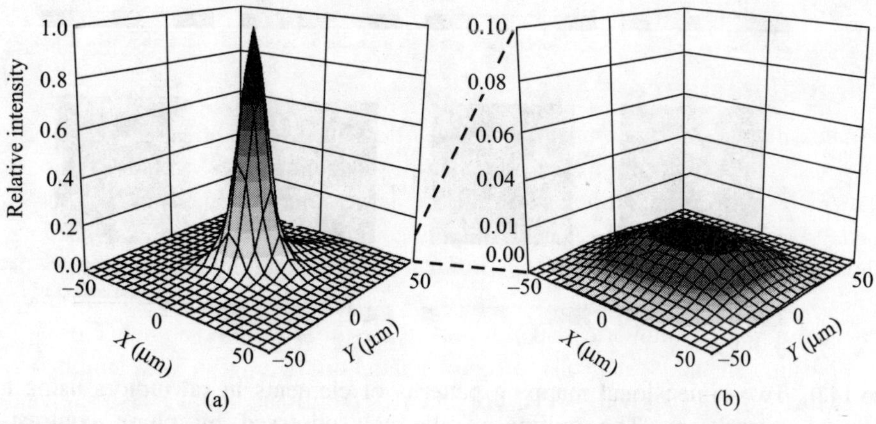

Figure 14.8 The intensity distribution of a microbeam for an infrared scanning microscope (a) at SPring-8 and (b) using a Grobar lamp (by courtesy of SPring-8).

14.2.4 Potential uses of synchrotron radiation microscopy in physiology

Metallic elements such as Fe, Zn, Mn, Cu, Co, function as active centers in enzymes and play crucial roles in biological function. The changes in quantities of these metallic elements in organisms may have a considerable influence on their metabolism. Metallic ions such as K^+, Na^+, Mg^{2+}, Ca^{2+}, participate fundamentally in energy metabolism in cells and in information transfer. Therefore the investigation on the behavior of the ions is extremely important to understand the changes in energy states, the motion of molecules and also information flow, in cells. It is known that some special trace metallic elements may accumulate abnormally in brain cells of patients suffering from brain diseases. These elements include Al in Alzheimer's disease (dementia), Cu in Wilson's disease, Fe and Cu in Parkinson's disease, *etc.* Therefore the mapping of such elements may provide the first direct information to offer us some clarification about the physiological functions of living organisms and the mechanisms of diseases.

Using X-ray fluorescence analyses with an synchrotron radiation microbeam will allow element mapping for most trace elements in biological samples to be undertaken without incurring damage to the samples. One example presents mapping results from particle induced X-ray emission (PIXE) using a proton microbeam. In this case, the samples were irradiated in air with proton microbeam of less than 1 μm in diameter and the X-ray fluorescence was detected. The sample was scanned in the *XY*-plane and the two-dimensional mapping with respect to the specific elements was performed.[32)] The samples were prepared from a tumor-bearing rat that was treated with anti-cancer drug (BrdU). After killing the rat, the tumor was excised and frozen, and a 10 μm thick slice was cut and mounted on a 4 μm thick Myler film and freeze-dried. The sample was attached to a target holder and irradiated by 2.6 MeV proton microbeam. Figure 14.9 shows the results of two-dimensional mapping with respect

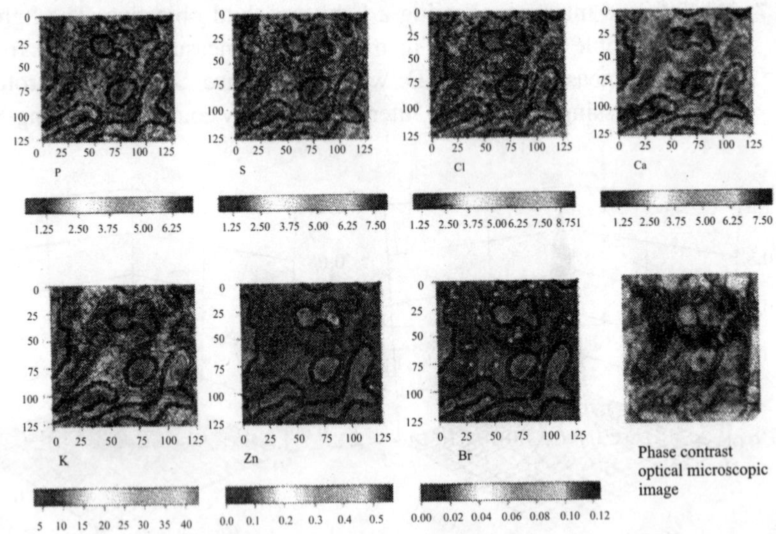

Figure 14.9 Two-dimensional mapping patterns of elements in rat tumors using a proton microbeam. The outline of the cell observed by phase contrast optical microscope is overlapped on each image (45 μm × 45 μm) (© World scientific 2002).[32)]

to the elements P, S, Cl, Ca, K, Zn and Br in the tumor tissue. The different distribution patterns for each element are clear and the element of Br, that is, the administrated BrdU seems to accumulate in the nucleus of the cell, although is not wholly conclusive because of the small quantities involved. It is clear that such measurements could be made more precisely using synchrotron radiation.

Since synchrotron radiation in the infrared is at least 100 times more brilliant than the infrared radiation emitted from conventional, blackbody light sources it is possible to create an infrared microbeam yielding excellent spatial resolution combined with a good S/N ratio. Infrared can be used to identify biochemical composition since every molecular bond has its own quite specific absorption energy. Figure 14.10 shows the infrared spectrum of hemoglobin as an example.[31] Similarly, one can obtain the spectra which characterize lipids, proteins, nuclear acids, *etc.* By choosing the energy of the microbeam, one can achieve two-dimensional mapping with respect to every chemical bond.[31] Therefore information can be obtained which will augment that obtained from X-ray fluorescence analysis and other techniques in studies on the relationships between biological structures and their functions. There is even the possibility of high resolution space-resolved mapping of chemical materials of not only at cellular levels, but even at sub-cellular levels, such as in the nucleus and cytoplasm, *etc.* All of these techniques may also be explored throughout the ns to ps time domains by taking advantage of the unique time modulation of all radiation produced by any synchrotron radiation light source. This will make it possible in principle to undertake time-resolved mapping observations of chemical reactions in real time.

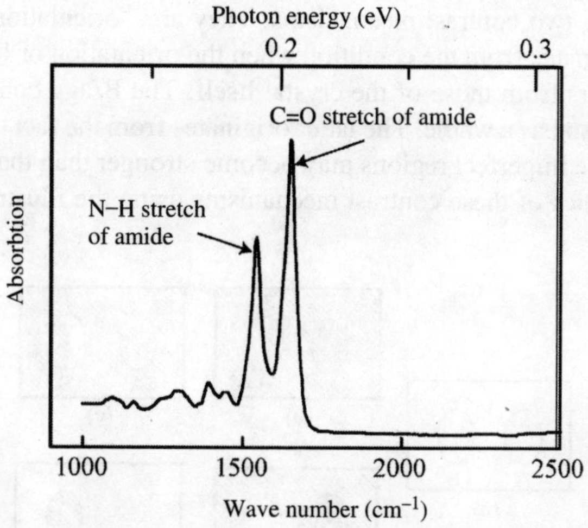

Figure 14.10 The infrared spectrum of hemoglobin (by courtesy of L. M. Miller, © Taylor and Francis 2000).[31]

14.3 X-RAY TOPOGRAPHY

Any distorted region in a crystal will modify the diffraction behavior of an X-ray beam. X-ray topography is an imaging technique for the identification of defects in such regions, since the contrast will originate from any spatial variation in intensity of the diffracted X-ray beam. Currently, the minimum spatial resolution in X-ray topography is roughly from 2 to 3 microns and is governed by the degree of parallelism of the X-ray beam, not the grain size of nuclear emulsion plate. Since the transmission of X-rays beam through any material is large in comparison with that of an electron beam, we can observe defects spreading in a bulk crystal using this method even up to 1 mm in thickness. The method is particularly appropriate in the investigation of semiconductors where defects will affect the device functionality, for example in the study of defect distributions, the characterization of defect properties in grown and processed crystals such as Si, GaAs and so forth. Even defects in heavy-metallic crystals can be measured.

The brilliance of synchrotron radiation has substantially extended the range of X-ray topographic applications and the beam is sufficiently parallel and brilliant for the resolution of topographic images to be reduced to about 1 μm. Defects in wider regions of bulk crystals can now be observed non-destructively with short exposure times. By using *in situ* techniques, the generation and structural modification of defects progressing during crystal growth, or under the effects of external applied fields can be quantitatively investigated in real time. Previously such kinds of investigations were difficult to conduct. As in electron microscopy, the defect nature given by Burgers vector (a vector characterizing the displacement caused by defects) can be established by selecting diffraction conditions suitably. Even slight distortions and mis-orientations within any crystal can also be easily detected.

14.3.1 The principles

The contrast in X-ray topographic diffraction images of a crystal with distorted or imperfect regions is defined by the following two contrast mechanisms. They are, 'orientation contrast' and 'extinction contrast'. The former originates from the condition when the orientation or lattice spacing in particular regions in the crystal differs from those of the crystal itself. The Bragg conditions in the regions will differ from those in the crystal as a whole. The latter originates from the fact that the integrated intensity of X-ray diffracted from the imperfect regions may become stronger than that from the perfect regions. We discuss the characteristics of these contrast mechanisms using the illustrations in Figs. 14.11(a) to 14.11(e).[33]

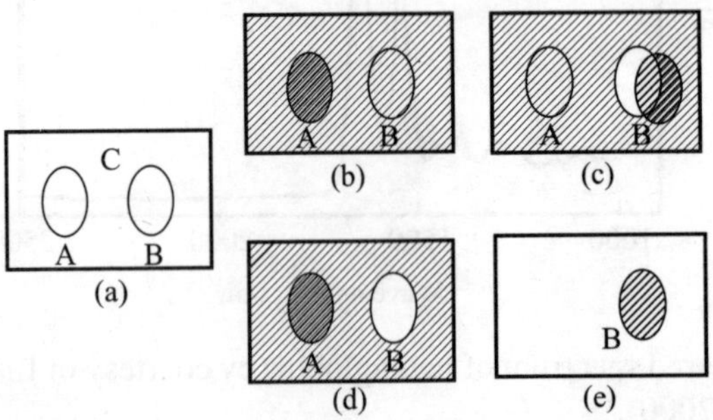

Figure 14.11 The generation of contrast by an X-ray beam diffracted by atomic distortions in a crystal (by courtesy of A. R. Lang, © Elsevier 1993).[33] See details in text.

In Fig. 14.11(a), suppose that the crystal consists of a perfect matrix C including two distorted regions, A and B. In region A, the lattice is imperfect, but its orientation is the same as that of C, while in region B, the lattice is perfect, but its orientation differs slightly from that of C. We now suppose that continuous (white) X-rays are used and that a nuclear plate is placed close to the crystal back. The region A and the matrix C can then be observed under the same Bragg conditions. Since the reflected X-ray intensity from the region A will be stronger than that from the region C by extinction contrast, the distinction between regions A and C is easily made. Contrarily, region B will reflect X-rays under different conditions from regions A and C, but the regions cannot be identified, because the X-ray intensity reflected in region B is the same as that for region C. The topographic image in this case is illustrated in Fig. 14.11(b). If the distance from the crystal to the X-ray plate is increased, then the direction of the reflected X-rays from the region B will differ from those of the reflected X-rays from the region C and the region B can be recorded with a finite displacement as shown in Fig. 14.11(c). Suppose that monochromatic collimated X-rays are used and the crystal is oriented such that the region C satisfies Bragg condition. The topographic image will then be as in Fig. 14.11(d). Region A will now resemble the previous case by extinction contrast. However, region B will be identified as an area with no image on the nuclear plate since there is no X-ray reflection. An appropriate rotation of the crystal could set region B to satisfy Bragg law leading to a topographic image, but then there will be no reflections from A or C (see Fig. 14.11(e)). From the set angles of the crystal, one can then measure differences between lattice spacings and orientations of region B from those of region C. No significant change in contrast will result from increase in distance between the crystal and the nuclear plate when using perfectly parallel monochromatic X-rays, but a consequence of such an increase would be a loss of resolution in the topographic image due to the band width and lack of parallelism of the incident X-rays. Orientation contrast is ideal for detection of lattice distortions and to observe the development of sub-grains in deformed crystals. Extinction contrast is used mainly to detect the distribution of lattice defects as dislocations.

14.3.2 The observation of defects using X-ray topography

X-ray topography can be divided into two primary modes, that is, the 'reflection' and 'transmission' modes. The reflection mode is applied to the observation of defects existing in the surface regions of an epitaxial film grown on a substrate. For example, Fig. 14.12 illustrates the Berg-Barrett technique in which monochromatic parallel X-ray beam from an X-ray line source is incident on the surface region and the reflected X-ray beam is then recorded in the nuclear plate. An asymmetrical arrangement where

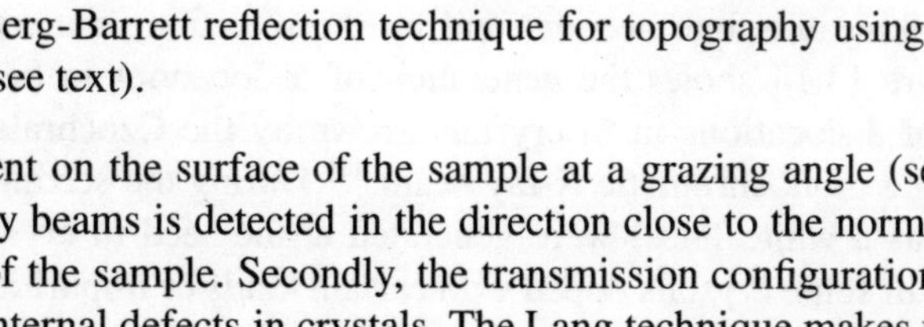

Figure 14.12 The Berg-Barrett reflection technique for topography using monochromatic X-rays (see text).

the X-ray beam is incident on the surface of the sample at a grazing angle (see Fig. 5.16) and where the set of reflecting X-ray beams is detected in the direction close to the normal of the surface gives a maximum illumination of the sample. Secondly, the transmission configuration is normally associated with the observation of internal defects in crystals. The Lang technique makes use of a monochromatic X-ray beam as shown in Fig. 14.13. A thin parallel beam of monochromatic X-rays generated by an

X-ray point source is passed through the crystal along the path OA. The crystal is set to fulfill the Bragg condition. The incident X-rays are reflected along the line OA and information about defects in the region △OAB is recorded in the nuclear plate as intensity variations in the reflected X-ray beam. This is called 'section topography'. Defects at particular depths within the crystal can be recorded by inserting a slit between the crystal and the nuclear plate as shown in Fig. 14.13(b). This is called 'tomography'. In the figure the method is also shown giving an image with directly transmitted X-rays. In order to obtain the distribution of defects across a wide area of the crystal, the Lang technique can be used in which the crystal and the plate are moved in parallel, synchronously against the incident X-ray beam as shown in Fig. 14.13(b). Thc mcthod is also known as 'projcction topography'.

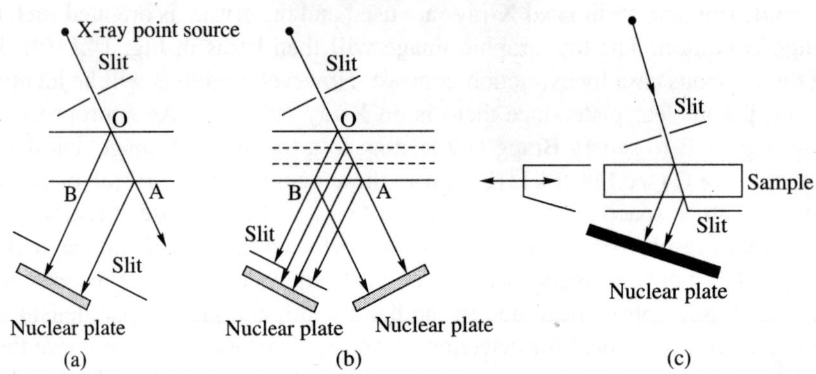

Figure 14.13 Techniques for transmission topography using monochromatic X-rays (details in text).

Monochromatic X-ray beams at short wavelengths are used because of their high level of transmission through materials and often the Ag Kα_1 or Mo Kα_1 lines are used. For synchrotron radiation of course the wavelength of the incident X-ray is not restricted. X-ray beams enable us in principle to investigate defects in quite thick crystals. As the beam passes through the material, the intensity will reduce exponentially with thickness. Usually, for thickness d of a crystal in transmission, the observation is chosen to roughly fulfill the condition $\mu d \approx 1$, where μ is the X-ray absorption coefficient of the crystal. For characteristic lines such as Ag Kα_1 and Mo Kα_1 the values of μ of Si are 7.938 and 15.58 cm^{-1}, respectively, while those for Ge are 181 and 344.7 cm^{-1}, respectively. The latter values are very close to those of GaAs. In these cases, a suitable thickness would be in the range from several tens of microns to about one mm. A crystal can be examined even in the condition up to $\mu d \approx 2$. In a symmetrical arrangement such as one in which the diffraction plane defined to be the plane including both incident and diffracted X-rays, is perpendicular to the crystal surface, the distortion of the topographic image from its original shape can be minimized. The sample crystal should be positioned taking this into account and by the selection of a reflecting plane that giving strong diffraction. Figure 14.14 shows the generation of dislocations in Si diffraction. Figure 14.14 shows the generation of dislocations in Si crystals grown by the Czochralski technique revealed by X-ray topography using a monochromatic X-ray beam.[34] During the seeding stage of crystal growth, dislocations (recorded as a white lines) were generated at the seed to crystal interface (S/C) due to thermal shock stresses. In some crystals doped with certain kinds of impurities, the dislocations cannot penetrate into the crystals. From such work, suppression effects on dislocations with selected impurities have been found and applied to grow crystals largely free from dislocations. An intense X-ray source is essential to minimize plate exposure times. This might enable an X-ray television (TV) observation system to be used instead of the nuclear plate.

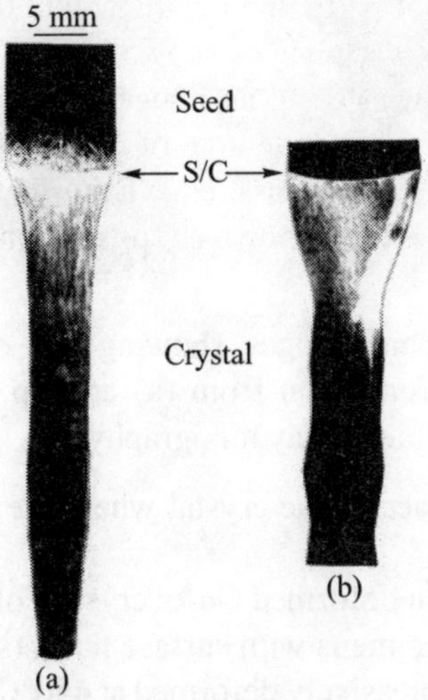

Figure 14.14 The observation of dislocation generation at the seed to crystal interface due to thermal shock during the seeding stage of Si crystal growth, obtained using X-ray topography.[34)] The Si seed used was prepared (a) from high purity Si and (b) from heavily B-doped Si.

Such a system would permit *in situ* observations to be made for the generation, multiplication and motion of dislocations in a crystal under the application of stress and at elevated temperatures. For example, *in situ* observations of the growth and modification of solid helium crystals in a high pressure apparatus and at ultra-low temperatures have been carried out using synchrotron radiation at the PF. Figure 14.15 is a white X-ray topographic image showing the martensitic transformation of solid ^{4}He, grown under a pressure at 3.5 MPa, from an hcp structure into a bcc structure.[35)] These *in situ* X-ray topographic observation experiments could be extended to study dislocation activities under stress, their impurity effects and other types of deformation behavior in crystals.[36)]

An X-ray topographic image will become optimal in the transmission configuration under the condition of $\mu d \approx 1$ for an absorption coefficient μ and crystal thickness d. In addition, if such a crystal has a very high degree of crystallographic perfection, then the defects in a thick crystal can be examined even with the condition of μd of from 10 to 20 according to the phenomenon of X-ray anomalous transmission, or the 'Borrmann effect'. This effect can be understood as follows. The incident and diffracted X-ray beams create two standing waves in the crystal. In one of the standing waves, the electric field creates a minimum at the lattice planes and in another, a maximum. In the case of the former, the X-ray beam satisfying the Bragg condition is transmitted anomalously and with very low loss through the crystal, in contrast to those X-rays which do not meet the Bragg condition. In those regions distorted with defects such anomalous transmission does not occur. Thus, any X-ray beam passing through such regions loses intensity, leading to a low intensity in the topographic image. In this

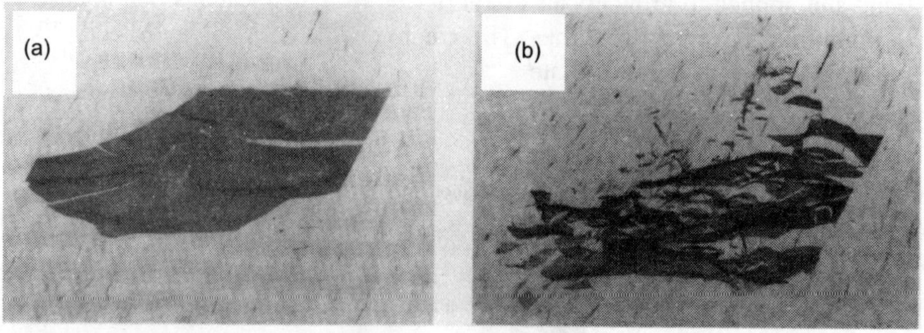

Figure 14.15 Section topographic images showing the changes in solid ^{4}He during the martensitic transformation from (a) an hcp structure into (b) a bcc structure observed using white X-ray topography.[35]

method, distortions at the back surface of the crystal where the X-ray exits can be detected as images with a strong contrast.

Figure 14.16 shows dislocations in deformed GaAs crystals observed by X-ray topography using the Borrmann effect.[37] Rectangular specimens with surface flaws (scratches) were drawn using a diamond stylus at room temperature and compressively deformed at 450°C, and subsequently 0.5 mm thick plates were cut parallel to the active primary slip plane from the samples. Dislocations, identified as black lines in the images, are generated preferentially from the scratches. In III-V compound semiconductor crystals with zinc blend structures such as GaAs, three types of dislocations can be identified namely, α, β and screw (designated by (s)) dislocations. Figure 14.16(c) shows schematically the types of dislocations which can be generated in a deformed crystal according to the crystallographic geometry. Different types of dislocations reveal differing dynamic activities and susceptibilities from impurities. In Si-doped GaAs, α dislocations are generated and they move faster than other types of dislocations

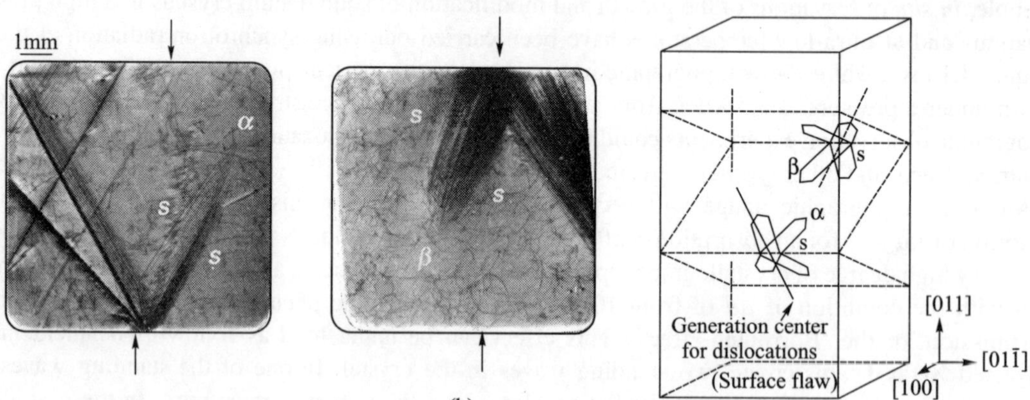

Figure 14.16 The observation of dislocations in X-ray topography using the Borrmann effect (a) in Si-doped GaAs (b) in Zn-doped GaAs.[37] (c) illustrates the geometry of the hexagonal half-loops of dislocations which develop from preferential generation centers.

to elongate screw dislocations as seen in Fig. 14.16(a). In Zn-doped GaAs, β dislocations are generated and they also move faster than other types of dislocation to elongate screw dislocations as shown in Fig. 14.16(b).

A parallel and monochromatic X-ray beam is indispensable for the accurate measurement of lattice distortions and the lattice constants of any crystal. The combination of a plane wavefront synchrotron radiation X-ray beam used in conjunction with from 2 to 4 monochromatizing crystals, with a high degree of crystallinity, greatly enhances the techniques capabilities when used at a high order of reflection. In such an apparatus, the measurement and determination of lattice distortions and mis-orientation around lattice defects can be achieved with an extremely high sensitivity of around ppb. The technique is called plane wave topography or ultra-plane wave topography which has been applied in the detection of micro defects and in the precise measurement of the lattice constants of crystals.[38]

Recently X-ray topographic studies have been further developed and extended, for example, to the observation of large crystal area,[39] three-dimensional observation (tomography)[40] and time-resolved observations.[41]

REFERENCES

1. Int. Conf. on X-Ray Microscopy has been held every 2~3 years. The proceedings are published in J. de Phys. IV **104** (2003), IPAP Conf. Ser. **7** (2006) , J. Phys.: Conf. Ser., **186** (2009), AIP. Conf. Proc. **1365** (2011) *et al.*
2. S. Suehiro, H. Miyaji and H. Hayashi: Nature **352** (1991) 385.
3. A. Snigirev, V. Kohn, I. Snigireva and B. Lengeler: Nature **384** (1996) 49.
4. Y. Ohishi: J. Jpn. Soc. Synch. Rad. Res. **14** (2001) 287, in Japanese.
5. B. Nohammer, J. Hoszowska, A. K. Freund and C. David: J. Synch. Rad. **10** (2003) 168.
6. Y. Kohmura, M. Awaji, Y. Suzuki, T. Ishikawa, Yu I. Dudchik, N. N. Kolchevsky and F. F. Komarov: AIP Conf. Proc. **507** (2000) 549.
7. H. Mimura, S. Handa, T. Kimura, H. Yumoto, D. Yamakawa, H. Yokoyama, S. Matsuyama, K. Inagaki, K. Yamamura, Y. Sano, K. Tamasaku, Y. Nishino, M. Yabashi, T. Ishikawa and K. Yamauchi: Nature Physics **6** (2009) 122.
8. D. H. Bilderback and D. J. Thiel: Rev. Sci. Instr. **66** (1995) 2059.
9. W. Jark, S. Di Fonzo, S. Lagomarsino, A. Cedola, E. di Fabrizio, A. Bram and C. Riekel: J. Appl. Phys. **80** (1996) 4831.
10. A. Smith, M. R. F. Siggel-King, G. M. Holder, A. Cricenti, M. Luce, P. Harrison, D. S. Martin, M. Surman, T. Craig, S. D. Barrett, A. Wolski, D. J. Dunning, N. R. Thompson, Y. Saveliev, D. M. Pritchard, A. Varro, S. Chattopadhyay and P. Weightman: Appl. Phys. Lett. **102** (2013) 53701.
11. A. Momose, T. Takeda, Y. Itai and K. Hirano: Nature Medicine **2** (1996) 473.
12. N. Yagi, Y. Suzuki, K. Umetani, Y. Kohmura and K. Yamasaki: Med. Phys. **26** (1999) 2190.
13. D. Chapman, W. Thomlinson, R. E. Johnston, D. Washburn, E. Pisano, N. Gmür, Z. Zhong, R. Menk, F. Arfelli and D. Sayer: Phys. Med. Biol. **42** (1997) 2015.

14. T. Takeda, Y. Itai, K. Hyodo, M. Ando, T. Akatsuka and C. Uyama: J. Synch. Rad. **5** (1998) 326.

15. B. Thomlinson: Synch. Rad. News **24** (2011) No.2, 2.

16. B. P. Tonner, D. Dunham, T. Droubay, J. Kimura and J. Delinger: J. Electr. Spectr. Rel. Phenom. **78** (1996) 13.

17. A. Ito, K. Shinohara, Y. Mizukami, H. Nakano, K. Yada, T. Uehara and T. Honda: J. Synch. Rad. **5** (1998) 1099.

18. D. M. Paganin: "Coherent X-Ray Optics" (Oxford University Press, 2006).

19. J. Miao, T. Ishikawa, B. Johnson, R. H. Anderson, B. Lai and K. O. Hodgson: Phys. Rev. Lett. **89** (2002) 88303.

20. N. Kamijo, Y. Suzuki, M. Awaji, A. Takeuchi, K. Uesugi, M. Yamamoto, S. Tamura, Y. Kohmura, A. Duevel, D. Rudolph and G. Schmahl: Nucl. Instr. Meth. A **467-468** (2001) 868.

21. F. Barbo, M. Bertolo, A. Bianco, G. Cautero, S. Fontana, T. K. Johal, S. La Rosa, G. Margaritondo and K. Kaznacheyev: Rev. Sci. Instr. **71** (2000) 5.

22. M. Toyoda, Y. Shitani, M. Yanagihara, T. Ejima, M. Yamamoto and M. Watanabe: Jpn. J. Appl. Phys. **39** (2000) 1926.

23. G. Denbeaux, E. Anderson, W. Chao, T. Eimüller, L. Johnson, M. Köhler, C. Larabell, M. Legros, P. Fischer, A. Pearson, G. Schütz, D. Yager and D. Attwood: "Soft X-Ray microscopy to 25 nm with application to biology and magnetic materials", Nucl. Instr. Meth. A **467-468** (2001) 841.

24. A. Yamamoto, R. Masaki, P. Guttmann, G. Schmahl and H. Kihara: J. Synch. Rad. **5** (1998) 1105.

25. N. Watanabe, K. Yamamoto, H. Takano, T. Ohigashi, H. Yokosuka, T. Aota and S. Aoki: "X-Ray fluorescence microtomography with a Wolter mirror system", Nucl. Instr. Meth. A **467-468** (2001) 837.

26. S. Hayakawa, S. Tohno, K. Takagawa, A. Hamamoto, Y. Nishida, M. Suzuki, Y. Sato and T. Hirokawa: Anal. Sci. **17** (2001) 115.

27. A. Iida and T. Noma: "Synchrotron X-ray microprobe and its application to human hair analysis", Nucl. Instr. Meth. B **82** (1993) 129.

28. J. V. Gilfrich, N. L. Gilfrich, E. F. Skelton, J. F. Kirkland, S. B. Quadri and D. J. Nagel: X-Ray Spectrom. **20** (1991) 203.

29. K. W. Jones, W. J. Berry, D. J. Borsay, H. T. Cline, W. C. Conner and C. S. Fullmer: X-Ray Spectrom. **26** (1997) 350.

30. S. Kimura, T. Nanba, T. Sada, M. Okuno, M. Matsunami, K. Shinoda, H. Kimura, T. Moriwalki, M. Yamagata, Y. Kondo, Y. Yoshimatsu, T. Takahashi, K. Fukui, T. Kawamoto and T. Ishikawa: Nucl. Instr. Meth. A **467-468** (2001) 893.

31. L. M. Miller, G. L. Carr, M. Jackson, P. Dumas and G. P. Williams: "The impact of infrared synchrotron radiation in biology: past, present and future", Synch. Rad. News **13** (2000) No. 5, 31.

32. A. Tanaka, K. Ishii, Y. Komori, S. Matsuyama, H. Yamazaki, Y. Oishi, M. Rodoriguez, Ts. Amartaivan, K. Kubota, H. Fukuda, T. Kamiya, T. Satoh, T. Sakai, M. Oikawa, K. Arakawa and M. Saidoh: "In-air micro-PIXE analysis of tissue samples", Int. J. PIXE **12** (2002) 79.

33. A. R. Lang: "Recent application of X-Ray topography" in "Modern Diffraction and Imaging Techniques in Materials Science", S. Amelinckx, R. Gevers, R. Remaut and J. van Landuyt eds. (North Holland, 1970) p. 407.

34. I. Yonenaga, T. Taishi, X. Huan and K. Hoshikawa: Mater. Sci. Eng. B **91-92** (2002) 192.

35. H. Suzuki, I. Iwasa, T. Nakajima and I. Yonenaga: Mater. Trans. JIM **33** (1992) 302.

36. M. Imai and K. Sumino: Philos. Mag. A **47** (1983) 599.

37. I. Yonenaga and K. Sumino: J. Crystal Growth **126** (1993) 19.

38. T. Kitano, T. Ishikawa and J. Matsui: Philos. Mag. A **63** (1991) 95.

39. A. N. Danilewsky, J. Wittge, A. Rack, T. Weitkamp, R. Simon and P. McNally: J. Mat. Sci.: Materials in Electronics **19** (2008) Suppl. 1, 269.

40. T. Mukaide, K. Kajiwara, T. Noma and K. Takada: J. Synch. Rad. **13** (2006) 484.

41. B. Capelle, J. Détaint and Y. Epelboin: J. Appl. Cryst. **34** (2001) 625.

Appendices

Appendix A: Theorems on the focusing of light

a. The sine theorem and the Lagrange invariant: Figure A.1 shows a conceptual diagram for image formation of an object by light reflection from a spherical mirror. At near normal incidence the ratio of object size x_1 to image size x_2 is given by

$$\frac{x_2}{x_1} = \frac{R - r'}{r - R} = \frac{\sin\theta_1}{\sin\delta}\frac{\sin\delta}{\sin\theta_2} = \frac{\sin\theta_1}{\sin\theta_2}. \tag{A.1}$$

Here θ_1 and θ_2 are angles of divergent and convergent light, and δ is the reflection angle. Thus the sine theorem

$$x_1 \sin\theta_1 = x_2 \sin\theta_2 \tag{A.2}$$

is given. For small θ_1 and θ_2,

$$x_1\theta_1 = x_2\theta_2. \tag{A.3}$$

The value, Eq. (A.3), is the 'Lagrange invariant'.

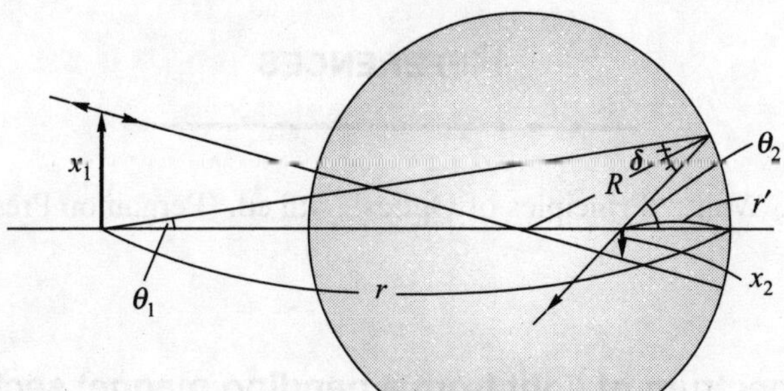

Figure A.1 Near normal incidence object and image reflection from a spherical mirror. r : incident light pass length, r': exit light pass length. x_1: object size, x_2: image size.

b. Abbe's sine condition: For a spherical mirror, using the sine theorem Eq. (A.1), the magnification factor m

$$m = \frac{x_2}{x_1} = \frac{\sin\theta_1}{\sin\theta_2}. \tag{A.4}$$

At near normal incidence, m is approximately constant for paraxial rays, and the image is not deformed significantly using a spherical mirror. To obtain a good image, m should be constant and independent of θ, called 'Abbe's sine condition'. This condition describes any optical system which is free from coma and spherical aberration. To achieve this condition, combinations of more than two spherical mirrors are required. The situation is similar for aspherical mirrors. Note that $m \approx r'/r$.

c. The numerical aperture (*NA*) and the resolution (the resolving power) of a light microscope: When light at the edge of the mirror of diameter D converges onto the optical axis, with the convergent angle 2θ, the numerical aperture is equal to sin θ. Let r' be exit ray pass length, then $NA = \sin\theta \approx D/2r'$. When r' is equal to the focal length ($f = R/2$), parallel light converges to a focal point and $NA = D/2f = D/R$. (f/D is called the 'F value' and $1/F$ expresses the acceptance angle.) NA is a measure of the resolution of the image which is determined by the diffuseness of the edges of the image due to diffraction. When a beam of parallel light is focused at the focal plane by a mirror and the beam is then reduced in width, the radius of the mirror will define the circular aperture for Fraunhofer diffraction. For simplicity, consider the case in which a beam of parallel light passes through an aperture with parallel edges and slit width a. The light intensity distribution is proportional to a function $\sin^2\alpha x/x^2$ in the focal plane, where $\alpha = 2\pi a/\lambda f$ and x is the position in the focal plane. The function has a shape similar to the curves given in Fig. 2.8, and has a main maximum at $x = 0$ and a first minimum at $x = \pm\lambda f/2a$. In order to distinguish two points separated by distance Δx in the focal plane by the beam passing through the slit aperture, the two main maxima must be separated by about $\Delta x = \lambda f/2a$. In the case of a circular aperture the separation of the two maxima is given by

$$\Delta x = 0.61\frac{\lambda}{NA} \tag{A.5}$$

using Rayleigh's criterion. Eq. (A.5) is used to define the resolution (also called the 'resolving power') of any microscope.[1)]

REFERENCES

1. M. Born and E. Wolf : "Principles of Optics", 4th ed. (Pergamon Press, 1970).

Appendix B: The spectrum of light from a bending magnet section and the mean photon energy

Consider the interaction between a moving electron and the electric field generated by the electron. For electric current $\boldsymbol{j}$ and electric field $\boldsymbol{E}$, the total power received by the current is $\int \boldsymbol{j} \cdot \boldsymbol{E} d\boldsymbol{r}$. Accordingly using Eq. (2.17), the total emitted power $\hat{P}$ will be given by

$$\hat{P} = -\int \boldsymbol{j} \cdot \boldsymbol{E} d\boldsymbol{r} = \int \boldsymbol{j} \cdot \left(\frac{\partial \boldsymbol{A}}{\partial t} + \text{grad}\,\Phi\right) d\boldsymbol{r}. \tag{B.1}$$

Radiated power P is given as total power expressed by Eq. (B.1) minus the time derivative of the particle energy $\frac{d}{dt}\int(\rho\,\Phi)\,d\boldsymbol{r}$. Using the equation $\operatorname{div}\boldsymbol{j}+\frac{\partial\rho}{\partial t}=0$, P is given by[1)]

$$P=\int\left\{\boldsymbol{j}\left(\boldsymbol{r},t\right)\cdot\frac{\partial\boldsymbol{A}_{\rm ret}}{\partial t}-\rho\left(\boldsymbol{r},t\right)\frac{\partial\Phi_{\rm ret}}{\partial t}\right\}d\boldsymbol{r}. \tag{B.2}$$

Here the subscript ret stands for retarded time and $\Phi_{\rm ret}$ given by Eq. (2.29) and $\boldsymbol{A}_{\rm ret}$ given by Eq. (2.30) are respectively the retarded scalar and vector potentials generated by a moving charge (see Fig. 2.1). Using $d\delta\left(-t\right)/dt=-\delta'\left(-t\right)$ and $\varepsilon_0\mu_0=c^{-2}$, Eq. (B.2) becomes

$$P=\frac{1}{4\pi\varepsilon_0}\int\left\{\rho\left(\boldsymbol{r}_{\rm P},t\right)\rho\left(\boldsymbol{r}_{\rm P'},t'\right)-\frac{1}{c^2}\boldsymbol{j}\left(\boldsymbol{r}_{\rm P},t\right)\boldsymbol{j}\left(\boldsymbol{r}_{\rm P'},t'\right)\right\}\frac{\delta'\left(t'-t+\dfrac{|\boldsymbol{r}_{\rm P}-\boldsymbol{r}_{\rm P'}|}{c}\right)}{|\boldsymbol{r}_{\rm P}-\boldsymbol{r}_{\rm P'}|}d\boldsymbol{r}_{\rm P}d\boldsymbol{r}_{\rm P'}dt'. \tag{B.3}$$

The total radiated power is given by $P=\int_0^\infty P\left(\omega,t\right)d\omega$, where $P(\omega,t)$ is the angular frequency component of the radiated power. Using Eqs. (2.31) and (2.32) and Fourier transform of derivative of delta function

$$\delta'\left(t'-t+\frac{|\boldsymbol{r}_{\rm P}-\boldsymbol{r}_{\rm P'}|}{c}\right)=-\frac{1}{\pi}\int_0^\infty\omega\sin\omega\left(t'-t+\frac{|\boldsymbol{r}_{\rm P}-\boldsymbol{r}_{\rm P'}|}{c}\right)d\omega, \tag{B.4}$$

$$P(\omega,t)=-\frac{e^2\omega}{4\pi^2\varepsilon_0}\int_{-\infty}^{\infty}\frac{1-\dfrac{\boldsymbol{v}\left(t\right)\cdot\boldsymbol{v}\left(t+\tau\right)}{c^2}}{|\boldsymbol{r}_{\rm P'}\left(t+\tau\right)-\boldsymbol{r}_{\rm P'}\left(t\right)|}\sin\omega\left(\tau+\frac{|\boldsymbol{r}_{\rm P'}\left(t+\tau\right)-\boldsymbol{r}_{\rm P'}\left(t\right)|}{c}\right)d\tau, \tag{B.5}$$

where $\boldsymbol{r}_{\rm P'}\left(t\right)$ and $\boldsymbol{r}_{\rm P'}\left(t'\right)$ are the electron positions at times t and t' respectively, and $t'-t=\tau$. Under the condition that a point charge moves in a circular orbit of R radius with relativistic velocity $(E/mc^2\gg 1)$,

$$\boldsymbol{v}\cdot\dot{\boldsymbol{v}}=0,\quad\frac{d^2\boldsymbol{v}^2}{dt^2}=2\frac{d}{dt}\boldsymbol{v}\cdot\dot{\boldsymbol{v}}=2\dot{\boldsymbol{v}}^2+2\boldsymbol{v}\cdot\ddot{\boldsymbol{v}}=0,\quad|\dot{\boldsymbol{v}}|=\frac{\boldsymbol{v}^2}{R}\approx\frac{c^2}{R}. \tag{B.6}$$

Using Eq. (B.6), $\boldsymbol{v}$ and $\boldsymbol{r}_{\rm p'}$ terms in Eq. (B.5) respectively take the forms

$$\boldsymbol{v}(t)\cdot\boldsymbol{v}(t+\tau)\approx\boldsymbol{v}\left(t\right)\cdot\left\{\boldsymbol{v}\left(t\right)+\dot{\boldsymbol{v}}\left(t\right)\tau+\frac{1}{2}\ddot{\boldsymbol{v}}\left(t\right)\tau^2\right\} \tag{B.7}$$

$$=\boldsymbol{v}^2+\frac{1}{2}\boldsymbol{v}\cdot\ddot{\boldsymbol{v}}\tau^2=\boldsymbol{v}^2-\frac{c^4}{2R^2}\tau^2 \tag{B.8}$$

$$|\boldsymbol{r}_{\rm P'}\left(t+\tau\right)-\boldsymbol{r}_{\rm P'}\left(\tau\right)|\approx\left|\boldsymbol{v}\tau+\frac{1}{2}\dot{\boldsymbol{v}}\tau^2+\frac{1}{6}\ddot{\boldsymbol{v}}\tau^3\right|$$

$$\approx\sqrt{\boldsymbol{v}^2\tau^2+\frac{1}{3}\boldsymbol{v}\cdot\ddot{\boldsymbol{v}}\tau^4+\frac{1}{4}\dot{\boldsymbol{v}}^2\tau^4}\approx v\,|\tau|\left(1-\frac{c^2\tau^2}{24R^2}\right). \tag{B.9}$$

Using Eqs. (B.7) to (B.9) and paying attention to sign of τ, we can change the integral range to (0, ∞) as

$$\rho(\omega,t) \approx -\frac{e^2\omega}{2\pi^2\varepsilon_0}\int_0^\infty \frac{1-\beta^2+\dfrac{c^2\tau^2}{2R^2}}{c\tau}\cos\omega\tau\sin\omega\tau\left(\beta-\frac{c^2\tau^2}{24R^2}\right)d\tau. \tag{B.10}$$

Furthermore, using the approximation

$$\cos\omega\tau\sin\omega\tau\left(\beta-\frac{c^2\tau^2}{24R^2}\right) \approx \frac{1}{2}\left\{\sin 2\omega\tau - \sin\omega\tau\left(1-\beta+\frac{c^2\tau^2}{24R^2}\right)\right\} \tag{B.11}$$

and the following change of variables

$$\tau = \frac{2R}{c}\left(1-\beta^2\right)^{\frac{1}{2}}x = \frac{2Rx}{c\gamma}, \tag{B.12}$$

$$\xi = \frac{2\omega R}{3c}\left(1-\beta^2\right)^{\frac{3}{2}} = \frac{2\omega R}{3c\gamma^3} = \frac{\omega}{\omega_c}\ ,\quad \omega_c = \frac{3}{2}\frac{c}{R}\gamma^3, \tag{B.13}$$

angular frequency component of power is given by

$$P(\omega,t) = \frac{e^2\omega}{4\pi^2\gamma^2\varepsilon_0 c}\left\{\int_0^\infty \left(1+2x^2\right)\sin\frac{3}{2}\xi\left(x+\frac{1}{3}x^3\right)\frac{dx}{x} - \frac{\pi}{2}\right\}. \tag{B.14}$$

Here we have employed the well-known integral relation

$$\int_0^\infty \frac{\sin ax}{x}dx = \frac{\pi}{2}\quad (a>0). \tag{B.15}$$

The integral contained in Eq. (B.14) can be rewritten using relation between the Airy integral and modified Bessel functions of the second kind given by[2)]

$$\frac{1}{\sqrt{3}}K_{1/3}(\xi) = \int_0^\infty \cos\frac{3}{2}\xi\left(x+\frac{1}{3}x^3\right)dx, \tag{B.16}$$

$$\frac{1}{\sqrt{3}}K_{2/3}(\xi) = \int_0^\infty x\sin\frac{3}{2}\xi\left(x+\frac{1}{3}x^3\right)dx. \tag{B.17}$$

The equation

$$\frac{1}{\sqrt{3}}K_{1/3}(\xi) = \frac{d}{d\xi}\int_0^\infty \sin\frac{3}{2}\xi\left(x+\frac{1}{3}x^3\right)\frac{dx}{x}$$

can be derived by taking into account the property of the delta function. From this equation we obtain

$$\frac{1}{\sqrt{3}}\int_\xi^\infty K_{1/3}(\eta)\,d\eta = \frac{\pi}{2} - \int_0^\infty \sin\frac{3}{2}\xi\left(x+\frac{1}{3}x^3\right)\frac{dx}{x}. \tag{B.18}$$

Using Eqs. (B.17) and (B.18), and the recursion formula for modified Bessel functions of the second kind

$$2\frac{d}{d\xi}K_{2/3}(\xi) + K_{1/3}(\xi) = -K_{5/3}(\xi), \tag{B.19}$$

Eq. (B.14) then becomes

$$P(\omega,t)=\frac{3^{3/2}}{16\pi^2\varepsilon_0}\frac{e^2\gamma^4}{R^2}\frac{c\omega}{\omega_c^2}\int_{\omega/\omega_c}^{\infty}K_{5/3}(\eta)\,d\eta. \tag{B.20}$$

Equation (B.20) gives the component of power emitted by a relativistic electron per unit angular frequency, that is the spectrum of the light from bending sections.

Let the energy of the emitted electromagnetic radiation at frequencies ω and ω_c be $u=\hbar\omega$ and $u_c=\hbar\omega_c$, respectively. Let $n(u)du$ be the number of photons emitted per unit time with energies between u and $u+du$. The power emitted by these photons is $un(u)du$ and should be equal to power emitted in the frequency interval $d\omega$. Thus

$$un(u)du=P(\omega)\,d\omega=u\frac{P}{u_c^2}F(\xi)\,du, \tag{B.21}$$

where

$$F(\xi)=\frac{9\sqrt{3}}{8\pi}\int_{\xi}^{\infty}K_{5/3}(\eta)\,d\eta \tag{B.22}$$

and P and ξ are defined by Eqs. (2.65) and (B.13), respectively. The mean number of photons emitted per second by an electron is given by

$$N_p=\int n(u)\,du=\frac{P}{u_c^2}\frac{9\sqrt{3}}{8\pi}\int_0^{\infty}du\int_{u/u_c}^{\infty}K_{5/3}(\eta)\,d\eta. \tag{B.23}$$

Using the integral form for modified Bessel functions the integral in Eq. (B.23) can be calculated to be

$$\frac{1}{u_c}\int_0^{\infty}du\int_{u/u_c}^{\infty}d\eta\int_0^{\infty}e^{-\eta\cosh t}\cosh(5/3)\,tdt=\int_0^{\infty}\frac{\cosh(5/3)\,t}{\cosh^2 t}dt$$

$$=B\left(\frac{1}{6},\frac{11}{6}\right)=\frac{\Gamma\left(\frac{1}{6}\right)\Gamma\left(\frac{11}{6}\right)}{\Gamma(2)}=\frac{5}{3}\pi. \tag{B.24}$$

The calculation is performed by replacing the hyperbolic function with an exponential function and also the substitution $y=e^{-2t}$ is made. Here B represents beta function and Γ, the gamma function. The relations $\Gamma(2)=1$, $\Gamma(z+1)=z\Gamma(z)$ and $\Gamma(z)\Gamma(1-z)=\pi/\sin\pi z$ are used. Substituting Eq. (B.24) for Eq. (B.23), the mean photon number N_p and mean photon energy $\langle u\rangle$ are calculated to be

$$N_p=\frac{15\sqrt{3}}{8}\frac{P}{u_c},\quad \langle u\rangle=\frac{8}{15\sqrt{3}}u_c. \tag{B.25}$$

Similarly, the mean square of photon energy $\int u^2 n(u)\,du$ and total value of the mean square are given by

$$\langle u^2\rangle=\frac{11}{27}u_c^2,\quad N_p\langle u^2\rangle=\frac{55}{24\sqrt{3}}u_cP. \tag{B.26}$$

References

1. J. Schwinger: Phys. Rev. **75** (1949) 1912.
2. G. N. Watson: "Bessel Functions" (The Macmillan Company, New York, 1945).

Appendix C: The angular distribution of light emitted from electrons moving on the arc of a circle

In calculation of the power of the emitted light from Eq. (2.55), the quantity $\boldsymbol{E}_{(r)}(t)$ is defined to be

$$\boldsymbol{E}_{(r)}(t) = \sqrt{\frac{1}{\mu_0 c}}\,\{r\boldsymbol{E}(r_{\mathrm{p}}, t)\}_{\mathrm{ret}}\,. \tag{C.1}$$

The Fourier transform of $\boldsymbol{E}_{(r)}(t)$ is represented by

$$\boldsymbol{E}_{(r)}(\omega) = \int_{-\infty}^{\infty} \boldsymbol{E}_{(r)}(t)\, e^{i\omega t} dt, \tag{C.2}$$

$$\boldsymbol{E}_{(r)}(t) = \frac{1}{2\pi}\int_{-\infty}^{\infty} \boldsymbol{E}_{(r)}(\omega)\, e^{-i\omega t} d\omega. \tag{C.3}$$

The total energy radiated per unit solid angle is given by

$$\begin{aligned}\frac{dW}{d\Omega} &= \int_{-\infty}^{\infty} \left|\boldsymbol{E}_{(r)}(t)\right|^2 dt \\ &= \frac{1}{(2\pi)^2}\int_{-\infty}^{\infty} \boldsymbol{E}_{(r)}(\omega)\, e^{-i\omega t} d\omega \int_{-\infty}^{\infty} \boldsymbol{E}^*_{(r)}(\omega')\, e^{i\omega' t} d\omega' \int_{-\infty}^{\infty} dt \\ &= \frac{1}{2\pi}\int_{-\infty}^{\infty} \left|\boldsymbol{E}_{(r)}(\omega)\right|^2 d\omega.\end{aligned} \tag{C.4}$$

The time integral is the delta function $\delta(\omega - \omega')$, where the delta function is defined to be

$$\delta(\omega) = \frac{1}{2\pi}\int_{-\infty}^{\infty} e^{i\omega t} dt. \tag{C.5}$$

The energy radiated per unit angular frequency is given by[1)]

$$\frac{d^2 W(\omega)}{d\omega d\Omega} = \frac{1}{2\pi}\left\{\left|\boldsymbol{E}_{(r)}(\omega)\right|^2 + \left|\boldsymbol{E}_{(r)}(-\omega)\right|^2\right\} = \frac{1}{\pi}\left|\boldsymbol{E}_{(r)}(\omega)\right|^2. \tag{C.6}$$

For relativistic electrons, Eq. (C.2) is given using Eq. (2.52a) by

$$\boldsymbol{E}_{(r)}(\omega) = \frac{-e}{4\pi\sqrt{\varepsilon_0 c}}\int_{-\infty}^{\infty} e^{i\omega t}\left[\frac{\boldsymbol{n}\times\{(\boldsymbol{n}-\boldsymbol{\beta})\times\dot{\boldsymbol{\beta}}\}}{(1-\boldsymbol{n}\cdot\boldsymbol{\beta})^3}\right]_{\mathrm{ret}} dt \tag{C.7}$$

The variable of integration t can be changed to $t' = t - |\boldsymbol{r}_{\mathrm{P}} - \boldsymbol{r}_{\mathrm{P'}}|/c$, where $|\boldsymbol{r}_{\mathrm{P}} - \boldsymbol{r}_{\mathrm{P'}}| = \boldsymbol{n}\cdot\boldsymbol{r}_{\mathrm{P}} - \boldsymbol{n}\cdot\boldsymbol{r}_{\mathrm{P}}$, is the observation point and $\boldsymbol{n}$ is the unit vector in the direction of observation (see Figs. 2.1 and 2.3). Since the observation point is significantly remote from the charge, $\boldsymbol{n}$ can be assumed to be independent of t'. Therefore in the integration of Eq. (C.2), $\boldsymbol{n}\cdot\boldsymbol{r}_{\mathrm{P}}$ gives only the phase factor. Since $1 = dt/dt' + \boldsymbol{n}\cdot\boldsymbol{\beta}$ (see Eq. (2.42)), Eq. (C.7) takes the form

$$\boldsymbol{E}_{(r)}(\omega) = \frac{-e}{4\pi\sqrt{\varepsilon_0 c}}\int_{-\infty}^{\infty} e^{i\omega\left(t' - \frac{\boldsymbol{n}\cdot\boldsymbol{r}_{\mathrm{P'}}}{c}\right)}\left[\frac{\boldsymbol{n}\times\{(\boldsymbol{n}-\boldsymbol{\beta})\times\dot{\boldsymbol{\beta}}\}}{(1-\boldsymbol{n}\cdot\boldsymbol{\beta})^2}\right]_{\mathrm{ret}} dt'. \tag{C.8}$$

In Eq. (C.8), the integrand excluding the exponential function is a total derivative,

$$\frac{\boldsymbol{n}\times\{(\boldsymbol{n}-\boldsymbol{\beta})\times\dot{\boldsymbol{\beta}}\}}{(1-\boldsymbol{n}\cdot\boldsymbol{\beta})^2} = \frac{d}{dt'}\frac{\boldsymbol{n}\times(\boldsymbol{n}\times\boldsymbol{\beta})}{(1-\boldsymbol{n}\cdot\boldsymbol{\beta})} \tag{C.9}$$

Integration by parts of Eq. (C.8) and its substitution for Eq. (C.6) leads to the intensity distribution

$$\frac{d^2W(\omega)}{d\omega d\Omega} = \frac{e^2\omega^2}{16\pi^3\varepsilon_0 c}\left|\int_{-\infty}^{\infty}\boldsymbol{n}\times(\boldsymbol{n}\times\boldsymbol{\beta})\,e^{i\omega\left(t' - \frac{\boldsymbol{n}\cdot\boldsymbol{r}_{\mathrm{P'}}}{c}\right)}dt'\right|^2. \tag{C.10}$$

For any electron executing circular motion, we assume that the time when the electron is at the starting point on the circular orbit is $t' = 0$. The z-axis represents velocity direction. The orbit plane is the xz-plane and the direction vertical to the orbit plane is the y-axis. The observation direction $\boldsymbol{n}$ can be chosen to be in the yz-plane. $\hat{\theta}$ is the angle between $\boldsymbol{n}$ and xz-plane and ψ is the rotation angle of an electron describing a circular orbit from $t' = 0$ (see Fig. 1.1). In this coordinate system, $\boldsymbol{n} = \left(0, \sin\hat{\theta}, \cos\hat{\theta}\right)$, $\boldsymbol{\beta} = \beta(\sin\psi, 0, \cos\psi)$ and $\psi = vt'/R$. Let us define unit vectors of directions x and y to be $\boldsymbol{e}_{\parallel}$ and $\boldsymbol{e}_{\perp}$. Since the radiation is observed for an infinitesimally small time and $\hat{\theta}$ is very small as mentioned in Sect. **2.3.1**, the vector product in Eq. (C.10) can be written as

$$\boldsymbol{n}\times(\boldsymbol{n}\times\boldsymbol{\beta}) \approx \beta\left(-\boldsymbol{e}_{\parallel}\sin\frac{vt'}{R} + \boldsymbol{e}_{\perp}\cos\frac{vt'}{R}\sin\hat{\theta}\right) \approx -\boldsymbol{e}_{\parallel}\frac{c}{R}t' + \boldsymbol{e}_{\perp}\hat{\theta}. \tag{C.11}$$

In Eq. (C.8), the exponential function has an independent variable which is

$$\omega\left(t' - \frac{\boldsymbol{n}\cdot\boldsymbol{r}_{\mathrm{P'}}}{c}\right) \approx \omega\left(t' - \frac{R}{c}\sin\frac{vt'}{R}\cos\hat{\theta}\right). \tag{C.12}$$

By expanding Eq. (C.12) around $t' = 0$ and $\hat{\theta} = 0$, and using $1 - \beta \approx 1/2\gamma^2$, one obtains

$$\omega\left(t' - \frac{\boldsymbol{n}\cdot\boldsymbol{r}_{\mathrm{P'}}}{c}\right) \approx \frac{\omega}{2}\left\{\left(\frac{1}{\gamma^2} + \hat{\theta}^2\right)t' + \frac{c^2}{3R^2}t'^3\right\}. \tag{C.13}$$

Substitution of Eqs. (C.11) and (C.12) for Eq. (C.10) yields the radiated energy distribution as

$$\frac{d^2W(\omega)}{d\omega d\Omega} = \frac{e^2\omega^2}{16\pi^3\varepsilon_0 c}\left|-\boldsymbol{e}_{\parallel}K_{\parallel}(\omega) + \boldsymbol{e}_{\perp}K_{\perp}(\omega)\right|^2, \tag{C.14}$$

where

$$K_{\parallel}(\omega) = \frac{c}{R}\int_{-\infty}^{\infty} t'\exp\left[i\frac{\omega}{2}\left\{\left(\frac{1}{\gamma^2} + \hat{\theta}^2\right)t' + \frac{c^2t'^3}{3R^2}\right\}\right]dt',$$

$$K_{\perp}(\omega) = \hat{\theta}\int_{-\infty}^{\infty} \exp\left[i\frac{\omega}{2}\left\{\left(\frac{1}{\gamma^2}+\hat{\theta}^2\right)t' + \frac{c^2t'^3}{3R^2}\right\}\right] dt'. \tag{C.15}$$

By making the variable change of

$$x = \frac{ct'}{R\left(\dfrac{1}{\gamma^2}+\hat{\theta}^2\right)^{1/2}} \tag{C.16}$$

and introduction of the parameter

$$\xi = \frac{\omega R}{3c}\left(\frac{1}{\gamma^2}+\hat{\theta}^2\right)^{3/2}, \tag{C.17}$$

Eqs. (C.15) and (C.16) take the forms

$$K_{||}(\omega) = \frac{R}{c}\left(\frac{1}{\gamma^2}+\hat{\theta}^2\right)\int_{-\infty}^{\infty} x\exp\left[i\frac{3}{2}\xi\left(x+\frac{1}{3}x^3\right)\right] dx, \tag{C.18}$$

$$K_{\perp}(\omega) = \frac{R}{c}\hat{\theta}\left(\frac{1}{\gamma^2}+\hat{\theta}^2\right)^{1/2}\int_{-\infty}^{\infty} \exp\left[i\frac{3}{2}\xi\left(x+\frac{1}{3}x^3\right)\right] dx. \tag{C.19}$$

Using Eqs. (B.16) and (B.17), the angular intensity distribution of the light from bending sections is given by

$$\frac{d^2W(\omega)}{d\omega d\Omega} = \frac{e^2}{12\pi^3\varepsilon_0 c}\left(\frac{\omega R}{c}\right)^2\left(\frac{1}{\gamma^2}+\hat{\theta}^2\right)^2\left\{K_{2/3}^2(\xi) + \frac{\hat{\theta}^2}{\dfrac{1}{\gamma^2}+\hat{\theta}^2}K_{1/3}^2(\xi)\right\}, \tag{C.20}$$

where $d\Omega \approx \theta d\theta d\phi \approx d\hat{\theta}d\psi$. Here in the curly bracket, the first term corresponds to radiation with the electric field polarized in the orbit plane and second term to that in the plane perpendicular to the orbit plane.

The phase difference between above-mentioned electric fields is $\pi/2$ and the radiation emitted in any direction other than horizontal direction is elliptically polarized for any wavelength.[2)]

REFERENCES

1. J. D. Jackson : "Classical Electrodynamics", 2nd ed. (John Wiley & Sons, 1975).
2. É. G. Mychelkin: Sov. Astron . – AJ **12** (1968) 326.

Appendix D: The spectrum of light from a planar undulator

Undulator radiation is generated by relativistic electrons ($\gamma \gg 1$), traversing a periodic magnetic structure. The radiation is extremely bright, quasi-monochromatic and is emitted in a narrow cone.[1] In this appendix, we have calculated the spectrum of light from a planar undulator. The orbit of the electron beam has been assumed to be sine-wave-like with the period λ_u. (Fig. 2.7 shows the coordinate system.) Here it is assumed that the electron moves in the z direction and that the undulation of the electron beam is within the xz-plane. To define the direction of radiated light, a polar coordinate around the z-axis is adopted. $\boldsymbol{i}$, $\boldsymbol{j}$, $\boldsymbol{k}$ are unit vectors in the x, y, z directions and the magnetic field is assumed to be in the y direction. When the deflection parameter of the undulator $K \leq 1$, the deflection angle $\psi_0 \ll 1$. Differentiating Eq. (2.88), expressing $\boldsymbol{\beta} = \boldsymbol{v}/c$ and by neglecting the third term of z, one can obtain

$$\boldsymbol{\beta} = \boldsymbol{i}\frac{K}{\gamma}\beta \cos \omega_0 t' + \boldsymbol{k}\overline{\beta}_z, \tag{D.1}$$

where $\omega_0 = 2\pi c\overline{\beta}_z/\lambda_u$. The retarded time t' is related to the time scale of the observer through the relation $dt = (1 - \boldsymbol{n} \cdot \boldsymbol{\beta})\, dt'$. The observation position is remote from the emitting point and the amplitude of undulation is so small that the starting point can represent the emitting point. The unit vector, directed to the observer from the emitting point is given by

$$\hat{\boldsymbol{n}} = \boldsymbol{i}\theta \cos \phi + \boldsymbol{j}\theta \sin \phi + \boldsymbol{k}, \tag{D.2}$$

where θ is the polar angle between $\hat{\boldsymbol{n}}$ and z direction. The angle ϕ is the azimuth angle between $\hat{\boldsymbol{n}}$ and the xz-plane. For $K \leq 1$, the deflection angle of electron beam ψ_0 is as small as γ^{-1}, thus $\theta \ll 1$. Calculating $1 - \hat{\boldsymbol{n}} \cdot \boldsymbol{\beta}$ using Eqs. (D.1) and (D.2), one can obtain t and t' by integration as

$$t = \frac{\omega_0}{\omega_1}t' - \beta\frac{K\theta}{\gamma\omega_0}\cos\phi \sin \omega_0 t' + \frac{r - \hat{\boldsymbol{n}} \cdot \boldsymbol{r}_{P'}}{c}, \tag{D.3}$$

$$\omega_1 = \frac{2\gamma^2\omega_0}{1 + K^2/2 + \gamma^2\theta^2}. \tag{D.4}$$

The first term in Eq. (D.3) is the constant of integration. $r - \hat{\boldsymbol{n}} \cdot \boldsymbol{r}_{P'}$ gives the distance from the emitting point to the observer. Here, r is the distance from the starting point and $\boldsymbol{r}_{P'}$ (x', y', z') is the electron position during undulation. $\hat{\boldsymbol{n}} \cdot \boldsymbol{r}_{P'}$ is given by

$$\hat{\boldsymbol{n}} \cdot \boldsymbol{r}_{P'} = x'\theta \cos \phi + y'\theta \sin \phi + z'. \tag{D.5}$$

The spectrum of light from an undulator is calculated using the Lienard-Wiechert potential, Eq. (2.38) to be[2]

$$\boldsymbol{A} = \frac{-\mu_0 ec}{4\pi r}\left[\frac{\boldsymbol{\beta}}{1 - \hat{\boldsymbol{n}} \cdot \boldsymbol{\beta}}\right]_{\text{ret}}. \tag{D.6}$$

The vector potential Eq. (D.6) can be expressed using the Fourier expansion of the fundamental wave Eq. (D.4) with n-th order higher harmonics as

$$\boldsymbol{A}(t) = \sum_n \boldsymbol{A}_n e^{-in\omega_1 t}, \tag{D.7}$$

$$\boldsymbol{A}_n = \frac{-\mu_0 ec}{4\pi r}\frac{\omega_1}{2\pi}\int_0^{\frac{2\pi}{\omega_1}} \frac{\boldsymbol{\beta}(t')}{1-\hat{\boldsymbol{n}}\cdot\boldsymbol{\beta}(t')} e^{in\omega_1 t} dt. \tag{D.8}$$

To calculate $\boldsymbol{A}_n$ let us substitute Eq. (D.3) for Eq. (D.8) and perform the integration with respect to t'. Using the formula given below, where J_p is the Bessel function of order p,

$$e^{iz\sin\psi} = \sum_{p=-\infty}^{\infty} J_p(z)\, e^{ip\psi} \tag{D.9}$$

and assuming $\overline{\beta}_z \approx 1$, one can obtain $\boldsymbol{A}_n$ as

$$\boldsymbol{A}_n = \frac{-\mu_0}{4\pi}\frac{ec}{r}\frac{\gamma\zeta}{n}\left\{\boldsymbol{i}K\left(\hat{S}_1+\hat{S}_{-1}\right)+\boldsymbol{k}2\gamma\hat{S}_0\right\} e^{i\frac{n\omega_1}{c}(r-\hat{\boldsymbol{n}}\cdot\boldsymbol{r}_{\mathrm{P}'})}, \tag{D.10}$$

where

$$\hat{S}_q = \sum_{p=-\infty}^{\infty} J_{n+2p+q}(2\gamma\theta K\zeta\cos\phi)\, J_p\left(K^2\zeta/4\right), \quad q=-1,0,1,$$

$$\zeta = \frac{n}{1+K^2/2+\gamma^2\theta^2}. \tag{D.11}$$

For an undulator with finite length the Fourier transform with respect to ω is given by

$$\boldsymbol{A}(\omega) = \sum_n \boldsymbol{A}_n \int_{-\frac{N_u\pi}{\omega_1}}^{\frac{N_u\pi}{\omega_1}} e^{i(\omega-n\omega_1)t} dt \tag{D.12}$$

$$= \sum_n \frac{2\boldsymbol{A}_n \sin N_u\pi(\omega/\omega_1-n)}{\omega-n\omega_1} = \frac{2\pi}{\omega_1}\sum_n \boldsymbol{A}_n h_n(\omega), \tag{D.13}$$

$$h_n = \frac{\sin N_u\pi(\omega/\omega_1-n)}{\pi(\omega/\omega_1-n)}. \tag{D.14}$$

The magnetic flux density can be calculated from $\boldsymbol{B} = \mathrm{rot}\,\boldsymbol{A}$ to be

$$\boldsymbol{B} = \mathrm{rot}\,\boldsymbol{A} = \mathrm{rot}\,\frac{1}{2\pi}\int_{-\infty}^{\infty}\boldsymbol{A}(\omega)\, e^{-i\omega t} d\omega$$

$$= \frac{1}{\omega_1}\sum_n \int_{-\infty}^{\infty} h_n(\omega)\, e^{-i\omega t}\mathrm{rot}\,\boldsymbol{A}_n d\omega = \frac{1}{2\pi}\sum_n \int_{-\infty}^{\infty}\boldsymbol{B}_n e^{-i\omega t} d\omega. \tag{D.15}$$

At the observer's position (x, y, z), rot $\boldsymbol{A}_n$ can be expressed in terms of the first power of θ as

$$\mathrm{rot}\,\boldsymbol{A}_n = \chi_n(\boldsymbol{r}_{\mathrm{P}'})\left[\boldsymbol{i}2\gamma\hat{S}_0\theta\sin\phi + \boldsymbol{j}\left\{K\left(\hat{S}_1+\hat{S}_{-1}\right)-2\gamma\hat{S}_0\theta\cos\phi\right\}\right.$$
$$\left. -\boldsymbol{k}K\left(\hat{S}_1+\hat{S}_{-1}\right)\theta\sin\phi\right], \tag{D.16}$$

$$\chi_n(\boldsymbol{r}_{\mathrm{P}'}) = -i\frac{\mu_0}{4\pi}\frac{ec}{r}\frac{\gamma\zeta\omega_1}{c} e^{i\frac{n\omega_1}{c}(r-\hat{\boldsymbol{n}}\cdot\boldsymbol{r}_{\mathrm{P}'})} \tag{D.17}$$

In the above calculation, the derivative of $\exp\{i\omega_1 (r - \hat{\boldsymbol{n}} \cdot \boldsymbol{r}_{\mathrm{P'}})/c\}$ with respect to x, y, z, was obtained by differentiation of it with respect to x', y', z' followed by change of sign. The calculation of $\boldsymbol{E}_n = c\boldsymbol{B}_n \times \hat{\boldsymbol{n}}$ yields $E_{n\|}$, the $\boldsymbol{i}$ component of $\boldsymbol{E}_n$ and $E_{n\perp}$, the $\boldsymbol{j}$ component of $\boldsymbol{E}_n$ as

$$E_{n\|} = -\left\{2\gamma\hat{S}_0\theta\cos\phi - K\left(\hat{S}_1 + \hat{S}_{-1}\right)\right\}\frac{2\pi c h_n \chi_n}{\omega_1}, \tag{D.18}$$

$$E_{n\perp} = -\left(2\gamma\hat{S}_0\theta\sin\phi\right)\frac{2\pi c h_n \chi_n}{\omega_1}. \tag{D.19}$$

Expressing { ... ... } in (D.18) and (... ...) in (D.19) by $\hat{E}_{n\|}$ and $\hat{E}_{n\perp}$, respectively, one can express the radiation intensity from Eq. (2.55) by

$$\begin{aligned} P_{n\|,\perp} d\omega d\Omega &= \frac{e^2\gamma^2\zeta^2}{4\pi\varepsilon_0 c}|\hat{E}_{n\|,\perp}|^2 \frac{\sin^2 N_{\mathrm{u}}\pi(\omega/\omega_1 - n)}{\pi^2(\omega/\omega_1 - n)^2} d\omega d\Omega \\ &= \frac{e^2\gamma^2\zeta^2}{4\pi\varepsilon_0 c}|\hat{E}_{n\|,\perp}|^2 H_n(\omega/\omega_1)\, d\omega d\Omega, \end{aligned} \tag{D.20}$$

where $d\Omega \approx \theta d\theta d\phi$. For the calculation of $|\hat{E}_{n\|,\perp}|^2$, the calculation method used in Eq. (C.4) to (C.6) is followed.

On axis $\theta = 0$, $E_\perp = 0$ from Eq. (D.19), while $E_\| \neq 0$, because the second term of Eq. (D.18) is not 0 for $n + 2p + 1 = 0$ and $n + 2p - 1 = 0$, though the first term is 0. Therefore the intensity function on axis, $F_n^{(\mathrm{p})}(K)$ is given by

$$F_n^{(\mathrm{p})}(K) = \frac{n^2K^2}{(1 + K^2/2)^2}\left\{J_{(n+1)/2}(Y_0) - J_{(n-1)/2}(Y_0)\right\}^2, \quad Y_0 = \frac{nK^2}{4 + 2K^2}. \tag{D.21}$$

In the above calculation, the formula $J_{-p}(z) = (-1)^p J_p(z)$ is used. As for helical undulators, see Ref. 3.

References

1. D. F. Alferov, Yu. A. Bashmakov and E. G. Bessonov: Sov. Phys. Tech. Phys. **18** (1974) 1336.
2. H. Kitamura: Jpn. J. Appl. Phys. **19** (1980) L185.
3. S. Yamamoto and H. Kitamura: Jpn. J. Appl. Phys. **26** (1987) L1613.

Appendix E: The Free Electron Laser (FEL)

a. The 'FEL interaction': Radiation can also be produced by an exchange of energy between the relativistic electron beam and the light traveling in the same direction (z) in an undulator. Here we describe this FEL interaction in a planar undulator which is widely used under the conditions of relatively small current density corresponding to the cavity-FEL (see Chaps. **1** and **3**). The rate of change of the electron beam energy under an electric field of $\boldsymbol{E}$ is given by the equation of motion Eq. (2.12) and can be rewritten in terms of the normalized electron velocity $\boldsymbol{\beta}$ and energy γ as

$$\frac{d\gamma}{dt} = -\frac{e}{mc}\boldsymbol{\beta}\cdot\boldsymbol{E} \tag{E.1}$$

For simplicity, the time in the electron system t' is denoted t. For a planar undulator, assuming the magnetic field to be in the vertical direction (y), the electron motion will be in the horizontal (xz-) plane. The electric field of linearly-polarized light traveling in the direction of the electron beam is given by

$$E_x = E_{x0}\cos\left(kz - \omega t + \phi_0\right), \tag{E.2}$$

where ϕ_0 is the initial phase at the entrance to the undulator. Using wave number $k_\mathrm{u} = 2\pi/\lambda_\mathrm{u}$ for the undulator with period λ_u, the velocity of electron in the undulator in the x direction can be expressed by

$$\beta_x = -\frac{eB_0\lambda_\mathrm{u}}{2\pi mc\gamma}\cos k_\mathrm{u} z\,. \tag{E.3}$$

Therefore, Eq. (E.1) can be rewritten as

$$\frac{d\gamma}{dt} = \frac{eE_{x0}K}{2mc\gamma}\left[\cos\left\{\left(k_\mathrm{u}+k\right)z - \omega t + \phi_0\right\} + \cos\left\{\left(k_\mathrm{u}-k\right)z + \omega t - \phi_0\right\}\right]. \tag{E.4}$$

In the above cosine functions, the arguments are $k_\mathrm{u}z \pm (kz - \omega t) \approx k_\mathrm{u}c\left(\overline{\beta}_z \mp 1\right)t$, where $\overline{\beta}_z$ is average velocity of an electron in the longitudinal direction and $z = c\overline{\beta}_z t$. The first term changes gradually, while the second term oscillates with high frequencies. Therefore, for the estimation of rate of change of energy, only the first term is required. Therefore,

$$\frac{d\gamma}{dt} = \frac{eE_{x0}K}{2mc\gamma}\cos\phi,\ \phi = \left(k_\mathrm{u}+k\right)z - \omega t + \phi_0\,, \tag{E.5}$$

where K is the deflection parameter. Using the approximations described in Sect. **2.4.1.b**, the rate of the change of phase in the z direction can be derived from Eq. (E.5) as

$$\frac{d\phi}{dz} = k_\mathrm{u} + k\left(1 - \frac{1}{\overline{\beta}_z}\right) = k_\mathrm{u} - \frac{k}{2\gamma^2}\left(1 + \frac{K^2}{2}\right). \tag{E.6}$$

At resonance energy γ_0, $d\phi/dz = 0$, therefore

$$k_\mathrm{u} = \frac{k}{2\gamma_0^2}\left(1 + \frac{K^2}{2}\right). \tag{E.7}$$

Assuming that the energy differs slightly from the resonance energy, $\gamma = \gamma_0 + \Delta\gamma$, rate of change of phase is

$$\frac{d\phi}{dz} = k_\mathrm{u}\left\{1 - \left(\frac{\gamma_0}{\gamma}\right)^2\right\} = 2k_\mathrm{u}\frac{\Delta\gamma}{\gamma_0} = \frac{4\pi}{\lambda_\mathrm{u}}\frac{\Delta\gamma}{\gamma_0}. \tag{E.8}$$

By transforming the time-derivative in Eq. (E.5) to the distance-derivative and setting $\overline{\beta}_z \approx 1$, the differential equation of $\Delta\gamma$ is rewritten as

$$\frac{d\,(\Delta\gamma/\gamma_0)}{dz} = \frac{eE_{x0}K}{2mc^2\gamma_0^2}\cos\phi. \tag{E.9}$$

Equations (E.8) and (E.9) describe the FEL interaction. Eliminating $\Delta\gamma/\gamma_0$ one obtains the second order differential equation,

$$\frac{d^2\phi}{dz^2} = k_\phi^2\cos\phi,\quad k_\phi^2 = \frac{2\pi eKE_{x0}}{mc^2\lambda_{\mathrm{u}}\gamma_0^2}. \tag{E.10}$$

This equation describes the electron motion in phase space, which resembles the equation for one kind of pendulum motion.

Consider the case in which electrons with initial energies equal to the resonance energy spread throughout phase space. Electrons with negative initial phase will be accelerated and with positive phase and then decelerated by the FEL interaction. The electrons will gradually coalesce around 0° phase and the average change of energy is 0, resulting in zero exchange of energy. However when the initial energy deviates from the resonant energy, an imbalance is generated between the number of electrons being accelerated and decelerated, leading to an exchange of energy between the radiation and the electron beam as a whole. The electrons will then redistribute around a particular phase except at 0°. The electron density is therefore modulated at the wavelength of the radiation resulting in the formation of microbunches separated by the wavelength of the radiation. The decelerated electron bunches emit light with almost the same phase. However when the electric field of the light becomes very large, the phase can change so as to destroy the microbunch of electrons, when the net FEL amplification factor will then decrease. The intensity at this time is called the 'saturated intensity'.

b. The amplification factor for an FEL: As a consequence of the FEL interaction, the change in the electron energy $\delta\gamma$ causes an alteration in the energy of the light. Consider light amplification for the case where the change of intensity of the light is negligibly small, when an electron traverses the undulator once, (the 'small signal regime'). Using the average velocity $\overline{\beta}_z$ in the z direction, Eq. (E.5) can be rewritten as

$$\frac{d\gamma}{dt} = \frac{eE_{x0}K}{2mc\gamma}\cos\left(\Psi t + \phi_0\right),$$
$$\Psi = \omega_0 - \omega\left(1 - \overline{\beta}_z\right),\quad \omega_0 = k_{\mathrm{u}}c\overline{\beta}_z. \tag{E.11}$$

By integrating Eq. (E.11) from $t = 0$ to $t = z/c\overline{\beta}_z = z/\overline{v}_z$, one can obtain the change of energy of electron $\delta\gamma$ as

$$\delta\gamma(z,\phi_0) = \frac{eE_{x0}K}{2mc\gamma\Psi}\left\{\sin\left(\frac{\Psi z}{\overline{v}_z} + \phi_0\right) - \sin\phi_0\right\}. \tag{E.12}$$

The bunch length in accelerators is much larger than that of the wavelength of light, in general, so that the initial phase can be considered to distribute from 0 to 2π. Therefore the average of Eq. (E.12) with respect to ϕ_0 is 0. That is, if FEL interactions only for those electrons of average velocity are taken into account, the change of electron energy is 0, so that exchange of energy does not occur. However, as seen from Eq. (E.10), the electrons are in motion such that their energies change dependent on the initial phases and are subsequently perturbed in their positions by the radiation (light) field. To express the electric field of light which acts on the electrons, we rewrite z in Eq. (E.2) as

$$z(t) = \int_0^t v_z(t')\,dt' = \int_0^t \{\overline{v}_z + c\,\delta\beta_z(t')\}\,dt'. \tag{E.13}$$

For the motion of electrons in the undulator, the following relation is derived from Eq. (E.6),

$$\delta\beta_z = \frac{1+K^2/2}{\gamma^3}\delta\gamma. \tag{E.14}$$

Using the above equation we can calculate Eq. (E.12). The change of the velocity depending on the initial phase is

$$\delta\beta_z = \frac{eKE_{x0}\,(1+K^2/2)}{2\gamma^4 mc\Psi}\{\sin(\Psi t+\phi_0) - \sin\phi_0\}. \tag{E.15}$$

By integrating Eq. (E.13), one can obtain the following relations,

$$z(t) = \overline{v}_z t - cD\left\{\frac{\cos(\Psi t+\phi_0) - \cos\phi_0}{\Psi} + t\sin\phi_0\right\}, \tag{E.16}$$

$$D \equiv \frac{eKE_{x0}\,(1+K^2/2)}{2\gamma^4 mc\Psi} \tag{E.17}$$

The velocity of the electron modulated by its interaction with light and the electric field Eq. (E.2) of the light acting on the electrons changes the electric field as given by

$$E_x(t) = E_{x0}\cos(k\overline{v}_z t + \Delta\phi - \omega t + \phi_0), \tag{E.18}$$

$$\Delta\phi = -\frac{\omega D}{\Psi}\{\cos(\Psi t+\phi_0) - \cos\phi_0 + \Psi t\sin\phi_0\}. \tag{E.19}$$

Thus, the change of phase $\Delta\phi$ appears in FEL interaction. Assuming this value is small, the change of energy is given by

$$\frac{d\gamma}{dt} = \frac{eKE_{x0}}{2\gamma mc}\left[\cos(\Psi t+\phi_0) + \frac{\omega D}{\Psi}\sin(\Psi t+\phi_0)\{\cos(\Psi t+\phi_0) - \cos\phi_0 + \Psi t\sin\phi_0\}\right]. \tag{E.20}$$

Taking an average of the above equation over whole initial phase yields

$$\langle\frac{d\gamma}{dt}\rangle_{\phi_0} = \frac{eKE_{x0}\omega D}{2\gamma mc\Psi}\left(-\frac{\sin\Psi t}{2} + \frac{\Psi t\cos\Psi t}{2}\right). \tag{E.21}$$

If τ is the required time for an electron to traverse the undulator, by integrating Eq. (E.21), one can obtain the change in energy due to the FEL interaction for an electron as

$$\langle\delta\gamma\rangle_{\phi_0} = -\frac{eKE_{x0}\omega D}{4\gamma mc\Psi^2}(2 - 2\cos\Psi\tau - \Psi\tau\sin\Psi\tau). \tag{E.22}$$

Providing $\overline{v}_z \approx c$, Eq. (E.22) can be rewritten as

$$\langle \delta\gamma \rangle_{\phi_0} = \frac{\pi e^2 K^2 E_{x0}^2 N_u^3 \lambda_u^2}{4\gamma^3 m^2 c^4} \frac{d}{d(\Psi\tau)} \left\{ \frac{\sin(\Psi\tau/2)}{(\Psi\tau/2)} \right\}^2, \quad \text{(E.23)}$$

where the total length of the undulator is $L = N_u\lambda_u$. Here the change of energy is expressed by the derivative of a function with variable $\Psi\tau$.

The energy of the electric field of light, W, in a volume V is $W = \varepsilon_0 E_{x0}^2 V/2$. If the change in energy due to the FEL interaction of N electrons is converted to the energy of light, the change in energy of the electric field is $\Delta W = -Nmc^2 \langle \delta\gamma \rangle_{\phi_0}$. Therefore the single pass gain $g = \Delta W/W$ is expressed by

$$g = \frac{\pi e^2 K^2 N_u^3 \lambda_u^2 \rho}{2\varepsilon_0 \gamma^3 m c^2} \frac{d}{d(\Psi\tau)} \left\{ \frac{\sin(\Psi\tau/2)}{(\Psi\tau/2)} \right\}^2, \quad \text{(E.24)}$$

where the electron density is $\rho = N/V$ and the sign of the equation is neglected. In Eq. (E.24) expressing the amplification factor for the FEL, $\Psi = \omega_0 - \omega\left(1 - \overline{\beta}_z\right)$. If $\beta \approx 1$, on z-axis, $1 - \overline{\beta}_z \approx \omega_0/\omega_1$ from Eqs. (2.86) and (2.95). Therefore, $\Psi\tau = 2N_u\pi(\omega_1 - \omega)/\omega_1$, so that the function to be differentiated in Eq. (E.24) is the same as the shape function for undulator radiation Eq. (2.97). Expressing the function by $f_{\text{spont}}(\omega)$, one can obtain

$$g \propto \frac{d}{d\omega} f_{\text{spont}}(\omega). \quad \text{(E.25)}$$

The gain is proportional to $df_{\text{spont}}(\omega)/d\omega$, which is called 'Madey's theorem'. By numerical calculation of the steepest gradient of $f(\omega)$, one can obtain the gain using practical units as[1)]

$$g = 1.75 \times 10^{-20} \left\{ \frac{L\,(\text{m})}{E\,(\text{GeV})} \right\}^3 \{B_0\,(\text{T})\}^2 \rho\left(\text{m}^{-3}\right) \lambda_u\,(\text{m})\, F. \quad \text{(E.26)}$$

In any real case, the overlap of the electron beam and light should be taken into account by means of the filling factor F. The effective amplification factor must include both the gain and the loss due to the mirror reflection and the light extraction. FEL oscillations occurs when such effective amplification factor is greater than 0. In the case of the SASE-FEL (see Chaps. **1** and **3**) the gain is extremely large because of the large current density and there are some parts of the amplification mechanism which may differ from the explanation given above.

When an electron beam traverses an undulator with periodic length λ_u, it is bunched at the wavelength of the fundamental. If this beam traverses an undulator having period length which is an odd integer fraction of the period of the fundamental ($\lambda_u/3$ for instance), one can obtain the higher harmonics of the odd integer.

REFERENCES

1. S. Takano, H. Hama, G. Isoyama, A. Lin and N. A. Vinokulov: Jpn. J. Appl. Phys. **31** (1992) 2621. The expression of the gain in this paper is corrected as Eq. (E.26).

Appendix F: The solution of the wave function of an electron in a Coulomb field and free space using polar coordinates

In a central field, the solution of the wave equation of a particle can be written as a product of radial part $\chi_l(r)$ and angular part $Y_{lm}(\theta, \phi)$. The wave function for energy eigenvalue E can be written as

$$\left\{\frac{p_r^2}{2m}+\frac{l^2}{2mr^2}+V(r)\right\}\chi_l(r)\,Y_{lm}(\theta,\phi)=E\chi_l(r)\,Y_{lm}(\theta,\phi), \tag{F.1}$$

$$\left\{\frac{-\hbar^2}{2mr}\frac{d^2}{dr^2}r+\frac{l(l+1)\hbar^2}{2mr^2}+V(r)\right\}\chi_l(r)=E\chi_l(r). \tag{F.2}$$

The wave equation for radial part is

$$\left\{-\frac{d^2}{dr^2}+\frac{l(l+1)}{r^2}+U(r)-F\right\}y_l(r)=0, \tag{F.3}$$

$$U(r)=\frac{2mV(r)}{\hbar^2},\quad F=\frac{2mE}{\hbar^2}, \tag{F.4}$$

where the radial part is rewritten as $\chi_l(r)=y_l(r)/r$.

The Coulomb potential of a point charge Ze for an electron at the origin is $V=-Ze^2/4\pi\varepsilon_0 r$. The radial part χ_{nl} is expressed by the Laguerre bi-polynomial where n is called principal quantum number and l, orbital quantum number. Its normalization condition is

$$\int_0^\infty\{\chi_{nl}(r)\}^2 r^2dr=1. \tag{F.5}$$

The number l = 0, 1, 2, 3, ... are often designated as s, p, d, f, ... The explicit expressions of $\chi_{nl}(r)$ are

$$\chi_{1s}(r)=2\left(\frac{Z}{a}\right)^{3/2}e^{-\frac{Zr}{a}},$$

$$\chi_{2s}(r)=\left(\frac{Z}{2a}\right)^{3/2}\left(2-\frac{Zr}{a}\right)e^{-\frac{Zr}{2a}},$$

$$\chi_{2p}(r)=\left(\frac{Z}{2a}\right)^{3/2}\frac{Zr}{a\sqrt{3}}e^{-\frac{Zr}{2a}},\quad a=\frac{\hbar^2}{me^2}. \tag{F.6}$$

The energy eigen value is given by

$$E_n=-\frac{Z^2me^4}{2\hbar^2}\frac{1}{n^2}. \tag{F.7}$$

The angular part of the wave function, $Y_{lm}(\theta,\phi)$, is the spherical harmonic function,

$$Y_{lm}(\theta,\phi)=N_l{}^m P_l{}^m(\cos\theta)\,e^{im\phi}, \tag{F.8}$$

here P_l^m is Legendre bi-function, whose mother function is the Legendre polynomial, P_l. When $m=0$,

$$P_l{}^0(\mu)=P_l(\mu). \tag{F.9}$$

For special cases,

$$P_l(1) = 1, \quad P_l(-1) = (-1)^l, \tag{F.10}$$

$$P_l^m(1) = P_l^m(-1) = 0, \quad m \neq 0, \tag{F.11}$$

$$P_0(\mu) = 1, \quad P_1(\mu) = \mu, \quad P_2(\mu) = \frac{1}{2}\left(3\mu^2 - 1\right). \tag{F.12}$$

For specific cases, Y_{lm} is

$$Y_{00} = \left(\frac{1}{4\pi}\right)^{1/2}, \quad Y_{10} = \left(\frac{3}{4\pi}\right)^{1/2} \cos\theta, \quad Y_{11} = -\left(\frac{3}{8\pi}\right)^{1/2} e^{i\phi} \sin\theta. \tag{F.13}$$

In free space, $U(r) = 0$ and $E = (\hbar k)^2/2m$. The wave equation for the radial part is given by

$$\left\{\frac{d^2}{d\rho^2} + \frac{2}{\rho}\frac{d}{d\rho} + 1 - \frac{l(l+1)}{\rho^2}\right\} \chi_l(\rho) = 0, \tag{F.14}$$

where $\rho = kr$. The solutions of Eq. (F.14) are spherical Bessel functions of $j_l(\rho)$, $h_l^+(\rho)$, $n_l(\rho)$ and $h_l^-(\rho)$. $h_l^\pm$ are related to real functions $n_l(\rho)$ and $j_l(\rho)$ as

$$h_l^\pm(\rho) = n_l(\rho) \pm i j_l(\rho). \tag{F.15}$$

$h_l^{(1)}$ and $h_l^{(2)}$ are spherical Hankel functions of the first and the second kind respectively. $h_l^\pm$ are also expressed using $h_l^{(1)}$ and $h_l^{(2)}$ as

$$h_l^+ = i h_l^{(1)}, \quad h_l^- = -i h_l^{(2)}. \tag{F.16}$$

Under $\rho \to \infty$, an asymptotic solution of is

$$h_l^+(\rho) \sim \frac{1}{\rho} e^{i\left(\rho - \frac{l\pi}{2}\right)}. \tag{F.17}$$

Appendix G: The Fano effect

For the bound state φ and the continuum state $\psi_{E'}$ the elements of the energy matrix are

$$\langle \varphi |H| \varphi \rangle = E_\varphi, \tag{G.1a}$$

$$\langle \psi_{E'} |H| \varphi \rangle = V_{E'}, \tag{G.1b}$$

$$\langle \psi_{E''} |H| \psi_{E'} \rangle = E' \delta(E'' - E'). \tag{G.1c}$$

For the eigenvalue E, the eigenfunction can be represented by

$$\Psi_E = a\varphi + \int b_{E'} \psi_{E'} dE'. \tag{G.2}$$

Using Eqs. (G.1a) to (G.1c), relations of a and $b_{E'}$ are obtained as

$$E_\varphi a + \int V^*_{E'} b_{E'} dE' = Ea, \tag{G.3a}$$

$$V_{E'} a + E' b_{E'} = E b_{E'}. \tag{G.3b}$$

Using Eq. (G.3b), $b_{E'}$ can be expressed as

$$b_{E'} = \left[\frac{1}{E - E'} + z(E)\delta(E - E') \right] V_{E'} a. \tag{G.4}$$

The states $\psi_{E'}$ have an asymptotic form $\sin[k(E')r + \delta]$. Their superposition with the coefficients Eq. (G.4) yields

$$\int b_{E'} \psi_{E'} dE' \propto \sin[k(E)r + \delta + \Delta], \tag{G.5}$$

in which

$$\Delta = -\arctan[\pi / z(E)] \tag{G.6}$$

is the phase shift due to the configuration interaction of $\Psi_{E'}$ with state φ. Substituting Eq. (G.4) for Eq. (G.3a), one obtains

$$z(E) = \frac{E - E_\varphi - F(E)}{|V_E|^2}, \tag{G.7}$$

$$F(E) = \mathcal{P} \int dE' \frac{|V_{E'}|^2}{E - E'}, \tag{G.8}$$

where $\mathcal{P}$ indicates principal value. Substituting Eq. (G.7) for Eq. (G.6), one obtains

$$\Delta = -\arctan \frac{\pi |V_E|^2}{E - E_\varphi - F(E)}. \tag{G.9}$$

The coefficient a in Eq. (G.2) is determined by the normalization condition

$$\langle \Psi_{\bar{E}} | \Psi_E \rangle = \delta(\bar{E} - E) \tag{G.10}$$

as

$$|a(E)|^2 = \frac{1}{|V_E|^2 [\pi^2 + z^2(E)]} = \frac{|V_E|^2}{[E - E_\varphi - F(E)]^2 + \pi^2 |V_E|^4}. \tag{G.11}$$

Thus the bound states φ are diluted with half width, $\pi |V_E|^2$ through a band of continuum states. Using the phase shift Eq. (G.9), a and $b_{E'}$ can be rewritten as

$$a = \sin \Delta / \pi V_E, \tag{G.12}$$

$$b_{E'} = \frac{V_{E'}}{\pi V_E} \frac{\sin \Delta}{E - E'} - \cos \Delta \, \delta(E - E'). \tag{G.13}$$

The transition moment between an initial state i and the state Ψ_E is given by

$$\langle \Psi_E|T|i\rangle = \frac{1}{\pi V_E^*}\langle \Phi|T|i\rangle \sin\Delta - \langle \psi_E|T|i\rangle \cos\Delta, \qquad (G.14)$$

$$\Phi = \varphi + \mathcal{P}\int \frac{V_{E'}\psi_{E'}}{E-E'}\,dE'. \qquad (G.15)$$

This indicates that the state Φ becomes modified by the initial state φ through an admixture of the continuum states. Thus, in the Fano effect, the ratio of the transition probability, $|\langle \Psi_E|T|i\rangle|^2$, to the probability of unperturbed continuum state, $|\langle \psi_E|T|i\rangle|^2$ is expressed by

$$\frac{|\langle \Psi_E|T|i\rangle|^2}{|\langle \psi_E|T|i\rangle|^2} = \frac{(q+\varepsilon)^2}{1+\varepsilon^2}, \qquad (G.16)$$

where

$$\varepsilon = -\cot\Delta = \frac{E-E_F-F(E)}{\pi|V_E|^2}, \qquad (G.17)$$

$$q = \frac{\langle \Phi|T|i\rangle}{\pi V_E^*\langle \psi_E|T|i\rangle} \qquad (G.18)$$

The ratio can be represented by a single family of curves. In the text $\pi|V_E|^2 = \Gamma$ is used.

References

1. U. Fano: Phys. Rev. **124** (1961) 1866.

Appendix H: The Born-Oppenheimer approximation and the Frank-Condon factor

a. The Born-Oppenheimer approximation: The position vectors of the j-th electron and the k-th nucleus (with respect to the center of mass) in a molecule are defined as $\boldsymbol{r}_j$ and $\boldsymbol{R}_k$, respectively. The sets of positions of the electrons and nuclei can be represented as $r \equiv \{\boldsymbol{r}_j\}$ and $R \equiv \{\boldsymbol{R}_k\}$, respectively. Here we assume that only Coulombic interactions exist between the electrons and the nuclei and that no external force is applied to the molecule. The total molecular Hamiltonian is given by

$$H = T_R + T_r + V(r,R). \qquad (H.1)$$

Here

$$T_R = -\frac{\hbar^2}{2}\sum_k \frac{\partial^2}{M_k \partial R_k^2}, \qquad (H.2)$$

$$T_r = -\frac{\hbar^2}{2m}\sum_j \frac{\partial^2}{\partial r_j^2}, \qquad (H.3)$$

where T_R and T_r are the kinetic energy of nuclear and electron systems, respectively and $V(r, R)$ is the potential energy of the whole system. In the Born-Oppenheimer approximation, the nuclear kinetic energy, T_R, is regarded as a small perturbation, so that the Hamiltonian can be written as

$$H = H_0 + T_R, \tag{H.4}$$

$$H_0 = T_r + V(r, R). \tag{H.5}$$

Here the electron-nucleus interaction is represented as the Coulomb potential of the nuclei clamped at certain positions in space. For each position set, R, the wave equation

$$[H_0 - \varepsilon_n(R)]\psi_n(R, r) = 0 \tag{H.6}$$

is solved. $\varepsilon_n(R)$, which is the energy eigenvalue of position set R, is called the 'adiabatic potential'. The total eigenfunction $\Psi(R, r)$ can be represented by a linear combination of $\psi_n(R, r)$, the eigenfunctions of H_0 as

$$\Psi(R, r) = \sum_n \Phi_n(R)\psi_n(R, r). \tag{H.7}$$

Substituting Eq. (H.7) for $(H - E)\Psi = 0$, multiplying $\psi_m^*(R, r)$, integrating with respect to r and neglecting the higher order terms, the following equation is obtained.

$$[T_R + \varepsilon_m(R)]\Phi_m(R) = E_m\Phi_m(R). \tag{H.8}$$

From this equation, the motion of nuclei can be understood to be as motion in an adiabatic potential.

b. The Frank-Condon factor: Let us consider a diatomic molecule. Under Born-Oppenheimer approximation, the cross section for photoabsorption by the molecule due to a transition from a ground state i with vibrational quantum number 0, to an excited state f with vibrational quantum number v' is given by

$$\sigma_{fv'} \sim \left| \int X_{fv'}^*(R)D(R)X_{i0}(R)dR \right|^2, \tag{H.9}$$

where $X_{i0}(R)$ and $X_{fv'}(R)$ are the wave functions for the ground state, $i0$ and the excited state, fv', respectively. Here electronic transition probability $D(R)$ is given by

$$D(R) = \int \psi_f^*(R, r)r\psi_i(R, r)dr. \tag{H.10}$$

Under the assumption that $D(R)$ does not change with any change in internuclear distance R, the photoabsorption cross section can be written as

$$\sigma_{fv'}(E) \sim |D(R_e)|^2 F(v'0), \tag{H.11}$$

where R_e is the equilibrium internuclear distance in the ground state and

$$F(v'0) = \left| \int X_{fv'}^*(R)X_{i0}(R)dR \right|^2, \tag{H.12}$$

which is called the 'Frank-Condon factor'. This description can be applied not only to those transitions to valence-excited states, but also to core-excited states. The intensities of the vibrational transitions redistribute according to $F(v'\,0)$.

In the linear-coupling model, the initial and the final states can be represented by harmonic oscillators of the same angular frequency with different equilibrium internuclear distances. In this case, the two wave functions in Eq. (H.12) can be two Hermite functions of which position coordinates have different origins. Then, the Frank-Condon factor is obtained using a Poisson distribution as[1]

$$I(0 \rightarrow v') = e^{-S}\frac{S^{v'}}{v'!}. \tag{H.13}$$

Here S is the expectation value for vibrational excitation, which is expressed using K, the spring constant of the harmonic oscillator and ΔR_e, the difference in equilibrium internuclear distances as

$$S = \frac{K}{2\hbar\omega}(\Delta R_e)^2. \tag{H.14}$$

where ω is the angular frequency of the harmonic oscillator. In any real system, there is a small change in frequency between the initial and final states. For diatomic molecules, the change in the equilibrium internuclear distance, ΔR_e is

$$\Delta R_e = \left(\frac{2S\hbar}{\omega}\frac{m_1 + m_2}{m_1 m_2}\right)^{1/2}\frac{\omega}{\omega'}, \tag{H.15}$$

where ω and ω' are respectively the angular frequencies of the initial and final states, and m_1 and m_2, masses of the two atoms. The sign of ΔR_e is negative, if $\omega < \omega'$ and positive, if $\omega > \omega'$.[2]

REFERENCES

1. For example, T. Keil: Phys. Rev. **140** (1965) A601.
2. S. J. Osborne, S. Sundin, A. Ausmees, S. Svensson, L. J. Saethre, O. Svaeren, S. L. Sorensen, J. Végh, J. Karvonen, S. Aksela and A. Kikas: J. Chem. Phys. **106** (1997) 1661.

Appendix I: Density of states, joint density of states

First we calculate the density of states $J(E)$ for a homogeneous periodic three-dimensional system (crystal). The following example is for the conduction band, of which energy dispersion is given by

$$E(\boldsymbol{k}) = a(k_x^2 + k_y^2 + k_z^2) = ak^2. \tag{I.1}$$

$J(E)$ is defined to be the number of states per unit volume and unit energy. The summation with respect to (k_x, k_y, k_z), $\sum_{k_x,k_y,k_z}$ can be replaced by integral in $\boldsymbol{k}$-space and $J(E)$ can be written as

$$J(E) = \frac{2}{(2\pi)^3}\int_{\mathrm{BZ}} \delta\left(E\left(\boldsymbol{k}\right) - E\right) d\boldsymbol{k}, \tag{I.2}$$

where the integration is over the Brillouin zone BZ. The factor 2 accounts for two directions of spin for each electron and it arises from the integration of spin variables. In general

$$\int f(\boldsymbol{k})\, d\boldsymbol{k} = \int f(\boldsymbol{k})\, dS_{\boldsymbol{k}} dn_{\boldsymbol{k}} = \int f(\boldsymbol{k})\, dS_{\boldsymbol{k}} \frac{df}{|\nabla_{\boldsymbol{k}} f|}. \tag{I.3}$$

Since $f(\boldsymbol{k}) = f\{E(\boldsymbol{k})\}, df/|\nabla_{\boldsymbol{k}} f| = dE/|\nabla_{\boldsymbol{k}} E|$ and f is delta function, the above integral is transformed to a surface integral in $\boldsymbol{k}$-space. Therefore $J(E)$ is given by

$$J(E) = \frac{2}{(2\pi)^3} \int_{E(\boldsymbol{k})=E} \frac{dS_{\boldsymbol{k}}}{|\nabla_{\boldsymbol{k}} E(\boldsymbol{k})|}, \tag{I.4}$$

where $dS_{\boldsymbol{k}}$ is the area element in $\boldsymbol{k}$-space. Since $|\nabla_{\boldsymbol{k}} E| = 2ak$ and $dS_{\boldsymbol{k}} = k^2 d\Omega$, $J(E)$ is obtained as

$$J(E) = \frac{2}{(2\pi)^3} \int \frac{k^2 d\Omega}{2ak} = \frac{k}{2\pi^2} = \frac{1}{2\pi^2} \sqrt{\frac{E}{a^3}}, \tag{I.5}$$

which is proportional to $\sqrt{E}$ as shown in Fig. 7.34.

In the case of two-dimensional systems, of which energy dispersion is given by

$$E(\boldsymbol{k}) = a(k_x^2 + k_y^2) = ak^2, \tag{I.6}$$

$J(E)$ is given by

$$J(E) = \frac{2}{(2\pi)^2} \oint_{E(\boldsymbol{k})=E} \frac{dl_{\boldsymbol{k}}}{|\nabla_{\boldsymbol{k}} E(\boldsymbol{k})|} = \frac{2}{(2\pi)^2} \int \frac{k d\phi}{2ak} = \frac{1}{2\pi a}, \tag{I.7}$$

which is constant and $J(E)$ shows a step. If the system is not completely two-dimensional, but it has also energy states dependent on z direction, $J(E)$ becomes multi step function, whose steps correspond to energy states dependent on z direction as shown in Fig. 7.34.

In the case of one-dimensional system, of which energy dispersion is given by

$$E(\boldsymbol{k}) = ak_x^2 = ak^2, \tag{I.8}$$

$J(E)$ is given by

$$J(E) = \frac{2}{2\pi} \sum_{E(\boldsymbol{k})=E} \frac{1}{|\nabla_{\boldsymbol{k}} E(\boldsymbol{k})|} = \frac{2}{2\pi} \frac{2}{2ak} = \frac{1}{\pi ak} = \frac{1}{\pi\sqrt{aE}}, \tag{I.9}$$

which shows divergent and asymmetric shape around $E = 0$. If the system has also energy states dependent on y and z directions, such shapes are found around the energies depending on y and z directions as shown in Fig. 7.34.

Next we consider the joint density of states. For a system with many oscillators, with oscillator number j and oscillator strength $\hat{f}_j$, the Lorentz model can be used to calculate the dielectric constant (see Sect. **4.2**). In crystals, j corresponds to a wave number $\boldsymbol{k}$. $\hat{\rho} \sum_j$ in the Lorentz model can be replaced by $\sum_{k_x,k_y,k_z}$ in $\boldsymbol{k}$-space and this sum can be replaced by the integral

$$\hat{\rho} \sum_j \rightarrow \frac{2}{(2\pi)^3} \int_{\text{BZ}} dk_x dk_y dk_z = \frac{2}{(2\pi)^3} \int_{\text{BZ}} d\boldsymbol{k}. \tag{I.10}$$

Using the relation between $\hat{f}_j$ and dipole moment obtained in Sect. **4.2**, imaginary part of dielectric constant $\varepsilon_2(\omega)$ is given by

$$\varepsilon_2(\omega) = \frac{e^2}{\varepsilon_0}\sum_{m,n}\frac{2\omega_{mn}}{3\hbar}\int_{\mathrm{BZ}}\frac{2}{(2\pi)^3}\frac{|\boldsymbol{r}_{mn}|^2\,\Gamma_{mn}\omega}{(\omega_{mn}^2-\omega^2)^2+\Gamma_{mn}^2\omega^2}d\boldsymbol{k}. \tag{I.11}$$

Using the characteristic of the delta function

$$\delta(x) = \frac{1}{\pi}\lim_{\sigma\to 0}\frac{\sigma}{x^2+\sigma^2}, \tag{I.12}$$

$\varepsilon_2(\omega)$ is rewritten as

$$\varepsilon_2(\omega) = \frac{\pi e^2}{3\hbar\varepsilon_0}\sum_{m,n}\int_{\mathrm{BZ}}\frac{2}{(2\pi)^3}|\boldsymbol{r}_{mn}|^2\,\delta\left(\omega-\omega_{mn}(\boldsymbol{k})\right)d\boldsymbol{k}. \tag{I.13}$$

This expression relates the band structure to the optical properties. For the case, where the dipole moment does not depend on band dispersion, the joint density of states is defined to be

$$J_{mn}(\hbar\omega) = \frac{2}{(2\pi)^3}\int_{\mathrm{BZ}}\delta\left(\hbar\omega_{mn}(\boldsymbol{k})-\hbar\omega\right)d\boldsymbol{k}, \tag{I.14}$$

where $\hbar\omega_{mn}(\boldsymbol{k}) = E_m(\boldsymbol{k}) - E_n(\boldsymbol{k})$. $E_m(\boldsymbol{k})$ and $E_n(\boldsymbol{k})$ correspond respectively to the conduction band and valence band or core level. This expression can be also obtained by quantum theory of solids to calculate the transition probability by similar way given in Sect. **4.3.1**. The above integral is transformed to a surface integral in $\boldsymbol{k}$-space and $J_{mn}(\hbar\omega)$ is given by

$$J_{mn}(\hbar\omega) = \frac{2}{(2\pi)^3}\int_{\hbar\omega_{mn}(\boldsymbol{k})=\hbar\omega}\frac{dS_{\boldsymbol{k}}}{|\nabla_{\boldsymbol{k}}\hbar\omega_{mn}(\boldsymbol{k})|}, \tag{I.15}$$

where $dS_{\boldsymbol{k}}$ is the area element at $\hbar\omega_{mn}(\boldsymbol{k}) = \hbar\omega$ in $\boldsymbol{k}$-space. In Eq. (I.15), the strong variations (singularities) in $J_{mn}(\hbar\omega)$ are expected when $\nabla_{\boldsymbol{k}}\hbar\omega_{mn}(\boldsymbol{k}) = 0$. Singularities of $J_{mn}(\hbar\omega)$ appear for particular values of photon energy $E = \hbar\omega_{mn}$, and these are called 'critical point energies'. The analytic behavior of $J_{mn}(\hbar\omega)$ near a singularity can be found by expanding ω_{mn} in a Taylor series with respect to $\boldsymbol{k}$ around the point in $\boldsymbol{k}$-space which gives critical point energies. If such point in $\boldsymbol{k}$-space is located at $\boldsymbol{k} = 0$ and the expansion is limited up to quadratic terms,

$$\hbar\omega_{mn}(\boldsymbol{k}) = \hbar\omega_0 + a_x k_x^2 + a_y k_y^2 + a_z k_z^2. \tag{I.16}$$

Thus the following relations are obtained.

$$|\nabla_{\boldsymbol{k}}\hbar\omega_{mn}(\boldsymbol{k})| = 2\sqrt{a_x^2k_x^2 + a_y^2k_y^2 + a_z^2k_z^2}, \tag{I.17}$$

$$dS_{\boldsymbol{k}} = \sqrt{1+\left(\frac{\partial k_z}{\partial k_x}\right)^2+\left(\frac{\partial k_z}{\partial k_y}\right)^2}\,dk_x dk_y. \tag{I.18}$$

Integration of J_{mn} Eq. (I.15) performed around $\hbar\omega = \hbar\omega_0$ results in a singularity (called the 'van Hove singularity') corresponding to a discontinuity in the joint density of states in a crystal. A detailed analysis shows that there are four types of van Hove singularities in crystals, depending on whether or not the band structure goes through a local maximum, a local minimum or a saddle point.[1]) When the signs of a_x, a_y and a_z are (+, +, +), (+, +, –), (+, –, –) and (–, –, –), the spectral shape becomes respectively M_0 (minimum), M_1 (saddle point), M_2 (saddle point) and M_3 (maximum) types, as shown in Fig. I.1. In this figure, E_0, E_1, E_2 and E_3 respectively stand for critical point energies of M_0, M_1, M_2 and M_3 types.

The spectral shape of lower dimensional systems can be also obtained by the similar way.

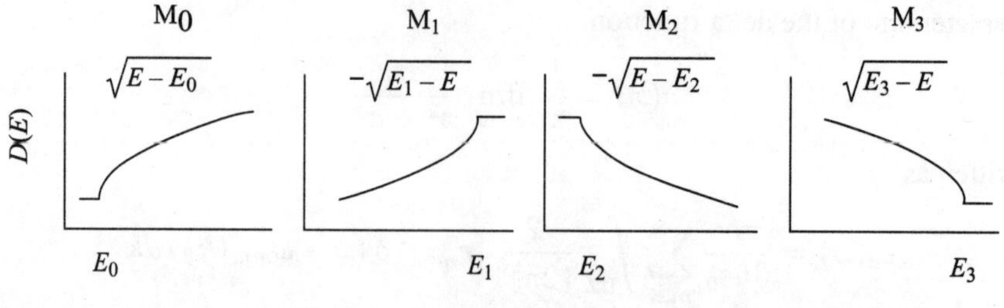

Figure I.1 Schematic of a van Hove singularity in a three-dimensional crystal.

REFERENCES

1. F. Bassani and G. Pastori Parravicni: "Electronic States and Optical Transitions in Solids" (Pergamon Press, 1975).

Appendix J: The magnetooptical effect

There are typically two kinds of experiments which exploit the magnetooptical effect.[1] The first is magnetic circular dichroism, in which the difference in absorption between right and left circularly-polarized light is measured. The second is a magnetooptical rotation experiment in which the rotation of the electric vector of linearly-polarized light, as it passes through the medium or is reflected by the medium is measured. Let z be the direction of propagation of light and x and y be the two orthogonal directions perpendicular to z direction with unit vectors, $\boldsymbol{i}$ and $\boldsymbol{j}$ (see Fig. J.1). Any right-handed rotation of circularly-polarized light is defined to be clockwise when viewed along the propagation direction of the light. For simplicity, assume that $n = 1$ and at $t = 0$ and $z = 0$, the electric vector is parallel to $\boldsymbol{i}$, that is $\phi = 0$. The meaning of right-handed rotation is that at $z = 0$, as t increases, the electric vector rotates clockwise. At $z = \Delta z > 0$, at $t = 0$, the rotation angle of the electric vector ϕ is negative and when time goes as $t = k\Delta z/\omega$, the electric vector becomes parallel to $\boldsymbol{i}$. Therefore right circularly-polarized light is described by

$$\boldsymbol{E}_1 = E_0\{\boldsymbol{i}\cos(\omega t - kz) + \boldsymbol{j}\sin(\omega t - kz)\}. \tag{J.1}$$

Since the wave equation is a second order differential equation, another solution for right circularly-polarized light will be given by

$$\boldsymbol{E}_2 = E_0\{-\boldsymbol{i}\sin(\omega t - kz) + \boldsymbol{j}\cos(\omega t - kz)\}. \tag{J.2}$$

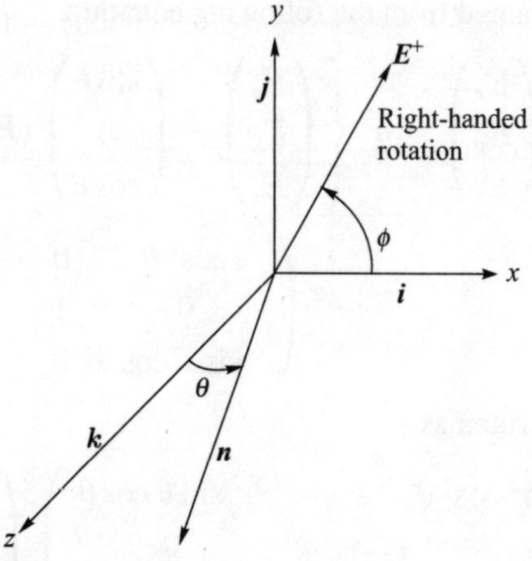

Figure J.1 Definition of rotational direction of circularly-polarized light.

The useful expression for right circularly-polarized light is

$$\begin{aligned} \boldsymbol{E}^+ &= \boldsymbol{E}_1 + i\boldsymbol{E}_2 \\ &= E_0\left[\boldsymbol{i}\left\{\cos(\omega t - kz) - i\sin(\omega t - kz)\right\} + \boldsymbol{j}\left\{\sin(\omega t - kz) + i\cos(\omega t - kz)\right\}\right] \\ &= E_0\left(\boldsymbol{i} + i\boldsymbol{j}\right)e^{-i(\omega t - kz)} . \end{aligned} \tag{J.3}$$

Similarly, that for left circularly-polarized light is

$$\boldsymbol{E}^- = E_0\left(\boldsymbol{i} - i\boldsymbol{j}\right)e^{-i(\omega t - kz)} . \tag{J.4}$$

First let us consider circular magnetic dichroism in an isotropic medium under a magnetic field which is applied along the z direction. Any isotropic medium has 2-fold and 4-fold rotational symmetry around the x-, y- and z-axes, and inversion symmetry. After magnetization, the medium becomes uniaxially anisotropic, so that the dielectric tensor is given by

$$\varepsilon = \varepsilon_0\tilde{\varepsilon} = \varepsilon_0\begin{pmatrix} \varepsilon_{xx} & \varepsilon_{xy} & 0 \\ -\varepsilon_{xy} & \varepsilon_{xx} & 0 \\ 0 & 0 & \varepsilon_{xx} \end{pmatrix} . \tag{J.5}$$

In general $\varepsilon_{zz} \neq \varepsilon_{xx}$, but we assume that $\varepsilon_{zz} = \varepsilon_{xx}$. The electric and magnetic fields in medium can then be expressed by

$$\boldsymbol{E} = \boldsymbol{E}_0 e^{-i\omega t} e^{i\tilde{n}\boldsymbol{k}\cdot\boldsymbol{r}}, \quad \boldsymbol{H} = \boldsymbol{H}_0 e^{-i\omega t} e^{i\tilde{n}\boldsymbol{k}\cdot\boldsymbol{r}} \tag{J.6}$$

(see Sect. **4.1.1.a**). The relation of electric flux density to the electric field is obtained from Maxwell's equations as

$$\boldsymbol{D} = \varepsilon_0\tilde{\varepsilon}\boldsymbol{E}_0 = \varepsilon_0\tilde{n}^2\left\{\boldsymbol{E} - \boldsymbol{n}\left(\boldsymbol{n}\cdot\boldsymbol{E}\right)\right\} , \tag{J.7}$$

where $\boldsymbol{n}$ is the unit vector in the direction of light propagation. In Fig. J.1, $\boldsymbol{n} = (\sin\theta,\, 0,\, \cos\theta)$. The complex refractive index is obtained from the following equation,

$$\begin{pmatrix} \varepsilon_{xx} & \varepsilon_{xy} & 0 \\ -\varepsilon_{xy} & \varepsilon_{xx} & 0 \\ 0 & 0 & \varepsilon_{xx} \end{pmatrix} \begin{pmatrix} E_x \\ E_y \\ E_z \end{pmatrix} = \tilde{n}^2 \left\{ \begin{pmatrix} E_x \\ E_y \\ E_z \end{pmatrix} - \begin{pmatrix} \sin\theta \\ 0 \\ \cos\theta \end{pmatrix} (E_x \sin\theta + E_z \cos\theta) \right\}$$

$$= \tilde{n}^2 \begin{pmatrix} \cos^2\theta & 0 & -\sin\theta\cos\theta \\ 0 & 1 & 0 \\ -\sin\theta\cos\theta & 0 & \sin^2\theta \end{pmatrix} \begin{pmatrix} E_x \\ E_y \\ E_z \end{pmatrix}. \tag{J.8}$$

The above equation can be rewritten as

$$\begin{pmatrix} \varepsilon_{xx} - \tilde{n}^2\cos^2\theta & \varepsilon_{xy} & \tilde{n}^2 \sin\theta\cos\theta \\ -\varepsilon_{xy} & \varepsilon_{xx} - \tilde{n}^2 & 0 \\ \tilde{n}^2\sin\theta\cos\theta & 0 & \varepsilon_{xx} - \tilde{n}^2\sin^2\theta \end{pmatrix} \begin{pmatrix} E_x \\ E_y \\ E_z \end{pmatrix} = 0. \tag{J.9}$$

Calculation of eigenvalue of the matrix gives $\tilde{n}_\pm^2 \approx \varepsilon_{xx} \pm i\varepsilon_{xy}\cos\theta_\pm$. In the case of normal incidence, $\tilde{n}_\pm^2 = \varepsilon_{xx} \pm i\varepsilon_{xy}$. Substitution of the eigenvalue for $\tilde{n}^2$ at normal incidence into Eq. (J.8) yields $\pm\, iE_x = E_y$. Therefore, the electric field of circularly-polarized light can be represented by

$$\boldsymbol{E}^\pm = (E_0 \boldsymbol{i} \pm iE_0 \boldsymbol{j})\, e^{-i(\omega t - \tilde{n}_\pm kz)}, \tag{J.10}$$

where $|E_x| = |E_y| = E_0$. Comparing Eqs. (J.3) and (J.4) with Eq. (J.10) indicates that $\pm$ corresponds to right and left circularly-polarized light and that circularly-polarized light incident on a medium remains circularly-polarized within the medium and also after passing through it. Circular dichroism originates from the imaginary part of $\tilde{n}_\pm$. Here the complex refractive index before application of the magnetic field is defined to be $\tilde{n} = n + i\kappa$ and the non-diagonal part of dielectric tensor, $\varepsilon_{xy} = \varepsilon'_{xy} + i\varepsilon''_{xy}$. Since $\tilde{n}_\pm^2 = \tilde{n}^2 \pm i\varepsilon_{xy}$, $\tilde{n}_\pm \approx \tilde{n}\,(1 \pm i\varepsilon_{xy}/2\tilde{n}^2)$. After further calculation,

$$\tilde{n}_\pm \approx (n + i\kappa)\left(1 \pm \frac{i\varepsilon'_{xy} - \varepsilon''_{xy}}{2\,(n^2 - \kappa^2 + 2in\kappa)}\right), \tag{J.11}$$

$$Im\,(\tilde{n}_\pm) \approx \kappa \pm \frac{n\varepsilon'_{xy} + \kappa\varepsilon''_{xy}}{2\,(n^2 + \kappa^2)}. \tag{J.12}$$

Setting $\Delta\kappa = \kappa^+ - \kappa^-$, the extinction coefficient defining magnetic circular dichroism is derived to be

$$\Delta\kappa = \frac{n\varepsilon'_{xy} + \kappa\varepsilon''_{xy}}{n^2 + \kappa^2}. \tag{J.13}$$

Multiplying $\Delta\kappa$ by $4\pi/\lambda$ gives the absorption coefficient. Similarly by setting $\Delta n = n^+ - n^-$, one obtains the following equation

$$\Delta n = \frac{\kappa\varepsilon'_{xy} - n\varepsilon''_{xy}}{n^2 + \kappa^2}. \tag{J.14}$$

Next we discuss the Faraday effect which describes the rotation of the electric vector of the linearly-polarized light during its passage through a magnetized medium. Assume that the linearly-polarized light $\boldsymbol{E}_1 = \boldsymbol{i}E_0 e^{-i(\omega t - \tilde{n}_1 \boldsymbol{k}\cdot\boldsymbol{r})}$ is incident from isotropic medium 1 to medium 2 with dielectric tensor described by Eq. (J.5). The incident linearly-polarized light is expressed by sum of right and left circularly-polarized light by

$$\boldsymbol{E}_1 = \boldsymbol{E}^+ + \boldsymbol{E}^- = \frac{E_0}{2}\{(\boldsymbol{i}+i\boldsymbol{j}) + (\boldsymbol{i}-i\boldsymbol{j})\}\, e^{-i(\omega t - \tilde{n}_1 \boldsymbol{k}\cdot\boldsymbol{r})}. \tag{J.15}$$

As previously shown, complex refractive indices are different for right and left circularly-polarized and when light propagates by distance z in medium 2, the electric field can be expressed by

$$\boldsymbol{E}_2 = \frac{E_0}{2} e^{-i\omega t}\left\{e^{i\tilde{n}_2^+ kz}(\boldsymbol{i}+i\boldsymbol{j}) + e^{i\tilde{n}_2^- kz}(\boldsymbol{i}-i\boldsymbol{j})\right\}. \tag{J.16}$$

After calculation of $\boldsymbol{i}$ and $\boldsymbol{j}$ components, by setting $\tilde{n}_2^+ = \tilde{n}_2 + \Delta\tilde{n}/2$ and $\tilde{n}_2^- = \tilde{n}_2 - \Delta\tilde{n}/2$,

$$\boldsymbol{E}_2 = E_0 e^{-i(\omega t - \tilde{n}_2 kz)}\left(\boldsymbol{i}\cos\frac{\Delta\tilde{n}kz}{2} - \boldsymbol{j}\sin\frac{\Delta\tilde{n}kz}{2}\right). \tag{J.17}$$

Finally by setting $\Delta\tilde{n} = \Delta n + i\Delta\kappa$, using $\cos ix = \cosh x$, $\sin ix = i\sinh x$ and approximating the hyperbolic function while assuming $\Delta\kappa kz$ is small,

$$\begin{aligned}\boldsymbol{E}_2 = E_0 e^{-i(\omega t - \tilde{n}_2 kz)} &\left\{\boldsymbol{i}\left(\cos\frac{\Delta nkz}{2} - i\frac{\Delta\kappa kz}{2}\sin\frac{\Delta nkz}{2}\right)\right. \\ &\left. -\boldsymbol{j}\left(\sin\frac{\Delta nkz}{2} + i\frac{\Delta\kappa kz}{2}\cos\frac{\Delta nkz}{2}\right)\right\}.\end{aligned} \tag{J.18}$$

When the coordinate rotates around z-axis by $\phi = -\dfrac{\Delta nkz}{2}$, unit vectors $\boldsymbol{i}'$ and $\boldsymbol{j}'$ in the new coordinates are given by

$$\begin{pmatrix}\boldsymbol{i}'\\ \boldsymbol{j}'\end{pmatrix} = \begin{pmatrix}\cos\phi & \sin\phi\\ \sin\phi & \cos\phi\end{pmatrix}\begin{pmatrix}\boldsymbol{i}\\ \boldsymbol{j}\end{pmatrix} \tag{J.19}$$

Using $\boldsymbol{i}'$ and $\boldsymbol{j}'$, the electric vector in medium 2 is rewritten as

$$\boldsymbol{E}_2 = E_0 e^{-i(\omega t - \tilde{n}_2 kz)}\left\{\boldsymbol{i}' - i\left(\frac{\Delta\kappa kz}{2}\right)\boldsymbol{j}'\right\} \tag{J.20}$$

From this equation, it is found that if $\Delta\kappa = 0$, linearly-polarized light rotates by ϕ. Therefore Faraday rotation angle ϕ_F is given by

$$\phi_F = -\frac{\Delta nkz}{2} = -\frac{\pi z \Delta n}{\lambda}. \tag{J.21}$$

Note that when the rotation direction is defined to be opposite (along the direction from the observer) the sign becomes opposite. If $\Delta\kappa \neq 0$, the linearly-polarized light becomes elliptically-polarized with the long axis rotated by ϕ_F. The ratio of lengths of the long and short axes of elliptically-polarized light describes the 'Faraday ellipticity', which is given by

$$\eta_F = -\frac{\Delta\kappa kz}{2} = -\frac{\pi z \Delta\kappa}{\lambda}. \tag{J.22}$$

Lastly we consider the magnetic Kerr effect where the electric vector of the linearly-polarized light rotates after reflection by a magnetized medium. After calculation of x, y and z components of the electric and magnetic fields, one can obtain the Fresnel equations for the magnetized medium using the continuity conditions of the components at the interface. In Faraday rotation, the case with an angle of incidence = 0 was explained, while in the following, the polar configuration Kerr rotation (magnetization $\|$ z-axis) with an angle of incidence ≥ 0 is discussed. In the magnetized medium, light propagates along two directions (called birefringence). Assume that the refractive index of the medium $\tilde{n}$ becomes $\tilde{n}_{\pm}$ after the application of magnetic field. Applying Snell's law, $\tilde{n}_{\pm} \sin\theta_{\pm} = \tilde{n} \sin\theta$, one can obtain the following equation

$$\cos\theta_{\pm} \approx \cos\theta \pm \frac{Q}{2} \sin^2\theta, \tag{J.23}$$

and express refractive index as

$$\tilde{n}_{\pm} = \tilde{n}\left(1 \pm \frac{Q}{2}\cos\theta\right), \tag{J.24}$$

where $Q = i\varepsilon_{xy}/\varepsilon_{xx}$. The relations between the electric fields are

$$\begin{aligned} &\varepsilon_{xx} E_x^{\pm} + \varepsilon_{xy} E_y^{\pm} = \tilde{n}_{\pm}^2 E_x^{\pm} \cos^2\theta_{\pm} - \tilde{n}_{\pm}^2 E_z^{\pm} \sin\theta_{\pm} \cos\theta_{\pm}, \\ &-\varepsilon_{xy} E_x^{\pm} + \varepsilon_{xx} E_y^{\pm} = \tilde{n}_{\pm}^2 E_y^{\pm}, \\ &\varepsilon_{xx} E_z^{\pm} = -\tilde{n}_{\pm}^2 E_x^{\pm} \sin\theta_{\pm} \cos\theta_{\pm} + \tilde{n}_{\pm}^2 E_z^{\pm} \sin^2\theta_{\pm}. \end{aligned} \tag{J.25}$$

From these equations, $E_x^{\pm}$ and $E_z^{\pm}$ can be obtained in terms of $E_y^{\pm}$ as

$$\begin{aligned} E_x^{\pm} &= \mp\, iE_y^{\pm} \cos\theta_{\pm}, \\ E_z^{\pm} &= \pm\, iE_x^{\pm} \sin\theta_{\pm}\left(1 \pm \frac{Q}{\cos\theta_{\pm}}\right). \end{aligned} \tag{J.26}$$

The electric field of the incident light at any angle of incidence θ can be written as

$$\begin{aligned} E_x &= E_{\mathrm{p}} \cos\theta\, e^{-i\{\omega t - \tilde{n}_1 k(x\sin\theta + z\cos\theta)\}}, \\ E_y &= E_{\mathrm{s}}\, e^{-i\{\omega t - \tilde{n}_1 k(x\sin\theta + z\cos\theta)\}}, \\ E_z &= -E_{\mathrm{p}} \sin\theta\, e^{-i\{\omega t - \tilde{n}_1 k(x\sin\theta + z\cos\theta)\}}, \end{aligned} \tag{J.27}$$

where E_{p} and E_{s} are respectively the p- and s-polarization components. As for reflected light,

$$\begin{aligned} E'_x &= -E'_{\mathrm{p}} \cos\theta e^{-i\{\omega t - \tilde{n}_1 k(x\sin\theta - z\cos\theta)\}} \\ E'_y &= E'_{\mathrm{s}} e^{-i\{\omega t - \tilde{n}_1 k(x\sin\theta - z\cos\theta)\}}, \\ E'_z &= -E'_{\mathrm{p}} \sin\theta\, e^{-i\{\omega t - \tilde{n}_1 k(x\sin\theta - z\cos\theta)\}}. \end{aligned} \tag{J.28}$$

Let θ_2 be the angle of refraction and assume linearly-polarized light is represented as combination of left and right circularly-polarized light. The electric field of refracted light can then be expressed by

$$E''_x = -iE''_+ \cos\theta_2^+ e^{-i\left\{\omega t - \tilde{n}_2^+ k\left(x\sin\theta_2^+ + z\cos\theta_2^+\right)\right\}} + iE''_- \cos\theta_2^- e^{-i\left\{\omega t - \tilde{n}_2^- k\left(x\sin\theta_2^- + z\cos\theta_2^-\right)\right\}},$$

$$E''_y = E''_+ e^{-i\left\{\omega t - \tilde{n}_2^+ k\left(x\sin\theta_2^+ + z\cos\theta_2^+\right)\right\}} + E''_- e^{-i\left\{\omega t - \tilde{n}_2^- k\left(x\sin\theta_2^- + z\cos\theta_2^-\right)\right\}},$$

$$E''_z = iE''_+ \sin\theta_2^+ \left(1 + \frac{Q}{\cos\theta_2^+}\right) e^{-i\left\{\omega t - \tilde{n}_2^+ k\left(x\sin\theta_2^+ + z\cos\theta_2^+\right)\right\}}$$

$$-iE''_- \sin\theta_2^- \left(1 - \frac{Q}{\cos\theta_2^-}\right) e^{-i\left\{\omega t - \tilde{n}_2^- k\left(x\sin\theta_2^- + z\cos\theta_2^-\right)\right\}}, \tag{J.29}$$

where $n_\pm = n^\pm$, $\theta_\pm = \theta^\pm$, $E^\pm = E_\pm$. In the case of $Q \to 0$, $\tilde{n}_2^\pm \to \tilde{n}_2$, E''_s becomes $E''_+ + E''_-$, E''_p, $-i\left(E''_+ - E''_-\right)$. Using $\boldsymbol{H} = -\frac{1}{\mu_0}\int \text{rot}\,\boldsymbol{E}dt$, one can calculate $\boldsymbol{H}$ for incident, reflected and refracted light from Eq. (J.29). From the continuity condition of electric and magnetic fields at the interface, the Fresnel formula is obtained from the following matrix equation,

$$\begin{pmatrix} E'_s \\ E'_p \end{pmatrix} = \begin{pmatrix} r_{ss} & r_{sp} \\ r_{ps} & r_{pp} \end{pmatrix} \begin{pmatrix} E_s \\ E_p \end{pmatrix}. \tag{J.30}$$

The terms r_{ss} and r_{pp} are, respectively the same quantities as r_s, Eq. (4.30) and r_p, Eq. (4.31). By setting $\theta = \theta_1$, the non-diagonal term is given by

$$r_{sp} = r_{ps} = \frac{iQ\tilde{n}_1\tilde{n}_2\cos\theta_1}{\left(\tilde{n}_1\cos\theta_1 + \tilde{n}_2\cos\theta_2\right)\left(\tilde{n}_1\cos\theta_2 + \tilde{n}_2\cos\theta_1\right)}. \tag{J.31}$$

The Kerr rotation angles for s- and p-polarized light are given by[2)]

$$\phi_K^s = r_{ps}/r_{ss}\,, \quad \phi_K^p = r_{sp}/r_{pp}\,. \tag{J.32}$$

It is the absolute values of these angles that one can measure.

REFERENCES

1. A. K. Zvezdin and V. A. Kotov: "Modern Magnetooptics and Magnetooptical Materials" (CRC Press, 2010).
2. C.-Y. You and S.-C. Shin: J. Appl. Phys. **84** (1998) 541.

Appendix K: Magnetic moment

The orbital magnetic moment $\langle L_z \rangle$ per single hole in the valence band is given from an integrated MCD signal by[1)]

$$\frac{\displaystyle\int_{j_+ + j_-} (\mu^+ - \mu^-)\, d\omega}{\displaystyle\int_{j_+ + j_-} (\mu^+ + \mu^- + \mu_0)\, d\omega} = \frac{1}{2}\,\frac{l\,(l+1) + 2 - c\,(c+1)}{l\,(l+1)\,(4l+2-n)}\,\langle L_z \rangle\,. \tag{K.1}$$

Here μ^+ and μ^- are absorption coefficients for left and right circularly-polarized light, c and l are angular momenta of the core hole and valence band and, j_+ and j_- are total angular momenta when core levels split into two levels due to spin-orbit splitting. The integral attached by j_+ and j_- means integration of the absorption from these levels. The relation between spin magnetic moment $\langle S_z \rangle$ and magnetic dipole moment $\langle T_z \rangle$ is given by[1)]

$$\frac{\displaystyle\int_{j_+} (\mu^+ - \mu^-)\, d\omega - \{c\,(c+1)/c\} \int_{j_-} (\mu^+ - \mu^-)\, d\omega}{\displaystyle\int_{j_+ + j_-} d\omega\,(\mu^+ + \mu^- + \mu_0)} = \frac{l\,(l+1) - 2 - c\,(c+1)}{3c\,(4l+2-n)}\,\langle S_z \rangle$$

$$+\,\frac{l\,(l+1)\,\{l\,(l+1) + 2c\,(c+1) + 4\} - 3\,(c-1)^2\,(c+2)^2}{6lc\,(l+1)\,(4l+2-n)}\,\langle T_z \rangle\,. \tag{K.2}$$

For $L_{2,3}$ absorption in transition metals, $\langle T_z \rangle$ is small in comparison with $\langle S_z \rangle$.[1)]

References

1. P. Carra, B. Thole, M. Altarelli and X. Wang: Phys. Rev. Lett. **70** (1993) 694.

Appendix L: The energy loss function

We first derive the energy loss function for a high velocity electron beam passing through matter. These electrons not only lose energy $\Delta E = \hbar\omega$, but also give a part of their momentum to electrons within the material. This phenomenon is understood to be the dielectric loss due to the electric field $\boldsymbol{E}(\boldsymbol{r}, t)$ induced by electrons. The total energy loss W is given by

$$W = Re \int \boldsymbol{E} \cdot \frac{\partial \boldsymbol{D}}{\partial t} d\boldsymbol{r} dt, \tag{L.1}$$

where $\boldsymbol{D}(\boldsymbol{r}, t)$ is the electric flux density. $\boldsymbol{D}(\boldsymbol{r}, t)$ is Fourier-transformed with respect to ω and $\boldsymbol{q}$ as

$$\boldsymbol{D}(\omega, \boldsymbol{q}) = \frac{1}{(2\pi)^4} \int \boldsymbol{D}(\boldsymbol{r}, t) e^{i(\omega t + \boldsymbol{q}\cdot\boldsymbol{r})} d\boldsymbol{r} dt. \tag{L.2}$$

Similarly if $\boldsymbol{E}(\boldsymbol{r}, t)$ is also Fourier-transformed to $E(\omega, \boldsymbol{q})$ the following relationship

$$\boldsymbol{D}(\omega, \boldsymbol{q}) = \varepsilon_0 \varepsilon(\omega, \boldsymbol{q}) \boldsymbol{E}(\omega, \boldsymbol{q}) \tag{L.3}$$

is obtained. From the condition that the fields are real, the following condition

$$\boldsymbol{D}(-\omega, -\boldsymbol{q}) = \boldsymbol{D}^*(\omega, \boldsymbol{q}) \tag{L.4}$$

is satisfied. Using above equations, the total energy loss is derived as

$$\begin{aligned} W &= Re \int \frac{1}{\varepsilon_0 \varepsilon(\omega, q)} \boldsymbol{D}(\omega, \boldsymbol{q}) e^{-i(\omega t + \boldsymbol{q}\cdot\boldsymbol{r})} d\omega d\boldsymbol{q} \cdot \int \boldsymbol{D}(-i\omega', \boldsymbol{q}') e^{-i(\omega' t + \boldsymbol{q}'\cdot\boldsymbol{r})} d\omega' d\boldsymbol{q}' d\boldsymbol{r} dt \\ &= Re \int \frac{-i\omega'(2\pi)^4}{\varepsilon_0 \varepsilon(\omega, \boldsymbol{q})} \delta(\omega + \omega') \delta(\boldsymbol{q} + \boldsymbol{q}') \boldsymbol{D}(\omega, \boldsymbol{q}) \cdot \boldsymbol{D}(\omega', \boldsymbol{q}') d\omega d\omega' d\boldsymbol{q} d\boldsymbol{q}' \\ &= \frac{(2\pi)^4}{\varepsilon_0} \int \omega Im \left\{ \frac{-1}{\varepsilon(\omega, \boldsymbol{q})} \right\} \boldsymbol{D}(\omega, \boldsymbol{q}) \cdot \boldsymbol{D}^*(\omega, \boldsymbol{q}) d\omega d\boldsymbol{q}. \end{aligned} \tag{L.5}$$

$\boldsymbol{D}(\boldsymbol{r}, t)$ is the flux density caused by $\boldsymbol{E}(\boldsymbol{r}, t)$ of electron, so that it is expressed as

$$\boldsymbol{D}(\boldsymbol{r}, t) = -\frac{e}{4\pi} \text{grad} \frac{1}{|\boldsymbol{r} - \boldsymbol{v}t|}. \tag{L.6}$$

Using

$$\frac{1}{r} = \frac{1}{(2\pi)^3} \int \left(\frac{4\pi}{q^2} \right) e^{-i\boldsymbol{q}\cdot\boldsymbol{r}} dq, \tag{L.7}$$

one can obtain

$$\begin{aligned} \boldsymbol{D}(\boldsymbol{r}, t) &= -\frac{e}{4\pi} \frac{1}{(2\pi)^3} \int \frac{4\pi}{q'^2} \text{grad}\, e^{-i\boldsymbol{q}'\cdot(\boldsymbol{r} - \boldsymbol{v}t)} d\boldsymbol{q}' \\ &= \frac{ie}{(2\pi)^3} \int \frac{\boldsymbol{q}' e^{-i\boldsymbol{q}'\cdot(\boldsymbol{r} - \boldsymbol{v}t)}}{q'^2} d\boldsymbol{q}'. \end{aligned} \tag{L.8}$$

By using this equation, Eq. (L.2) is written as

$$\begin{aligned} \boldsymbol{D}(\omega, \boldsymbol{q}) &= \frac{1}{(2\pi)^4} \int \frac{ie}{(2\pi)^3} \frac{\boldsymbol{q}' e^{-i\boldsymbol{q}'\cdot(\boldsymbol{r} - \boldsymbol{v}t)}}{q'^2} e^{i(\omega t + \boldsymbol{q}\cdot\boldsymbol{r})} d\boldsymbol{q}' d\boldsymbol{r} dt \\ &= \frac{ie}{(2\pi)^3} \int \frac{\boldsymbol{q}' \delta(\boldsymbol{q}' - \boldsymbol{q}) \delta(\omega + \boldsymbol{q}'\cdot\boldsymbol{v})}{q'^2} d\boldsymbol{q}' = \frac{ie}{(2\pi)^3} \frac{\boldsymbol{q} \delta(\omega + \boldsymbol{q}\cdot\boldsymbol{v})}{q^2} \end{aligned} \tag{L.9}$$

By calculating the inner product of $\boldsymbol{D}^*(\omega, \boldsymbol{q})$ and $\boldsymbol{D}(\omega, \boldsymbol{q})$, Eq. (L.5) can be expressed by

$$W = \frac{1}{(2\pi)^2 \varepsilon_0} \int \omega Im \left\{ \frac{-1}{\varepsilon(\omega, \boldsymbol{q})} \right\} \frac{e^2}{q^2} \delta(\omega + \boldsymbol{q}\cdot\boldsymbol{v}) d\omega d\boldsymbol{q}. \tag{L.10}$$

Here $\dfrac{1}{\hbar^2 \omega} \dfrac{\partial W}{\partial \omega}$ are number of oscillators with resonant energy $\hbar\omega$, which receive the energy from high velocity electrons. This value is normalized by unit energy. $-Im\{1/\varepsilon(\omega, q)\}$ is the energy loss function.

This energy loss function is not only valid for high velocity electrons, but also for the dielectric response to the charge introduced from outside.[1)] Assume that the density distribution of electrons

$$\rho_{\text{ext}}(\boldsymbol{r},t) = \sum_{\omega,\boldsymbol{q}} \rho_{\text{ext}}(\omega,\boldsymbol{q})\, e^{-i(\omega t+\boldsymbol{q}\cdot\boldsymbol{r})} \tag{L.11}$$

is introduced from outside. This density of the electrons gives div $\boldsymbol{D}$. Densities of electrons from outside and the induced density in the matter defined to be $\langle\rho(\omega,\boldsymbol{q})\rangle$ give div $\boldsymbol{E}$. The Fourier transforms of $\boldsymbol{D}$ and $\boldsymbol{E}$ like Eq. (L.2) and executions of operation div yield

$$\begin{aligned} i\boldsymbol{q}\cdot\boldsymbol{D}(\omega,\boldsymbol{q}) &= i\varepsilon_0\varepsilon(\omega,\boldsymbol{q})\,\boldsymbol{q}\cdot\boldsymbol{E}(\omega,\boldsymbol{q}) = e\rho_{\text{ext}}(\omega,\boldsymbol{q}) \\ i\boldsymbol{q}\cdot\boldsymbol{E}(\omega,\boldsymbol{q}) &= e\rho_{\text{ext}}(\omega,\boldsymbol{q}) + e\langle\rho(\omega,\boldsymbol{q})\rangle. \end{aligned} \tag{L.12}$$

From the above equations, one can obtain

$$\frac{1}{\varepsilon_0\varepsilon(\omega,\boldsymbol{q})} = 1 + \frac{\langle\rho(\omega,\boldsymbol{q})\rangle}{\rho_{\text{ext}}(\omega,\boldsymbol{q})}. \tag{L.13}$$

In order to calculate the dielectric response, $\langle\rho(\omega,\boldsymbol{q})\rangle$ is required which is calculated quantum-mechanically as follows. The spatial electron distribution in matter can be given by

$$\rho(\boldsymbol{r}) = \sum_i \rho_i(\boldsymbol{r}) = \sum_i \delta(\boldsymbol{r}-\boldsymbol{r}_i). \tag{L.14}$$

$\rho_{\boldsymbol{q}}$ is defined to be

$$\rho_{\boldsymbol{q}} = \sum_i \int \delta(\boldsymbol{r}-\boldsymbol{r}_i)\, e^{i\boldsymbol{q}\cdot\boldsymbol{r}} d\boldsymbol{r} = \sum_i e^{i\boldsymbol{q}\cdot\boldsymbol{r}_i}. \tag{L.15}$$

The interaction between the external electron and electrons in the matter in unit volume at t can be calculated to be

$$\begin{aligned} &\frac{e^2}{4\pi\varepsilon_0}\sum_i \int \frac{\rho_i(\boldsymbol{r})\,\rho_{\text{ext}}(\boldsymbol{r}',t)\,d\boldsymbol{r}d\boldsymbol{r}'}{|\boldsymbol{r}-\boldsymbol{r}'|} \\ &= \frac{e^2}{\varepsilon_0}\sum_{i,\boldsymbol{q}} \int \frac{1}{q^2} e^{-i\boldsymbol{q}\cdot(\boldsymbol{r}-\boldsymbol{r}')}\delta(\boldsymbol{r}-\boldsymbol{r}_i) \sum_{\omega,\boldsymbol{q}'} \rho_{\text{ext}}(\omega,\boldsymbol{q}')\, e^{-i(\omega t+\boldsymbol{q}'\cdot\boldsymbol{r}')} d\boldsymbol{r}d\boldsymbol{r}' \\ &= \frac{e^2}{\varepsilon_0}\sum_{\omega,\boldsymbol{q}} \frac{\rho_{\boldsymbol{q}}^*\rho_{\text{ext}}(\omega,\boldsymbol{q})\, e^{-i\omega t}}{q^2}. \end{aligned} \tag{L.16}$$

Here Eq. (L.7) is used after replacing integration by summation and $\rho_{\boldsymbol{q}}^* = \rho_{-\boldsymbol{q}}$. Define the time dependent perturbation to be

$$H' = \frac{e^2}{\varepsilon_0}\lim_{\sigma\to 0}\sum_{\omega,\boldsymbol{q}} \frac{1}{q^2}\rho_{\boldsymbol{q}}^*\rho_{\text{ext}}(\omega,\boldsymbol{q})\, e^{-i\omega t}e^{\sigma t} \quad (\sigma > 0). \tag{L.17}$$

The Schrödinger equation is given by

$$i\hbar\frac{\partial\Psi(\boldsymbol{r},t)}{\partial t} = (H+H')\,\Psi(\boldsymbol{r},t). \tag{L.18}$$

Its solution is expressed as

$$\Psi(\boldsymbol{r},t)=\sum_{n}a_n\left(t\right)\psi_n\left(\boldsymbol{r}\right)e^{-iE_nt/\hbar}. \tag{L.19}$$

Let us consider the first order perturbation. Denoting an initial state by i and a final state by f, we can calculate $a_f(t)$ to be

$$a_f(t)=\frac{e^2}{\varepsilon_0\hbar q^2}\left\{\frac{\rho_{\text{ext}}\left(\omega,\boldsymbol{q}\right)\left\langle f|\rho_{\boldsymbol{q}}^*|i\right\rangle e^{-i\left(\omega-\omega_{fi}+i\sigma\right)t}}{\omega-\omega_{fi}+i\sigma}\right\}, \tag{L.20}$$

where $\omega_{fi}=\left(E_f-E_i\right)/\hbar$ and $a_f\left(t\right)$ represents only leading term when $\omega\approx\omega_{fi}$. The expectation value of $\rho_{\boldsymbol{q}}$ can be regarded as the induced charge density at time t after the external electron is introduced, can be calculated as

$$\int\left\{\psi_i^*e^{iE_it/\hbar}+\sum_{f}a_f^*\left(t\right)\psi_f^*e^{iE_ft/\hbar}\right\}\rho_{\boldsymbol{q}}\left\{\psi_ie^{-iE_it/\hbar}+\sum_{f'}a_{f'}\left(t\right)\psi_{f'}e^{-iE_{f'}t/\hbar}\right\}d\boldsymbol{r}$$

$$\approx\rho_{\boldsymbol{q}}+\sum_{f}\left\{\left\langle i\left|\rho_{\boldsymbol{q}}\right|f\ \right\rangle a_f\left(t\right)e^{-i\omega_{fi}t}+\left\langle f\left|\rho_{\boldsymbol{q}}\right|i\right\rangle a_f^*\left(t\right)e^{i\omega_{fi}t}\right\}. \tag{L.21}$$

Therefore, the expectation value of $\rho_{\boldsymbol{q}}$ is given by

$$\left\langle\rho_{\boldsymbol{q}}\left(t\right)\right\rangle=\frac{e^2}{\varepsilon_0\hbar q^2}\rho_{\text{ext}}\left(\omega,\boldsymbol{q}\right)e^{-i(\omega+i\sigma)t}\sum_{f}\frac{\left|\left(\rho_{\boldsymbol{q}}^*\right)_{fi}\right|^2}{\omega-\omega_{fi}+i\sigma}. \tag{L.22}$$

Since the time dependence of the perturbation is proportional to $e^{-i\omega t}e^{\sigma t}$, the term responding to it (only the first term in $\sum_f$ in Eq. (L.21)) is extracted. In the right side of the above equation, the factor multiplied by $e^{-i\omega t}$, can be regarded as $(\omega,\boldsymbol{q})$ component of induced charge density, that is $\left\langle\rho\left(\omega,\boldsymbol{q}\right)\right\rangle$. By substituting this factor for Eq. (L.13), one can obtain the following equation

$$\frac{1}{\varepsilon_0\varepsilon\left(\omega,\boldsymbol{q}\right)}-1=\frac{e^2}{\varepsilon_0\hbar q^2}\sum_{f}\left|\left(\rho_{\boldsymbol{q}}^*\right)_{fi}\right|^2\frac{1}{\omega-\omega_{fi}+i\sigma}. \tag{L.23}$$

Using the characteristics of the delta function

$$\lim_{\sigma\to 0}\frac{1}{x+i\sigma}=\mathcal{P}\frac{1}{x}-i\pi\delta(x),\quad \mathcal{P}:\ \text{principal value}, \tag{L.24}$$

and making σ infinitesimally small ($\sigma\to 0$), one can obtain the following equation from (L.23),

$$Im\frac{-1}{\varepsilon\left(\omega,\boldsymbol{q}\right)}=\frac{\pi e^2}{\hbar q^2}\sum_{f}\left|\left(\rho_{\boldsymbol{q}}^*\right)_{fi}\right|^2\delta\left(\omega-\omega_{fi}\right). \tag{L.25}$$

From comparison of Eq. (L.25) with Eq. (12.1), it is found that the cross section for the inelastic scattering of light by matter is proportional to $Im\left\{-1/\varepsilon\left(\omega,\ \boldsymbol{q}\right)\right\}$. This means that electrons scattered by incident X-rays behave as though they were introduced electrons, $\rho_{\text{ext}}\left(\omega,\boldsymbol{q}\right)$.

REFERENCES

1. D. Pines: "Elementary Excitations in Solids. Lectures of Phonons, Electrons and Plasmons" (W. A. Benjamin, INC., 1964).

INDEX

I

J

K

L

M

N

O

P

Q

R

About the Editors

Makoto WATANABE: Professor Emeritus, Tohoku University, Japan. Dr.Sc. Research specialities are solid state spectroscopy in the VUV and soft X-ray regions, and engineering for synchrotron radiation sources and their optical systems. B.Sc. Kyoto University, 1963. In 1967 started research using an electron synchrotron and from 1973 took part in the construction of SOR-RING at The University of Tokyo. In 1979, moved to the Institute for Molecular Science, Okazaki to assist in construction of the UVSOR Facility. From 1993 to 2004, working at Tohoku University. Subsequently from 2004 to 2010, Appointed Special Professor and Researcher, at Saga University and Guest Professor, Shanghai Dianji University, China.

Shigeru SATO: Professor Emeritus, Tohoku University, Japan. Dr.Sc. Research fields are spectroscopy and photoelectron spectroscopy on solids using VUV and soft X-rays, and the engineering for magnets and beamlines at synchrotron radiation sources. M.Sc. Tohoku University in 1966 and began research using synchrotron radiation. From 1973 a member of the SOR-RING construction team at the University of Tokyo, then from 1979 participated in the construction of the Photon Factory. From 1989 to 2004 working at Tohoku University. Dean of the School of Science for three years. From 2004 to 2008, Professor Tohoku University of Technology and from 2011 to 2012, Director, Tsukuba Institute of Science and Technology (TIST). Currently Chief Researcher and Manager, Comprehensive Research Organization for Science and Society (CROSS).

Ian MUNRO: Emeritus Professor, University of Manchester, Honorary Scientist, Science and Technologies Facilities Council UK. FInst.P., CPhys. B.Sc., Ph.D., 1962/4, Manchester University. Lecturer Manchester University, 1965. Coordinator of UK Synchrotron Radiation Science Programme at Daresbury Laboratory, 1972, organized the worlds first symposium on synchrotron radiation research, 1973. Atomic and organic molecular spectroscopy at SRF (NINA electron synchrotron). NRC Ottawa, Canada, 1976/7, Stanford Medical Centre USA, 1977/8 constructed a fluorescence lifetimes beamline on SPEAR for work on protein mobility. Involved in design and construction of SRS, the first storage ring to be purpose-built as an X-ray Users Facility. In 1985, Honorary Professor, Aberdeen University. Head of Daresbury Laboratory from 1993 to 1997. Initiated the project of DIAMOND. Guest Professor, Institute for Molecular Science, Okazaki, Japan in 1997.

Gyanendra LODHA: Retired in March 2015 as Head, Indus Synchrotrons Utilization Division at Raja Ramanna Centre for Advanced Technology (RRCAT), Indore, India. Presently Professor at Homi Bhabha National Institute (HBNI). M.Sc. Rajasthan University, 1977. Ph.D. Kashmir University, 1987. UVSOR and ISAS, Japan, 1991 and Nagoya University, Japan, 1996-97. At RRCAT responsible to design and construct beamlines for Indus-1 and Indus-2 synchrotron sources and to coordinate their utilization program for users. At HBNI involved in coordinating the "Concepts in X-ray Physics", "Accelerator Physics" and "Materials Science" courses designed for postgraduate students. Current research field is refractive, diffractive and reflective X-ray optics for high brightness synchrotron sources.